The Second World Ocean Assessment

WORLD OCEAN ASSESSMENT II

Volume II

United Nations

Cover photo: Yung-Sen Wu
United Nations World Oceans Day Photo Competition

United Nations publication
Sales no.: E.21.V.5
ISBN: 978-92-1-130422-0
eISBN: 978-92-1-604006-2

Printed at the United Nations, New York

Contents

Volume I

Volume II

Chapter 8
Trends in the state of human society in relation to the ocean

Chapter lead member: Alan Simcock.

Chapter 8A
Coastal communities and maritime industries

Contributors: Alan Simcock (chapter lead member and convener), Austin Becker, Marcelo Bertellotti, Anthony Charles, Leandra Gonçalves, Miguel Iñíguez, Osman Keh Kamara (co-lead member), Paula Keener, Jenna Lamphere, Candace May, Ishmael Mensah, Essam Yassin Mohammed (co-lead member), Tanya O'Gara, Christina Pita, Jean Edmond Randrianantenaina, Maria Sahib, Regina Salvador, Anastasia Strati (co-lead member) and Jean-Claude Tibe.

Keynote points

- About 40 per cent of the world's population lives in the coastal zone, that is, within 100 km of the coast. The proportion is increasing.
- Coastal communities play a key role in supporting all components of the ocean economy, as well as a range of social and cultural values, and all forms of coastal and marine management and governance. While coastal communities often have to deal with physical and social vulnerabilities, they are crucial contributors to conservation, to marine hazard responses and to climate mitigation and adaptation.
- The ocean supports a wide range of economic activities, including the harvesting of food, shipping, seabed mining, offshore hydrocarbon exploration and exploitation, tourism and recreation, use of marine genetic resources, production of fresh water by desalinization and production of salt. The various economic activities are steadily growing in scale. Separate chapters in part 5 of the present *Assessment*, on trends in pressures on the marine environment, give more detail on areas not discussed in depth here.
- Shipping carries about 90 per cent by volume of international trade, which makes it fundamental to the global economy. It is still recovering from the economic crisis of the period 2008–2011.
- Globally, tourism continues to grow at about 6 per cent per year. Coastal tourism represents a substantial proportion of overall economic activity for many countries, especially small island developing States and archipelagic States.
- Shipping and tourism have been seriously dislocated by the COVID-19 pandemic.
- Desalinization continues to grow in importance, in particular in the Middle East, North Africa and small island States and archipelagic States. Sea salt production also continues at a generally steady level, but accounts for only about one eighth of total salt production.

1. Introduction

The present chapter contains an overview of the relationship between humans, their economic activities and the ocean. It starts with a description of the way in which the human population is concentrated to a growing extent around the coasts. It then provides an overview of the communities in which those costal populations live, followed by an overview of the main economic activities that involve the ocean: harvesting food from the ocean; shipping; tourism and recreation; seabed mining; offshore hydrocarbon exploration and exploitation; the use of marine genetic resources; the production of fresh water by desalinization; and the production of salt. It is intended to provide, as far as possible, information on levels of economic activity, levels of employment, gender perspectives and the safety aspects of the activities. Some of the industries are discussed in detail in part 5 with regard to the pressures they impose. Therefore, the present chapter contains cross references to chapters in part V in order to avoid duplication. For shipping and tourism, however, more detail is given in the present chapter. The pressures from shipping are dealt with in chapter 10 on nutrient pollution, chapter 11 on liquid and atmospheric inputs and chapter 12 on solid waste. Tourism infrastructure is considered in chapter 14 on marine infrastructure, and the effects of tourism on species and habitats are considered in chapters 6 and 7 on the state of species and habitats. Where appropriate, pressures from those industries are noted in the present chapter to the extent that they are not covered elsewhere.

Coastal communities are crucial components of economic activity on the coast, as home to the people who work or are involved in all kinds of maritime industries, but also in terms of the social and cultural aspects of the coast, with a range of artistic endeavours, traditional practices and communal involvement with the sea. Coastal communities also play a key role in supporting the many decision-making, management and governance activities on the coast and for the sea. In view of that link, the present chapter also provides an overview of coastal communities.

2. Coastal communities

In chapter 1 of the first *World Ocean Assessment* (United Nations, 2017a), it was noted that 38 per cent of the world's population lives within 100 km of the shore, 44 per cent lives within 150 km, 50 per cent within 200 km, and 67 per cent within 400 km (Small and Cohen, 2004). A more detailed analysis was carried out in chapter 18 of the first Assessment (United Nations, 2017b) on the location and level of activity of the world's ports, but a more general analysis of the status of coastal communities was not carried out, since the focus of discussion on human activities was sectoral.

2.1. Coastal population and size of coastal communities

Although there have been calls for regular monitoring and assessment of the process of change in coastal areas (see, for example, Shi and Singh, 2003), they have largely been at the national or regional levels. Little, if anything, has been published about the total global coastal population since the early 2000s. Because of the significance of the impacts of sea level rise, studies since then have concentrated, in particular, on low-elevation coastal zones, which have a narrower scope (for example, Neumann and others, 2015).

Studies in the early 2000s showed that, globally, there is a major concentration of population in the coastal zones. Figure I, which is drawn from a 2005 study conducted under the aegis of the Food and Agriculture Organization of the United Nations (FAO) on the basis of national population data, shows an estimate of global population density in 2015 based on trends observed over the five years preceding the study. The project uses night-time satellite data of observed light sources to identify urban areas and reallocates census count data within administrative boundaries. The resulting map (figure I) shows that the global coastal population is concentrated mostly in East, South-East and South Asia. The evidence suggests that concentration in the coastal zone is increasing as a proportion of the total global population (Merkens and others, 2016). Nevertheless, access to the ocean, in particular for maritime transport, remains important for landlocked States.

Urban areas near the coast reinforce the concentration: 40 per cent of the population within 100 km of the coast lives in 4 per cent of the land area within that distance (Small and Nicholls, 2003). Much of the concentration (about 90 per cent) is in coastal cities with populations of over 1 million. An analysis of such cities as recorded in *The World's Cities in 2018* (United Nations, Department of Economic and Social Affairs (UNDESA), 2018) is shown in table 1.

Figure I
Global population density estimate, 2015

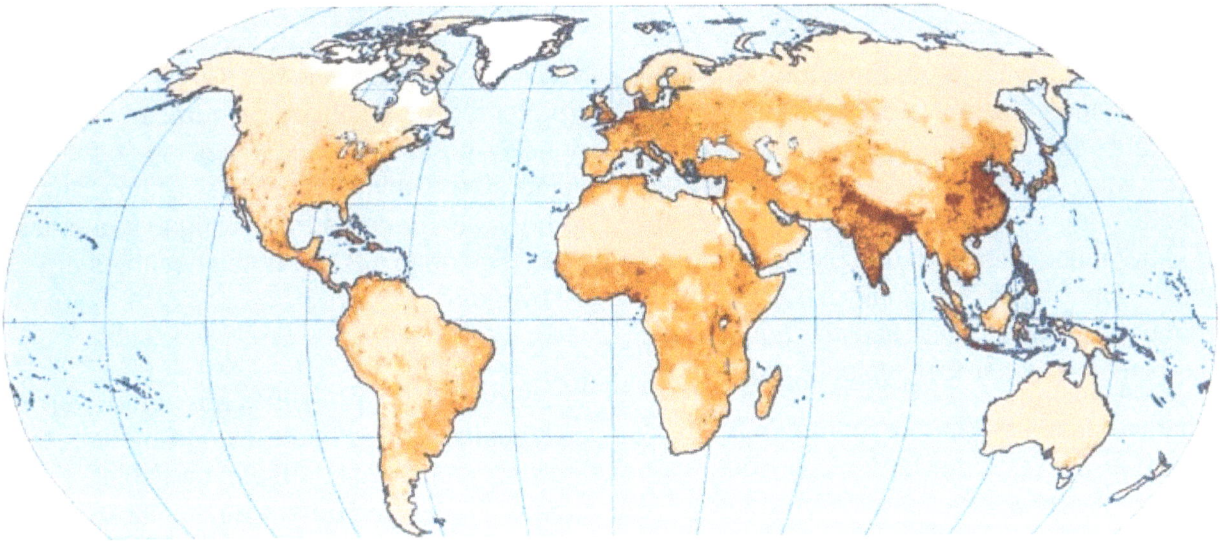

Persons per square kilometre: 1–5; 5.1–25; 25.1–50; 50.1–100; 100.1–250; 250 or more

Source: Center for International Earth Science Information Network and FAO, 2005.

Table 1
Coastal cities with populations of over 1 million in 2018

Region	Number of coastal cities of over 1 million population in 2018	Total population of those cities in 2018 (millions)	Range of annual average growth rates of those cities, 2000–2018
Sub-Saharan Africa	21	54.6	6.6 – 0.4
North Africa	6	16.1	3.5 – 0.7
East Asia	60	258.7	6.3 – 0.1
South Asia	12	86.3	5.6 – 1.2
South-East Asia	20	74.4	6.8 – 0.6
West Asia	14	44.8	5.2 – 1.3
Europe	19	48.1	1.5 – (-0.1)
Latin America and the Caribbean	28	94.2	2.7 – (-0.1)
North America	15	66.5	2.7 – 0.2
Oceania	5	16.8	2.1 – 0.9
Total	200	760.5	

Source: United Nations, Department of Economic and Social Affairs, 2018.

The analysis thus shows that the main concentrations of urban coastal population are in East, South and South-East Asia, and that the most rapid rates of growth of such populations are in those regions and sub-Saharan Africa.

At the other end of the scale are tens of thousands of smaller coastal communities around the world. The number of, and populations in, such communities are unknown. It seems likely, however, that the number of such communities along the coasts of the world is high, and that official local government units often contain many more than one community. For example, in Nova Scotia, Canada, a recent assessment indicates that, while there are about 50 official municipalities, there are approximately 1,000 separate coastal communities (Charles, 2020). Accordingly, there is great diversity among coastal communities across the globe, notably in differences between the big cities noted above and rural communities, where such economic activities as fishing, aquaculture, shipping and tourism are typically prominent.

Whatever the size of the community, it often plays a role in stewardship of the coast. Indeed, the role of coastal communities in conservation is being increasingly recognized and valued, in terms of many local initiatives in ocean conservation, around the world, that often succeed both in improving livelihoods and protecting communities (Charles, 2017; Charles and others, 2020).

The role of coastal communities in conservation is being increasingly valued. Many coastal communities around the world and their small-scale fishers have undertaken a large number of local initiatives in ocean conservation, often with considerable success. The successes of those communities are often based on local knowledge, structures and cooperation (Charles, 2017).

The vulnerability of coastal communities to the impacts of climate change is of increasing concern. It is relevant to the planning of tourism development, in particular in small island developing States with economies that are dependent on tourism, and fisheries management. The Intergovernmental Panel on Climate Change concludes that, under current trends of the increasing exposure and vulnerability of coastal communities to climate change, the risks of erosion and land loss, flooding, salinization and cascading impacts owing to mean sea level rise and extreme weather events, among others, are projected to increase significantly throughout the present century (Intergovernmental Panel on Climate Change (IPCC), 2019). Coastal communities located in the Arctic, in low-lying (often deltaic) States, such as Bangladesh and Guyana, in paths frequented by cyclones or hurricanes and in densely populated megacities are especially vulnerable. On the other hand, there appear to be health benefits from living in the coastal zone (see chap. 8B on human health as affected by the ocean).

Small coastal communities are not just physically vulnerable to climate change impacts; they are also socially vulnerable, in particular in rural areas (Charles and others, 2019). Rural coastal communities are vulnerable to weather events and flooding as a result of geographic location and limited access to health care, goods, transportation and other services. Sensitivity to market fluctuations from their dependence on natural resources, and poverty, limited economic opportunities and losses of populations, create problems when trying to adapt (Armitage and Tam, 2007; Amundsen, 2015; Bennett and others, 2016; Metcalf and others, 2015; May, 2019c). Such factors strain material assets, as well as the social and moral foundations that facilitate collective problem-solving (Amundsen, 2015; May, 2019a). Communities are more likely to mobilize collective resources in response to threats when people actively care about each other and the place they live (Amundsen, 2015; May, 2019b; Wilkinson, 1991). That may be a function of attachment to the history, culture or environmental context of a place and/or the people in a place. Those attachments can become potential sources of resistance to change in contexts of low social diversity and slow population change, or the basis for conflict in contexts of high social diversity and fast population change (Graham and others, 2018; May, 2019b, 2019c). The combined effect of physical and social vulnerability on community capacities is particularly challenging at a time when collective action efforts for mitigation and adaptation are more important than ever (May, 2019b, 2019c).

The Intergovernmental Panel on Climate Change warns that, for our most vulnerable communities, many of which are coastal, transformative mitigation and adaptation is necessary to assuage the worst impacts of climate change. Incremental change is no longer seen as a possibility by most States: more radical action is thought to be needed to reduce the impacts of and adapt to a changing climate. Responses to threats from climate change are varied and include a mix of hard and soft coastal defences. Built infrastructure, such as sea walls or dykes, is widely used but tends to be more costly and maintenance-dependent than ecosystem-based measures, such as marshes, mangroves, reefs or seagrass (see also sect. 7.3). Having limited data inhibits estimates of the cost effectiveness of both hard and soft measures, especially across geographies and scales (Oppenheimer and others, 2019), although State-level estimates exist (see, for example, Environment Agency of the United Kingdom, 2015). The World Bank estimated that, without concrete climate and development action, over 143 million people could be forced to move within their own countries to escape the slow-onset impacts of climate change by 2050 in just three regions: sub-Saharan Africa, South Asia and Latin America (Rigaud and others, 2018). To address those problems, in coastal areas, integrated coastal zone management is widely regarded as an effective approach to climate change and other drivers (Nicholls and Klein, 2005; Nicholls and others, 2007; see also chap. 27 on management approaches).

3. Capture fisheries, shellfish harvesting and aquaculture

Food from the sea represents the largest maritime industry in terms of the numbers of people involved. In 2017, the total first sale value of total production was estimated at $221 billion, of which $95 billion was from marine aquaculture production (including fish, shellfish and seaweed). Those figures include small proportions of production not used for food (FAO, 2019). Further details are given in chapter 15 on capture fisheries, chapter 16 on aquaculture and chapter 17 on seaweed harvesting.

The world fishing fleet consisted of about 4.5 million vessels in 2017, a number that has been relatively stable since 2008. Globally, just under one third of the fishing fleet is still composed of unpowered vessels, which reflects the large proportion of small-scale and subsistence fisheries. Only 2 per cent of the total fleet consists of vessels of 24 or more m in length overall, and about 36 per cent of vessels are less than 12 m in length overall (FAO, 2019).

In 2017, an estimated 135 million people were involved in capture fisheries and marine aquaculture: some 120 million in capture fisheries and some 15 million in marine aquaculture. Employment in capture fisheries (as opposed to subsistence fishing) amounts to about 40.4 million, and employment in marine aquaculture is about 15.6 million. In addition, there is a slightly smaller workforce engaged in post-harvest processing. About 13 per cent of that employed workforce are women. Including subsistence fishing, about 50 per cent of those engaged in that group of activities are women (FAO, 2019; World Bank and others, 2012). There have been no recent surveys of death and injuries in the fishing industry. However, the most recent survey shows that those engaged in the industry suffer much higher levels of death and injury at work than in other industries: about 18–40 times higher than the average in a range of developed countries for which statistics were available (Petursdottir and others, 2001).

Apart from subsistence fisheries, fisheries and aquaculture depend on substantive supply chains from producer to consumer. The problems caused by the COVID-19 pandemic are challenging fishing industries, especially in relation to international trade of products, and disrupting the supply chains. Fishing operations have also been affected, with effort reduced by an estimated 6.5 per cent in March and April 2020. In some areas (e.g., the Mediterranean and the Black Sea), small-scale fisheries have been halted. In the future, COVID-19-compliant practices will lead to restrictions on working practices both on the water and in post-harvest handling (FAO, 2020).

Further information on capture fisheries, aquaculture and seaweed harvesting can be found in chapters 15, 16 and 17, respectively.

4. Shipping

4.1. Situation as shown in the first *World Ocean Assessment*

When the first *World Ocean Assessment* was written, international shipping was still recovering from the financial crisis that occurred from 2008 to 2011. Shipping is conventionally reckoned to represent 90 per cent of international trade, although one estimate in the first Assessment put it nearer to 75 per cent by volume and about 60 per cent by value (United Nations, 2017f).

4.2. Cargo traffic

Until 2020, recovery of the world's economy after 2011 has been reflected in the growth of world trade and, consequently, in the tonnage of cargo carried by international shipping (figure II). When the distances over which the cargoes were carried are taken into account, the growth in ton-miles is even larger (United Nations Conference on Trade and Development (UNCTAD), 2019). Recovery is still in progress and has been seriously affected by the massive drop in world trade caused by the COVID-19 crisis.

Such growth, however, has occurred against a weak competitive background for the international shipping industry. The economic crisis that took place from 2008 to 2011 occurred during a time when world shipping had commissioned a large increase in tonnage to meet the increased freight demand of the preceding years. The additional tonnage was delivered at a time when demand had started to reduce, with the result that, during the 2010s, the shipping industry was operating against a background of oversupply, which had the consequence of depressing freight rates. As measures to further control the pollutant emissions from ships take effect (from 2020), further pressures associated with implementing modifications to fleets will be placed on the shipping industry. To meet the new requirements (as detailed in chap. 11), ships must either purchase bunkers with a lower sulfur content (which may have a higher price, since the traditional ships' bunkers have been the high-sulfur oils for which there was less demand) or retrofit scrubbers to clean the ships' exhaust. Further economic pressures of that kind are described in chapter 11. The combined effect of continuing overcapacity and higher operating costs remains unclear (UNCTAD, 2019).

For many years, the quantities of cargo loaded in ports in developing countries were smaller than those unloaded in those countries, marking an imbalance in seaborne trade. By the time of the first Assessment, the quantities, on average, were nearly in balance and, since then, the quantities loaded in developing countries now exceed those unloaded. Even excluding China, as the single largest developing country importer/exporter, there is still an excess of unloading in developing countries (UNCTAD, 2019).

Figure II
International seaborne trade by commodity type, 2013–2018

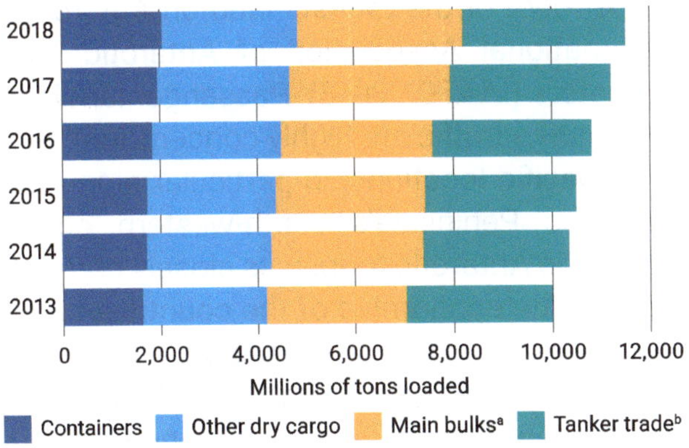

Source: UNCTAD, 2019.
a "Main bulks" are iron ore, grain and coal.
b "Tanker trade" covers crude oil, refined petroleum products, gas and chemicals.

Container traffic continues to be focused on the main East-West arteries across the northern hemisphere (Asia-Europe, trans-Pacific and trans-Atlantic), which account for 40 per cent of all container shipping. Of the remaining 60 per cent, 27 per cent is intraregional, 13 per cent occurs across the other East-West routes in the northern hemisphere, 12 per cent is associated with traffic between southern hemisphere countries, and 8 per cent is associated with North-South traffic (UNCTAD, 2019). At the same time, there is a growing tendency to consolidate container shipping, so the combined market share of the top 10 container shipping lines increased from 68 per cent in 2014 to 90 per cent in 2019. That is combined with a returning interest in container shipping lines integrating their operations with traffic between originators and ports and between ports and the ultimate destinations. Those developments have the ability to undermine competition and thus to result in higher transit costs (UNCTAD, 2019).

The total world fleet of ships carrying all that cargo amounted to 96,295 ships in early 2019, accounting for 1.97 billion dead-weight tons of capacity. Bulk carriers and oil tankers maintained the largest market shares of vessels that dominated the world fleet, at 42.6 per cent of all vessels and 28.7 per cent of dead-weight tons, respectively. A large proportion of the world's tonnage continues to be registered in a relatively small number of registries. Nearly 70 per cent of the world's tonnage is registered in seven registries: Panama (17 per cent), Marshall Islands (12 per cent), Liberia (12 per cent), Hong Kong Special Administrative Region of China (10 per cent), Singapore (7 per cent), Malta (6 per cent) and China (5 per cent). No other registry is responsible for more than 4 per cent of the world's tonnage (UNCTAD, 2019).

Likewise, ownership and control of shipping continues to be concentrated in the hands of firms in a relatively small number of countries. In 2019, five economies accounted for more than 50 per cent of the world tonnage: Greece, Japan, China, Singapore and Hong Kong, China. Between 2015 and 2019, Greece, Singapore, China and Hong Kong, China have increased the proportion that they own/control (UNCTAD, 2019).

The construction of new ships still remains very concentrated in China, Japan and the Republic of Korea, which together represent 90 per cent of all cargo ship construction activity. The demolition of ships that have reached the end of their useful life likewise continues to be concentrated in the same countries as reported in the first Assessment. In 2018, 47.2 per cent of the total reported tonnage of propelled seagoing vessels of 100 gross tons and above that were sold for demolition were demolished in Bangladesh, 25.6 per cent in India, 21.5 per cent in Pakistan, 2.3 per cent in Turkey and 2 per cent in China, leaving 1.4 per cent for the rest of the world. The share of the market held by China, India and Turkey has been declining (UNCTAD, 2019).

In 2020, the COVID-19 pandemic has been disrupting global trade extensively. Demand for the transport of raw materials and finished goods has dropped significantly, while demand for the transport of health-related goods has risen (United Nations Coordinating Committee on Statistical Activities (UNCCSA), 2020). Overall, cargo shipping activity has dropped significantly: for example, trade from the European Union to China and the United States dropped in the first 31 weeks of 2020 by 47 per cent and 25 per cent, respectively, compared with 2019; trade in the reverse directions has dropped by 26 per cent and 38 per cent, respectively (European Maritime Safety Agency (EMSA), 2020).

4.3. Passenger traffic

Passenger traffic is almost entirely carried on local ferries or on cruise ships. The pattern of ferry traffic remains as described in the first Assessment, but the level of traffic has grown steadily (International Shipping Economics and Logistics (ISL), 2017).

The activities of cruise ships have also continued to grow steadily with the increased global market for cruising: the number of passengers is increasing at an average of about 5 per cent per year (figure III). The size of individual cruise ships is also growing steadily (figure IV). The overall market remains dominated by passengers from the United States (about 50 per cent of the total market) and the global distribution of cruising remains largely as described in the first Assessment, with the major focuses being the Caribbean and the Mediterranean, which together accounted for a little over half of all traffic in 2017 (Cruise Lines International Association (CLIA), 2018).

The first Assessment noted the relatively recent, but rapid, growth of tourism to Antarctica, in particular with regard to cruise ships – from 27,324 cruise ship passengers in the 2003–2004 season to 37,044 in the 2013–2014 season, which is an increase of 35 per cent. The growth has continued, reaching 51,700 in the 2017–2018 season (an increase of a further 40 per cent), with a forecast of further growth to 55,750 in the 2018–2019 season. Over 80 per cent of the tourists land on Antarctica (International Association of Antarctic Tour Operators (IAATO), 2018). Passenger landings and marine traffic are highly concentrated at a few specific locations, in particular along the Antarctic Peninsula's south-western coast. Growth in Antarctic tourism is closely correlated with the economies of the countries sending the most visitors to the region: 60 per cent of the tourists come from the United States (33 per cent), China (16 per cent) and Australia (11 per cent). The proportion of tourists from China increased significantly between 2013 and 2014 and between 2017 and 2018. Markets for Antarctic travel are probably far from saturated, and demand is therefore likely to continue to grow (Bender and others, 2016). Apart from some categories, such as private yachts, that shipping traffic is covered by the new mandatory Polar Code (International Maritime Organization (IMO), 2015).

Figure III
Numbers of passengers on cruise ships, 2006–2017 (millions)

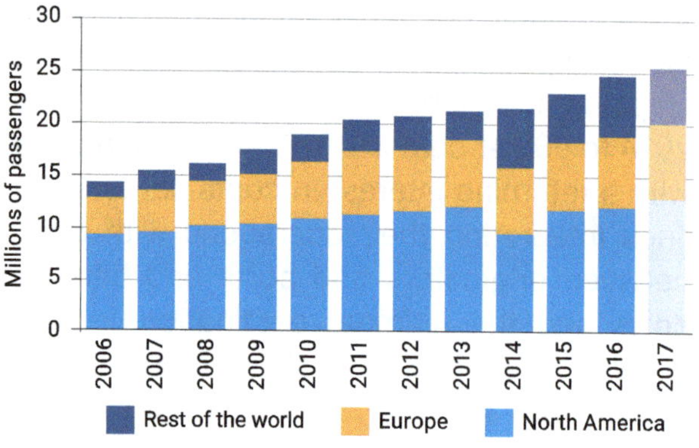

Source: ISL, 2017.

Note: 2017 statistic estimated.

Figure IV
Numbers of cruise ships and their gross tonnage

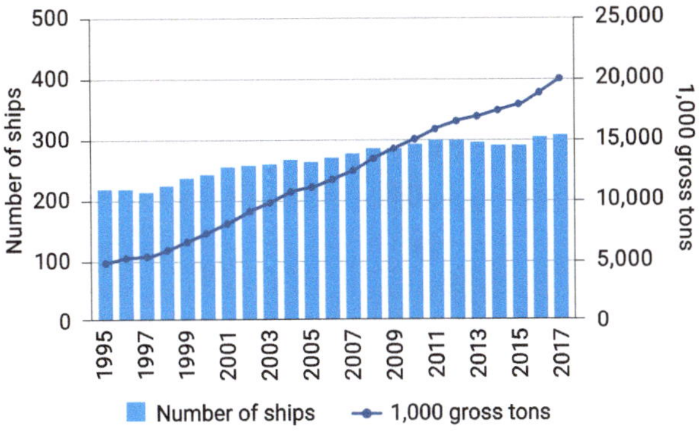

Source: ISL, 2017.

Tourism is also increasing rapidly in the Arctic: summer tourism quadrupled and winter tourism increased by over 600 per cent between 2006 and 2016, although large areas remain unaffected. The increase is likely to have an impact on Arctic ecosystems and communities, especially as new parts of the Arctic open up with less sea ice, new airports and continued promotion of the area (Runge and others, 2020).

In 2020, passenger traffic on ferries dropped significantly early in the year as a result of the COVID-19 pandemic, but, by August 2020, it was beginning to recover (e.g., EMSA, 2020). Cruise ship activity has plummeted for the same reason: in August 2019, there were 1.8 million persons on board cruise ships; in August 2020, there were only a small number of crew (EMSA, 2020).

4.4. Seafarers

The number of seafarers serving on international merchant ships was estimated in 2015 at 1,647,500, of which 774,000 were officers and 873,500 ratings. A new survey will be carried out in 2020. China, the Philippines, Indonesia, the Russian Federation and Ukraine were estimated to be the five largest supply countries for all seafarers. For officers, China was reported to be the largest supplier, followed by the Philippines, India, Indonesia and the Russian Federation. For ratings, the Philippines was the largest supplier, followed by China, Indonesia, the Russian Federation and Ukraine. In 2015, there was thought to be a shortage of about 16,500 officers and a surplus of about 119,000 ratings. While the global supply of officers is forecast to increase steadily, the trend is expected to be outpaced by increasing demand (Baltic and International Maritime Council and the International Chamber of Shipping (BIMCO/ICS), 2016). The important international instruments for the protection of seafarers were described in the first Assessment.

The best estimate of the proportion of seafarers who are women remains at about 2 per cent, mainly in the cruise ship sector (International Transport Workers Federation (ITF), 2019).

Travel and border restrictions imposed in 2020 to control the spread of COVID-19 have created a major crisis for seafarers. In July 2020, there were estimated to be 600,000 seafarers affected: approximately 300,000 seafarers kept working aboard ships owing to problems related to changing crews, and an equal number of unemployed seafarers were waiting ashore to join their ships (ITF, 2020).

4.5. Piracy and armed robbery against ships

There was a slight decline in the total number of attempted and actual cases of piracy and armed robbery against ships between 2015 and 2019 (table 2). The most significant areas in which piracy and armed robbery occur remain those in South-East Asia and West Africa.

Table 2
Attempted and actual cases of piracy and armed robbery against ships, 2015–2019

Region	2015	2016	2017	2018	2019
East Asia	31	16	4	7	5
South-East Asia	147	68	76	60	53
South Asia	24	17	15	18	4
East Africa, the Red Sea and the Gulf of Aden	3	6	13	5	4
West Africa and the Mediterranean	32	57	45	82	67
South America	8	22	24	25	24
Rest of the world	1				
Total	246	191	180	201	162

Source: International Maritime Bureau of the International Chamber of Commerce, 2020.

4.6. Environmental impacts

Discharges and emissions from ships and sewage are discussed along with other liquid and atmospheric pollution in chapter 11, with garbage derived from ships considered in chapter 12 and noise inputs to the ocean from ships covered in chapter 20.

Environmental impacts associated with the growth of shipping in the Arctic Ocean are considered in chapter 7K. Steps are being taken to prepare sustainably for such traffic, with the International Maritime Organization adopting the International Code for Ships Operating in Polar Waters (Polar Code),[1] which is mandatory under both the International Convention for the Safety of Life at Sea[2] and the International Convention for the Prevention of Pollution from Ships[3] (IMO, 2015). The Arctic Council has also set up arrangements for emergency prevention, preparedness and response for shipping incidents and, in 2011, it adopted a legally binding Agreement on Cooperation on Aeronautical and Maritime Search and Rescue in the Arctic (Arctic Council, 2011).

5. Seabed mining

There are two distinct aspects to the seabed mining industry. One is the long-established mining of relatively shallow deposits by a number of countries within their own waters. The other is the potential development of deep seabed mining for which commercial operations have not yet commenced. The established mining undertakings include, among others, aggregates (sand and gravel) in many Western European countries; placer diamond mining in Namibia; placer tin mining in several South-East Asian countries; and, most recently, ironsand mining in New Zealand. There are also projects related to mining for phosphorite under development in Mexico, Namibia and New Zealand. Details of both established and potential activities are given in chapter 18 on seabed mining.

1 International Maritime Organization, document MEPC 68/21/Add.1, annex 10.
2 United Nations, *Treaty Series*, vol. 1184, No. 18961.
3 International Maritime Organization, document MEPC 62/24/Add.1, annex 19, resolution MEPC.203(62).

The established mining activities are disparate, since they involve very different countries and situations. No overview of the economics of such activities is available, and there have been no surveys of employment, of the occurrence of death and injury to workers or of pay across the field.

6. Offshore hydrocarbons

In 2016, approximately 27 per cent of the global production of oil, and 30 per cent of that of natural gas, was offshore. Offshore oil is produced in more than 50 different countries, including Brazil, Mexico, Norway, Saudi Arabia and the United States (International Energy Agency (IEA), 2018). For natural gas, Australia, Iran (Islamic Republic of), Norway and Qatar were the main offshore producers in 2017. The offshore industry had an estimated annual global investment capital expenditure of $155 billion in 2018, which is projected to reach $200 billion by 2021. Further details are given in chapter 19 on hydrocarbon exploration and extraction.

Chapter 21 of the first Assessment (United Nations, 2017c) provided a survey of the social aspects of the offshore hydrocarbon industry. In general, that description remains accurate. Employment numbers inevitably fluctuate significantly, depending on the international price of crude oil and the planned capital expenditure by oil and gas companies. The workforce draws heavily from a global talent pool.

7. Tourism and recreation

7.1. Situation as shown in the first *World Ocean Assessment*

Chapter 27 of the first Assessment (United Nations, 2017d) assessed the full range of aspects of tourism and recreational activities affecting the ocean. They included the scale, showing rapid growth over several decades; the social and economic aspects, showing the economic importance for many countries (especially small island developing States); the demands for built environments; and the many pressures that tourists and their activities impose on the marine environment. Exceptionally, cruising was treated as part of chapter 17 on shipping.

In the present Assessment, tourism-related infrastructure and development is considered in chapter 14, and the problems associated with atmospheric, liquid and solid wastes resulting from tourist activities are considered in chapters 11 and 12. The present section, therefore, deals with the social and economic aspects of tourism.

The picture has recently changed substantially because of the COVID-19 pandemic. The World Tourism Organization projects that the number of international tourist arrivals in 2020 is likely to drop by between 58 per cent and 78 per cent compared with 2019, depending on what happens with travel restrictions imposed through efforts to control COVID-19 in the second half of the year. In March 2020, arrivals dropped by 60 per cent compared with 2019. (UNCCSA, 2020). The countries most affected are those that rely substantially on tourism, including island nations in the Pacific Ocean, the Indian Ocean and the Atlantic Ocean (Pacific Community, 2020; UNCCSA, 2020).

7.2. Scale and distribution of tourism

Tourism affecting the ocean, other than cruising, is predominantly located in the coastal zone. Statistics are not available globally to show the scale of tourism in the coastal zone. Because of their geography, some countries with large tourism industries, such as Greece, inevitably have a very large proportion of that industry in coastal areas. Elsewhere, evidence from different regions of the world continues to show that coastal tourism remains a major component of overall tourism. For example, in addition to the evidence quoted in the first Assessment:

(a) In the countries of the European Union, four of the five regions with the highest levels of tourist activity in 2016 (Canary Islands, Catalonia, Adriatic Croatia and Balearic Islands) were coastal regions (the other region was Île-de-France, around Paris) (European Commission, 2018);

(b) The percentage of tourists in the Republic of Korea who visited the coastal zone increased from 49.5 per cent in 2000 to 69.1 per cent in 2010, and the total number of beach visitors in 2014 was 69 million (Chang and Yoon, 2017);

(c) Destinations in the four coastal provinces of Northern Cape, Western Cape, Eastern Cape and KwaZulu-Natal in South Africa accounted for 28 per cent of the total tourism trips and 40 per cent of total tourism spending in 2015. Overall, coastal destinations were dominated substantially by domestic tourists: 9.8 million domestic tourism trips as compared with 1.6 million international tourist trips; tourism activity is particularly concentrated around Cape Town and in the eThekwini Metropolitan Municipality (which includes Durban), which in 2015 together accounted for 75 per cent of total tourism spending in South African coastal areas (Rogerson and Rogerson, 2018, 2019).

International travel and associated tourism play a major role in many parts of the world, in particular in the "sun, sea and sand" type of tourism. The relatively rapid rate of growth in international travel observed in the first Assessment continued throughout the 2010s (table 3) and between 2011 and 2017. Throughout the world as a whole, the rate of growth in the numbers of international tourists continued between 2011 and 2017 at above the long-term rate, reaching an annual average rate of 5.7 per cent, slightly higher than that reported in the first Assessment. The estimated income derived from international tourism has continued to grow globally, at an annual average rate of 4.0 per cent, but not in line with the number of tourists. That implies that, on average, tourists are spending less. However, the global growth in tourist numbers is sufficient to more than offset the decline, and the share of tourism in export earnings globally has continued to increase (World Bank, 2019).

Global patterns in numbers of tourists and expenditure vary significantly between regions (table 4). The absolute scale of tourism in different regions also varies significantly. Collectively, some of the countries in South Asia and South-East Asia (Bangladesh, India, Maldives, Myanmar and Pakistan) achieved a 119 per cent increase in inbound international tourist numbers between 2011 and 2017 (although from a relatively low base), far outstripping other regions. Other regions have, in general, experienced growth rates of less than 10 per cent (table 4). Nevertheless, Caribbean States, such as the Dominican Republic and Jamaica, have had growth rates of around 25 per cent, well above the regional average (World Bank, 2019). The Middle East and North Africa has experienced relatively low growth in tourist numbers, but a substantial growth in tourist income, suggesting that the tourist industry is offering more upmarket experiences (World Bank, 2019).

Table 3
Inbound international tourism by global region

Area	Inbound international tourists (millions)		Average annual increase, 2011–2017 (percentage)	Inbound international tourism expenditure (billions of dollars)		Average annual increase, 2011–2017 (percentage)	Regional average of inbound international tourism spending (percentage of total exports)	
	2011	2017		2011	2017		2011	2017
World	997.7	1 341.5	5.7	1 231.0	1 525.7	4.0	5.5	6.7
East Asia and the Pacific	206.8	300.6	7.6	291.2	373.0	4.7	4.5	5.2
Europe and Central Asia	512.8	669.5	5.1	534.6	594.5	1.9	5.7	6.3
Latin America and the Caribbean	75.9	112.4	8.0	70.9	101.8	7.3	5.1	7.8
Middle East and North Africa	75.2	89.2	3.1	74.0	112.5	8.7	5.5	10.8
North America	79.1	98.0	4.0	208.1	272.3	5.1	7.8	9.5
South Asia	10.4	22.8	119.2	23.0	37.9	10.8	4.4	6.5
Sub-Saharan Africa	33.1	42.4	4.7	29.0	34.4	3.1	5.8	9.2

Source: Compiled from World Bank, 2019.

Table 4
Share of international tourist arrivals by global region

Region	International tourist arrivals 2017 (percentage)
World	100
East Asia and the Pacific	22.5
Europe and Central Asia	49.9
Latin America and the Caribbean	8.4
Middle East and North Africa	6.7
North America	7.4
South Asia	1.3
Sub-Saharan Africa	3.3

Source: Compiled from World Bank, 2019.

Domestic tourism dominates the tourist market in most major economies (figure V), with 73 per cent of expenditure on tourism and travel derived from domestic sources globally (World Tourism and Travel Council (WTTC), 2018). While it will include much tourism and travel that does not have an impact on the marine environment, coastal tourism is, as noted above, a major component of total tourism. Domestic tourism has grown generally in line with total tourism, and growth rates are estimated at over 10 per cent per year in many Asia-Pacific countries, such as China, Malaysia and the Philippines, over the period 2011–2017 (WTTC, 2018).

Figure V
Relative importance of domestic and international tourism and travel expenditure in 31 countries (percentage of travel and tourism spending)

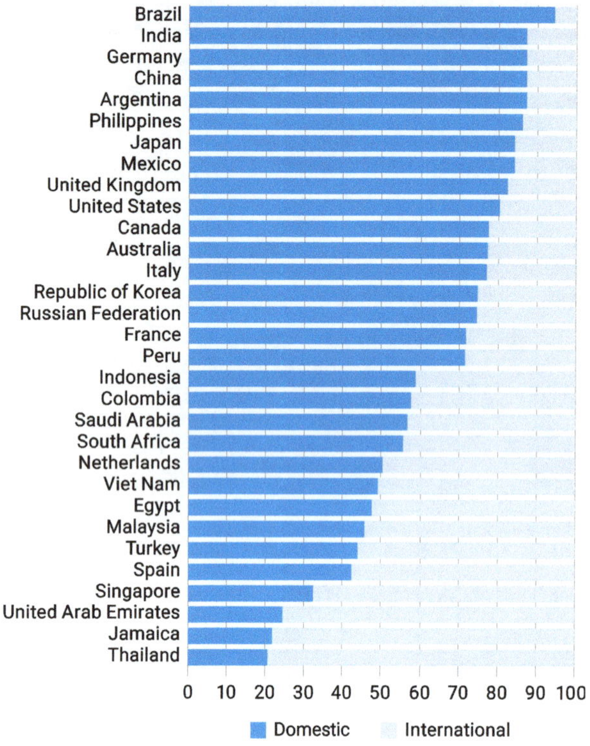

Source: WTTC, 2018.

7.3. Impacts on the marine environment

Throughout all tourist areas, the major impact on the marine environment comes from coastal development, including the proportion of land covered by buildings, such as hotels, restaurants and retail shops, and transport infrastructure, including ports, airports and train terminals, and the need for hard built coastal defences, street lighting and sewerage (see also chap. 14). Where such development is not subject to effective planning and management, impacts on marine flora and fauna can be disastrous. For example, at Vlora Bay in Albania, unplanned development over 15 years has resulted in the disappearance of 50 per cent of the seagrass meadows and a substantial reduction in macroalgae (Fraschetti and others, 2011).

In tourist regions, beach feeding or beach nourishment, which is the replacement of sand on beaches which have had sand removed by coastal currents or extreme weather events, can have considerable economic benefits (Klein and Osleeb, 2010). For example, in the Republic of Korea, an evaluation of the economic benefits of the restoration of the Songdo beach at Busan after typhoon damage in 2003 put the benefits at about $230 million (Chang and Yoon, 2017).

The management of beaches is a significant element in managing the impacts of coastal tourism on the marine environment. Beach cleaning and the building of sea walls are generally done to give "sun, sea and sand" tourists surroundings that they find more attractive, and they have significant effects on the local flora and fauna, as recorded in the first Assessment. Studies continue to show that beaches used extensively for tourism support ecosystems that are less rich than those of comparable beaches in the same vicinity that are in protected areas, for example, along the New Jersey coast in the United States, (Kelly, 2014) and near Cadiz, Spain (Reyes-Martínez and others, 2015), and that seawalls supported 23 per cent less biodiversity and 45 per cent fewer organisms than natural shorelines (Gittman and others, 2016).

Other interventions to attract tourists to beaches have included the creation of artificial surfing reefs. The limited success of such structures was recorded in the first Assessment, but there is now a report of a new venture based on an inflatable artificial reef at Bunbury, Australia (West Australian, 2019). National legislation to promote public access to coasts and beaches can also be significant.

7.4. Enjoyment of marine wildlife

7.4.1. Diving

Snorkelling and scuba diving continue to be a significant element in marine tourism, focused on enabling tourists to enjoy underwater wildlife. The substantial growth (about 25 per cent) in the levels of the activity recorded in the period from 2000 to 2013 and reported in the first Assessment has now slowed down but still continues. Based on the statistics of the Professional Association of Diving Instructors, between 2013 and 2019, there was about 6 per cent growth in the number of establishments offering diving training (about 6,600 in 2019), about 1 per cent growth in the number of individual trainers (about 137,000 in 2019) and about an 11 per cent increase in the number of people trained annually (about 1 million in 2019) (Professional Association of Diving Instructors (PADI), 2019).

The main interest in diving lies in areas endowed with coral reefs – the corals and other reef biota are spectacular and attract large numbers of tourists who want to see them. In some areas, as recorded in the first Assessment, studies suggest that it is possible to manage coral reef tourism (e.g., by limiting the number of divers in an area, specifying divers' behaviour and generally increasing divers' awareness of the problems) compatibly with sustaining the condition and health of the reef. In other areas, however, studies continue to suggest that the interaction of divers with coral is damaging the reefs. A recent study of the coral reefs around the island of Bonaire in the Caribbean part of the Netherlands showed that diving is at levels probably at least twice those considered to be the upper limit beyond which damage is likely to occur (see Hawkins and Roberts, 1997), and that damage, albeit largely unintentional, is occurring but could be controlled by better management measures (Jadot and others, 2016).

As part of the decommissioning of offshore installations, significant numbers of disused installations are being used to create artificial reefs. In the Gulf of Mexico alone, 532 installations had, by 2018, been used as artificial reefs (Bureau of Safety and Environmental Enforcement of the United States (BSEE), 2020). In 2016, it was estimated that some 600 offshore installations would be decommissioned between 2017 and 2021. Not all of them were intended as places for divers to explore, but a substantial proportion are being used in that way (Van Elden and others, 2019).

A new area of interest for scuba diving is emerging in the form of diving over muddy substrates, known as "muck diving", which focuses on finding rare, cryptic species that are seldom seen on coral reefs. A recent study investigated the value of "muck diving", its participant and employee demographics and potential threats to the industry. Results indicate that "muck diving" tourism is worth more than $150 million annually in Indonesia and the Philippines combined. It employs over 2,200 people and attracts more than 100,000 divers per year (De Brauwer and others, 2017).

7.4.2. Wildlife watching

Birdwatching ("avitourism") continues to be a significant element in coastal tourism, but coastal birdwatching can rarely be disaggregated from other birdwatching. Increased efforts are being made to promote birdwatching generally as a basis for tourism. The Netherlands Centre for the Promotion of Imports from Developing Countries (international tourism, of course, counts as an export from the country where it takes place), has identified India, Kenya, Namibia and the United Republic of Tanzania as significant destinations for avitourism, and Brazil, Costa Rica, Ecuador, Morocco, South Africa and Sri Lanka as emerging destinations (Netherlands Enterprise Agency (NEA), 2019). Statistical evidence is sparse; it seems, however, that in some areas, the market may be becoming saturated: in the United States, the National Survey on Recreation and the Environment reported that, in 2012, the number of people taking avitourism trips, including to domestic locations, stood at 19.9 million, but that in 2016, the numbers had declined to 17.6 million (United States National Survey of Fishing, Hunting and Wildlife-Associated Recreation (USNSFHWAR), 2016).

Whale watching, reported in the first Assessment as an activity with a global turnover of about $2.1 billion, continues to be a significant tourist activity: an estimated 13 million people engaged in whale watching across the globe in 2017; in Iceland, the activity was reported to have been growing by 20 per cent per year since 2015 (Hoyt, 2009, 2017), and in Peru, it grew from nought to $3 million between 2008 and 2018 (Guidino, 2020). Whale watching may benefit conservation by changing attitudes towards wild animals and natural habitats (Argüelles and others, 2016), especially if commercial tour operators educate tourists about related long-term sustainable benefits (Wearing and others, 2014). Species that live in coastal environments are the most utilized as tourist attractions because of easy access to them. If conducted properly, whale watching is relatively benign (Argüelles and others, 2016). However, uncontrolled whale watching may disturb whales, thus causing changes in their natural behaviour that could, in turn, modify their distribution, reproduction and survival (Williams and others, 2006; Lusseau and others, 2006). The International Whaling Commission, Governments and non-governmental organizations have attempted to reduce the impact of the activity worldwide by developing guidelines and codes of conduct that are aimed both at reducing the negative effects of the activity and at giving an educational opportunity to visitors (Garrod and Fennel, 2004; Cole, 2007; Argüelles and others, 2016; International Whaling Commission (IWC), 2019).

The first Assessment cited an estimate of $300 million per year as the global revenue from shark watching. A survey of shark watching in Australia supports estimates of that order, since it evaluates annual expenditure on shark watching in Australia alone at $28.5 million per year (Huveneers and others, 2017).

7.4.3. Recreational boating

In chapter 27 of the first Assessment (United Nations, 2017d), a sustained growth in recreational boating was recorded for the countries for which statistics were available over the preceding 50 years, but it was noted that, in the United States, there had been a slight reduction between 2012 and 2013, the latest date for which information was available. In the United States, the growth has more or less halted: in 2018, the number of registered recreational boats, some of which are in inland waters, is still just under 12 million, as in 2013 (National Marine Manufacturers Association (NMMA), 2018). Similarly, in the European Union, the number of recreational boats has remained roughly constant at about 6 million, while the age of those involved in boating has increased substantially, suggesting that younger people are not taking up the activity. On the other hand, outside those areas, there appears to be an active market for new boats (Ecorys, 2015).

8. Marine genetic resources

Most commercial activity with respect to marine genetic resources continues to be concentrated in a comparatively small number of countries. Some idea of the scale of activity in the sector can be gained from the fact that 28 candidates are currently in clinical trials and a further 10 drugs derived from marine natural products have already gained regulatory approval, and that 76 publicly available cosmeceutical ingredients derived from marine natural products have been marketed. The investigation of marine genetic resources is not a separate sector from pharmaceutical and industrial research generally, and the economic and social aspects of the marine component are limited in scale and cannot yet be separated. More detail is given in chapter 23 on marine genetic resources.

9. Marine renewable energy

Energy from offshore wind, wave and tidal power, that is, marine renewable energy, is increasingly feeding national distribution systems in a number of countries, although not in Africa or, to any large extent, in the Americas. Of those power sources, offshore wind technology is the most mature and technically advanced, providing a capacity of about 28.3 MW in 18 countries (International Renewable Energy Agency (IRENA), 2020c). For further information, see chapter 21 on renewable energy sources.

Total employment in the onshore and offshore wind energy sector represented about 1.2 million jobs in 2018, of which perhaps 20 per cent (240,000) related to offshore activities. Women account for about 21 per cent of persons employed in the wind energy sector as a whole (IRENA, 2020a, 2020b).

10. Desalinization

10.1. Situation as shown in the first *World Ocean Assessment*

In chapter 28 of the first Assessment, it was shown that the global installed capacity for desalinization of seawater to produce fresh water had increased from negligible amounts in 1965 to about 86.5 million m3 per day in 2015 (United Nations, 2017e). Of the two techniques predominantly used in desalinization, 71 per cent of capacity was based on membrane processes, and the remaining 29 per cent of capacity for desalinization used thermal processes. About 27 per cent of the total global capacity was found in States in the Persian Gulf area, overwhelmingly (96 per cent of the total capacity in the area) in the six States members of the Gulf Cooperation Council (Bahrain, Kuwait, Oman, Qatar, Saudi Arabia and United Arab Emirates). Significant sea-related capacities also existed in Algeria, Australia, China, Israel, Japan, Spain, the United States and islands such as Malta and Singapore, as well as many Caribbean islands.

The environmental impacts of desalinization plants noted in the first Assessment included the emission of greenhouse gases, the intake of feedwater and the discharge of brine. The impact of intakes on marine biota above microscopic sizes and the effects of discharges (which can contain significant levels of chlorine, copper and antiscalants) can be minimized by proper design.

In the first Assessment, it was also noted that growth in the population of States with shortages of fresh water, and the effects of climate change, would most likely lead to desalinization being increasingly considered as an adaptation measure for communities suffering increased and related water stress.

10.2. Current desalinization capacity and processes

The world's desalinization capacity has continued to grow. From an installed capacity of 86.5 million m3 per day in 2015, it reached 97.4 million m3 per day in 2018, with 48 per cent of that capacity in the Middle East and North Africa (International Desalinization Association (IDA), 2019; Jones and others, 2019).

Membrane processes remain dominant in desalinization (more than 65 per cent of production), although multi-stage distillation is still important in the States members of the Gulf Cooperation Council, where it is linked to power generation from oil or gas and provides about 60 per cent of capacity (IDA, 2019; Mogielnicki, 2020).

New demands for desalinated seawater seem likely from the mining industry. For example, substantial new growth in desalinization output is proposed in Chile in connection with copper mining, where about 1 million m3 per day of desalinated water are expected to be needed by 2027 for the copper mining industry, an increase of nearly 200 per cent over 2016 levels (Comisión Chilena del Cobre (CCC), 2016).

Global statistics for employment in desalinization operations are not available. However, it has been estimated that, between 2010 and 2030, a further 50,000 technicians at different skill levels would be needed to service the desalinization industry in the Middle East and North Africa. If the projected increase in output in that region translated to staff required is consistent around the world, it would imply a total global current workforce in desalinization of about 400,000 people (Ghaffour, 2009).

10.3. Potential pressures on the ocean

As noted above, the predominant view of waste discharge from desalinization plants has been that proper design can minimize adverse impacts on the ocean. However, a recent study of the impact of desalinization on the ocean has argued that the amount of brine discharged to the ocean from desalinization has been underestimated, together with its potential impact on the marine environment (Jones and others, 2019). It estimates that the amount of brine discharged daily stands at 142 million m3, of which 48 per cent is discharged in the Persian Gulf area. It also argues that the high-salinity water can have a serious adverse impact on the seabed flora and fauna. On the other hand, reports from Australia, based on seven years of observation of the site where discharges are released from a large desalinization plant serving Sydney, have been mixed, with adverse impacts observed on some marine invertebrates within 100 m of the discharges, while barnacles increased in numbers (Clark and others, 2018) and, at the same time, a threefold increase in fish numbers in the area was observed (Kelaher and others, 2020). Six years of monitoring brine discharges from two large desalinization plants in Israel observed almost no impact on seawater quality (Kress and others, 2020).

11. Salt production

11.1. Situation as shown in the first *World Ocean Assessment*

Salt production was only briefly considered in the first Assessment, in relation to its importance in the cultural aspects of food. It was noted that, although salt production by evaporation of seawater was still important, most salt was produced from rock salt and brine deposits in the ground. It was also noted that sea salt production was still important for some countries, such as Brazil, India and Spain (United Nations, 2017f).

11.2. Current situation

Salt production from evaporation of seawater is still a significant source of salt around the world. However, comprehensive statistics at a global level remain unavailable. The British Geological Survey, in its overview of world mineral production, identifies production of about 35 million tons of salt from seawater out of a reported total world production of 265 million tons (table 5), but it does not identify the source of salt for many countries, and it notes that salt is also produced in a number of countries for which data are not available (Brown and others, 2019). In most regions where reports are available, salt production from seawater has remained relatively stable, with the notable exception of a 34 per cent increase in India (table 5). The size of the workforce involved in sea salt production is unknown.

Table 5
Salt production from seawater (thousands of tons)

Country or territory	Sea salt production, 2013	Sea salt production, 2017
Albania	49[a]	47[a]
Montenegro	10[a]	10[a]
Portugal	91	115
Spain	1 221	1 111
Algeria	172	160*
Brazil	5 926	6 000[a]
Colombia	113	165
Bangladesh	1 439	1 496
India	17 517	23 500[a]
Pakistan	297	209
Mauritius	4	1
Mozambique	150	140[a]
Bonaire (Netherlands)	400[a]	400[a]
El Salvador	100[a]	100[a]
Guatemala	60[a]	60[a]
Nicaragua	30[a]	30[a]
Philippines	992	993[a]
Total	28 571[a]	34 537[a]

Source: Adapted from Brown and others, 2019.
[a] Estimated.

12. Key knowledge and capacity-building gaps

In relation to coastal communities, better information on their state, the threats they face and their economic and social situation is needed, especially for communities of indigenous peoples, given the crucial roles they play in maritime industries, in social and cultural aspects, and in ocean conservation.

In relation to maritime industries, knowledge and capacity-building gaps are identified in the following chapters: for harvesting food from the sea (chaps. 15, 16 and 17); for seabed mining (chap. 18); for offshore hydrocarbons (chap. 19); for marine renewable energy (chap. 21); and for marine genetic resources (chap. 23).

For shipping, the main knowledge gaps concern the social aspects. For example, better information is needed on the rates of injury and death of seafarers and other aspects of their welfare. Capacity-building gaps exist in some regions in terms of the training and development of seafarers: Africa and South America provide fewer seafarers than their share of the global population would support. Given the projected shortages in the supply of officers, there is clearly scope for expanding training in such areas.

For tourism, there is limited information on the scale of coastal and marine tourism and its growth, as compared with tourism generally. Equally, there is a lack of global information on the social and economic aspects of coastal and marine tourism. In particular, there is a lack of knowledge on the extent to which host countries benefit from their coastal and marine tourism industries, and on the status of employment in those industries.

For desalinization, there is scope for further examination of the relationship between discharge designs and impacts on the marine environment.

13. Outlook

The chapters on specific industries (chaps. 15, 16, 17, 18, 19, 21 and 23) describe the outlook for the industries concerned.

The outlook for shipping is closely linked to development of the global economy. The shipping industry has largely overcome the problems resulting from the economic crisis that occurred from 2008 to 2011, but challenges in controlling air pollution remain and increased concentration of cargo shipping seems likely. The future of the cruise industry is also closely linked to the development of disposable income in major economies.

The level of activity in the tourism industry, including coastal and marine tourism, is governed by the levels of available discretionary disposable income. The outlook for coastal and marine tourism is, therefore, dependent on maintaining current levels of expenditure from tourists from the regions and countries that are now the principal sources of tourists and increasing the interest in coastal and marine tourism from other countries as their discretionary disposable income increases.

References

Amundsen, Helene (2015). Place attachment as a driver of adaptation in coastal communities in Northern Norway. *Local Environment*, vol. 20, No. 3, pp. 257–276.

Arctic Council (2011). *Agreement on Cooperation on Aeronautical and Maritime Search and Rescue in the Arctic*. https://oaarchive.arctic-council.org/handle/11374/531.

Argüelles, María Belén, and others (2016). Impact of whale-watching on the short-term behavior of Southern right whales (*Eubalaena australis*) in Patagonia, Argentina. *Tourism Management Perspectives*, vol. 18, pp. 118–24. https://doi.org/10.1016/j.tmp.2016.02.002.

Armitage, Derek, and Chui-Ling Tam (2007). A political ecology of sustainable livelihoods in coastal Sulawesi, Indonesia. *Canadian Journal of Development Studies/Revue Canadienne d'études Du Développement*, vol. 28, No. 1, pp. 39–57.

Baltic and International Maritime Council and the International Chamber of Shipping (BIMCO/ICS) (2016). *Manpower Report: The Global Supply and Demand for Seafarers in 2015*. Bagsværd, Denmark: BIMCO.

Bender, Nicole A., and others (2016). Patterns of tourism in the Antarctic Peninsula region: a 20-year analysis. *Antarctic Science*, vol. 28, No. 3, pp. 194–203.

Bennett, Nathan James, and others (2016). Communities and change in the anthropocene: understanding social-ecological vulnerability and planning adaptations to multiple interacting exposures. *Regional Environmental Change*, vol. 16, No. 4, pp. 907–926.

Brown, T., and others (2019). *World Mineral Production 2013–2017*. Nottingham, United Kingdom: British Geological Survey.

Bureau of Safety and Environmental Enforcement of the United States (BSEE) (2020). *Rigs to Reefs*. www.bsee.gov/what-we-do/environmental-focuses/rigs-to-reefs.

Center for International Earth Science Information Network and Food and Agricultural Organization of the United Nations (2005). *Mapping Global Urban and Rural Population Distributions: Estimates of Future Global Population Distribution to 2015*, Environment and Natural Resources Series, No. 24. Rome: FAO, annex.

Chang, Jeong-In, and Sungsoon Yoon (2017). Assessing the Economic Value of Beach Restoration: Case of Song-do Beach, Korea. *Journal of Coastal Research*, vol. 79, No. sp1, pp. 6–10. https://doi.org/10.2112/SI79-002.1.

Charles, A. (2017). Chapter 21 – The big role of coastal communities and small-scale fishers in ocean conservation. In *Conservation for the Anthropocene Ocean*, Phillip S. Levin and Melissa R. Poe, eds., pp. 447–61. Academic Press. https://doi.org/10.1016/B978-0-12-805375-1.00021-0.

________(2020). *Looking to the Future in Nova Scotia's Coastal Communities*. Halifax, Canada: Saint Mary's University.

Charles, A., and others (2019). *Addressing the Climate Change and Poverty Nexus: A Coordinated Approach in the Context of the 2030 Agenda and the Paris Agreement. Rome; 2019*. Rome: FAO.

Charles, A., and others. (2020). *Looking to the Future in Nova Scotia's Coastal Communities*. Halifax, Canada: Saint Mary's University.

Clark, Graeme F., and others (2018). First large-scale ecological impact study of desalination outfall reveals trade-offs in effects of hypersalinity and hydrodynamics. *Water Research*, vol. 145, pp. 757–768.

Cole, Stroma (2007). Implementing and evaluating a code of conduct for visitors. *Tourism Management*, vol. 28, No. 2, pp. 443–51. https://doi.org/10.1016/j.tourman.2006.03.010.

Comisión Chilena del Cobre (CCC) (2016). *Proyección de Consumo de Agua En La Minería Del Cobre 2016–2027*. Santiago, Chile.

Cruise Lines International Association (CLIA) (2018). *2018 Global Passenger Report*. https://cruising.org/-/media/research-updates/research/clia-global-passenger-report-2018.pdf.

De Brauwer, Maarten, and others (2017). The economic contribution of the muck dive industry to tourism in Southeast Asia. *Marine Policy*, vol. 83, pp. 92–99.

Ecorys (2015). *Study on the Competitiveness of the Recreational Boating Sector*. Rotterdam: European Consortium for Sustainable Industrial Policy.

Environment Agency of the United Kingdom (2015). *Cost Estimation for Coastal Protection – Summary of Evidence*. Bristol, United Kingdom: Environment Agency.

European Commission (2018). *Eurostat News, Coastal Regions: Popular Tourist Destinations*. https://ec.europa.eu/eurostat/web/products-eurostat-news/-/EDN-20180927-1.

European Maritime Safety Agency (EMSA) (2020). *COVID-19 Impact on Shipping*. www.emsa.europa.eu/newsroom/covid19-impact/item/4037-august-2020-covid-19-impact-on-shipping-report.html.

Food and Agriculture Organization of the United Nations (FAO) (2019). *Fishery and Aquaculture Statistics 2017*. Rome. www.fao.org/fishery/static/Yearbook/YB2017_USBcard/index.htm.

________(2020). *Summary of the impacts of the COVID-19 pandemic on the fisheries and aquaculture sector: Addendum to the State of World Fisheries and Acquaculture*. Rome. https://doi.org/10.4060/ca9349en.

Fraschetti, Simonetta, and others (2011). Effects of Unplanned Development on Marine Biodiversity: A Lesson from Albania (Central Mediterranean Sea). *Journal of Coastal Research*, vol. 2011, No. 10058, pp. 106–115. https://doi.org/10.2112/SI_58_10.

Garrod, Brian, and David A. Fennell (2004). An analysis of whalewatching codes of conduct. *Annals of Tourism Research*, vol. 31, No. 2, pp. 334–52. https://doi.org/10.1016/j.annals.2003.12.003.

Ghaffour, Noreddine (2009). The challenge of capacity-building strategies and perspectives for desalination for sustainable water use in MENA. *Desalination and Water Treatment*, vol. 5, Nos. 1–3, pp. 48–53.

Gittman, Rachel K., and others (2016). Ecological consequences of shoreline hardening: a meta-analysis. *BioScience*, vol. 66, No. 9, pp. 763–73. https://doi.org/10.1093/biosci/biw091.

Graham, Sonia, and others (2018). Local values and fairness in climate change adaptation: insights from marginal rural Australian communities. *World Development*, vol. 108, No. C, pp. 332–43. https://doi.org/10.1016/j.worlddev.2017.12.008.

Guidino, Chiara, and others (2020). Whale-watching in Northern Peru: An economic boom?, *Tourism in Marine Environments,* January 2020. https://doi.org/10.3727/154427320X15819596320544.

Hawkins, Julie P., and C.M. Roberts (1997). Estimating the carrying capacity of coral reefs for scuba diving. In *Proceedings of the 8th International Coral Reef Symposium*, 2: pp. 1923–1926. Smithsonian Tropical Research Institute Panama.

Hoyt, Erich (2009). Whale watching. In *Encyclopedia of Marine Mammals*, William F. Perrin, and others, eds., 2nd ed., pp. 1223–27. Academic Press.

________ (2017). *The Global Status and True Value of Whale Watching: A Presentation to the Conference Organised by the Secretariat of the Pacific Regional Environment Programme on Whales in a Changing Ocean.* www.sprep.org/attachments/Publications/Presentation/whale-conference/global-status-and-true-value-of-whale-watching.pdf.

Huveneers, Charlie, and others (2017). The economic value of shark-diving tourism in Australia. *Reviews in Fish Biology and Fisheries*, vol. 27, No. 3, pp. 665–80. https://doi.org/10.1007/s11160-017-9486-x.

Intergovernmental Panel on Climate Change (IPCC) (2019). Summary for policymakers. In *IPCC Special Report on the Ocean and Cryosphere in a Changing Climate*, Hans-Otto Pörtner and others, eds., Intergovernmental Panel on Climate Change.

International Association of Antarctic Tour Operators (IAATO) (2018). *IAATO Overview of Antarctic Tourism: 2017–18 Season and Preliminary Estimates for 2018–19 Season.* Antarctic Treaty Consultative Meeting XLI, Information Paper 071. www.ats.aq/devAS/Meetings/DocDatabase?lang=e.

International Desalinization Association (IDA) (2019). *Dynamic Growth for Desalinization and Water Reuse in 2019.* https://idadesal.org/dynamic-growth-for-desalinization-and-water-reuse-in-2019.

International Energy Agency (IEA) (2018). *Offshore Energy Outlook. World Energy Outlook Series.* www.iea.org/reports/offshore-energy-outlook-2018.

International Maritime Bureau of the International Chamber of Commerce (2020). *Piracy and Armed Robbery against Ships: Report for the Period 1 January to 31 December 2019. ICC IMB.* London.

International Maritime Organization (IMO) (2015). *International Code for Ships Operating in Polar Waters (Polar Code).* IMO Document MEPC 68/21/Add.1, annex 10.

International Renewable Energy Agency (IRENA) (2020a). *Renewable Capacity Statistics.* Abu Dhabi: IRENA.

________ (2020b). *Renewable Energy and Jobs: Annual Review 2019.* Abu Dhabi: IRENA.

________ (2020c). *Wind Energy: A Gender Perspective.* Abu Dhabi: IRENA.

International Shipping Economics and Logistics (ISL) (2017). *Shipping Statistics and Market Review 2017.* Bremen, Germany: ISL.

International Transport Workers Federation (ITF) (2019). *Women Seafarers.* www.itfseafarers.org/en/issues/women-seafarers.

________ 2020. *Press release: 300,000 seafarers trapped at sea.* www.itfglobal.org/en/news/300000-seafarers-trapped-sea-mounting-crew-change-crisis-demands-faster-action-governments.

International Whaling Commission (IWC) (2019). *Whale Watching Handbook.* https://wwhandbook.iwc.int/en.

Jadot, Catherine, and others (2016). Intentional and Accidental Diver's Contact to Reefs at Popular Locations in the Dutch Caribbean. *Diving for Science 2016*, p. 74.

Jones, Edward, and others (2019). The state of desalination and brine production: a global outlook. *Science of the Total Environment*, vol. 657, pp. 1343–1356.

Kelaher, Brendan P., and others (2020). Effect of desalination discharge on the abundance and diversity of reef fishes. *Environmental Science & Technology.*

Kelly, Jay F. (2014). Effects of human activities (raking, scraping, off-road vehicles) and natural resource protections on the spatial distribution of beach vegetation and related shoreline features in New Jersey. *Journal of Coastal Conservation*, vol. 18, No. 4, p. 383.

Klein, Yehuda L., and Jeffrey Osleeb (2010). Determinants of coastal tourism: a case study of Florida beach counties. *Journal of Coastal Research*, vol. 26, No. 6, pp. 1149–1156.

Kress, Nurit, and others (2020). Seawater quality at the brine discharge site from two mega size seawater reverse osmosis desalination plants in Israel (Eastern Mediterranean). *Water Research*, vol. 171, art. 115402.

Lusseau, David, and others (2006). An individual-based model to infer the impact of whalewatching on cetacean population dynamics.

May, Candace K. (2019a). Governing resilience through power: explaining community adaptations to extreme events in coastal Louisiana. *Rural Sociology*, vol. 84, No. 3, pp. 489–515.

________ (2019b). Political ecology of culture clash: Amenity-led development, vulnerability, and risk in coastal North Carolina. *Journal of Rural and Community Development*, vol. 14, No. 3, pp. 24–48.

________(2019c). Resilience, vulnerability, & transformation: Exploring community adaptability in coastal North Carolina. *Ocean & Coastal Management*, vol. 169, pp. 86–95.

Merkens, Jan-Ludolf, and others (2016). Gridded population projections for the coastal zone under the shared socioeconomic pathways. *Global and Planetary Change*, vol. 145, pp. 57–66.

Metcalf, Sarah J., and others (2015). Measuring the vulnerability of marine social-ecological systems: a prerequisite for the identification of climate change adaptations. *Ecology and Society*, vol. 20, No. 2. https://doi.org/10.5751/ES-07509-200235.

Mogielnicki, R. (2020). *Water Worries: The Future of Desalination in the UAE*. Washington, D.C.: Arab Gulf States Institute in Washington. https://agsiw.org/wp-content/uploads/2020/03/Mogielnicki_Desalination_ONLINE.pdf.

National Marine Manufacturers Association (NMMA) (2018). *Recreational Boating Statistical Abstract*. Chicago: NMMA.

Netherlands Enterprise Agency (NEA), Centre for Promoting Imports from Developing Countries (2019). *Bird-Watching Tourism from Europe*. www.cbi.eu/node/752/pdf.

Neumann, Barbara, and others (2015). Future coastal population growth and exposure to sea-level rise and coastal flooding – a global assessment. *PloS One*, vol. 10, No. 3, e0118571.

Nicholls, R.J., and others (2007). Coastal systems and low-lying areas. In *Climate Change 2007: Impacts, Adaptation and Vulnerability*, M.L. Parry and others, eds., pp. 315–356. Cambridge: Cambridge University Press.

Nicholls, R.J., and R.J.T. Klein (2005). Climate change and coastal management on Europe's coast. In *Managing European Coasts: Past, Present and Future*, J.E. Vermaat and others, eds., pp. 199–226. Environmental Science Monograph Series. Heidelberg, Germany: Springer.

Oppenheimer, M., and others (2019). Sea level rise and implications for low-lying islands, coasts and communities. In *IPCC Special Report on the Ocean and Cryosphere in a Changing Climate*, H-O. Pörtner and others, eds., in press.

Pacific Community (2020). The economic and social impact of the COVID-19 pandemic on the Pacific Island economies. https://sdd.spc.int/news/2020/04/29/economic-and-social-impact-covid-19-pandemic-pacific-island-economies.

Petursdottir, Gudrun, and others (2001). *Safety at Sea as an Integral Part of Fisheries Management*. Rome: FAO.

Professional Association of Diving Instructors (PADI) (2019). *Worldwide Corporate Statistics*. www.padi.com/sites/default/files/documents/2019-02/2019%20PADI%20Worldwide%20Statistics.pdf.

Reyes-Martínez, Ma. José, and others (2015). Human pressure on sandy beaches: implications for trophic functioning. *Estuaries and Coasts*, vol. 38, No. 5, pp. 1782–1796.

Rigaud, Kanta Kumari, and others (2018). *Groundswell: Preparing for Internal Climate Migration*. World Bank.

Rogerson, Christian M., and Jayne M. Rogerson (2019). Emergent planning for South Africa's blue economy: evidence from coastal and marine tourism. *Urbani Izziv*, vol. 30, pp. 24–36.

________ (2018). Africa's tourism economy: uneven progress and challenges. In *The Routledge Handbook of African Development*, T. Binns, and others, eds., pp. 545–560. Abingdon, United Kingdom: Routledge.

Runge, C.A., and others (2020). Quantifying tourism booms and the increasing footprint in the Arctic with social media data. *PLoS ONE* 15(1). https://doi.org/10.1371/journal.pone.0227189.

Shi, Hua, and Ashbindu Singh (2003). Status and interconnections of selected environmental issues in the global coastal zones. *AMBIO: A Journal of the Human Environment*, vol. 32, No. 2, pp. 145–152.

Small, C., and Cohen, J.E. (2004). Continental physiography, climate, and the global distribution of human population. *Current Anthropology*, vol. 45, No. 2.

Small, C., and Nicholls, R.J. (2003). A global analysis of human settlement in coastal zones. *Journal of Coastal Research*, vol. 19, No. 3, pp. 584–599.

United Nations (2017a). Chapter 1: Introduction – planet, oceans and life. In *The First Global Integrated Marine Assessment: World Ocean Assessment I*. Cambridge: Cambridge University Press.

________ (2017b). Chapter 18: Ports. In *The First Global Integrated Marine Assessment: World Ocean Assessment I*. Cambridge: Cambridge University Press.

________ (2017c). Chapter 21: Offshore hydrocarbon industries. In *The First Global Integrated Marine Assessment: World Ocean Assessment I*. Cambridge: Cambridge University Press.

________ (2017d). Chapter 27: Tourism and recreation. In *The First Global Integrated Marine Assessment: World Ocean Assessment I*. Cambridge: Cambridge University Press.

________ (2017e). Chapter 28: Desalinization. In *The First Global Integrated Marine Assessment: World Ocean Assessment I*. Cambridge: Cambridge University Press.

________ (2017f). *The First Global Integrated Marine Assessment: World Ocean Assessment I*. Cambridge: Cambridge University Press.

United Nations Conference on Trade and Development (UNCTAD) (2019). *Review of Maritime Transport 2019*. New York: United Nations.

United Nations Coordinating Committee on Statistical Activities (UNCCSA) (2020). *How COVID-19 is changing the world: a statistical perspective*. https://unstats.un.org/unsd/ccsa/documents/covid19-report-ccsa.pdf.

United Nations, Department of Economic and Social Affairs, Population Division (UNDESA) (2018). *The World's Cities in 2018—Data Booklet*. New York: United Nations.

United States National Survey of Fishing, Hunting and Wildlife-Associated Recreation (USNSFHWAR) (2016). Washington, D.C.: United States Department of the Interior, United States Department of Commerce, United States Census Bureau.

Van Elden, Sean, and others (2019). Offshore oil and gas platforms as novel ecosystems: a global perspective. *Frontiers in Marine Science*, vol. 6, art. 548. https://doi.org/10.3389/fmars.2019.00548.

Wearing, Stephen Leslie, and others (2014). Whale watching as ecotourism: how sustainable is it? *Cosmopolitan Civil Societies: An Interdisciplinary Journal*, vol. 6, No. 1, pp. 38–55.

West Australian (2019). *Artificial Surfing Reef for Bunbury.* https://thewest.com.au/news/south-western-times/artificial-surfing-reef-for-bunbury-ng-b881227223z.

Wilkinson, Kenneth P. (1991). *The Community in Rural America*. Westport, Connecticut, United States of America: Greenwood Publishing Group.

Williams, Rob, and others (2006). Estimating relative energetic costs of human disturbance to killer whales (*Orcinus orca*). *Biological Conservation*, vol. 133, No. 3, pp. 301–11. https://doi.org/10.1016/j.biocon.2006.06.010.

World Bank (2019). *World Bank World Development Indicators*. Table 6.14. http://wdi.worldbank.org/table/6.14.

World Bank, and others (2012). *Hidden Harvest: The Global Contribution of Capture Fisheries*. Worldbank; WorldFish.

World Tourism and Travel Council (WTTC) (2018). *Domestic Tourism Importance and Economic Impact*. London: World Tourism and Travel Council.

Chapter 8B
Human health as affected by the ocean

Contributors: Michael Moore (convener), Martin Edwards, Bella S. Galil, Alan Simcock (chapter lead member) and Dick Vethaak.

Keynote points

- There are both health benefits and risks to living near the sea. The advantages can include enhanced air quality, exercise opportunities, novel marine-derived pharmaceuticals and ready access to food from the sea, which itself has health benefits (as a source of protein and essential micronutrients), although seafood is also traded inland; as well as sources of renewable energy.
- The ocean presents health risks from tsunamis, storms and tropical cyclones. Humans are also subject to increased risks from contaminated food from the sea, sea level rise and storms and cyclones from climate change.
- Chemical contaminants (including air pollution particulates), harmful or toxic algal blooms and pathogens pose health risks, in particular in estuarine and coastal waters where there is adjacent urbanization and/or recreational usage.
- Novel pollutants, such as antibiotics, hormones, nanomaterials (e.g., fullerenes, carbon nanotubes, metallic nanoparticles and nanoplastics) and microplastics, are a cause for concern. Combustion nanoparticles (e.g., $PM_{2.5}$) as a major component of air pollution, are well established as contributing to cardiovascular disease and lung cancer.

1. Introduction

In the first *World Ocean Assessment* (United Nations, 2017), various adverse impacts on human health were noted from sewage discharge, disease vectors linked to seawater (especially from sewage discharge), nanomaterials and microplastics, especially from plastic waste. Nanomaterials include both materials intentionally manufactured, for use in cosmetics, for example, and those resulting from the breakdown of plastic waste. Some benefits to human health were also noted, especially from fish and seaweed as food elements, marine pharmaceuticals and marine nutraceuticals, and the recreational effects of time spent by the seaside. There was no comprehensive discussion on the relationship between human health and the ocean. The present chapter, therefore, seeks to give an overview of all aspects of the relationship between human health and the ocean.

2. General aspects of the relationship between human health and the ocean

The marine environment brings both benefits and risks to human health, especially for people who live near it (see figure below; Depledge and others, 2013; Moore and others, 2013, 2014). Health has been defined as a state of complete physical, mental and social well-being and not merely the absence of disease or infirmity (World Health Organization-Regional Office for Europe (WHO-Europe), 1984). However, people live in an interdependent existence with the totality of the living world. Hence, human health cannot be separated from the health of our total planetary biodiversity and has now been redefined as the ability of a body to adapt to new threats and infirmities (Lancet-Editorial, 2009). The complex interactions between the seas and oceans and human health and well-being have been viewed primarily within a risk framework, for example, the adverse impacts of extreme weather, chemical pollution (from domestic and industrial effluents, aquaculture, offshore industries, air pollutants and road dust run-off, and black carbon in the Arctic) and, increasingly, climate change (Borja and others, 2020; Depledge and others, 2017, 2019; Fleming and others, 2019; Pleijel and others, 2013; Tornero and Hanke, 2016; Valotto and others, 2015; Walker and others, 2019; Winiger and others, 2019). However, new research is expanding our concept of the "health" of the "global ocean", with a broader recognition of its essential and beneficial contribution to the current and future health and well-being of humankind (Borja and others, 2020; Depledge and others, 2019; Ercolano and others, 2019; Lindequist, 2016; see table below).

The marine environment contributes significantly to human health through the provision and quality of the air we breathe, the food we eat, the water we drink and marine-derived pharmaceuticals, as well as providing health-enhancing economic and recreational opportunities (see chaps. 5 and 8A; Ercolano and others, 2019; Lindequist, 2016). The coastal environment can also have a calming effect (White and others, 2013) and provide important cultural benefits (see chap. 28, sect. 1.4). However, at the same time, the marine environment is under pressure from such human activities as transport, industrial processes, fishing, agricultural and waste management practices, climate change-related impacts associated with rising sea levels and coastal erosion, and biological invasions. The figure below summarizes the links between the degradation of the marine environment and human health.

The assessment and management of the impacts on marine ecosystems and on human health resulting from the pressures on those ecosystems have largely been undertaken separately under the umbrella of different disciplines and, frequently, with little or no obvious collaborative interaction (Depledge and others, 2013; Moore and others, 2013, 2014). Consequently, many of our perceptions of the interactions between the marine environment and human health are limited and still relatively unchallenged, leaving an opportunity to address critical knowledge gaps to further inform science-based policies for the sustainable use of marine resources and environmental and human health protection (see figure below and Moore and others, 2014).

The complex nature of the interactions between the marine environment and human health was reviewed by the European Marine Board (Moore and others, 2013, 2014) and others (Borja and others, 2020; Depledge and others, 2013, 2017, 2019; Fleming and others, 2014, 2019). The reviews have emphasized the need for an interdisciplinary approach to address all levels of organization, from genes to ecosystems.

There are five key scientific challenges to improving our understanding of the linkages between the marine environment and human health (Galloway and others, 2017; Moore and others, 2014):

(a) To improve the measurement and monitoring of the distribution of marine pollutants, including algal toxins, nanoparticles as contributing factors to cardiovascular disease and lung cancer (Chang and others, 2020; Liu and others, 2016; Moore, 2020; Mossman and others, 2007; Numan and others, 2015; Stapleton, 2019), microparticles and plastic marine litter as a vector, as well as pathogens and non-indigenous species as potential health hazards at required time and spatial scales (Galil, 2018; Vezzulli and others, 2016);

(b) To improve knowledge of processes and models of the dynamics of transport and transformation in the environment of marine pollutants, pathogens and non-indigenous species that present health hazards;

(c) To improve the assessment of marine pollutant, pathogen and non-indigenous species health hazard exposure and risk to humans (Galil, 2018; Moore and others, 2013, 2014; Vezzulli and others, 2016);

(d) To understand the impacts of waste management activities on the marine environment and human health;

(e) To find explanations for the association between the marine environment and observed human health benefits, described as the "Blue Gym" effect (Depledge and Bird, 2009; Robinson and others, 2020; White and others, 2013; Wyles and others, 2019), including socioeconomic influences (Li and Zhu, 2006; Sachs and others, 2001).

Summary of the interconnectivity of the key adverse processes between the marine environment and human health

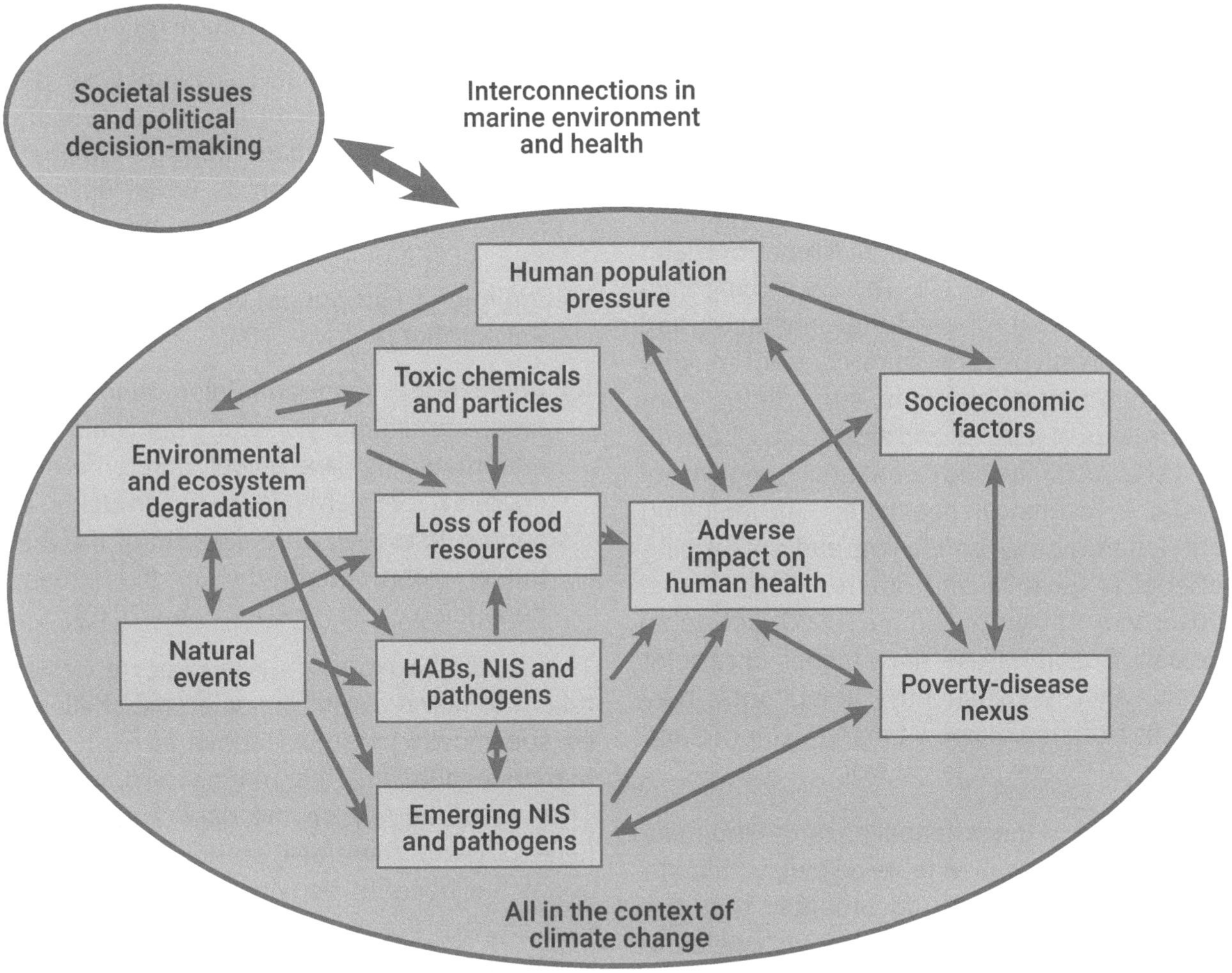

Source: Original diagram partly adapted from Moore and others, 2014.
Note: "Toxic chemicals and particles" includes air pollution particulates, nanoparticles and microplastics.
Abbreviations: HABs, toxic or harmful algal blooms; NIS (poisonous and venomous) non-indigenous species.

The potential benefits for human health from living in proximity to the sea (see table below), such as novel pharmaceuticals (e.g., antimicrobial, antitumour, antidiabetic, anticoagulant, antioxidant, anti-inflammatory, antiviral, antimalarial, antitubercular, anti-ageing and antiprotozoal) derived from marine organisms and essential micronutrients in seafood, have often been overlooked in the past (see table below; Borja and others, 2020; Depledge and others, 2019; Ercolano and others, 2019; Fleming and others, 2019; Gascon and others, 2017; Hosomi and others, 2012; Lindequist, 2016; Wheeler and others, 2012; White and others, 2014; and Wyles and others, 2019). However, it is becoming well established that there are various health benefits to be gained from living by the sea (Giles, 2013). The reason why that should be is less clear and has so far eluded an overall scientific explanation. However, several hypotheses have been proposed: psychological stress reduction owing to pleasant surroundings (Gascon and others, 2017; White and others, 2014); improved immunoregulation from exposure to bacteria and parasites with which we co-evolved (Rook, 2013); and exposure to bioactive natural products (biogenics), such as harmful or toxic algal toxins (Berdalet and others, 2016, 2017). The third (biogenic) hypothesis has proposed that inhalation and ingestion (with upper respiratory tract mucus) of certain natural products, such as low concentrations of aerosolized algal toxins, have direct effects on the body's molecular regulatory systems, resulting in health benefits, including anti-inflammatory, anticancer and anti-ageing effects (Asselman and others, 2019; Moore, 2015; Van Acker and others, 2020; see table below). Coastal areas have higher ultraviolet levels and, consequently, inhabitants may benefit from increased vitamin D (Cherrie and others, 2015; see table below).

With regard to the potential hazards and risks for human health (see table below), which are more comprehensively documented than the benefits (Borja and others, 2020; Depledge and others, 2013, 2017, 2019; Fleming and others, 2014, 2019; Moore and others, 2013, 2014), the European Environment and Health Process, coordinated by the World Health Organization (WHO), has identified five general "key environment and health challenges of our time". Their particular focus in relation to the marine environment includes:

(a) Health and environmental impacts of climate change (for example, tropical cyclones);

(b) Health risks to children and other vulnerable groups posed by poor environmental, working and living conditions, especially the lack of water and sanitation (e.g., contaminated food from the sea);

(c) Socioeconomic and gender inequalities in the human environment and health (e.g., the poor injury record of fishers and seafarers and limited access to health care for women as a result of cultural traditions);

(d) The burden of non-communicable diseases, in particular to the extent that the burden can be reduced through adequate policies in areas such as urban development, transport, food safety and nutrition, and living and working environments (e.g., the role of fish protein in providing essential nutrients);

(e) Persistent, endocrine-disrupting and bioaccumulating harmful chemicals and nanomaterials; and novel and emerging chemical problems (e.g., the impacts of such substances on the health of the marine environment and thus on the humans depending on it) (WHO-Europe, 2010).

The marine aspects of those policy priorities reflect to some extent the scientific challenges specifically identified above in relation to human health and the marine environment. They focus largely on the risks and tend to leave out and, therefore, do not take into account the benefits deriving from the marine environment. Furthermore, both gender differences and gender inequalities can give rise to inequities between men and women in health status and access to health care. However, gender norms and values are not fixed and can evolve over time, can vary substantially from place to place and are subject to change (WHO, 2014). Nevertheless, there are a number of threats to human health arising from the marine environment that have now been identified:

(a) Increase in the spread of pathogens related to climate warming (e.g., *Vibrio*). Also, there is some evidence related to an increase in some harmful algal bloom species related to climate warming is some regions (Hinder and others, 2012; Vezzulli and others, 2016);

(b) Recently, non-indigenous species, sometimes called invasive alien species, have started to be considered as one of the major threats to global marine ecosystems through impacts on the ecosystems' structure, function and services (Galil, 2018). A small number of poisonous or venomous marine non-indigenous species represent potential threats to human health. Intensification of anthropogenic activities, coupled with rapidly increasing coastal urbanization, drive complex and fundamental changes in coastal waters, including increases in alien species. Some of the alien venomous and poisonous species have attracted the attention of scientists, managers, the media and the public for their conspicuous human health impacts. In the Mediterranean alone, 10 non-indigenous species are considered human health hazards, running the gamut from nuisance to lethal (Galil, 2018). Human health hazards of non-indigenous species are expected to worsen as a result of climate change. The poleward influx of warm water biota enables them to spread to regions as yet uncolonized;

(c) A further, recently identified health threat is the potential role of plastic marine litter as a vector for opportunistic human pathogens and antibiotic-resistant microorganisms (Barboza and others, 2018; Harrison and others, 2018; Imran and others, 2019). Various pathogenic bacteria bind, in particular and strongly, to plastic litter (for example, *Vibrio cholerae* and some strains of *Escherichia coli*). Such human pathogens can colonize plastic surfaces in stable biofilms. The scientific and medical understanding of that health threat of plastic pollution is inadequate but the threat is dealt with as a further aspect of the problem of marine litter discussed in chapter 12. A severe problem could arise in areas that are highly polluted as a result of natural disasters, climate crises or occurring epidemics, or in conflict zones (Vethaak and Leslie, 2016; Keswani and others, 2016; Galloway and others, 2017; Leonard and others, 2018a, 2018b; Moore and others, 2014).

In a general context, some new multinational, interdisciplinary projects are now addressing some of those issues, including:

(a) The Seas, Oceans and Public Health in Europe project, funded by the European Union (European Union, 2020), which has developed a "research road map" to help scientists to gather evidence and inform policies that enhance and protect both human health and the health of the marine environment;

(b) The Blue Communities programme, a research capacity-building programme for marine planning in East and South-East Asia, which includes a project to assess the benefits and risks of coastal living associated with environmental, demographic and climate change.[1]

[1] See www.blue-communities.org/About_the_programme.

Summary of benefits and of hazards and risks associated with living in proximity to the sea

Benefits	Hazards and risks
Improvements in the length and quality of life (Gascon and others, 2017)	Chemical and radionuclide pollutants, including toxic airborne particulates (both land-derived and from shipping) and coastal ozone (Moore and others, 2014; Pleijel and others, 2013; Valotto and others, 2015; Vom Saal and others, 2007; Walker and others, 2019; Wan and others, 2016)
Improved physical and mental health (Gascon and others, 2017; White and others, 2014; Wyles and others, 2019)	Nanomaterials and microplastics (Chang and others, 2020; Galloway and others, 2017; Moore and others, 2014; Mossman and others, 2007; Numan and others, 2015)
Increased vitamin D (Cherrie and others, 2015)	Pathogens and public health consequences from sewage, agricultural run-off and flooding (Leonard and others, 2018a; Moore and others, 2013, 2014; Vezzulli and others, 2016)
Reduced behavioural problems in children (Gascon and others, 2017)	Environmental impacts on food security and safety, such as the collapse of fisheries and the contamination of food resources (Moore and others, 2014)
Low concentrations of airborne aerosolized algal toxins may have beneficial hormetic effects on health (anti-inflammatory and anticancer effects) (Asselman and others, 2019; Moore, 2015; Van Acker and others, 2020)	Harmful or toxic algal blooms and algal toxins (Berdalet and others, 2016)
Benefits from consumption of seafood that is high in protein and essential micronutrients (Hosomi and others, 2012)	Poisonous or venomous indigenous and non-indigenous species, such as silverstripe blaasop (produces tetrodotoxin), nomadic jellyfish and lionfish (Galil, 2018)
Marine-derived pharmaceuticals (Ercolano and others, 2019; Lindequist, 2016)	Adverse natural events (volcanic eruptions, earthquakes, tsunamis, tropical cyclones and flooding) (Moore and others, 2014; Powell and others, 2019; Ruskin and others, 2018)
	Transmission of antimicrobial resistance and pathogens through natural bacterial ecosystems (Leonard and others, 2018b; Imran and others, 2019)
	Plastic marine litter as an emerging potential vector for pathogens and their possible global transport (Vethaak and Leslie, 2016; Keswani and others, 2016); as well as possible collisions with large pieces of plastic litter at sea
	Increased risks from overcrowding as coastal population increases (Moore and others, 2014)

3. Health of coastal communities relative to inland communities

Studies comparing the health of coastal communities to that of inland communities have, so far, largely been confined to developed countries. The evidence differs between physical health and mental health. For physical health, there is evidence from Australia (Ball and others, 2007), New Zealand (Witten and others, 2008), the United States (Gilmer and others, 2003) and the United Kingdom (White and others, 2013) that living in a coastal setting encourages greater levels of recreational physical activity. Although there is some evidence that the extra activity may translate into healthier weight, for the most part, even among children living at the coast (Wood and others, 2016), the evidence is equivocal (Bell and others, 2019). A re-examination of responses to a question in the 2001 Census of England and Wales showed that a significantly higher proportion of people in coastal areas said that they enjoyed good health. The effect may be greater for more socioeconomically deprived groups (Wheeler and others, 2012). In Belgium, a recent survey concluded that people living less than 5 km from the coast report themselves as enjoying better general health than do those living 50–100 km from the coast (Hooyberg and others, 2020).

As regards mental health, an increasing amount of evidence suggests that living in coastal settings, visiting them frequently or simply having a coastal view from home is associated with increased life satisfaction (Brereton and others, 2008) and a decreased risk of anxiety and depression (Nutsford and others, 2016; White and others, 2013; Wyles and others, 2019).

Differences in human health between coastal and inland areas can be attributed to causes other than the proximity of the sea. Socioeconomic status has a major effect on health generally (Marmot and Wilkinson, 2005); and where there are differences in economic prosperity between coastal areas and inland areas, differences in human health between those areas may be attributable in part to those economic differences, rather than to direct health benefits of being close to the ocean (Li and Zhu, 2006). However, interpreting the very complicated relationship between economic prosperity and health is often difficult owing to the plethora of potential interacting factors (Sachs and others, 2001).

A key challenge is to determine how each coastal community can improve its resilience to sociodemographic change and the increasing number of extreme weather events and environmental threats. Evidence shows that there are advantages to policies that offer a range of benefits to both the environment and health. However, any policy response is complicated by the fact that the diversity of coastal communities means that there is unlikely to be any "one size fits all" solution (Depledge and others, 2017; Li and Zhu, 2006; Sachs and others, 2001).

4. Effects of exposure to contaminated seawater

Many of the main activities related to coastal tourism and recreation involve contact with seawater, with paddling, swimming, boating, surfing, recreational fishing and diving being among the most common. Fishers and seafarers also come into contact with seawater as part of their work. Such contact brings with it the risk of exposure to pathogens, including algal toxins, in the water or in marine aerosols. For a long period after the disposal of municipal wastewater into the sea became common, there was little concern about the effect of pathogens in the wastewater on human health – the scale of dispersal of the wastewater into the much greater volumes of seawater was thought to minimize risk through dilution (Sullivan, 1971). However, eventually, concern did grow and led to the adoption of measures, in Europe, for example, such as the Bathing Waters Directive (European Economic Community, 1975).

Studies in many places have quantified the scale of the risk to human health from contact with seawater containing pathogens, such as some strains of *Escherichia coli* – bacteria commonly found in the gut of warm-blooded animals (Zmirou and others, 2003; Wade and others, 2006). For example, in Hong Kong, China, a major epidemiological study was conducted in 1992 in which 25,000 beachgoers were interviewed in order to establish the health effects of exposure to bathing water. The results indicated that the total incidence of swimming-related illness symptoms was 41 per 1,000 interviewees, higher than the 30 per 1,000 found earlier, in 1987. Eye, skin and respiratory symptoms were 2–20 times more prevalent in swimmers than in non-swimmers (Kueh, 1995).

Likewise, in Santander, Spain, a study over the main holiday season in 1998 showed that 7.5 per cent of the 1,858 bathers studied reported fever or respiratory, gastrointestinal, eye or ear symptoms within seven days – and that was in waters that met the regulatory standards in force (Prieto, 2001). A similar study of 654 surfers was carried out in the winter seasons from 2013 to 2015 in San Diego, California, United States, where the quality of coastal waters was adversely affected after heavy rainfall (which usually leads to increased run-off or discharge of contaminants). The study examined the incidence of gastrointestinal illness, sinus infections, ear infections and infected wounds within three days of over 10,000 surfing sessions. It found that the incidence of those conditions rose by between 26 and 105 per cent (varying among types of complaint) after surfing sessions in dry weather, as compared with periods when the persons studied were not surfing. After heavy rainfall, and a consequent increase in surface run-off, the incidence of post-surfing diseases rose by a further 26 to 102 percentage points as compared with non-surfing periods (Arnold and others, 2017). Sewage-contaminated seawater contains a range of microbial pathogens, and exposed individuals may experience various disease symptoms, such as skin rashes, conjunctivitis, sinus infections and, in particular, gastroenteritis (Harder-Lauridsen and others, 2013). With a predicted increasing frequency of heavy rainfall associated with climate change in some regions, the future implications for human health worldwide could be considerable, in particular in those areas without well-functioning sewage systems or where current sewage systems are unable to contain the excess run-off and raw sewage is discharged (Harder-Lauridsen and others, 2013). Climate change-related increases in the frequency and severity of riverine and coastal flooding leading to the release of raw sewage and run-off of vector animal faeces may also represent a health problem through the transmission of emerging infectious microbial agents, such as in the COVID-19 pandemic (Seneviratne and others, 2012).

The global impact of poor water quality was examined in a study by the Joint Group of Experts on the Scientific Aspects of Marine Environmental Protection (GESAMP) and WHO. Based on global estimates of the number of tourists who go swimming, and WHO estimates of the relative risks at various levels of contamination, the study estimated that bathing in polluted seas causes some 250 million cases of gastroenteritis and upper respiratory disease every year and that some of those people affected would be disabled over the longer term. Measured by adding up the total years of healthy life that are lost through disease, disability and death, the worldwide burden of disease incurred by bathing in contaminated seawater is some 400,000 disability-adjusted life-years (a standard measure of time lost owing to premature death and time spent disabled by disease), comparable to the global impacts of diphtheria and leprosy. GESAMP and WHO estimated that the cost to society, worldwide, amounted to about $1.6 billion per year (GESAMP, 2001). Furthermore, harmful or toxic algal blooms can induce serious neurological disease and also have major financial impacts (Bechard, 2020; Diaz and others, 2019).

The most common pollutants tend to come from one of two places: humans or animals. Human faecal matter in water bodies constitutes the greatest public health threat because humans are reservoirs for many bacteria, parasites and viruses that are dangerous to other humans and can lead to a variety of illnesses. The cause of many problems can often be traced back to sewage overflows or leaky residential septic systems. Run-off from agricultural land can also represent a serious health concern, as faecal waste from farmed animals can contain pathogens, including various viruses, cryptosporidium, *Escherichia coli* and salmonella, while pet waste on beaches can also pose health threats to humans (Food and Agriculture Organization of the United Nations (FAO), 2017; Moore and others, 2014; Woods Hole Oceanographic Institution (WHOI), 2020).

Exposure to contaminated seawater thus affects the health of those enjoying recreation by the sea and adversely affects coastal tourism and recreation. Drawing together the scientific work in the field, in 2003, WHO published *Guidelines for Safe Recreational Water Environments: Coastal and Fresh Waters* (WHO, 2003). More recently, WHO, with the support of the European Union, prepared recommendations on scientific, analytical and epidemiological developments relevant to the parameters for bathing-water quality, with special reference to Europe (WHO, 2018). WHO has indicated that the recommendations will inform the revision of the 2003 Guidelines (WHO, 2020). However, achievement of such standards requires adequate planning and infrastructure. Even where, as in some parts of India, strenuous efforts are being made to install properly operating sewage treatment systems, problems persist. For example, in Goa, a major tourist location, faecal coliform bacteria exceeded the relevant standards at all 10 of the beaches monitored (Goa State Pollution Control Board (GSPCB), 2019).

The monitoring of bathing water will not achieve its aim of improving public health without improvements in the communication to the public of the findings so that they are readily understandable. The current European Union legislation on bathing water (European Union, 2006) provides for standardized ways of publicizing the results of the monitoring that is required. Similar systems are found in various Australian States (New South Wales Department of Planning, Industry and Environment (NSW-DPIE), 2020; South Australia Environment Protection Agency (SA-EPA), 2020) and in the United States (WHOI, 2020).

Climate change may be influencing the prevalence of microbial infections (Deeb and others, 2018; Konrad and others, 2017). For example, increases in *Vibrio vulnificus* and *Vibrio parahaemolyticus* infections, both topical, and infections from ingesting seafood (oysters), have been described in relation to climate change, with rises in cases overall, as well as new cases found in high latitude areas that were previously not affected, as they are having more days over the minimum temperature threshold (Vezzulli and others, 2016).

5. Problems for human health posed by food from the sea

Human health can be affected by many aspects of food from the sea. Some problems are the result of pollutants (such as mercury) or pathogens (often from sewage and ballast water) discharged into the sea and taken up by plants, fish and shellfish that are harvested for human consumption (Takahashi and others, 2008). Others are the result of toxins generated by, or viruses found in, various biota in the sea and taken up by some fish and shellfish (see chaps. 10 and 11).

According to WHO, mercury is one of the 10 most poisonous substances to human health (WHO, 2013). A principal form of mercury to which humans are exposed is organic methyl mercury (MeHg). The principal source of inorganic mercury in the sea is the burning of fossil fuels (see chap. 11). Such mercury is converted into MeHg by microbes in the aquatic environment, where it bioaccumulates in food webs. In humans, MeHg exposure occurs predominantly through the consumption of seafood. MeHg is a neurotoxin and is particularly harmful to fetal brain development. A large body of research has demonstrated a link between exposure to MeHg in the womb and developmental neurotoxicity (e.g., deficits in fine motor skills, language and memory) among populations that consume seafood regularly. A review of studies in 43 countries showed that pooled average biomarkers suggested an intake of MeHg that was:

(a) Several times above the Food and Agriculture Organization of the United Nations-WHO reference level for consumption in fish-consuming inhabitants of coasts and riverbanks living near small-scale gold-mining installations;[2]

(b) Well over the reference level in consumers of marine mammals in Arctic regions;

(c) Approaching the reference level in coastal regions in South-East Asia, the western Pacific and the Mediterranean.

Although the two former groups have a higher risk of neurotoxicity than the latter, the coastal regions of South-East Asia are home to very large populations. In all three areas, many of the samples showed levels of MeHg intake in excess of the reference value (Sheehan and others, 2014). Other experts, while recognizing the threat from MeHg, argue that it is important also to balance the benefits from fish-derived lipids with possible risks when considering fish as part of the diet of mothers and their children (Myers and others, 2015). Certain fish species have been identified as being at greater risk for MeHg exposure than others (e.g., MeHg biomagnifies in the aquatic food chain and larger predatory fish, such as shark, swordfish, king mackerel and certain species of tuna), so making appropriate choices in fish consumption can lead to increasing the benefits of eating seafood while decreasing the potential risk (Silbernagel and others, 2011).

Contamination of seafood by the presence of hormones, antibiotics, and persistent organic pollutants, such as polycyclic aromatic hydrocarbons and polychlorinated biphenyls continues to represent a hazard for human health (Binelli and Provini, 2003; Chen and others, 2015; Lu and others, 2018; European Commission, 2000). The recently recognized contamination of the ocean by nanomaterials and microplastics is of emerging concern, not only because of the potential ecological impacts, but also the potential to compromise food security, food safety and consequently human health. The presence of nanomaterials and microplastics in marine animals used for human food is now an emergent global phenomenon

2 The FAO/WHO reference level is 2.0 micrograms per gram, which is not considered as posing an appreciable risk (WHO, 2008).

that requires further research to determine whether there is human health risk (Chang and others, 2020; Galloway and others, 2017; Mossman and others, 2007; Numan and others, 2015; Sforzini and others, 2020; Smith and others, 2018; Stapleton, 2019; Stern and others, 2012; Vethaak and Leslie, 2016; Von Moos and others, 2012). Combustion nanoparticles are taken up into cells by endocytosis and accumulate in lysosomes, where overloading of the lysosomes results in membrane permeabilization, with resultant release of intralysosomal iron that causes oxidative cell injury leading to oxidative stress, with subsequent tissue and organ damage (Moore, 2020; Numan and others, 2015; Stern and others, 2012; Sforzini and others, 2020; Von Moos and others, 2012). There are now concerns that other nanoparticles including nano- and microplastics, may behave in a similar way (Boverhof and others, 2015; Von Moos and others, 2012).

Shellfish are the major vector of illnesses caused by pathogens discharged to the sea. Oysters, for example, can concentrate such pathogens up to 99 times the level in their surrounding water (Burkhardt and Calci, 2000; Morris and Acheson, 2003; Motes and others, 1994; Vezzulli and others, 2016). The most common viral pathogens involved were norovirus (83.7 per cent) and the hepatitis A virus (12.8 per cent) (Bellou and others, 2013). No global database exists on outbreaks of illness of that kind. However, a survey of outbreaks reported between 1980 and 2012 found records of about 368 shellfish-borne viral outbreaks. The majority were located in East Asia, with more than half in Japan, followed by Europe, the Americas, Oceania and Africa. In addition to sewage-borne pathogens, toxins (e.g., yessotoxins, brevetoxins and ciguatoxins) can be produced by toxic algae (e.g., dinoflagellates), often at relatively low concentrations (e.g., 200 cells/l of *Alexandrium* spp.) and not necessarily restricted to algal blooms (see chap. 10 for causes of such blooms; and United States Centers for Disease Control and Prevention databases). Algal toxins can enter the food web and are often present in shellfish and fish where they can cause illness as a result of their consumption as human food. The health impacts of algal toxins are not limited to illnesses and deaths caused by poisoning, but also include health impacts from the loss of shellfish and other fisheries that have to be closed to protect people from poisoning, and the disruption of ecosystems caused by deaths of fish and top predators that ingest the algae or the toxins that they produce. Many toxic algal bloom events are reported annually, from all parts of the world, and the number is growing. The increased numbers are, in part, attributable to improved observation and recording, but there is reliable evidence demonstrating a real increase in the incidence of such problem blooms as a result of the interaction of many factors, including rising sea temperatures, increased inputs of nutrients to the ocean, the transfer of non-indigenous species by shipping, and changes in the balance of nutrients in the sea (Hinder and others, 2012). Health warning systems could be implemented in higher risk areas by involving not only the public health authorities, but also community planners, utility managers and designers.

Effective monitoring and management programmes are, however, in place in some "at-risk" regions to prevent such toxins from being found in commercial seafood (Anderson, 2009; Anderson and others, 2001; see chap. 10). Such programmes are based on rigorous research on method development and validation, as well as the understanding of temporal and spatial patterns of toxic algae and knowledge of their transfer to humans.

Toxic algal blooms are complex phenomena, and many different disciplines need to be involved in finding a way to address the problems they cause, ranging from molecular and cell biology to large-scale field surveys, numerical modelling and remote sensing (Intergovernmental Oceanographic Commission of the United Nations Educational, Scientific and

Cultural Organization (UNESCO-IOC), 2017). Other biogenic toxins of health concern, which are not produced by algal blooms, include cyanotoxins produced by cyanobacteria, tetrodotoxins produced by symbiotic bacteria, which are used by metazoans as a defensive biotoxin to ward off predation or as both a defensive and predatory venom, and palytoxins, which are intense vasoconstrictors that pose risks to humans primarily through exposure to coral (Bane and others, 2014; Ramos and Vasconcelos, 2010; Zanchett and Oliveira-Filho, 2013). Humans who consume shellfish contaminated with brevetoxins, which are produced by some species of plankton, are at high risk of developing neurotoxic shellfish poisoning. There are also reports of skin ailments resulting from contact with brevetoxin-contaminated water and of respiratory illness from brevetoxin aerosols, in particular in vulnerable people with asthma (Hoagland and others, 2009). Shellfish metabolites of brevetoxin can also show different patterns of toxicity (Turner and others, 2015). Tetrodotoxins, produced by some species of bacteria, and ciguatoxins, produced by some species of plankton, can accumulate in fish and other seafood and are poisonous when consumed. Those types of biogenic toxins were previously associated with tropical waters but are now being found in temperate zones (Rodriguez and others, 2008; Silva and others, 2015a, 2015b). The social costs of all those illnesses can be huge, and the estimated costs related to illnesses from toxic algal blooms in just one single county in Florida, United States, amounted to between \$0.5 million and \$4.0 million (Hoagland and others, 2009).

6. Key remaining knowledge and capacity-building gaps

Knowledge gaps mainly relate to:

(a) Ways in which, and the extent to which, the ocean can produce health benefits through proximity to it, delivery of marine-derived pharmaceuticals and the development of novel seafoods;

(b) Extent to which health threats from the ocean affect human health in different parts of the world: for example, the ways in which marine vectors can deliver pathogens to humans; the scale and location of illness from swimming in contaminated water and from seafood; and the extent of contamination of fish and shellfish;

(c) Socioeconomic and gender inequalities in the human environment and health, including health risks to children and other vulnerable groups posed by poor environmental, working and living conditions (especially the lack of water and sanitation) (Moore and others, 2013, 2014; WHO, 2014);

(d) Burden of non-communicable diseases, in particular to the extent that it can be reduced through adequate policies in such areas as urban development, transport, food safety and nutrition, and living and working environments (Moore and others, 2013, 2014);

(e) Mechanisms by which novel health threats may arise from the ocean: for example, the role of nanomaterials (including combustion particulates) and nano- and microplastics, and the extent of human exposure to them (Galloway and others, 2017; Mossman and others, 2007; Numan and others, 2015; Sforzini and others, 2020; Stapleton, 2019; Stern and others, 2012; Vethaak and Leslie, 2016; Von Moos and others, 2012; Wright and Kelly, 2017); and the conditions under which algal blooms can become toxic (see chap. 10);

(f) Empirical assessment of the socioeconomic and health effects of marine protected areas is sparse. Ban and others (2019) reveal that most studies on well-being outcomes of marine protected areas focused on economic and governance aspects, whereas social, health and cultural aspects received only a cursory mention. Furthermore, the largest marine protected areas are situated far from human habitation (e.g., Marae Moana (Cook Islands) Ross Sea marine reserve (Antarctica), Papahānaumokuākea Marine National Monument (Hawaii), Pacific Remote Islands Marine National Monument (United States), Coral Sea Marine Park (Australia), whereas in the densely inhabited Mediterranean, fully protected marine protected areas, which conceivably may provide health benefits, constitute only 0.06 per cent of the countries' exclusive economic zone (Kersting and others, 2020).

(g) Health and environmental impacts of climate change (WHO, 2014).

Efforts to address those issues must include interdisciplinary research, which, in turn, requires the building of capacities to carry it out and apply the results. That necessitates both the training and retention of expert staff and the provision and financing of the necessary infrastructure. Efforts to tackle the causes of ill health linked to the ocean must also include the provision of adequate infrastructures and skilled personnel, in particular with regard to the environmentally sound management of chemicals and all wastes throughout their life cycle, integrated water resources management and the testing of harvested food (Sustainable Development Goal 12).

7. Outlook

Increased knowledge of the linkages between the ocean and human health will help to improve interventions to protect human health from threats and to increase the health benefits derived by humans from the sea. Improved capacities around the world, including in effectively managing marine protected areas (Organization for Economic Cooperation and Development (OECD), 2017), will enable the challenges posed by the sea to human health to be addressed more universally.

References

Anderson, D.M. (2009). Approaches to monitoring, control and management of harmful algal blooms (HABs). *Ocean & Coastal Management*, vol. 52, No. 7, 342. https://doi.org/10.1016/j.ocecoaman.2009.04.006.

Anderson, D.M., and others (2001). Monitoring and management strategies for harmful algal blooms in coastal waters, APEC #201-MR-01.1, Asia Pacific Economic Program, Singapore, and Intergovernmental Oceanographic Commission Technical Series No. 59, Paris.

Arnold, Benjamin F., and others (2017). Acute illness among surfers after exposure to seawater in dry- and wet-weather conditions. *American Journal of Epidemiology*, vol. 186, No. 7, pp. 866–875.

Asselman, Jana, and others (2019). Marine biogenics in sea spray aerosols interact with the mTOR signaling pathway. *Scientific Reports*, vol. 9, No.1, pp. 1–10.

Ball, Kylie, and others (2007). Personal, social and environmental determinants of educational inequalities in walking: a multilevel study. *Journal of Epidemiology & Community Health*, vol. 61, No. 2, pp. 108–114.

Ban, N., and others (2019). Well-being outcomes of marine protected areas. *Nature Sustainability*, vol. 2, No. 6, pp. 524–532.

Bane, V., and others (2014). Tetrodotoxin: chemistry, toxicity, source, distribution and detection. *Toxins*, vol. 6, No. 2, pp. 693–755.

Barboza, Luís Gabriel Antão, and others (2018). Marine microplastic debris: an emerging issue for food security, food safety and human health. *Marine Pollution Bulletin*, vol. 133, pp. 336–348.

Bechard, A. (2020). Harmful Algal Blooms and Tourism: The economic impact to counties in Southwest Florida. *Review of Regional Studies*. vol. 50, No. 2:12705.

Bell, S., and others (2019). The shadows of risk and inequality within salutogenic coastal waters. In: Foley, R., and others (ed.) *Hydrophilia Unbounded: Blue Space, Health and Place*. Routledge, Taylor & Francis, Milton Park, United Kingdom.

Bellou, M., and others (2013). Shellfish-borne viral outbreaks: a systematic review. *Food and Environmental Virology*, vol. 5, No.1, pp. 13–23.

Berdalet, Elisa, and others (2016). Marine harmful algal blooms, human health and wellbeing: challenges and opportunities in the 21st century. *Journal of the Marine Biological Association of the United Kingdom*, vol. 96, pp. 61–91.

Berdalet, E., and others (2017). Harmful algal blooms in benthic systems: Recent progress and future research. *Oceanography*, vol. 30, No. 1, pp. 36–45.

Binelli, A., and Provini, A. (2003). POPs in edible clams from different Italian and European markets and possible human health risk. *Marine Pollution Bulletin*, vol. 46, No. 7, pp. 879–886.

Borja, Angel, and others (2020). Moving toward an agenda on ocean health and human health in Europe. *Frontiers in Marine Science*, vol. 7, art. 37.

Boverhof, D.R., and others (2015). Comparative assessment of nanomaterial definitions and safety evaluation considerations. *Regulatory Toxicology and Pharmacology*, vol. 73, pp. 137–150.

Brereton, Finbarr, and others (2008). Happiness, geography and the environment. *Ecological Economics*, vol. 65, No. 2, pp. 386–396.

Burkhardt, William, and Kevin R. Calci (2000). Selective accumulation may account for shellfish-associated viral illness. *Applied and Environmental Microbiology*, vol. 66, No. 4, pp. 1375–1378.

Chang, X.R., and others (2020). Potential health impact of environmental micro- and nanoplastics pollution. *Journal of Applied Toxicology*, vol. 40, pp. 4–15.

Chen, H., and others (2015). Antibiotics in typical marine aquaculture farms surrounding Hailing Island, South China: occurrence, bioaccumulation and human dietary exposure. *Marine Pollution Bulletin*, vol. 90, Nos. 1–2, pp. 181–187.

Cherrie, M., and others (2015). Coastal climate is associated with elevated solar irradiance and higher 25(OH)D level. *Environment International*, vol. 77, pp. 76–84.

Deeb, R., and others (2018). Impact of climate change on *Vibrio vulnificus* abundance and exposure risk. *Estuaries and Coasts*, vol. 41, No. 8, pp. 2289–2303.

Depledge, M., and others (2013). Changing views of the interconnections between the oceans and human health in Europe. *Microbial Ecology*, vol. 65, No. 4, pp. 852–859.

Depledge, M., and others (2017). Future of the sea: health and wellbeing of coastal communities. United Kingdom Government Office for Science. 2017. https://ore.exeter.ac.uk/repository/handle/10871/31606.

Depledge, M., and others (2019). Time and tide: our future health and well-being depends on the oceans.

Depledge, M., and William J. Bird (2009). The blue gym: health and wellbeing from our coasts. *Marine Pollution Bulletin*, vol. 58, No. 7, p. 947.

Diaz, R.E., and others (2019). Neurological illnesses associated with Florida red tide (*Karenia brevis*) blooms. *Harmful Algae*, vol. 82, pp. 73–81.

Ercolano, G., and others (2019). New drugs from the sea: pro-apoptotic activity of sponges and algae derived compounds. *Marine Drugs*, vol. 17, No. 1, 31. https://doi.org/10.3390/md17010031.

European Commission (2000). Towards the establishment of a priority list of substances for further evaluation of their role in endocrine disruption. https://ec.europa.eu/environment/archives/docum/pdf/bkh_main.pdf.

European Economic Community (1975). *Council Directive 76/160/EEC of 8 December 1975 Concerning the Quality of Bathing Water.* https://eur-lex.europa.eu/legal-content/EN/ALL/?uri=celex:31976L0160.

European Union (2006). *Directive 2006/7/EC Concerning the Management of Bathing Water Quality and Repealing Directive 76/160/EEC.* https://eur-lex.europa.eu/legal-content/EN/TXT/PDF/?uri=celex:32006L0007&from=GA.

________ (2020). *About Seas, Oceans & Public Health in Europe*. https://sophie2020.eu/about.

Food and Agriculture Organization of the United Nations (FAO) (2017). *Water Pollution from Agriculture: A Global Review*. Rome: FAO and Colombo, Sri Lanka: International Water Management Institute on behalf of the Water Land and Ecosystems research programme.

Fleming, Lora E., and others (2014). Oceans and human health: a rising tide of challenges and opportunities for Europe. *Marine Environmental Research*, vol. 99, pp. 16–19.

Fleming, Lora E., and others (2019). Fostering human health through ocean sustainability in the 21st century. *People and Nature*, vol. 1, No. 3, pp. 276–283.

Gaibor, Nikita, and others (2020). Composition, abundance and sources of anthropogenic marine debris on the beaches from Ecuador – A volunteer-supported study. *Marine Pollution* Bulletin, vol. 154, 111068. https://doi.org/10.1016/j.marpolbul.2020.111068.

Galil, Bella (2018). Poisonous and venomous: marine alien species in the Mediterranean Sea and human health. In *Invasive Species and Human Health*, G. Mazza and E. Tricarico, eds., pp. 1–15. Wallingford, United Kingdom: CABI.

Galloway, Tamara S., and others (2017). Interactions of microplastic debris throughout the marine ecosystem. *Nature Ecology & Evolution*, vol. 1, No.5, p. 0116. https://doi.org/10.1038/s41559-017-0116.

Gascon, Mireia, and others (2017). Outdoor blue spaces, human health and well-being: a systematic review of quantitative studies. *International Journal of Hygiene and Environmental Health*, vol. 220, No. 8, pp. 1207–21. https://doi.org/10.1016/j.ijheh.2017.08.004.

Giles, Sarah (2013). Green space is great, but blue might be better.... 2013. http://blogs.royalsociety.org/in-verba/2013/04/09/blue_space.

Gilmer, Mary Jo, and others (2003). Youth characteristics and contextual variables influencing physical activity in young adolescents of parents with premature coronary heart disease. *Journal of Pediatric Nursing*, vol. 18, No. 3, pp. 159–168.

Goa State Pollution Control Board (GSPCB) (2019). *Annual Report 2017/18*. http://goaspcb.gov.in/Media/Default/Annual%20Report%20uploads/GSPCB_2017-2018.pdf.

Harder-Lauridsen, Nina Majlund, and others (2013). Gastrointestinal illness among triathletes swimming in non-polluted versus polluted seawater affected by heavy rainfall, Denmark, 2010–2011. *PloS One*, vol. 8, No. 11.

Harrison, Jesse P., and others (2018). Microplastic-associated biofilms: a comparison of freshwater and marine environments. In *Freshwater Microplastics*, M. Wagner and S. Lambert, eds., pp.181–201. Cham, Switzerland: Springer.

Hinder, S.L., and others (2012). Changes in marine dinoflagellate and diatom abundance under climate change. *Nature Climate Change*. https://doi.org/10.1038/NCLIMATE1388.

Hoagland, Porter, and others (2009). The costs of respiratory illnesses arising from Florida Gulf Coast *Karenia brevis* blooms. *Environmental Health Perspectives*, vol. 117, No. 8, pp. 1239–1243.

Hooyberg, Alexander, and others (2020). General health and residential proximity to the coast in Belgium: Results from a cross-sectional health survey. *Environmental Research*, vol. 184, art. 109225.

Hosomi, R., and others (2012). Seafood consumption and components for health. *Global Journal of Health Science*, vol. 4, No. 3, pp. 72–86.

Imran, Md., and others (2019). Co-selection of multi-antibiotic resistance in bacterial pathogens in metal and microplastic contaminated environments: an emerging health threat. *Chemosphere*, vol. 215, pp. 846–857.

Intergovernmental Oceanographic Commission of the United Nations Educational, Scientific and Cultural Organization (UNESCO-IOC) (2017). *Global Ocean Science Report: The Current Status of Ocean Science around the World*. Luis Valdés and others, eds. Paris: UNESCO Publishing.

Joint Group of Experts on the Scientific Aspects of Marine Environmental Protection (GESAMP) (2001). *Protecting the oceans from land-based activities – Land-based sources and activities affecting the quality and uses of the marine, coastal and associated freshwater environment*. Rep. Stud. GESAMP No. 71, UNEP Nairobi.

Kersting, D., and others (2020). The efficiency of full protection in MPAs. MedPAN. Marseille, France.

Keswani, Anisha, and others (2016). Microbial hitchhikers on marine plastic debris: human exposure risks at bathing waters and beach environments. *Marine Environmental Research*, vol. 118, pp. 10–19.

Konrad, S., and others (2017). Remote sensing measurements of sea surface temperature as an indicator of *Vibrio parahaemolyticus* in oyster meat and human illnesses. *Environmental Health*, vol. 16, No. 92. https://doi.org/10.1186/s12940-017-0301-x.

Kueh, C.S.W., and others (1995). Epidemiological study of swimming-associated illnesses relating to bathing-beach water quality. *Water Science and Technology*, vol. 31, Nos. 5–6, pp. 1–4.

Lancet-Editorial (2009). What is health? The ability to adapt. *Lancet,* vol. 373 (9666): 781. https://doi.org/10.1016/S0140-6736(09)60456-6.

Leonard, Anne F.C., and others (2018a). Exposure to and colonisation by antibiotic-resistant *E. coli* in UK coastal water users: Environmental surveillance, exposure assessment, and epidemiological study (Beach Bum Survey). *Environment International*, vol. 114, pp. 326–333.

Leonard, Anne F.C., and others (2018b). Is it safe to go back into the water? a systematic review and meta-analysis of the risk of acquiring infections from recreational exposure to seawater. *International Journal of Epidemiology*, vol. 47, No. 2, pp. 572–586.

Li, Hongbin, and Yi Zhu (2006). Income, income inequality and health: Evidence from China. WIDER Discussion Paper 2006/07. Helsinki: The United Nations University World Institute for Development Economics Research (UNU-WIDER). http://hdl.handle.net/10419/84654.

Lindequist, U. (2016). Marine-Derived Pharmaceuticals – Challenges and Opportunities. *Biomolecules & Therapeutics*, vol. 24, No. 6, pp. 561–571.

Liu, Huan, and others (2016). Health and climate impacts of ocean-going vessels in East Asia. *Nature Climate Change,* vol. 6, pp. 1037–1041.

Lu, J., and others (2018). Occurrence, distribution, and ecological-health risks of selected antibiotics in coastal waters along the coastline of China. *Science of the Total Environment*, vol. 644, pp. 1469–1476.

Marmot, Michael, and Richard Wilkinson, eds. (2005). *Social Determinants of Health*. 2nd ed. Oxford: Oxford University Press. https://doi.org/10.1093/acprof:oso/9780198565895.001.0001.

Moore, Michael N. (2015). Do airborne biogenic chemicals interact with the PI3K/Akt/mTOR cell signalling pathway to benefit human health and wellbeing in rural and coastal environments? *Environmental Research*, vol. 140, pp. 65–75. https://doi.org/10.1016/j.envres.2015.03.015.

________ (2020). Lysosomes, autophagy and hormesis in cell physiology, pathology and age-related disease. *Dose-Response*, vol. 18, No. 3. https://doi.org/10.1177/1559325820934227.

Moore, Michael N. and others (2013). Oceans and Human Health (OHH): a European Perspective from the Marine Board of the European Science Foundation (Marine Board-ESF). *Microbial Ecology*, vol. 65, No. 4, pp. 889–900. https://doi.org/10.1007/s00248-013-0204-5.

Moore, Michael N. and others (2014). *Linking Oceans and Human Health: A Strategic Research Priority for Europe*. Marine Board Position Paper 19. Ostend: European Marine Board.

Morris, J.G., Jr., and D. Acheson, (2003). Cholera and other types of vibriosis: a story of human pandemics and oysters on the half shell. *Clinical Infectious Diseases*, vol. 37, No. 2, pp. 272–280.

Mossman, B.T., and others (2007). Mechanisms of action of inhaled fibers, particles and nanoparticles in lung and cardiovascular diseases. *Particle and Fibre Toxicology,* vol. 4, No. 4. https://doi.org/10.1186/1743-8977-4-4.

Motes, M., and others (1994). Occurrence of toxigenic *Vibrio cholerae* O1 in oysters in Mobile Bay, Alabama: an ecological investigation. *Journal of Food Protection*, vol. 57, No. 11, pp. 975–980.

Myers, Gary J., and others (2015). Methylmercury exposure and developmental neurotoxicity. *Bulletin of the World Health Organization*, vol. 93, pp. 132A–132B.

New South Wales Department of Planning, Industry and Environment (NSW-DPIE) (2020). *Monitoring Beach Water Quality*. www.environment.nsw.gov.au/topics/water/beaches/monitoring-beach-water-quality.

Numan, M.S., and others (2015). Impact of air pollutants on oxidative stress in common autophagy-mediated aging diseases. *International Journal of Environmental Research & Public Health*, vol. 12, pp. 2289–2305.

Nutsford, Daniel, and others (2016). Residential exposure to visible blue space (but not green space) associated with lower psychological distress in a capital city. *Health & Place*, vol. 39, pp. 70–78.

Organization for Economic Cooperation and Development (OECD) (2017). Marine Protected Areas: Economics, Management and Effective Policy Mixes, OECD Publishing, Paris. https://doi.org/10.1787/9789264276208-en.

Pleijel, H., and others (2013). Surface Ozone in the Marine Environment—Horizontal Ozone Concentration Gradients in Coastal Areas. *Water Air Soil Pollution*, vol. 224, p. 1603. https://doi.org/10.1007/s11270-013-1603-4.

Powell, T.M., and others (2019). Stress and coping in social service providers after Superstorm Sandy: An examination of a postdisaster psychoeducational intervention. *Traumatology*, vol. 25, No. 2, pp. 96–103. https://doi.org/10.1037/trm0000189.

Prieto, M.D., and others (2001). Recreation in coastal waters: health risks associated with bathing in sea water. *Journal of Epidemiology & Community Health*, vol. 55, No. 6, pp. 442–447.

Ramos Vitor, and Vitor Vasconcelos (2010). Palytoxin and analogs: biological and ecological effects. *Marine Drugs*, vol. 8, No. 7, pp. 2021–37.

Robinson, Jake M., and others (2020). Let nature be thy medicine: a socioecological exploration of green prescribing in the UK. *International Journal of Environmental Research and Public Health*, vol. 17, No. 3460. https://doi.org/10.3390/ijerph17103460.

Rodriguez, Paula, and others (2008). First toxicity report of tetrodotoxin and 5, 6, 11-trideoxyTTX in the trumpet shell *Charonia lampas* in Europe. *Analytical Chemistry*, vol. 80, No. 14, pp. 5622–5629.

Rook, Graham A. (2013). Regulation of the immune system by biodiversity from the natural environment: an ecosystem service essential to health. *Proceedings of the National Academy of Sciences*, vol. 110, No. 46, pp. 18360–18367.

Ruskin, J., and others (2018). Lack of access to medical care during Hurricane Sandy and mental health symptoms. *Preventive Medicine Reports*, vol. 10, pp. 363–9.

Sachs, Jeffrey D., and others (2001). The geography of poverty and wealth. *Scientific American*, vol. 284, No. 3, pp. 70–75.

Seneviratne, Sonia I., and others (2012). Changes in climate extremes and their impacts on the natural physical environment. In: *Managing the Risks of Extreme Events and Disasters to Advance Climate Change Adaptation. A Special Report of Working Groups I and II of the Intergovernmental Panel on Climate Change (IPCC).* Cambridge, New York: Cambridge University Press, pp. 109–230.

Sforzini, S., and others (2020). Effects of fullerene C_{60} in blue mussels: role of mTOR in autophagy related cellular/tissue alterations. Chemosphere, vol. 246:125707. https://doi.org/10.1016/j.chemosphere.2019.125707.

Sheehan, Mary C., and others (2014). Global methylmercury exposure from seafood consumption and risk of developmental neurotoxicity: a systematic review. *Bulletin of the World Health Organization*, vol. 92, pp. 254–269.

Silbernagel, Susan M., and others (2011). Recognizing and preventing overexposure to methylmercury from fish and seafood consumption: information for physicians. *Journal of Toxicology*, vol. 2011, ID 983072. https://doi.org/10.1155/2011/983072.

Silva, Marisa, and others (2015a). Emergent toxins in North Atlantic temperate waters: A challenge for monitoring programs and legislation. *Toxins*, vol. 7, No. 3, pp. 859–885.

Silva, Marisa, and others (2015b). First report of ciguatoxins in two starfish species: *Ophidiaster ophidianus* and *Marthasterias glacialis*. *Toxins*, vol. 7, No. 9, pp. 3740–3757.

Smith, Madeleine, and others (2018). Microplastics in seafood and the implications for human health. *Current Environmental Health Reports*, vol. 5, No. 3, pp. 375–386.

South Australia Environment Protection Agency (SA-EPA) (2020). *Beach Alert*. www.epa.sa.gov.au/data_and_publications/water_quality_monitoring/beach_water_advice.

Stapleton P.A. (2019). Toxicological considerations of nano-sized plastics. *AIMS Environmental Science*, vol. 6, No. 5, pp. 367–378.

Stern, S.T., and others (2012). Autophagy and lysosomal dysfunction as emerging mechanisms of nanomaterial toxicity. *Particle and Fibre Toxicology*, vol. 9, No. 20. https://doi.org/10.1186/1743-8977-9-20.

Sullivan, A.J. (1971). Ecological effects of sewage discharge in the marine environment. *Proceedings of the Royal Society of London. Series B. Biological Sciences*, vol. 177, No. 1048, pp. 331–351.

Takahashi, C.K., and others (2008). Ballast water: a review of the impact on the world public health. *Journal of Venomous Animals and Toxins Including Tropical Diseases*, vol. 14, No. 3, pp. 393–408.

Tornero, V., and G. Hanke (2016). Chemical contaminants entering the marine environment from sea-based sources: a review with a focus on European seas. *Marine Pollution Bulletin*, vol. 112, pp. 17–38.

Turner, A.D., and others (2015). Potential threats posed by new or emerging marine biotoxins in UK waters and examination of detection methodology used in their control: brevetoxins. *Marine Drugs*, vol. 13, No. 3, pp. 1224–1254.

United Nations (2017). *The First Global Integrated Marine Assessment: World Ocean Assessment I*. Cambridge: Cambridge University Press.

Valotto, Gabrio, and others (2015). Environmental and traffic-related parameters affecting road dust composition: a multi-technique approach applied to Venice area (Italy). *Atmospheric Environment*, vol. 122, pp. 596–608.

Van Acker, Emmanuel, and others (2020). Aerosolizable marine phycotoxins and human health effects: in vitro support for the biogenics hypothesis. *Marine Drugs*, vol. 18, No. 1, art. 46.

Vethaak, A. Dick, and Heather A. Leslie (2016). Plastic debris is a human health issue. Environmental Science & Technology, vol. 50, No. 13, pp. 6825–26. https://doi.org/10.1021/acs.est.6b02569.

Vezzulli, L., and others (2016). Climate influence on Vibrio and associated human diseases during the past half-century in the coastal North Atlantic. *Proceedings of the National Academy of Sciences*, 201609157. https://doi.org/10.1073/pnas.1609157113.

Vom Saal, F.S., and others (2007). Chapel Hill bisphenol A expert panel consensus statement: integration of mechanisms, effects in animals and potential to impact human health at current levels of exposure. *Reproductive Toxicology*, vol. 24, No. 2, pp. 131–138.

Von Moos, N., and others (2012). Uptake and effects of microplastics on cells and tissue of the blue mussel *Mytilus edulis* L. after an experimental exposure. *Environmental Science & Technology*, vol. 46, pp. 11327–11335.

Wade, Timothy J., and others (2006). Rapidly measured indicators of recreational water quality are predictive of swimming-associated gastrointestinal illness. *Environmental Health Perspectives*, vol. 114, No. 1, pp. 24–28.

Walker, Tony R., and others (2019). Environmental effects of marine transportation. In: *World Seas: An Environmental Evaluation*, 2nd edition, C. Sheppard, ed., Academic Press, pp. 505–530, chap. 27.

Wan, Zheng, and others (2016). Three steps to a green shipping industry. *Nature*, vol. 530, pp. 275–277.

Wheeler, Benedict W., and others (2012). Does living by the coast improve health and wellbeing? *Health & Place*, vol. 18, No. 5, pp. 1198–1201.

White, Mathew P., and others (2013). Feelings of restoration from recent nature visits. *Journal of Environmental Psychology*, vol. 35, pp. 40–51.

White, Mathew P., and others (2014). Coastal proximity and physical activity: is the coast an under-appreciated public health resource? *Preventive Medicine*, vol. 69, pp. 135–140.

Winiger, P., and others (2019). Source apportionment of circum-Arctic atmospheric black carbon from isotopes and modeling. *Science Advances*, vol. 5, No. 2, eaau8052. http://doi.org/10.1126/sciadv.aau8052.

Witten, K., and others (2008). Neighbourhood access to open spaces and the physical activity of residents: a national study. *Preventive Medicine*, vol. 47, No. 3, pp. 299–303.

Wood, Sophie L., and others (2016). Exploring the relationship between childhood obesity and proximity to the coast: a rural/urban perspective. *Health & Place*, vol. 40, pp. 129–136.

Woods Hole Oceanographic Institution (WHOI) (2020). *Beach Closures*. www.whoi.edu/know-your-ocean/ocean-topics/pollution/beach-closures.

World Health Organization (WHO) (2003). *Guidelines for Safe Recreational Water Environments: Coastal and Fresh Waters*. vol. 1. Geneva.

________(2008). *Guidance for Identifying Populations at Risk from Mercury Exposure*. Geneva.

________(2014). Gender, climate change and health. Geneva.

________(2013). *Mercury and Health (Fact Sheet No. 361)*. Geneva.

________ (2018). *Guidelines for safe recreational water environments: Volume 1 – Coastal and Fresh Waters*, WHO, Geneva. www.who.int/water_sanitation_health/water-quality/recreational/guidelines-for-safe-recreational-environments/en.

________ (2020). *Water Safety and Quality – Bathing Waters.*

WHO-Regional Office for Europe (WHO-Europe) (1984). Health promotion: a discussion document on the concept and principles: summary report of the Working Group on Concept and Principles of Health Promotion, Copenhagen, 9–13 July 1984.

________ (2010). Parma Declaration on Environment and Health. In *Fifth Ministerial Conference on Environment and Health. Protecting Children's Health in a Changing Environment. Parma, Italy.* www.euro.who.int/__data/assets/pdf_file/0011/78608/E93618.pdf?ua=1.

Wright, Stephanie L., and Frank J. Kelly (2017). Plastic and human health: a micro issue? *Environmental Science & Technology*, vol. 51, No. 12, pp. 6634–6647.

Wyles, K.J., and others (2019). Are some natural environments more psychologically beneficial than others? The importance of type and quality on connectedness to nature and psychological restoration. *Environmental Behaviour*, vol. 51, pp. 111–143.

Zanchett, Giliane, and Eduardo C. Oliveira-Filho (2013). Cyanobacteria and cyanotoxins: from impacts on aquatic ecosystems and human health to anticarcinogenic effects. *Toxins*, vol. 5, No. 10, pp. 1896–1917.

Zmirou, Denis, and others (2003). Risks associated with the microbiological quality of bodies of fresh and marine water used for recreational purposes: summary estimates based on published epidemiological studies. *Archives of Environmental Health: An International Journal*, vol. 58, No. 11.

Part five
Trends in pressures on the marine environment

Chapter 9
Pressures from changes in climate and atmosphere

Contributors: Carlos Garcia-Soto (lead member and convener), Denise Breitburg, Monica Campillos, Patricia Castillo-Briceno, Sanae Chiba (co-lead member), Matthew Collins, Ganix Esnaola, Karen Evans (co-lead member), Louise B. Firth, Thomas Frölicher, Jason M. Hall-Spencer, David Halpern, Karen L. Hunter, Gabriel Ibarra, Sung-Yong Kim, Roxy M. Koll, Kathleen McInnes, Jon Saenz, Ca Thanh Vu (co-lead member), Bess Ward and Tymon Zielinski (co-lead member).

Keynote points

- **Extreme climate events.** Marine heatwaves and tropical cyclones are shown to be increasing in severity owing to human activities and are having an impact on nature and human societies. Extreme El Niño events have been observed but, because they occur infrequently, a human influence has not been detected. All three phenomena are projected to increase in the future, with the severity of impacts also increasing, but such increases can be reduced by climate change mitigation efforts.
- **Sea level rise.** The alarming observed pace of sea level rise, combined with increasing storminess and coastal urbanization, has resulted in the amplified susceptibility of coastal cities to erosion and flooding and increased the need for substantial investments in hard infrastructure and the restoration of natural barriers, such as reefs.
- **Ocean acidification and deoxygenation.** The accelerated increase of anthropogenic CO_2 in the atmosphere is creating an increase in the acidification and deoxygenation of the ocean. Under such conditions, both in nature and in the laboratory, marine organisms that support ecosystems and human livelihoods and nutrition typically respond poorly. Marine habitats experience a loss of diversity, many long-lived organisms die and a few resilient species proliferate. Less serious damage to life-supporting ecosystems would be possible under lower-emission scenarios.
- **Other physical and chemical properties.** Changes in ocean temperature and salinity induced by climate change and human activities are affecting marine ecosystems by changing the distribution of marine species, decreasing the ecological value of coastal ecosystems and changing marine primary production. Human well-being and the economy are consequently affected.

1. Introduction

The first part of the present chapter is based on three topics in the context of extreme climate events related to the ocean, namely, marine heatwaves, extreme El Niño Southern Oscillation events and tropical cyclones. Both physical aspects of the impact of climate change on the phenomena and potential impacts on natural and human systems are considered. The conclusions are based on a much more detailed assessment that can be found in chapter 6 of the *Special Report on Oceans and Cryosphere in a Changing Climate* of the Intergovernmental Panel on Climate Change (2019).

An extreme event is one that is rare at a particular place and time of year. Definitions of "rare" vary, but an extreme event is normally as rare as, or rarer than, the tenth or ninetieth percentile of a probability estimated from observations. By definition, the characteristics of what is called an extreme event may vary from place to place in an absolute sense. When a pattern of extreme weather persists for some time, such as a season, it may be classed as an extreme climate event, especially if it yields an average or total that is itself extreme (e.g., high temperature, drought or total rainfall over a season).

The second part of the chapter expands upon pressures from changes in ocean physical and chemical properties. Projected sea temperature increases of up to 1.5°C over pre-industrial levels by 2050 will continue to drive latitudinal abundance shifts in marine species, including those of importance for coastal livelihoods. Many large coastal cities are located in deltaic settings and are vulnerable to floods because of their proximity to rivers and the sea, general low elevations and land subsidence (Nicholls and others, 2008).

Carbon dioxide emissions and global warming are also causing ocean acidification and deoxygenation. Those changes have consequences for the people who depend on healthy marine ecosystems worldwide. At the time of the first *World Ocean Assessment* (United Nations, 2017), the chemistry of ocean acidification was well understood, yet the consequences for ecosystems and society were poorly known. The effects of declining oxygen on nutrient cycles and fish stocks were predicted to worsen, especially when climate change-driven oxygen depletion combines with coastal eutrophication. Reduced biodiversity and declines in fish populations were linked to falling oxygen levels across the world's oceans. New information is provided on marine organism and ecosystem responses to ocean acidification and deoxygenation and related capacity-building.

In the present chapter, in conjunction with chapter 5, the climate change aspects of the present Assessment are developed. The present chapter expands on the pressures on marine ecosystems and human populations of some of the physical and chemical changes caused by climate change. Some related aspects are also covered in chapter 7K and chapter 15.

2. Climate pressures: extreme climate events and pressures from changes in ocean physical and chemical properties

2.1. Extreme climate events

Marine heatwaves are periods of extremely high ocean temperatures that persist for days to months, that can extend up to thousands of km and can penetrate multiple hundreds of m into the deep ocean (Hobday and others, 2016). Over the past two decades, marine heatwaves have had a negative impact on marine organisms and ecosystems in all ocean basins, including critical foundation species such as corals, seagrasses and kelps (Hughes and others, 2018; Smale and others, 2019). Satellite observations reveal that marine heatwaves doubled in frequency between 1982 and 2016, and that they have also become longer lasting and more intense and extensive (Frölicher and others, 2018; Oliver and others, 2018). Between 2006 and 2015, 84 to 90 per cent of all globally occurring marine heatwaves were attributable to the temperature increase since the period 1850–1900 (Frölicher and others, 2018).

Marine heatwaves will further increase in frequency, duration, spatial extent and intensity under future global warming (Frölicher and others, 2018; Darmaraki and others, 2019), pushing some marine organisms, fish stocks and ecosystems beyond the limits of their resilience, with cascading impacts on economies and societies (Smale and others, 2019). Globally, the frequency of marine heatwaves is very likely to increase by a factor of about 50 times by the period 2081–2100 under the high-emission Representative Concentration Pathway (RCP) 8.5 scenario and by a factor of about 20 times under the low-emission RCP 2.6 scenario (Van Vuuren and others, 2011), relative to the reference period 1850–1900. Such future trends in marine heatwave frequency can largely be explained by increases in mean ocean temperature. The largest changes in the frequency of marine heatwaves are projected for the Arctic Ocean and the tropical oceans (figure I; Intergovernmental Panel on Climate Change (IPCC), 2019, chap. 6, figure 6.4).

Figure I
Locations of extreme events with an identified link to climate change caused by human activities

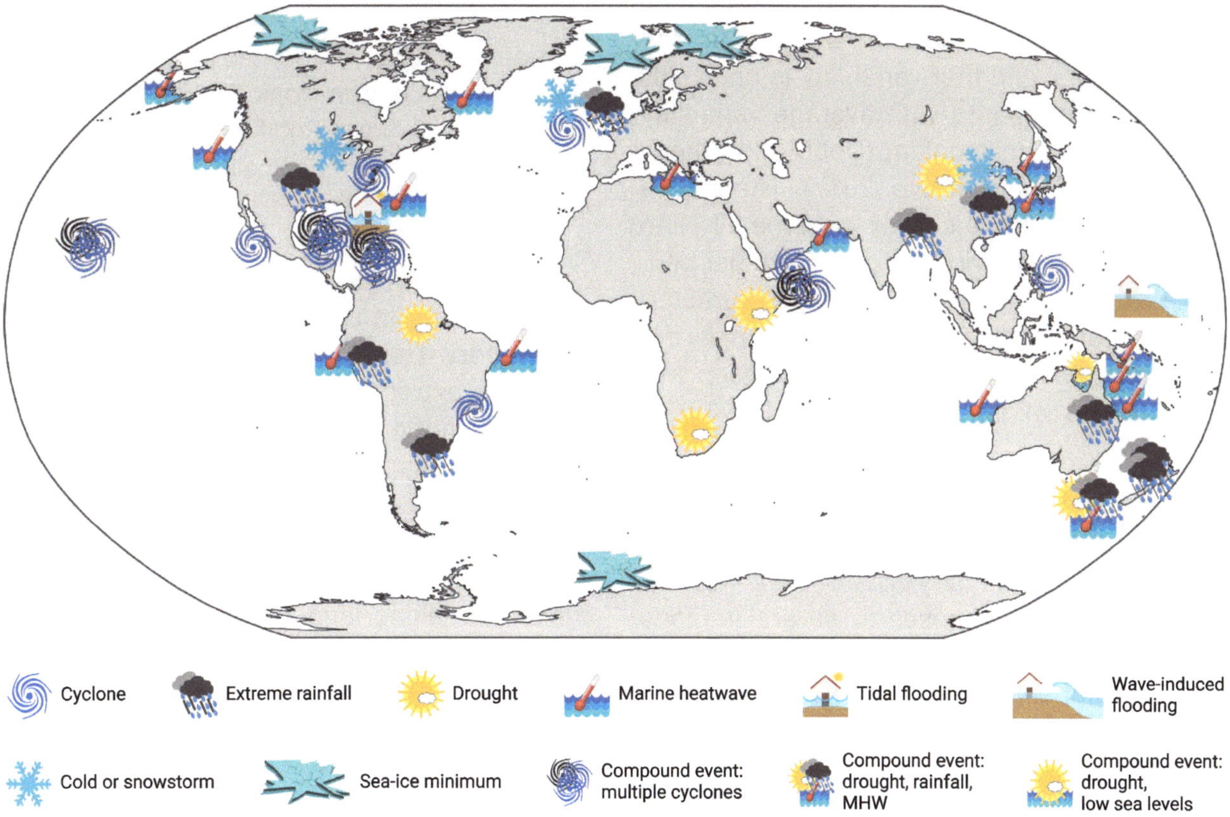

Source: Figure adapted from IPCC, 2019, figure 6.2.

Limiting global warming would reduce the risk of impacts of marine heatwaves, but critical thresholds for some ecosystems (e.g., kelp forests and coral reefs) will be reached even at relatively low levels of future global warming (King and others, 2017). Early warning systems, producing skilful forecasts of marine heatwaves, can further help to reduce vulnerabilities in fishing, tourism and conservation, but are yet unproven on a large scale (Payne and others, 2017; Tommasi and others, 2017).

One of the best data-rich examples of the impact of a marine heatwave on well-managed fisheries is of the Gulf of Alaska in the North Pacific. A prolonged warm ocean event weakened benthic ocean and surface mixing, in turn disrupting trophies, invertebrate and forage fish populations, and decimated the Pacific cod fishery, triggering a series of repeating mass marine mammal and seabird die-offs that had a ripple effect through coastal economies.

The El Niño Southern Oscillation is a coupled atmosphere-ocean phenomenon, identified by an oscillation between warm and cold ocean temperatures in the tropical central eastern Pacific Ocean and an associated fluctuation in the global-scale tropical and subtropical surface pressure patterns. Typically, it has a preferred timescale of about two to seven years. It is often measured by the surface pressure anomaly difference between Tahiti, French Polynesia, and Darwin, Australia, and/or the sea surface temperatures in the central and eastern equatorial Pacific (Rasmussen and Carpenter, 1982). It has climatic effects throughout the Pacific region and in many other parts of the world through global teleconnections. The warm phase of the Oscillation is called El Niño and the cold phase is called La Niña.

The strongest El Niño and La Niña events since the pre-industrial era have occurred during the past 50 years, and that variability is unusually high when compared with average variability during the last millennium (Cobb and others, 2013; Santoso and others, 2017). There have been three occurrences of extreme El Niño events during the modern observational period (1982/83, 1997/98, 2015/16), all characterized by pronounced rainfall in the normally dry equatorial East Pacific. There have been two occurrences of extreme La Niña (1988/89, 1998/99).

Extreme El Niño and La Niña events are likely to occur more frequently with global warming and are likely to intensify existing impacts, with drier or wetter responses in several regions across the globe, even at relatively low levels of future global warming (Cai and others, 2014; Cai and others, 2015; Power and Delage, 2018).

Sustained long-term monitoring and improved forecasts can be used in managing the risks of extreme El Niño and La Niña events associated with human health, agriculture, fisheries, coral reefs, aquaculture, wildfire, drought and flood management (L'Heureux and others, 2017).

A tropical cyclone is the general term for a strong, cyclonic-scale disturbance that originates over the tropical ocean. Based on one-minute maximum sustained wind speed, the cyclonic disturbances are categorized into tropical depressions (≤ 17 m/s), tropical storms (18–32 m/s) and tropical cyclones (≥ 33 m/s, category 1 to category 5) (Knutson and others, 2010). A tropical cyclone is called a hurricane, typhoon or cyclone, depending on geographic location.

Anthropogenic climate change has increased precipitation, winds and extreme sea level events associated with a number of observed tropical cyclones. For example, studies have shown that the rainfall intensity of tropical cyclone (Hurricane) Harvey increased by at least 8 per cent (8–19 per cent) owing to climate change (Risser and Wehner, 2017; Van Oldenborgh and others, 2017). Anthropogenic climate change may have contributed to a poleward migration of maximum tropical cyclone intensity in the western North Pacific in recent decades related to anthropogenically forced tropical expansion (Sharmila and Walsh, 2018). There is emerging evidence of a number of regional changes in tropical cyclone behaviour, such as an increase in the annual global proportion of category 4 or 5 tropical cyclones in recent decades, extremely severe tropical cyclones occurring in the Arabian Sea, cyclones making landfall in East and South-East Asia, an increase in frequency of moderately large storm surge events in the United States since 1923 and a decrease in frequency of severe tropical cyclones making landfall in eastern Australia since the late 1800s. There is low confidence that they represent detectable anthropogenic signals. Extreme wave heights, which contribute to extreme sea level events, coastal erosion and flooding, have increased in the Southern Ocean and the North Atlantic Ocean by about 1.0 cm per year and 0.8 cm per year over the period 1985–2018 (Young and Ribal, 2019).

An increase in the average intensity of tropical cyclones, and the associated average precipitation rates, is projected for a 2°C global temperature rise, although there is low confidence in future frequency changes at the global scale (Yamada and others, 2017). Rising sea levels will contribute to higher extreme sea levels associated with tropical cyclones in the future (Garner and others, 2017). Projections suggest that the proportion of category 4 and 5 tropical cyclones will increase (Knutson and others, 2015; Park and others, 2017). Such changes will affect storm surge frequency and intensity, as well as coastal infrastructure and mortality.

Investment in disaster risk reduction, flood management (ecosystem and engineered) and early warning systems decreases economic loss from tropical cyclones that occur near coasts and islands. However, such investments may be hindered by limited local capacities (e.g., ageing infrastructure and other non-climatic factors) that, for example, can lead to increased losses and mortality from extreme winds and storm surges in developing countries despite adaptation efforts. There is emerging evidence of increasing risks for locations affected by unprecedented storm trajectories. Management of risk from such changing storm trajectories and intensity proves challenging because of the difficulties of early warning and its receptivity by affected populations.

2.2. Sea level rise and cities

Cities located along coastlines and in archipelagic and island States are becoming increasingly susceptible to erosion and sea level rise (De Sherbinin and others, 2007; Hanson and others, 2011; Takagi and others, 2016). Many comprise large areas of reclaimed land (the gain of land from the sea, wetlands or other water bodies), which is retained and protected from erosion by hard engineered structures, such as sea walls and rock armouring (Sengupta and others, 2018). It is likely that many of such engineered coastlines will need to be adapted and upgraded to keep pace with rising sea levels. In highly urbanized environments that are often already heavily degraded, hard engineered structures are often the only option available and are considered to be successful options (Hallegatte and others, 2013; Hinkel and others, 2014), but there are a wide range of broader negative impacts of land reclamation and those structures on the surrounding environment (Dafforn and others, 2015). Globally, many regions (especially cities) are claiming that more than 50 per cent of their coastlines are armoured (e.g., Chapman, 2003; Burt and others, 2013), and that number will likely rise in the future in response to burgeoning economies, coastal populations and urbanization (e.g., see plans for the reclamation of the entire coastlines of two Malaysian states in Chee and others, 2017).

As an alternative to hard engineered coastal defences, construction of which is complex and expensive, where possible, natural coastal ecosystems such as mangroves and salt marshes should be used as natural barriers or combined with hard infrastructure using hybrid approaches (Temmerman and others, 2013). The use of such ecosystems can not only protect the land but also provide valuable ecosystem functions and services. As hard engineered coastal defences may be considered an effective short-term solution to coastal flooding, more investment will be needed owing to observed increasing storminess and sea level rise (Mendelsohn and others, 2012; Vitousek and others, 2017). By 2010, the global average sea level was calculated to be 52.4 mm above the 1993 level and, by 2018, it had risen to 89.9 mm above the 1993 level (National Oceanic and Atmospheric Administration (NOAA), 2019). The rate of change is also increasing. For the period 1993–2018, the rate of increase was calculated at 3.2 mm per year, while for the period 2010–2018, it was calculated to be much faster, at 4.7 mm per year. Despite significant uncertainties remaining, the Intergovernmental Panel on Climate Change predicts that sea level rise will continue for centuries, even if mitigation measures are put in place. The potential widespread collapse of ice shelves could lead to a larger twenty-first century sea level rise of up to several tenths of a metre (Church and others, 2013), which will have drastic consequences for coastal, archipelagic and small island cities, in particular those in low-lying areas.

Urbanization could, however, also provide opportunities for risk reduction, given that cities are engines of economic growth and centres of innovation, political attention and private sector investments (Garschagen and Romero-Lankao, 2015). Hallegatte and others (2013) conducted a global analysis of present and future losses in the 136 largest coastal cities. They predicted that global flood losses would increase from an average of $6 billion per year in 2005 to $1 trillion by 2050, with projected socioeconomic change, climate change and subsidence. Even if adaptation investments remain constant, flood probability, subsidence and sea level rise will increase global flood losses to $60 billion–$63 billion per year in 2050. The same study found that developing countries are particularly vulnerable to flood risk, with much lower investment in flood protection measures (Hallegatte and others, 2013).

Case study: Rotterdam

Low-lying cities in the Netherlands, a country that has long been a pioneer in both land reclamation and climate change adaptation, are taking a multipronged approach to the problem of sea level rise. For instance, Rotterdam's adaptation system is based on a flood and sea level rise defence system (C40 Cities, 2019) consisting of the Maeslantkering flexible storm surge barrier, permanent sand dunes along the coast, dykes along the rivers and a tailored "inner-dyke/outer-dyke" approach. The inner-dyke city, which is mostly below sea level, is formed by a system of polders drained by water outlets and pumps and protected by smaller secondary dykes. The outer-dyke city area (3–5.5 m above sea level), of 40,000 inhabitants, is vulnerable to rising sea level or smaller temporary floods. It is being adapted through the use of innovative technologies (e.g., floating buildings) and more traditional approaches (e.g., insulation of building facades and raising of electrical installations).

2.3. Pressures from changes in temperature

Ocean warming caused by anthropogenic climate change will continue for centuries after the anthropogenic forcing is stabilized (IPCC, 2019). It will affect marine ecosystems through increasing cumulative pressures owing to the changing climate and the intensity of human activities and is also interfering with other ocean properties, such as salinity and nutrient or carbon cycles, owing to the interconnection of all such processes.

Temperature-dependent biological sensitivity varies between species and is affected by other ocean properties. For example, for pelagic species, analysis of long-term trends in primary production has revealed that a rise in ocean temperatures, leading to enhanced stratification, nutrient limitation and shifts towards small phytoplankton, will have the greatest influence on decreasing the flux of particulate organic carbon to the deep ocean (Boyd and others, 2016; Fu and others, 2016). Reductions in particulate organic carbon flux are predicted at low and middle latitudes, but increases are possible at high latitudes, associated with a reduction in sea ice cover (Sweetman and others, 2017; Yool and others, 2017; FAO, 2018).

The special report entitled *Global Warming of 1.5°C* of the Intergovernmental Panel on Climate Change (2018) indicates that ocean ecosystems are already experiencing large-scale changes, and critical thresholds are expected to be reached at 1.5°C and higher levels of global warming. The changes to water temperatures are expected to drive some species (e.g., plankton and fish) to relocate to higher latitudes and cause novel ecosystems to assemble (Jonkers and others, 2019).

The increase in temperatures directly affects coastal communities, not only in terms of the effects on coastal marine ecosystems, but also on the ecosystem goods and services they deliver (Worm and others, 2006; Pendleton and others, 2016). They include, for example, the number of viable fisheries, the provision of nursery functions and the filtering services provided by coastal wetlands (Cochard and others, 2008; Barange and others, 2018). Coral reefs are one of coastal ecosystems heavily affected by ocean warming, and the coral bleaching phenomenon can affect not only marine life but also marine tourism.

Changes in temperature and salinity also have an impact on human well-being (food and health). With respect to food security, fish is one of the most consumed foods in the world and a major contributor to a healthy diet, owing to its proteins, fatty acids, vitamins and other elements that are essential for health (Hilmi and others, 2014). Climate change could decrease seafood availability (Golden and others, 2016) and, as a consequence, reduce protein supply to coastal communities, in general (Blanchard and others, 2017). That would have a strong impact on communities with high seafood dependence, including indigenous and other coastal communities.

An increased prevalence and transmission of diseases is also likely to occur with warmer ocean temperatures. Ocean warming could raise the risk of waterborne diseases and bloom algae toxins (see chap. 6a), affecting the populations and economies of affected areas. For example, the bacterial pathogen *Vibrio cholerae* is expected to grow faster owing to the increase in ocean temperatures (Semenza and others, 2017).

2.4. Pressures from changes in ocean chemistry

Ocean uptake of carbon dioxide emissions is rapidly changing seawater chemistry in a process known as ocean acidification (see chap. 5). As the partial pressure of carbon dioxide in seawater increases, it causes the carbonate saturation state to fall below levels suitable for globally important reef-forming taxa (Albright and others, 2018). Most coral reefs (shallow and deep) are vulnerable to rising CO_2 concentrations (Lam and others, 2019). Ocean acidification is causing the depth at which seawater is corrosive to carbonate to shoal, threatening deepwater coral reefs worldwide through dissolution and intensified bioerosion (Gómez and others, 2018). Ocean acidification combines with warming, rising sea level and more severe storms to reduce reef resilience on a global scale and augment reef destruction. In the Arctic, there has been a rapid expansion in the area where surface seawater is corrosive to calcareous organisms (Brodie and others, 2014).

Ocean acidification may affect all marine life, for example, through changes in gene expression, physiology, reproduction and behaviour (Riebesell and Gattuso, 2015; IPCC, 2019). Between 2005 and 2009, ocean acidification jeopardized a $270 million shellfish aquaculture industry that provided 3,200 jobs per year in Washington State, United States. Billions of oysters died in hatcheries because seawater had become corrosive to larval shells (Ekstrom and others, 2015). In addition to its negative impacts on calcifying phyto- and zooplankton, acidification can lower the nutritional value of seafood.

Ocean acidification also affects ecosystem properties, functions and services. Some groups of organisms do well in acidified conditions, but many taxa do not (Agostini and others, 2018). Many algae are resilient to the levels of ocean acidification projected under the Intergovernmental Panel on Climate Change RCP 8.5 scenario, yet shifts in community composition greatly alter seaweed habitats (Brodie and others, 2014; Enochs and others, 2015). Increased carbon availability stimulates primary production and can increase the standing stock of kelps and seagrasses (Russell and others, 2013; Linares and others, 2015; Cornwall and others, 2017), although microalgae and turf algae dominate acidified waters in exposed conditions (Agostini and others, 2018; Connell and others, 2018).

Research at natural marine CO_2 seeps has shown that there is about a 30 per cent decrease in macrofaunal biodiversity as average pH declines from 8.1 to 7.8 (Agostini and others, 2018; Foo and others, 2018), which is attributable to direct effects, such as increased metabolic costs of coping with hypercapnia, or indirect effects, such as increased susceptibility to predation (Sunday and others, 2017). Some corals grow well in seawater with elevated CO_2 concentrations, but the habitats they form lack diversity as reefs are degraded by ocean acidification owing to chemical dissolution and enhanced bioerosion, causing a shift to less diverse ecosystems. Chapter 7D also reviews the impacts of ocean acidification on coral reefs. The dual effects of increased CO_2 and decreased carbonate alter trophic interactions. Reductions in the abundance and size of calcareous herbivores contribute to the overgrowth of weedy turf algae and a simplification of food webs, with losses in functional diversity (Vizzini and others, 2017; Teixidó and others, 2018).

Damage from ocean acidification results in less coastal protection and less habitat for biodiversity and fisheries (Hall-Spencer and Harvey, 2019). Live coral cover on tropical reefs has nearly halved in the past 150 years, the decline accelerating over the past two decades owing to increased water temperature and ocean acidification exacerbating other drivers of coral loss. When combined with rising temperatures, sea level rise and increasing extreme climate events, ocean acidification further threatens the goods and services provided by coastal ecosystems. That is particularly important for those people who are heavily reliant on marine resources for protection, nutrition, employment and tourism (Lam and others, 2019).

Proposed actions to lessen the impacts of ocean acidification and to build resilience are primarily intended to reduce CO_2 emissions but also include: reduction of pollution and other stressors (such as overfishing and habitat damage); seaweed cultivation and seagrass restoration; water treatment, (e.g., for high-value aquaculture); adaptation of human activities such as aquaculture; and repair of damaged ecosystems (Cooley and others, 2016), for example, through the rewilding of the ocean.

Regarding deoxygenation, since the middle of the twentieth century, the ocean (including coastal waters, such as estuaries and semi-enclosed seas) has lost about 2 per cent, or over 150 billion tons, of its total oxygen content (Schmidtko and others, 2017), and more than 600 coastal water bodies have reported oxygen concentrations of less than 2 mg per l (Diaz and Rosenberg, 2008; Breitburg and others, 2018). Climate change is projected to cause more oxygen decline in many coastal systems where deoxygenation is currently driven primarily by an oversupply of anthropogenic nutrients. Such deoxygenation is of great concern because oxygen is fundamental to life in the oceans (figure II; Laffoley and Baxter, 2019). It constrains productivity and biodiversity, regulates global cycles of nutrients and carbon, and is required for the survival of individual organisms (Breitburg and others, 2018). When oxygen is sufficient, it does not limit or negatively affect the physiology, behaviour and ecological interactions of organisms dependent on aerobic (oxygen-utilizing) respiration. Waters are considered to be hypoxic when oxygen levels are insufficient and those processes are impaired. A threshold value of 2 mg dissolved oxygen/l is often used to define hypoxia, but the oxygen concentration or saturation at which life processes are impaired varies considerably among species, processes and habitats and is affected by temperature.

As the oxygen content of water declines, an increasing fraction of production is diverted to microbes (Diaz and Rosenberg, 2008; Wright and others, 2012). Food webs change because of altered encounter rates and the species-specific effects of low oxygen on the feeding efficiencies of predators and escape behaviours of prey. Energy transfer to tolerant animals, such as gelatinous species, can increase (Keister and Tuttle, 2013). The roles of vision (McCormick and Levin, 2017) and carnivory (Sperling and others, 2016) can decline within low oxygen areas because those activities are energy intensive. In contrast, predation can intensify above low oxygen zones as visual feeders are forced into shallower waters with higher light levels (Koslow and others, 2011).

Declining ocean oxygen is expected to negatively affect a wide range of biological and ecological processes. The magnitude of the effects will vary among species and processes, however, and whether the magnitude of responses will be directly proportional to the magnitude of oxygen decline is uncertain. Some effects of oxygen decline are dependent on direct exposure within low-oxygen waters, while others involve the movement of organisms and material (e.g., nutrients, organic matter, greenhouse gases) among locations that vary in oxygen content, and still other effects are primarily dependent on oxygen levels at particular locations that are critical for a species or life stage. Many responses involve threshold oxygen levels at which biological functions can no longer be maintained.

The biomass and diversity of eukaryotic organisms tend to decline and species composition changes as oxygen declines (Gallo and Levin, 2016). As low-oxygen waters expand, tolerant species can expand their depth range, while ranges of species that are more sensitive contract (Sato and others, 2017). The relative abundance of species within systems reflects variation in species' tolerances to low oxygen and other co-stressors (Koslow and others, 2018). Organisms, including crustaceans and fish adapted to low-oxygen environments, can reach very high densities in low-oxygen areas (Pineda and others, 2016; Gallo and others, 2019). However, in naturally low-oxygen habitats, such as oxygen minimum zones, even very small changes (representing less than 1 per cent of the oxygen content of well-oxygenated surface waters) can result in the exclusion of species that would otherwise be abundant (Wishner and others, 2018).

Figure II
Control of oxygen over biological and biogeochemical processes in the open ocean and coastal waters

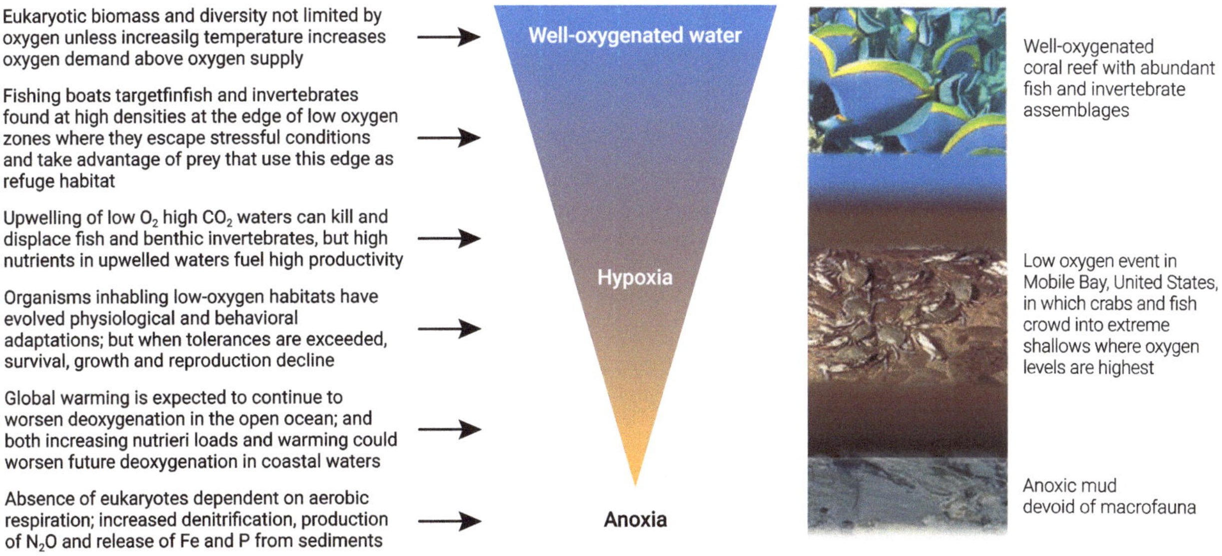

Source: Figure modified from Breitburg and others, 2018.

Note: Oxygen exerts a strong control over biological and biogeochemical processes in the open ocean and coastal waters. Whether oxygen patterns change over space, as with depth, or over time, as effects of nutrients and warming become more pronounced, biological diversity, biomass, and productivity decline with decreasing levels of oxygen.

Chronic exposure to suboptimal oxygen conditions can reduce growth (Thomas and others, 2019) and reproduction (Thomas and others, 2015). Numerical models indicate that those chronic effects can lead to population declines over time (Rose and others, 2018), even in the absence of direct low oxygen-induced mortality. Increased acquisition or progression of infections and decreased host immune responses resulting from exposure to low oxygen have been reported for a range of vertebrate and invertebrate hosts (Breitburg and others, 2019) and may increase the transmission of pathogens to humans through consumption of immunosuppressed hosts (Hernroth and Baden, 2018).

Microbes have evolved and adapted to exploit even the most extreme habitats on Earth, including those that contain no oxygen. Biogeochemical cycling of elements by microbes in the absence of oxygen leads to the production of greenhouse gases, including nitrous oxide and methane (Buitenhuis and others, 2018). The expansion of anoxic habitats could, therefore, lead to the increased release of greenhouse gases to the atmosphere, further increasing warming and stratification. That outcome is uncertain, however, because warming and stratification, both of which might increase greenhouse gas production, will also affect the rates and distribution of primary production upon which all other biological processes depend (Battaglia and Joos, 2018).

Ocean deoxygenation does not occur in isolation from other human-caused ocean stressors. With elevated ocean temperatures, microbes that are dependent on aerobic respiration and the vast majority of marine animals will need to consume more oxygen in order to survive (Pörtner, 2012). Elevated ocean temperatures therefore decrease the availability of suitable habitat both by increasing oxygen requirements and by inducing further oxygen loss. Predicted shifts in distribution poleward and into deeper, cooler waters, local extinctions and decreased maximum size of many fish species are attributed, at least in part, to increased oxygen requirements at warmer temperatures (Deutsch and others, 2015; Pauly and Cheung, 2018). The combined effects of ocean climate change stressors, namely, deoxygenation, warming and acidification, may also result in spatial, temporal and evolutionary mismatches between zooplankton and fish larvae that lead to altered larval fish growth and survival, and ultimately negative effects on fisheries (Dam and Baumann, 2017). More generally, the role of oxygen in converting food to energy means that oxygen supply can determine the amount of energy that is available to respond to other stressors (Sokolova, 2013).

Fisheries catches are often low in oxygen-depleted waters as a result of the avoidance behaviour of highly mobile species, as well as the mortality and recruitment failure of species that are sessile or have limited mobility (Breitburg and others, 2009; Rose and others, 2018). There is concern that low-oxygen areas and their expansion make fish and mobile shellfish more susceptible to overfishing (Craig, 2012; Purcell and others, 2017) by leading to high-density aggregations above and at the edge of low-oxygen waters (Craig, 2012; Stramma and others, 2012). For example, spatial shifts in fishing effort have been well documented in both the brown shrimp fishery in the Gulf of Mexico and the Dungeness crab fishery in Hood Canal, United States, whereby the spatial overlap between fishing fleets and target species increases as hypoxic zones increase on a seasonal basis or among years that vary in the spatial extent of hypoxia (Purcell and others, 2017; Froehlich and others, 2017). Fishing mortality may increase where such refuge locations are targeted and where shallower distributions increase catch rates (Purcell and others, 2017). Low-oxygen events have also been an important source of mortality in both finfish and shellfish aquaculture, causing substantial losses to local economies, with consequences to both human health and food security (Cayabyab and others, 2002; Rice, 2014).

3. Capacity-building: Global Ocean Acidification Observing Network and Global Ocean Oxygen Network

Sustainable Development Goal 14 addresses the need to "conserve and sustainably use the oceans, seas and marine resources for sustainable development", including by meeting target 14.3, to "minimize and address the impacts of ocean acidification, including through enhanced scientific cooperation at all levels".[1] Concern about the problem of deoxygenation was also noted in the "Our ocean, our future: call for action" declaration, the outcome of the United Nations Conference to Support the Implementation of Sustainable Development Goal 14: Conserve and sustainably use the oceans, seas and marine resources for sustainable development.[2]

The ability to attribute ecosystem impacts to changing ocean chemistry requires continued advances in ocean observation systems. Global initiatives in ocean research, such as Biogeochemical Argo, and the Global Ocean Acidification Observing Network and Global Ocean Oxygen Network of the Intergovernmental Oceanographic Commission are reducing barriers and building capacity in support of improved global understanding of ocean acidification and deoxygenation. The Global Ocean Acidification Observing Network and the Global Ocean Oxygen Network provide access to collaboration and mentoring in support of improving ocean observations of pH and oxygen through training sessions, partnerships and support for the creation of regional hubs. Currently, ocean acidification and deoxygenation observation and research efforts are concentrated in a relatively small number of countries, leaving large knowledge and capacity gaps around the world, especially in the southern hemisphere and in small island developing States and least developed countries (Global Ocean Acidification Observing Network (GOA-ON), 2019). Higher capacity to collect complex data and deliver better observations across the globe means that the predictive power of experiments and ecosystem models may improve as they replicate real-world scenarios more effectively to meet Goal 14.

Marine ecosystem services depend on which basic biotic functions are maintained (Connell and others, 2018), which ecosystem engineers and keystone species are retained (Sunday and others, 2017) and whether the spread of nuisance species is avoided (Hall-Spencer and Allen, 2015). Knowledge gaps for ecosystem responses to changes in ocean chemistry remain large. However, multi-stressor experiments and ecosystem models that incorporate advances in ecophysiology and genomics may better describe the scope of impact and reduce uncertainty about its extent. How deoxygenation is altering microbial pathways and rates of processes within the water column and the deep ocean needs to be better understood (Breitburg and others, 2018). The call by Riebesell and Gattuso (2015) for a shift towards multi-stressor and multispecies experiments to understand more specifically the ecological impacts of ocean acidification on marine communities has been taken up (Munday, 2017). Further advances will result from deepening and broadening the understanding of the relationships of ocean acidification and oxygen with other environmental drivers, how ecological processes and species interactions change under conditions that matter to them and how individual variation, plasticity and adaptation in response to ocean chemistry change shape impacts on marine ecosystems. Advancing research on those topics will support more

1 See General Assembly resolution 70/1.
2 See General Assembly resolution 71/312, annex; see also https://oceanconference.un.org/callforaction.

effective measures to mitigate the impacts of ocean acidification and deoxygenation, which may, as a result, have less serious consequences for the millions of people who are dependent on coastal protection, fisheries and aquaculture in lower-emission scenarios.

4. Summary

Marine heatwaves are shown to be increasing in frequency and intensity owing to climate change caused by human activities and are having a mostly negative impact on marine ecosystems. Marine heatwaves and their impacts are projected to increase in the future but those increases can be strongly limited by efforts to mitigate climate change. Forecasting systems may be employed in adapting to the effects of marine heatwaves.

Extreme El Niño and La Niña events have been observed but, because they occur infrequently, a human influence has not been detected. Nevertheless, models indicate an increase in the frequency of both phases of the oscillation under future scenarios of global warming. As in the case of marine heatwaves, forecasting systems, which already exist, may be employed in risk management and adaptation.

While changes in the frequency and spatial distribution of tropical cyclones are hard to detect in the observational record, studies of individual cyclones have shown a human influence on their intensity, in particular, the associated rainfall. Changes in intensity are projected to increase in the future, with associated impacts on storm surges and coastal infrastructure.

Although all coastal cities are already facing rising sea levels, low-lying cities and developing countries that lack the ability to invest in coastal defence measures and natural barrier restoration will suffer damage and losses of a higher degree. Global population studies suggest that people are relocating to coastal areas and will continue to do so, thereby putting more people at risk economically and socially. Although cities are typically centres for innovation and investment, key examples demonstrate the difficulty in solving such complex problems in vulnerable locations.

Damage and losses are also driven by existing vulnerabilities in coastal infrastructure and may not be solely attributed to rising sea levels. Rather, increasing sea levels may exacerbate existing issues, increasing risk.

The complex interactions of temperature and salinity with nutrients and chemical cycles of the ocean imply that variations in those variables owing to climate change and anthropogenic impact thus affect marine ecosystems, population, coastal communities and the related economy. Ocean warming is causing significant damage to marine ecosystems, and species are losing their habitats, forcing them to adapt or relocate to new temperatures or look for new feeding, spawning or nursery areas.

Ocean acidity and the availability of sufficient oxygen both underpin the provision of marine ecosystem services to human society. Rapid changes in ocean acidity and falling oxygen levels caused by climate change and anthropogenic CO_2 emissions are, however, now being observed, which is changing marine habitats and ecosystems worldwide. Warming is causing oxygen levels to fall, and acidification is rapidly changing the carbonate chemistry of surface ocean waters, which together are reducing the growth and survival of many organisms and degrading ecosystem resilience.

Closing knowledge gaps in ocean science by supporting capacity-building efforts that increase the understanding of how the ocean and its ecosystems are responding to changes in ocean physical and chemical properties is an important pathway to reducing the impacts of such changes and achieving Sustainable Development Goal 14.

References

Introduction

Intergovernmental Panel on Climate Change (IPCC) (2019). *IPCC Special Report on the Ocean and Cryosphere in a Changing Climate.*

Nicholls, R.J., and others (2008). Ranking Port Cities with High Exposure and Vulnerability to Climate Extremes, No. 1. https://doi.org/10.1787/011766488208.

United Nations (2017). *The First Global Integrated Marine Assessment: World Ocean Assessment I.* Cambridge: Cambridge University Press.

Extreme climate events

Cai, Wenju, and others (2014). Increasing frequency of extreme El Niño events due to greenhouse warming. *Nature Climate Change*, vol. 4, No. 2, p. 111.

Cai, Wenju, and others (2015). Increased frequency of extreme La Niña events under greenhouse warming. *Nature Climate Change*, vol. 5, No. 2, p. 132.

Cobb, Kim M., and others (2013). Highly variable El Niño – southern oscillation throughout the Holocene. *Science*, vol. 339, No. 6115, pp. 67–70.

Darmaraki, Sofia, and others (2019). Future evolution of Marine Heatwaves in the Mediterranean Sea. *Climate Dynamics*, pp. 1–22.

Frölicher, Thomas L., and others (2018). Marine heatwaves under global warming. *Nature*, vol. 560, No. 7718, p. 360.

Garner, Andra J., and others (2017). Impact of climate change on New York City's coastal flood hazard: Increasing flood heights from the preindustrial to 2300 CE. *Proceedings of the National Academy of Sciences*, vol. 114, No. 45, pp. 11861–11866.

Hobday, Alistair J., and others (2016). A hierarchical approach to defining marine heatwaves. *Progress in Oceanography*, vol. 141, pp. 227–238.

Hughes, Terry P., and others (2018). Global warming transforms coral reef assemblages. *Nature*, vol. 556, No. 7702, p. 492.

Intergovernmental Panel on Climate Change (IPCC) (2018). *Global Warming of 1.5°C. An IPCC Special Report on the Impacts of Global Warming of 1.5°C above Pre-Industrial Levels and Related Global Greenhouse Gas Emission Pathways, in the Context of Strengthening the Global Response to the Threat of Climate Change, Sustainable Development, and Efforts to Eradicate Poverty.* Valérie Masson-Delmotte and others, eds.

_________(2019). *IPCC Special Report on the Ocean and Cryosphere in a Changing Climate.*

King, Andrew D., and others (2017). Australian climate extremes at 1.5°C and 2°C of global warming. *Nature Climate Change*, vol. 7, No. 6, p. 412.

Knutson, Thomas R., and others (2010). Tropical cyclones and climate change. *Nature Geoscience*, vol. 3, No. 3, pp. 157–163.

Knutson, Thomas R., and others (2015). Global projections of intense tropical cyclone activity for the late twenty-first century from dynamical downscaling of CMIP5/RCP4. 5 scenarios. *Journal of Climate*, vol. 28, No. 18, pp. 7203–7224.

L'Heureux, Michelle L., and others (2017). Observing and predicting the 2015/16 El Niño. *Bulletin of the American Meteorological Society*, vol. 98, No. 7, pp. 1363–1382.

Oliver, Eric C.J., and others (2018). Longer and more frequent marine heatwaves over the past century. *Nature Communications*, vol. 9, No. 1, art. 1324.

Park, Doo-Sun R., and others (2017). Asymmetric response of tropical cyclone activity to global warming over the North Atlantic and western North Pacific from CMIP5 model projections. *Scientific Reports*, vol. 7, art. 41354.

Payne, Mark R., and others (2017). Lessons from the first generation of marine ecological forecast products. *Frontiers in Marine Science*, vol. 4, art. 289.

Power, Scott B., and François P.D. Delage (2018). El Niño – southern oscillation and associated climatic conditions around the world during the latter half of the twenty-first century. *Journal of Climate*, vol. 31, No. 15, pp. 6189–6207.

Rasmussen, E.M. and T.H. Carpenter (1982). Variations in tropical sea surface temperature and surface wind fields associated with the Southern Oscillation/El Niño. *Monthly Weather Review*, vol. 110, No. 5, pp. 354–384.

Risser, Mark D., and Michael F. Wehner (2017). Attributable human-induced changes in the likelihood and magnitude of the observed extreme precipitation during hurricane Harvey. *Geophysical Research Letters*, vol. 44, No. 24, pp. 12–457.

Santoso, Agus, and others (2017). The defining characteristics of ENSO extremes and the strong 2015/2016 El Niño. *Reviews of Geophysics*, vol. 55, No. 4, pp. 1079–1129.

Sharmila, S., and K.J.E. Walsh (2018). Recent poleward shift of tropical cyclone formation linked to Hadley cell expansion. *Nature Climate Change*, vol. 8, No. 8, p. 730.

Smale, Dan A., and others (2019). Marine heatwaves threaten global biodiversity and the provision of ecosystem services. *Nature Climate Change*, vol. 9, No. 4, p. 306.

Tommasi, Desiree, and others (2017). Managing living marine resources in a dynamic environment: the role of seasonal to decadal climate forecasts. *Progress in Oceanography*, vol. 152, pp. 15–49.

Van Oldenborgh, G.J., and others (2017). Attribution of extreme rainfall from Hurricane Harvey, August 2017. *Environment Research Letters*, vol. 12, No. 12, 124009, http://doi.org/10.1088/1748-9326/aa9ef2.

Vuuren, Detlef P., and others (2011). The representative concentration pathways: an overview. *Climatic Change*, vol. 109, No. 1, art. 5. https://doi.org/10.1007/s10584-011-0148-z.

Yamada, Yohei, and others (2017). Response of tropical cyclone activity and structure to global warming in a high-resolution global nonhydrostatic model. *Journal of Climate*, vol. 30, No. 23, pp. 9703–9724.

Young, Ian R., and Agustinus Ribal (2019). Multiplatform evaluation of global trends in wind speed and wave height. *Science*, vol. 364, No. 6440, pp. 548–552.

Sea level rise and cities

Burt, John A., and others (2013). Urban breakwaters as reef fish habitat in the Persian Gulf. *Marine Pollution Bulletin*, vol. 72, No. 2, pp. 342–350.

C40 Cities (2019). www.c40.org/other/the-future-we-don-t-want-staying-afloat-the-urban-response-to-sea-level-rise.

Chapman, M.G. (2003). Paucity of mobile species on constructed seawalls: effects of urbanization on biodiversity. *Marine Ecology Progress Series*, vol. 264, pp. 21–29.

Chee, Su Yin, and others (2017). Land reclamation and artificial islands: walking the tightrope between development and conservation. *Global Ecology and Conservation*, vol. 12, pp. 80–95.

Church, J.A., and others (2013). Sea level change. In *Climate Change 2013: The Physical Science Basis. Contribution of Working Group I to the Fifth Assessment Report of the Intergovernmental Panel on*

Climate Change, T.F. Stocker, and others, eds., pp. 1137–1216. Cambridge, New York: Cambridge University Press.

Dafforn, Katherine A., and others (2015). Marine urbanization: an ecological framework for designing multifunctional artificial structures. *Frontiers in Ecology and the Environment*, vol. 13, No. 2, pp. 82–90.

De Sherbinin, Alex, and others (2007). The vulnerability of global cities to climate hazards. *Environment and Urbanization*, vol. 19, No. 1, pp. 39–64.

Garschagen, Matthias, and Patricia Romero-Lankao (2015). Exploring the relationships between urbanization trends and climate change vulnerability. *Climatic Change*, vol. 133, No. 1, pp. 37–52.

Hallegatte, Stephane, and others (2013). Future flood losses in major coastal cities. *Nature Climate Change*, vol. 3, No. 9, p. 802.

Hanson, Susan, and others (2011). A global ranking of port cities with high exposure to climate extremes. *Climatic Change*, vol. 104, No. 1, pp. 89–111.

Hinkel, Jochen, and others (2014). Coastal flood damage and adaptation costs under 21st century sea-level rise. *Proceedings of the National Academy of Sciences*, vol. 111, No. 9, pp. 3292–3297.

Jonkers, L., and others (2019). Global change drives modern plankton communities away from the pre-industrial state. *Nature*, 570, pp. 372–375. https://doi.org/10.1038/s41586-019-1230-3.

Mendelsohn, Robert, and others (2012). The impact of climate change on global tropical cyclone damage. *Nature Climate Change*, vol. 2, No. 3, p. 205.

National Oceanic and Atmospheric Administration (NOAA) (2019). Sea Level Rise Viewer. 2019. https://coast.noaa.gov/digitalcoast/tools/slr.html.

Sengupta, Dhritiraj, and others (2018). Building beyond land: An overview of coastal land reclamation in 16 global megacities. *Applied Geography*, vol. 90, pp. 229–238.

Takagi, Hiroshi, and others (2016). Projection of coastal floods in 2050 Jakarta. *Urban Climate*, vol. 17, pp. 135–145.

Temmerman, Stijn, and others (2013). Ecosystem-based coastal defence in the face of global change. *Nature*, vol. 504 (7478), pp. 79–83.

Vitousek, Sean, and others (2017). Doubling of coastal flooding frequency within decades due to sea-level rise. *Scientific Reports*, vol. 7, No. 1, art. 1399.

Pressures from changes in temperature

Barange, M., and others (2018). Impacts of climate change on fisheries and aquaculture: FAO Fisheries and Aquaculture Technical Paper No. 627. Rome, FAO.

Blanchard, Julia L., and others (2017). Linked sustainability challenges and trade-offs among fisheries, aquaculture and agriculture. *Nature Ecology & Evolution*, vol. 1, No. 9, p. 1240.

Boyd, P.W., and others (2016). Physiological responses of a Southern Ocean diatom to complex future ocean conditions. *Nature Climate Change*, vol. 6, No. 2, p. 207.

Cochard, Roland, and others (2008). The 2004 tsunami in Aceh and Southern Thailand: a review on coastal ecosystems, wave hazards and vulnerability. *Perspectives in Plant Ecology, Evolution and Systematics*, vol. 10, No. 1, pp. 3–40.

Food and Agriculture Organization of the United Nations (FAO) (2018). *Deep Ocean Stewardship Initiative. Deep-Ocean Climate Change Impacts on Habitat, Fish and Fisheries*. Technical Paper 638. Rome.

Fu, Weiwei, and others (2016). Climate change impacts on net primary production (NPP) and export production (EP) regulated by increasing stratification and phytoplankton community structure in the CMIP5 models. *Biogeosciences*, vol. 13, No. 18, pp. 5151–70. https://doi.org/10.5194/bg-13-5151-2016.

Golden, Christopher D., and others (2016). Nutrition: Fall in fish catch threatens human health. *Nature News*, vol. 534, No. 7607, p. 317.

Hilmi, Nathalie, and others (2014). Exposure of Mediterranean countries to ocean acidification. *Water*, vol. 6, No. 6, pp. 1719–1744.

Intergovernmental Panel on Climate Change (IPCC) (2018). *Global Warming of 1.5°C. An IPCC Special Report on the Impacts of Global Warming of 1.5°C above Pre-Industrial Levels and Related Global Greenhouse Gas Emission Pathways, in the Context of Strengthening the Global Response to the Threat of Climate Change, Sustainable Development, and Efforts to Eradicate Poverty.* Valérie Masson-Delmotte and others, eds.

Pendleton, Linwood H., and others (2016). Has the value of global marine and coastal ecosystem services changed? *Marine Policy*, vol. 64, pp. 156–158.

Semenza, Jan C., and others (2017). Environmental suitability of *Vibrio* infections in a warming climate: an early warning system. *Environmental Health Perspectives*, vol. 125, No. 10, art. 107004.

Sweetman, Andrew K., and others (2017). Major impacts of climate change on deep-sea benthic ecosystems. *Elementa: Science of the Anthropocene*, vol. 5, No. 4.

Worm, Boris, and others (2006). Impacts of biodiversity loss on ocean ecosystem services. *Science*, vol. 314, No. 5800, pp. 787–790.

Yool, Andrew, and others (2017). Big in the benthos: Future change of seafloor community biomass in a global, body size-resolved model. *Global Change Biology*, vol. 23, No. 9, pp. 3554–3566.

Pressures from changes in ocean chemistry

Agostini, Sylvain, and others (2018). Ocean acidification drives community shifts towards simplified non-calcified habitats in a subtropical- temperate transition zone. *Scientific Reports*, vol. 8, No. 1, art. 11354.

Albright, Rebecca, and others (2018). Carbon dioxide addition to coral reef waters suppresses net community calcification. *Nature*, vol. 555, No. 7697, p. 516.

Battaglia, Gianna, and Fortunat Joos (2018). Marine N_2O emissions from nitrification and denitrification constrained by modern observations and projected in multimillennial global warming simulations. *Global Biogeochemical Cycles*, vol. 32, No. 1, pp. 92–121.

Breitburg, Denise L., and others (2009). Hypoxia, nitrogen, and fisheries: integrating effects across local and global landscapes. *Annual Review of Marine Science*, vol. 1, pp. 329–349.

Breitburg, Denise L., and others (2018). Declining oxygen in the global ocean and coastal waters. *Science*, vol. 359, No.6371, eaam7240.

Breitburg, Denise L., and others (2019). Multiple stressors – forces that combine to worsen deoxygenation and its effects. In *Ocean Deoxygenation: Everyone's Problem Causes, Impacts, Consequences and Solutions*. Loffely, D., and Baxter, J. eds. IUCN, pp. 225–247.

Brodie, Juliet, and others (2014). The future of the northeast Atlantic benthic flora in a high CO_2 world. *Ecology and Evolution*, vol. 4, No. 13, pp. 2787–2798.

Buitenhuis, Erik T., and others (2018). Constraints on global oceanic emissions of N_2O from observations and models. *Biogeosciences*, vol. 15, No. 7, pp. 2161–2175.

Cayabyab, R.R., and others (2002). *Histamine Fish Poisoning Following Massive Fishkill in Bolinao, Pangasinan, February 2002*. Regional Epidemiology and Surveillance Unit I Report 3. Philippines: Department of Health.

Connell, Sean D., and others (2018). The duality of ocean acidification as a resource and a stressor. *Ecology*, vol. 99, No. 5, pp. 1005–1010.

Cooley, Sarah R., and others (2016). Community-level actions that can address ocean acidification. *Frontiers in Marine Science*, vol. 2, art. 128.

Cornwall, Christopher E., and others (2017). Inorganic carbon physiology underpins macroalgal responses to elevated CO_2. *Scientific Reports*, vol. 7, art. 46297.

Craig, J. Kevin (2012). Aggregation on the edge: effects of hypoxia avoidance on the spatial distribution of brown shrimp and demersal fishes in the Northern Gulf of Mexico. *Marine Ecology Progress Series*, vol. 445, pp. 75–95.

Dam, Hans G., and Hannes Baumann (2017). Climate change, zooplankton and fisheries. *Climate Change Impacts on Fisheries and Aquaculture: A Global Analysis*, vol. 2, pp. 851–874.

Deutsch, Curtis, and others (2015). Climate change tightens a metabolic constraint on marine habitats. *Science*, vol. 348, No. 6239, pp. 1132–1135.

Diaz, Robert J., and Rutger Rosenberg (2008). Spreading dead zones and consequences for marine ecosystems. *Science*, vol. 321, No. 5891, pp. 926–929.

Ekstrom, Julia A., and others (2015). Vulnerability and adaptation of US shellfisheries to ocean acidification. *Nature Climate Change*, vol. 5, No. 3, p. 207.

Enochs, I.C., and others (2015). Shift from coral to macroalgae dominance on a volcanically acidified reef. *Nature Climate Change*, vol. 5, No. 12, p. 1083.

Foo, Shawna Andrea, and others (2018). The carbon dioxide vents of Ischia, Italy, a natural system to assess impacts of ocean acidification on marine ecosystems: an overview of research and comparisons with other vent systems. In *Oceanography and Marine Biology*, pp. 237–310. CRC Press.

Froehlich, Halley E., and others (2017). When does hypoxia affect management performance of a fishery? A management strategy evaluation of Dungeness crab (*Metacarcinus magister*) fisheries in Hood Canal, Washington, United States. *Canadian Journal of Fisheries and Aquatic Sciences*, vol. 74, No. 6, pp. 922–932.

Gallo, Natalya D., and others (2019). Home sweet suboxic home: remarkable hypoxia tolerance in two demersal fish species in the Gulf of California. *Ecology*, vol. 100, No. 3, e02539.

Gallo, N.D., and L.A. Levin (2016). Fish ecology and evolution in the world's oxygen minimum zones and implications of ocean deoxygenation. In *Advances in Marine Biology*, vol. 74, pp. 117–198. Elsevier.

Global Ocean Acidification Observing Network (GOA-ON) (2019). *Global Ocean Acidification Observing Network (GOA-ON) Implementation Strategy*. www.goa-on.org.

Gómez, Carlos E., and others (2018). Growth and feeding of deep-sea coral *Lophelia pertusa* from the California margin under simulated ocean acidification conditions. *PeerJ*, vol. 6, e5671.

Hall-Spencer, Jason M., and Ben P. Harvey (2019). Ocean acidification impacts on coastal ecosystem services due to habitat degradation. *Emerging Topics in Life Sciences*, vol. 3, No. 2, pp. 197–206.

Hall-Spencer, Jason M., and Ro Allen (2015). The impact of CO_2 emissions on "nuisance" marine species. *Research and Reports in Biodiversity Studies*, vol. 4, pp. 33–46.

Hernroth, Bodil E., and Susanne P. Baden (2018). Alteration of host-pathogen interactions in the wake of climate change – Increasing risk for shellfish associated infections? *Environmental Research*, vol. 61, pp. 425–438.

Intergovernmental Panel on Climate Change (IPCC) (2019). *IPCC Special Report on the Ocean and Cryosphere in a Changing Climate*.

Keister, Julie E., and Loren B. Tuttle (2013). Effects of bottom-layer hypoxia on spatial distributions and community structure of mesozooplankton in a sub-estuary of Puget Sound, Washington, U.S.A. *Limnology and Oceanography*, vol. 58, No. 2, pp. 667–80. https://doi.org/10.4319/lo.2013.58.2.0667.

Koslow, J. Anthony, and others (2011). Impact of declining intermediate-water oxygen on deepwater fishes in the California Current. *Marine Ecology Progress Series*, vol. 436, pp. 207–218.

Koslow, J. Anthony, and others (2018). The evolving response of mesopelagic fishes to declining midwater oxygen concentrations in the southern and central California Current. *ICES Journal of Marine Science*, vol. 76, No. 3, pp. 626–638.

Lam, Vicky W.Y., and others (2019). Dealing with the effects of ocean acidification on coral reefs in the Indian Ocean and Asia. *Regional Studies in Marine Science*, vol. 28, 100560.

Linares, Cristina, and others (2015). Persistent natural acidification drives major distribution shifts in marine benthic ecosystems. *Proceedings of the Royal Society B: Biological Sciences*, vol. 282, No. 1818, 20150587.

McCormick, Lillian R., and Lisa A. Levin (2017). Physiological and ecological implications of ocean deoxygenation for vision in marine organisms. *Philosophical Transactions of the Royal Society A: Mathematical, Physical and Engineering Sciences*, vol. 375, No. 2102, 20160322.

Munday, Philip L. (2017). New perspectives in ocean acidification research: editor's introduction to the special feature on ocean acidification. *Biology Letters*, vol. 13.

Pauly, Daniel, and William W.L. Cheung (2018). Sound physiological knowledge and principles in modeling shrinking of fishes under climate change. *Global Change Biology*, vol. 24, No. 1, pp. e15–e26.

Pineda, Jesús, and others (2016). A crab swarm at an ecological hotspot: patchiness and population density from AUV observations at a coastal, tropical seamount. *PeerJ*, vol. 4, e1770.

Pörtner, Hans-O. (2012). Integrating climate-related stressor effects on marine organisms: unifying principles linking molecule to ecosystem-level changes. *Marine Ecology Progress Series*, vol. 470, pp. 273–290.

Purcell, Kevin M., and others (2017). Fleet behavior is responsive to a large-scale environmental disturbance: Hypoxia effects on the spatial dynamics of the northern Gulf of Mexico shrimp fishery. *PloS One*, vol. 12, No. 8, e0183032.

Rice, Michael A. (2014). Extension programming in support of public policy for the management of aquaculture in common water bodies. *Aquacultura Indonesiana*, vol. 15, No. 1.

Riebesell, Ulf, and Jean-Pierre Gattuso (2015). Lessons learned from ocean acidification research. *Nature Climate Change*, vol. 5, No. 1, p. 12.

Rose, Kenneth A., and others (2018). Modeling the population effects of hypoxia on Atlantic croaker (*Micropogonias undulatus*) in the northwestern Gulf of Mexico: part 2—realistic hypoxia and eutrophication. *Estuaries and Coasts*, vol. 41, No. 1, pp. 255–279. https://doi.org/10.1007/s12237-017-0267-5.

Russell, Bayden D., and others (2013). Future seagrass beds: Can increased productivity lead to increased carbon storage? *Marine Pollution Bulletin*, vol. 73, No. 2, pp. 463–469.

Sato, Kirk N., and others (2017). Habitat compression and expansion of sea urchins in response to changing climate conditions on the California continental shelf and slope (1994–2013). *Deep Sea Research Part II: Topical Studies in Oceanography*, vol. 137, pp. 377–389.

Schmidtko, Sunke, and others (2017). Decline in global oceanic oxygen content during the past five decades. *Nature*, vol. 542, No. 7641, p. 335.

Sokolova, Inna M. (2013). Energy-limited tolerance to stress as a conceptual framework to integrate the effects of multiple stressors. *Integrative and Comparative Biology*, vol. 53, No. 4, pp. 597–608.

Sperling, Erik A., and others (2016). Biodiversity response to natural gradients of multiple stressors on continental margins. *Proceedings of the Royal Society B: Biological Sciences*, vol. 283, No. 1829, 20160637.

Stramma, Lothar, and others (2012). Expansion of oxygen minimum zones may reduce available habitat for tropical pelagic fishes. *Nature Climate Change*, vol. 2, No. 1, p. 33.

Sunday, Jennifer M., and others (2017). Ocean acidification can mediate biodiversity shifts by changing biogenic habitat. *Nature Climate Change*, vol. 7, No. 1, p. 81.

Teixidó, Nuria, and others (2018). Functional biodiversity loss along natural CO_2 gradients. *Nature Communications*, vol. 9, No. 1, art. 5149.

Thomas, Peter, and others (2015). Impaired gamete production and viability in Atlantic croaker collected throughout the 20,000 km^2 hypoxic region in the northern Gulf of Mexico. *Marine Pollution Bulletin*, vol. 101, No. 1, pp. 182–192.

Thomas, Y., and others (2019). Effects of hypoxia on metabolic functions in marine organisms: Observed patterns and modelling assumptions within the context of Dynamic Energy Budget (DEB) theory. *Journal of Sea Research*, vol. 143, pp. 231–242.

Vizzini, S., and others (2017). Ocean acidification as a driver of community simplification via the collapse of higher-order and rise of lower-order consumers. *Scientific Reports*, vol. 7, No. 1, art. 4018.

Wishner, Karen F., and others (2018). Ocean deoxygenation and zooplankton: Very small oxygen differences matter. *Science Advances*, vol. 4, No. 12, eaau5180.

Wright, Jody J., and others (2012). Microbial ecology of expanding oxygen minimum zones. *Nature Reviews Microbiology*, vol. 10, No. 6, p. 381.

Chapter 10
Changes in nutrient inputs to the marine environment

Contributors: Thomas C. Malone (convener), Archis Ambulker, Maria João Bebianno (co-lead member), Paula Bontempi, Michael Krom, Harri Kuosa, Joseph Montoya, Alice Newton, Yapo Ossey, João Sarkis Yunes, Walker Smith, Lars Sonesten, Georgios Sylaios, Juying Wang (lead member) and Kedong Yin.

Keynote points

- Inputs of nitrogen (N) and phosphorus (P) to coastal ecosystems through river runoff and atmospheric deposition increased rapidly during the twentieth century owing to anthropogenic inputs derived primarily from the use of synthetic fertilizer, combustion of fossil fuels, cultivation of legumes (N_2-fixation), production of manure by livestock and municipal wastes.
- Increases in anthropogenic nutrient inputs have fuelled a global increase in cultural eutrophication of the coastal ocean and now exceed inputs owing to natural processes.
- Ecological responses to the process of cultural eutrophication include increases in the severity and extent of coastal hypoxia, acidification and toxic algal events. Thus, cultural eutrophication is a serious threat to the health of coastal ecosystems and their capacity to provide services that are valued by society.
- It is projected that anthropogenic N and P production will increase by nearly a factor of two during the first half of the twenty-first century.
- Reducing anthropogenic inputs of N and P to the coastal ocean to minimize the extent and risk of coastal eutrophication during the course of the twenty-first century should be an international priority.

1. Introduction

During the course of the twenty-first century, increases in anthropogenic inputs of N and P to coastal ecosystems through river discharge became the primary cause of cultural eutrophication[1] and consequent ecosystem degradation of the coastal ocean worldwide (Rabalais and others, 2009a, 2009b; Paerl and others, 2014; Beusen and others, 2016; Ngatia and others, 2019), a trend that is arguably the most widespread anthropogenic threat to the health of coastal ecosystems (Rabalais and others, 2009b; IPCC, 2014).

Nixon (1995) defined eutrophication as an increase in the rate of supply of organic matter to an ecosystem and noted that increases in the supply of organic matter to coastal ecosystems have various causes, the most common being excess inputs of biologically active, inorganic N and P. Since phytoplankton net primary production in most coastal ecosystems is limited primarily by the availability of N (Howarth and Marino, 2006; Elser and others, 2007), phytoplankton biomass in the coastal ocean has increased accordingly (Howarth and others, 2011). Combined with additional anthropogenic inputs of organic nutrients from land-based sources, the resulting accumulation of organic matter has led to the cultural eutrophication of many coastal ecosystems worldwide (see figure below), a process that is arguably the most serious threat to the marine ecosystem services valued by society, for example, the provision of biodiversity, the production of oxygen, the mitigation of coastal flooding, fisheries and the sequestration of atmospheric CO_2 (Howarth and others, 2000; Bachmann and others, 2006; Martínez and others, 2007; Costanza and others, 2017).

The focus in the present chapter is on anthropogenic inputs of biologically reactive fixed

[1] Eutrophication driven by anthropogenic inputs of nutrients and organic matter that lead to undesirable changes in ecosystem health (Smith and others, 2006; Rabalais and others, 2009a, 2009b).

N (such as dissolved nitrate, nitrite, ammonium, urea and free amino acids) and P (PO_4^{-3}) (such as orthophosphate, polyphosphate and organically bound phosphates) to the coastal ocean as defined by the global network of large marine ecosystems.[2] In that context, the objectives of the present chapter are to: (a) document changes in anthropogenic inputs of N and P to selected coastal marine ecosystems; (b) assess the impacts of cultural eutrophication on those ecosystems; (c) project how those changes will likely affect the capacity of the coastal ecosystem to support ecosystem services during the course of the twenty-first century in the context of global climate change; and (d) identify gaps in current knowledge.

Global distribution of eutrophic coastal marine ecosystems

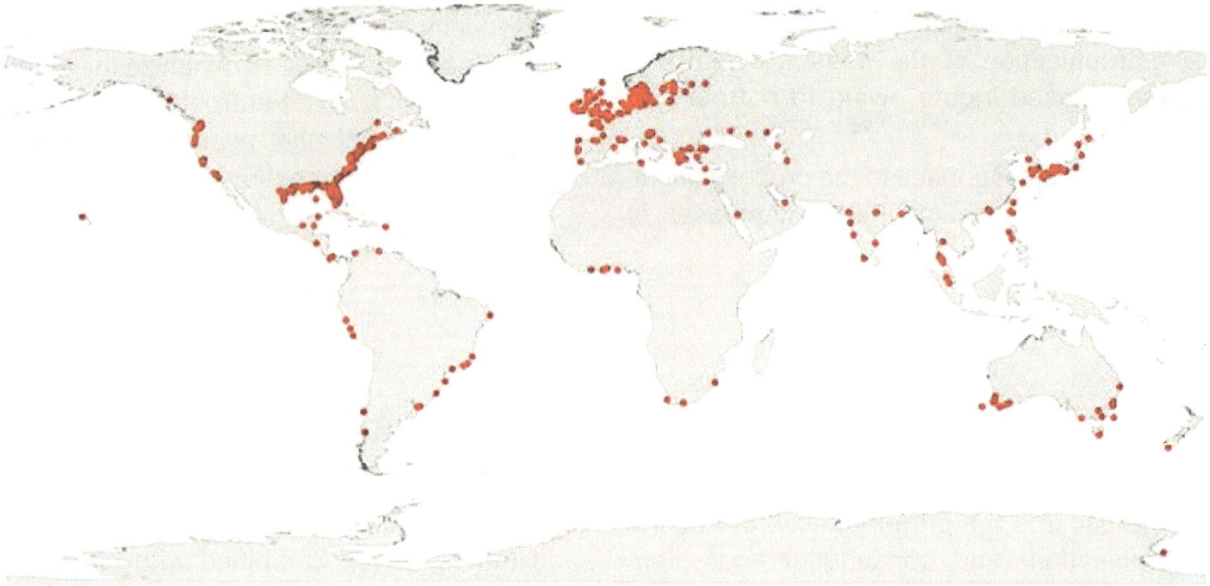

Source: Breitburg and others, 2018.
Note: Recent coastal surveys of the United States and Europe found that a staggering 78 per cent of the assessed continental United States coastal area and approximately 65 per cent of the Atlantic coast of Europe exhibit symptoms of eutrophication.

The information presented herein is relevant to a number of chapters in the present Assessment (chapters 4–9, 11–15, 22 and 28). Chapter 5 (trends in the physical and chemical state of the ocean), and chapter 6A (plankton diversity) are particularly relevant. The former is addressed in the present chapter to the extent that changes in nutrient inputs and eutrophication are related trends in physical and chemical environmental conditions (emphasis on climate-driven changes). The latter is addressed to the extent that changes in plankton diversity are relevant to the problem of coastal eutrophication.

2 The global network of large marine ecosystems includes coastal watersheds and the coastal ocean (estuaries and the open waters of the continental shelves (available at www.lmehub.net). Large marine ecosystems vary in size from about 200,000 km² to more than 1,000,000 km² and encompass areas of the coastal ocean where primary productivity is generally higher than in the open ocean.

2. Situation reported in the first *World Ocean Assessment*

Chapter 20 of the first *World Ocean Assessment* (United Nations, 2017) contained a review of coastal, riverine and airborne inputs of contaminants from land-based sources, with an emphasis on hazardous substances, endocrine disruptors, nutrients and waterborne pathogens, and radioactive substances. Those aspects that are related to anthropogenic nutrient inputs to the ocean, in general, and coastal ecosystems, in particular, are most relevant to the present chapter. In addition to the global view summarized below, chapter 20 of the first Assessment included a summary of inputs to and impacts of nutrients for different regions of the global ocean (Arctic Ocean and regions of the Atlantic Ocean, the Indian Ocean and the Pacific Ocean).

Major sources of anthropogenic nutrients include municipal wastewater, fertilizers used for agriculture, the combustion of fossil fuels and food-related industries. Transport routes to the ocean from those land-based sources include river run-off and atmospheric deposition. Controlling nutrient inputs from municipal wastes remains a challenge in the developing world. With regard to agriculture, the use of fertilizers has grown rapidly in recent decades, resulting in a 42 per cent increase globally between 2002 and 2012. However, fertilizer use in Latin America, southern Asia, eastern Asia and Oceania has more than doubled during the same period. Airborne inputs of nitrogen from the combustion of fossil fuels have also increased. In north-western Europe, over 25 per cent of nitrogen emissions to the atmosphere are from those sources. The precise consequences of excess nutrient loading depend on local environmental conditions, including the rate at which semi-enclosed bodies of water are flushed by currents and the strength of density stratification of the water column.

Land-based inputs of nutrients are not, in themselves, harmful but can cause problems when they are excessive. Inputs of anthropogenic N and P, which more than doubled in the past century, have affected the health of marine ecosystems worldwide. Increased inputs stimulate the growth of phytoplankton, resulting in excessive net primary production, which often leads to accumulations of phytoplankton biomass and eutrophication. That has led to the development of oxygen-depleted "dead zones", the loss of seagrass beds and increases in the occurrence of toxic phytoplankton blooms. The global spread of oxygen-depleted ("dead") zones in coastal waters has increased exponentially to over 400 systems since the 1960s and has reached a cumulative area of about 245,000 km^2 worldwide.

3. Global-scale patterns and trends

3.1. Anthropogenic inputs of biologically reactive nitrogen and phosphorus

3.1.1. Sources

During the twentieth century, the global supply of biologically reactive N and P doubled owing to anthropogenic activities (Beusen and others, 2016; Seitzinger and Mayorga, 2016). Over half of new[3] N and P loads to most coastal ecosystems (73 per cent of large marine ecosystems) are related to anthropogenic sources, current inputs of which have been estimated to be in the range of 210–223 x 10^9 kg N per year (Lee and others, 2016) and about 34 x 10^9 kg P per year (Harrison and others, 2005). Inputs of those nutrients to large marine ecosystems are derived from agricultural practices,[4] the combustion of fossil fuels and municipal wastes (Galloway and others, 2004; Howarth, 2008), as follows:

(a) The single largest source of anthropogenic N and P is synthetic fertilizers[5] (Vitousek and others, 1997; Mosier and others, 2004). The amount of synthetic fertilizer used for agriculture has grown exponentially from near zero in 1910 to about 118 x 10^9 kg N per year and 17.5 x 10^9 kg P per year in 2013 (Peñuelas and others, 2013; Lu and Tian, 2017). Hotspots of fertilizer use shifted from the United States and Western Europe in the 1960s to eastern Asia in the early twenty-first century. In 2013, East Asia, South Asia and South-East Asia accounted for 71 per cent of global fertilizer use, followed by North America (11 per cent), Europe (7 per cent) and South America (6 per cent) (Lu and Tian, 2017). Of the N loading, the volatilization of ammonia from agricultural fields emits an estimated 10 x 10^9 kg N per year into the atmosphere (Vitousek and others, 1997; Bouwman and others, 2013);

(b) The combustion of fossil fuels releases fixed N from long-term storage in geological formations back into the atmosphere in the form of nitrogen oxides (NO_X). Altogether, emissions from coal- and oil-fired power plants, automobiles and other combustion processes release on the order of 40 x 10^9 kg N per year (Peñuelas and others, 2013). The global distribution of NO_X emissions is not uniform, with Asia, Europe, North America and sub-Saharan Africa accounting for 30, 20, 17 and 12 per cent of emissions, respectively (Lamsal and others, 2011);

(c) As large areas of natural vegetation have been replaced with monocultures of legumes that support symbiotic N_2-fixing bacteria, anthropogenic input from biological N_2-fixation to coastal watersheds is estimated to be 33 x 10^9 kg per year (Boyer and Howarth, 2008);

(d) Livestock production of manure has increased rapidly over the past century. Current loads of manure N and P are estimated to be approximately 18 x 10^9 kg N per year and approximately 2.5 x 10^9 kg P per year, with hotspots in Western Europe, south-eastern Australia, north-eastern China, and India (Peñuelas and others, 2013; Zhang and others, 2017);

(e) Globally, 80 per cent of municipal wastewater is released into the environment

---

3 New N inputs are those coming from outside the ecosystem as opposed to those that are regenerated within the ecosystem as organic matter is decomposed.

4 Agricultural practices include the use of synthetic fertilizers, animal husbandry and the culture of legumes (biological N_2-fixation).

5 Synthetic fertilizers include ammonium nitrate, ammonium phosphate, superphosphate and urea.

untreated (World Water Assessment Programme (WWAP), 2017). Thus, the most prevalent urban source of nutrient pollution is human sewage, which is estimated to have released about 9 x 10^9 kg N and about 1.4 x 10^9 kg P into the environment in 2018 (extrapolated from Van Drecht and others, 2009). The percentage of treated[6] sewage varies regionally, from 90 per cent in North America, 66 per cent in Europe, 35 per cent in Asia and 14 per cent in Latin America and the Caribbean to less than 1 per cent in Africa (Selman and others, 2010).

Non-point (diffuse) source inputs (subparas. (a)–(d) above; 218 x 10^9 kg N/year) far exceed point source inputs from wastewater (subpara. (e) above; approximately 9 x 10^9 kg N/year) and are more difficult to control. Ultimately, most of those inputs are transported to the coastal ocean through river run-off and atmospheric deposition (Howarth, 2008; Spokes and Jickells, 2005; Jickells and others, 2017).[7] Both transport pathways are major routes of input for N, while atmospheric deposition of reactive P is negligible relative to riverine inputs. Thus, climate-driven acceleration of the global water cycle and associated increases in the magnitude and frequency of major rainfall events (Sinha and others, 2017) will accelerate nutrient inputs from diffuse sources (e.g., agriculture) to coastal waters (Howarth and others, 2012). In that context, it should be noted that reductions in N and P loads have come primarily from advanced wastewater treatment in developed countries, while efforts to reduce diffuse inputs from agricultural sources have, for the most part, been less effective (Boesch, 2019).

3.1.2. Transport of anthropogenic nutrients to the coastal ocean

Anthropogenic inputs to the coastal ocean through river run-off are fuelled by anthropogenic supplies to coastal watersheds, wet precipitation within watersheds and riverine transport from watersheds (Howarth and others, 1996; Green and others, 2004). Globally, there is a significant linear correlation between net anthropogenic N supplies to coastal watersheds and total river-borne N export to the coastal ocean (Boyer and Howarth, 2008). During the twentieth century, total riverine inputs of N and P to the coastal ocean increased from about 27 x 10^9 kg N per year to about 48 x 10^9 kg N per year and from about 2 x 10^9 kg P per year to about 4 x 10^9 kg P per year (Galloway and others, 2004; Beusen and others, 2016). Boyer and Howarth (2008) estimated riverine inputs of N to the ocean basins as follows: Atlantic (primarily from eastern North America and western Europe) 15–25 x 10^9 kg N per year; Pacific (primarily from East Asia) 10–14 x 10^9 kg N per year; Indian 7–8 x 10^9 kg N per year; and Arctic 2–4 x 10^9 kg N per year.

Atmospheric N compounds are derived from both agricultural sources (volatilization of ammonia) and fossil fuels (emission of NO_x). In contrast to river-borne nutrient loading, N inputs delivered through atmospheric deposition are fuelled by anthropogenic supplies to and emissions from coastal airsheds (which are generally much larger than watersheds), atmospheric transport from airsheds and wet precipitation directly over the coastal ocean (Valigura and others, 2001). As for river-borne N inputs, atmospheric deposition of N to the global ocean increased rapidly during the twentieth century from a pre-industrial rate of about 22 x 10^9 kg N per year to more than

6 Primary, secondary or tertiary treatment.

7 Groundwater discharge accounts for approximately 2.4 per cent of nutrient inputs to the coastal ocean globally (Luijendijk and others, 2020) and is not documented for most of the large marine ecosystems addressed in the present chapter. Inputs from aquaculture operations are also low; it is estimated that nutrients released to the coastal ocean annually by finfish aquaculture operations account for about 1 per cent of anthropogenic inputs worldwide (Hargrave, 2005). Thus, those input pathways are not considered in the present chapter.

45 x 10^9 kg N per year today (Dentener and others, 2006; Duce and others, 2008). Of that amount, it is estimated that atmospheric deposition to the coastal ocean is currently on the order of 8 x 10^9 kg N per year (Seitzinger and others, 2010; Ngatia and others, 2019). The relative importance of atmospheric deposition as a new N source varies among coastal ecosystems from 2–5 per cent in ecosystems with large riverine N inputs (e.g., the northern Gulf of Mexico, the continental shelf of Brazil) to as much as 40 per cent in ecosystems with relatively low riverine inputs (e.g., the Kiel Bight in the Baltic Sea and the Pamlico Sound in North Carolina, United States) (Paerl and others, 2002). Globally, atmospheric deposition of N accounts for approximately 4 per cent of anthropogenic inputs to the coastal ocean.

3.2. Documented impacts of anthropogenic nutrient inputs

3.2.1. Oxygen depletion and acidification

Since 1950, the number of coastal ecosystems experiencing hypoxia (dissolved oxygen (O_2) ≤ 2 mg/l or 63 millimoles (mmol)/l) has increased from about 50 in 1950 to more than 500 in 2015 as a consequence of anthropogenic nutrient loading and ocean warming (Diaz and Rosenberg, 2008; Kemp and others, 2009; Breitburg and others, 2018). In 2019, a further estimate suggests that the number was actually higher, namely, around 700 (Diaz and others, 2019). The spread of coastal hypoxia has not only resulted in the loss of oxygenated habitats for aerobic organisms; it also threatens the survival of coral reefs (Fabricius, 2011; Altieri and others, 2019). In addition, the global spread of hypoxia is amplifying ocean acidification as increases in biological oxygen demand produce CO_2 as a by-product of aerobic respiration (Wallace and others, 2014).

3.2.2. Toxic algal events

The production of toxins can cause mass mortalities of fishes and shellfish and cause harm to the health of people who consume contaminated fish and shellfish or are exposed to toxins through direct contact (Glibert and others, 2005). Globally, there have been more toxic algal events in coastal waters during the past decade than in previous decades (Heisler and others, 2008), largely as a consequence of anthropogenic nutrient inputs and changing N:P ratios (Glibert and Bouwman, 2012; Glibert and others, 2018), introductions of non-native toxic species, ocean acidification (Riebesell and others, 2018) and increases in water temperature and vertical stratification of the upper ocean[8] (Glibert and others, 2014).

3.2.3. Loss of critical, biologically engineered habitats

Coral reefs and seagrass meadows support a wide range of ecosystem services, including coastal protection, erosion control, the maintenance of biodiversity and fisheries (Barbier and others, 2011). At the same time, warm-water coral reefs and seagrass meadows are threatened by multiple anthropogenic stresses (e.g., ocean warming and acidification, eutrophication, overfishing and destructive fishing practices). Ocean warming has been affecting coral reefs for more than three decades through the bleaching and mortality of corals owing to heat stress (Heron and others, 2017), and the risk of bleaching has increased globally at a rate of 4 per cent per year, with 8 per cent of reefs per year being affected by bleaching in the 1980s and 31 per cent affected in 2016 (Hughes and others 2018), a trend that is expected to be exacerbated by coastal eutrophication (Wear and Thurber, 2015). The spatial extent of seagrass beds has been shown to be negatively affected by eutrophication and increases in water temperature

[8] Upper 1,000 m of the water column.

(Waycott and others, 2009; Mvungi and Pillay, 2019). Thus, seagrass meadows have declined in area by about 29 per cent since the beginning of the twentieth century, at an annual rate of about 1.5 per cent (Fourqurean and others, 2012).

4. Patterns and trends within regions

Many large marine ecosystems are hotspots of anthropogenic nutrient loading in both developed and developing countries. In order to provide regional and global perspectives on changing nutrient inputs to coastal systems throughout the world, an international working group developed a global watershed model that relates human activities and natural processes in watersheds to nutrient inputs to coastal systems globally (Seitzinger and others, 2005; Lee and others, 2016). Based on the contribution of anthropogenic dissolved inorganic N to the total dissolved inorganic N loads to large marine ecosystems (Lee and others, 2016), nine large marine ecosystems that represent a range of sizes and anthropogenic dissolved inorganic N inputs are highlighted in the table below.

Surface areas and anthropogenic nitrogen loads of the nine ecosystems addressed below

Ecosystem	Area (km²)	N loading (kg/year)
Baltic Sea	0.4×10^6	0.6×10^9
Bay of Bengal	3.7×10^6	7.1×10^9
Brazil Shelf	1.0×10^6	1.0×10^9
Guinea Current	2.0×10^6	1.0×10^9
Gulf of Mexico	1.5×10^6	1.3×10^9
Great Barrier Reef	1.3×10^6	0.1×10^9
East China Sea	1.0×10^6	2.0×10^9
North Sea	0.7×10^6	4.8×10^9
South China Sea	5.7×10^6	0.7×10^9

4.1. North Sea (Large Marine Ecosystem 22; 690,000 km²)

The North Sea encompasses two subregions: (a) shallow, eutrophic coastal waters along its south-eastern border; and (b) deeper, oligotrophic waters of the open sea. Nutrient inputs to the latter have remained virtually unchanged over the past 50 years, while coastal waters have experienced an increase in N load from about 2.9 to 4.8 x 10^9 kg N per year between 1950 and 1990; over the same period, the P-load increased from 0.44 to 0.64 x 10^9 kg P per year (Vermaat and others, 2008). River-borne inputs of N and P to the coastal subregion account for most of the anthropogenic loading, 75 per cent of which occurs via the Rhine and Elbe Rivers that discharge into the coastal waters of the south-eastern North Sea (Radach and Pätsch, 2007; Paramor and others, 2009). The discharge of N and P to those coastal waters increased rapidly during the period 1965–1985, as illustrated by the Rhine River, in which N and P increased fivefold and tenfold, respectively. As a result, the frequency and magnitude of blooms of *Phaeocystis pouchetii*[9] increased during that period (Lancelot and others, 1987; Lancelot, 1995). While summer hypoxia (< 2 mg O_2/l) occurs in some locations, it is limited to parts of the stratified open sea (Greenwood and others, 2010).

From 1990 to 2000, the P load decreased to the pre-eutrophication levels of the 1950s (Vermaat and others, 2008). The anthropogenic portion of the annual nutrient budget of the

9 *Phaeocystis* can produce large amounts of foam, which often affects coastlines and beaches, and can also produce dimethylsulphide, an aerosol that contributes to cloud formation and acid rain.

coastal North Sea is currently declining and is less than inputs from the benthos or open sea.

4.2. Baltic Sea (Large Marine Ecosystem 23; 400,000 km^2)

The Baltic Sea a is a brackish, shallow sea (mean depth, 55 m; maximum depth, 460 m) with limited water exchange with the North Sea. By virtue of its bathymetry and estuarine circulation regime,[10] the Baltic Sea is particularly vulnerable to eutrophication. Thus, it is host to the largest anthropogenically induced hypoxic zone in the world (Carstensen and others, 2014). The change from a healthy state without eutrophication problems began in the late 1950s and early 1960s.

Riverine inputs of N and P account for most inputs to the Baltic Sea from 1995 to 2015 (Sonesten and others, 2018). Inputs of N and P were generally higher during the period 1995–2002 (650–900 x 10^6 kg N/year and 33–43 x 10^6 kg P/year) compared with the period 2003–2015 (500–775 x 10^6 kg N/year and 22–35 x 10^6 kg P/year). Natural background loads of N and P made up about 33 percent of those inputs during the latter period (Sonesten and others, 2018). Atmospheric deposition also declined during the period from about 300 x 10^6 kg N per year in 1995 to 210 x 10^6 kg N per year in 2011. Low inputs during the period 2003–2015 were attributable, in part, to dry periods with low river flows (2003, 2014, 2015).

Over roughly the same period (1993–2016), the spatial extent of seasonal hypoxia-anoxia increased from about 5,000 km^2 (1.3 per cent of the Baltic) to more than 60,000 km^2 (> 16 per cent of the Baltic) (Limburg and Casini, 2018), in part because of increases in the strengths of the seasonal thermocline and halocline in the upper water column (< 100 m) (Liblik and Lips, 2019) and in part because episodes of deepwater ventilation in the basins have been less frequent and of shorter duration during the past two decades (Carstensen and others, 2014; Schmale and others, 2016). Seasonal hypoxia not only affects aerobic benthic life, it may also promote the development of more blooms of cyanobacteria. Massive surface accumulations of nitrogen-fixing cyanobacteria (largely *Nodularia* spp.) during the summer have intensified since 1982, a trend that is correlated with increases in the spatial extent of hypoxia and anthropogenic P loading (Pliński and others, 2007; Funkey and others, 2014). The enhanced downward flux of degradable organic matter from those blooms elevate oxygen demand and the regeneration of P in bottom waters, creating positive feedback between anthropogenic nutrient enrichment, cyanobacteria blooms and oxygen depletion. In addition, some species of cyanobacteria produce toxins that affect recreation and fisheries. Thus, although ocean warming and shifts in circulation patterns are important factors modulating the extent of hypoxia, further nutrient reductions in the Baltic Sea will be necessary to reduce the ecosystem impacts of deoxygenation.

Abatement of eutrophication in the Baltic Sea has received more concerted effort and sustained research than any other coastal region in the world (Boesch, 2019). Since the mid-1990s, statistically significant reductions in the anthropogenic loads of N and P have been achieved (Baltic Marine Environment Protection Commission (HELCOM), 2018; Sonesten and others, 2018). Relative to the reference period (1997–2003), flow-normalized riverine inputs of N and P have declined by 12 per cent and 25 per cent, respectively, and, compared with 1995, precipitation-normalized atmospheric deposition of N declined

10 A sill less than 20 m deep separates the Baltic Sea and its basins from the North Sea. Estuarine (density-driven) circulation consists of surface water flowing from the Baltic through the Danish Strait into the North Sea and bottom water flowing into the Baltic's basins through the Danish Strait from the North Sea (Szymczycha and others, 2019).

by 29 per cent. Following the introduction of nutrient abatement measures, recovery began in some basins during the late 1990s, while in others, it commenced early in the twenty-first century (Murray and others, 2019). However, given the sustained increase in vertical stratification and the associated isolation of deep water from oxygenated surface water (Liblik and Lips, 2019), the susceptibility of the Baltic to eutrophication will increase if the trend continues, which emphasizes the importance of achieving the maximum allowable inputs specified in the Baltic Sea Action Plan.[11] To that end, inputs of N and P from anthropogenic sources as a whole (riverine plus atmospheric sources) need to be further reduced by 12 per cent and 25 per cent, respectively, to ensure a healthy Baltic Sea.

4.3. Gulf of Mexico (Large Marine Ecosystem 5; 1,530,400 km^2)

Impacts of anthropogenic nutrient loading are greatest in the northern Gulf of Mexico. Inter-annual variations in nutrient loading are directly related to variations in the flow of the Mississippi and Atchafalaya Rivers (Rabalais and others, 2007). During the period 1980–2017, annual dissolved inorganic N inputs fluctuated around 1,000 x 10^6 kg per year, with a minimum of about 600 x 10^6 kg per year in 2000 and a maximum of about 1,800 x 10^6 kg per year in 1993.[12] As a consequence, during the summer, the northern Gulf of Mexico has the second largest coastal hypoxic zone in the world, the spatial extent of which has varied between less than 5,000 km^2 in 2000 to 22,720 km^2 in 2017, with a mean of 13,700 km^2 (Rabalais and others, 2007; Matli and others, 2018).

In addition to bottom-water hypoxia, increases in nutrient loading appear to be promoting toxic phytoplankton blooms. The abundance of *Pseudo-nitzschia* spp. has increased over the shelf since the 1950s, a trend that may be related to the long-term increase in nutrient loading (Dortch and others, 1997). Seasonal blooms develop when surface waters begin to warm in the spring and river discharge is increasing, but before seasonal peaks in flow and phytoplankton biomass occur (Bargu and others, 2016). Peaks in the abundance of potentially toxin-producing dinoflagellates (*Dinophysis* spp. and *Prorocentrum* spp.) have been observed to coincide with the seasonal peak in river flow (Bargu and others, 2016).

4.4. North Brazil Shelf (Large Marine Ecosystem 17; 1,034,600 km^2)

With a mean freshwater discharge of 120,000 m^3/s (seasonal maximum, about 240,000 m^3/s in May; minimum, 80,000 m^3/s in November), the Amazon River forms an extensive and dynamic surface plume of low-salinity, relatively nutrient-rich water that extends well offshore over the northern Brazil shelf. The river is the primary source of silicate (83–91 per cent), nitrate (62–76 per cent) and phosphate (48–65 per cent) to the North Brazil Shelf Large Marine Ecosystem (Demaster and Pope, 1996). The annual supply of river-borne N (mean, about 1,050 x 10^6 kg N/year) supports a eutrophic ecosystem (730 g carbon (C) m^2/year) in the mesohaline (salinity 30–35) waters of the coastal plume (Dagg and others, 2004; Santos and others, 2008; Coles and others, 2013).

Net primary production is nitrate limited, and extensive blooms of diatom-diazotroph associations[13] have been observed in the

11 Available at https://helcom.fi/baltic-sea-action-plan. .
12 Available at https://nrtwq.usgs.gov/mississippi_loads/#/GULF.
13 The diatoms *Hemiaulus hauckii* and *Rhizosolenia clevei*, containing the symbiotic cyanobacteria *Richelia* sp., formed about 28 per cent of the biomass in mesohaline waters of the plume.

mesohaline plume during both the spring and autumn (Gomes and others, 2018). Given the spread of plume water into the Caribbean and equatorial Atlantic (Coles and others, 2013), such blooms may be a significant source of new N to support primary production and the great Atlantic Sargassum belt (Wang and others, 2019) in nutrient-poor, tropical waters (Subramaniam and others, 2008; Yeung and others, 2012).

4.5. Guinea Current (Large Marine Ecosystem 28; 1,958,800 km^2)

Lying within the Guinea Current Large Marine Ecosystem (Heileman, 2008), the Gulf of Guinea receives freshwater discharges from 15 rivers, including the Congo (the second largest river on Earth) with an annual mean discharge of about 40,000 m^3 per s (Hopkins and others, 2013). It is also the world's second largest exporter of terrestrial organic carbon into the oceans (Spencer and others, 2012). The outflow of such a large volume of water into the south-eastern Atlantic produces a vast low salinity plume with a signature of high chlorophyll that can be detected as far as 700–800 km to the west and north from the river's mouth (Hopkins and others, 2013).

Most of the coastal cities bordering the Gulf lack basic infrastructure for sewage treatment, and substantial quantities of N and P from municipal and agricultural sources are transported to the Gulf through river run-off.[14] The current anthropogenic river-borne N loading is estimated to be between 600 and 1,000 x 10^6 kg per year, which places the region in the high-risk category for eutrophication (Seitzinger and Mayorga, 2016).

Consequently, the Gulf is characterized by high phytoplankton net primary production (356–438 g C m^2/year, 2003–2013) supported by nutrient input from both river run-off and coastal upwelling.[15] Nutrient pollution in coastal lagoon systems, in particular near urban centres, has caused increases in phytoplankton biomass and oxygen depletion, resulting in decreases in fish reproduction levels and increases in waterborne diseases (Scheren and others, 2002). In addition, while the phytoplankton community of coastal waters beyond the lagoons has been shown to be dominated by diatoms and cyanobacteria, potentially toxic dinoflagellate species (*Dinophysis caudata*, *Lingulodinium polyedrum* and *Prorocentrum* spp.) have been detected (Zendong and others, 2016).

4.6. Bay of Bengal (Large Marine Ecosystem 34; 3,657,500 km^2)

Freshwater inputs to the Bay of Bengal are high as a consequence of monsoonal rainfall and river run-off (Yaremchuk and others, 2005). A total of 5 of the world's 50 largest rivers flow into the Bay (Sengupta and others, 2006). Salinity is lowest in the northern Bay, off the Ganges River delta and off the Ayeyarwady River delta in the Gulf of Martaban, especially during the June–October monsoon season (Akhil and others, 2016). Rivers exported 35–45 per cent more N and P to the Bay of Bengal in 2000 than in 1970, largely as a consequence of increases in fertilizer use (Sattar and others, 2014). In 2000, rivers exported 7,100 x 10^6 kg N per year and 1,500 x 10^6 kg P per year to the Bay. Three rivers (Ganges, Godāvari and Ayeyarwady) account for 75–80 per cent of the total river input of N and P (Pedde and others, 2017). Atmospheric deposition has been estimated to be in the range of 100–3,100 x 10^6 kg N per year, with most estimates near the upper end of the range (Srinivas and Sarin, 2013). Thus, atmospheric deposition may be a major source of N in addition to riverine inputs. Ratios of N

14 See https://some.grida.no/media/23569/state-of-the-coastal-and-marine-ecosystems-in-gclme.pdf.
15 See http://onesharedocean.org/public_store/lmes_factsheets/factsheet_28_Guinea_Current.pdf.

and P to silicon (Si) have also been increasing, indicating an increasing risk for blooms of non-diatom species that may produce toxins and otherwise disrupt coastal ecosystems (Pedde and others, 2017).

A strong halocline limits nutrient enrichment from deep water, so the central Bay is oligotrophic (Kay and others, 2018). Coastal waters are much more productive[16] (> 300 g C m^2/year) as a consequence of riverine N and P inputs. Hotspots of coastal eutrophication occur off the Ganges River delta (Bangladesh) of the northern Bay and in the Gulf of Martaban off the Ayeyarwady River delta (Myanmar) of the eastern Bay (Kay and others, 2018; Monolisha and others, 2018). Phytoplankton biomass from those fertile areas that is not consumed in the euphotic zone sinks and decays at depth (150–600 m), leading to one of the largest hypoxic zones (60,000 km^2) in the global ocean (Bristow and others, 2017; Kay and others, 2018). In addition, potentially toxic species have been observed along the east coast of India (Mohanty and others, 2007; Sahu and others, 2014).

4.7. South China Sea (Large Marine Ecosystem 36; 5,661,000 km^2)

The South China Sea as a whole is considered to be moderately productive[17] (150–300 g C m^2/year) but has the "highest" risk of eutrophication (Seitzinger and Mayorga, 2016). Riverine inputs of fresh water and nutrients to coastal waters of the Sea are dominated by the rivers that flow into the Pearl River estuary (Harrison and others 2008; Chen and others, 2009). During the wet season (April–September) when 80 per cent of river discharge occurs (Yin and others, 2001), the two-layered estuarine circulation extends onto the inner shelf as the surface, nutrient-rich plume is transported along the coast and spreads at least 250 km into the interior of the Sea (Jilan, 2004; Chen and others, 2017).

In the late 1970s, the fertile river delta to the north of Hong Kong, China, was primarily used for agriculture. Since then, the Pearl River delta has been transformed from farmland into a large megalopolis. As a consequence, dissolved N and P inputs through the Pearl River delta increased by a factor of 2–5 during the 1980s and 1990s, largely owing to increases in urban waste discharges and nutrients released from aquaculture operations (Yin and Harrison, 2008). Inputs plateaued during the period 2006–2012, when concentrations remained in the range of 500–1,000 x 10^6 kg N per year and 20–40 x 10^6 kg P per year, with no inter-annual trend (Tong and others, 2015). Although atmospheric deposition of N over the South China Sea as a whole is estimated to be nearly an order of magnitude higher (about 9,200 x 10^6 kg N/year) than the riverine input (Luo and others, 2014), deposition is dispersed over the entire Sea with little impact on coastal eutrophication relative to river-borne inputs.

Overall, the impact of anthropogenic nutrient loading appears to be limited to the coastal margins of the Sea (Sun, 2017), with hotspots of seasonal hypoxia and toxic algal events located in the vicinity of major river deltas with substantial urban development (United Nations Environment Programme (UNEP) and others, 2005; Qian and others, 2018). The areas with the most severe eutrophication are associated with the estuaries of the main rivers. Among the most severely affected is the lower Pearl River estuary, which has experienced annual summer hypoxia in bottom waters. Oxygen depletion in bottom waters of the lower Pearl River estuary has occurred every summer for at least the past 25 years (Qian and others, 2018). During that period, the annual minimum dissolved oxygen concentration in bottom water decreased at a rate of about 2 ± 0.9 mmol

16 See http://lme.edc.uri.edu/LME/images/Content/LME_Briefs/lme_34.pdf.
17 See http://lme.edc.uri.edu/LME/images/Content/LME_Briefs/lme_36.pdf.

per l per year as a consequence of dissolved inorganic N loading, which increased at a rate of about1.4 ± 0.3 mmol N per l per year (Qian and others, 2018).

The frequency of toxic algal events in Chinese coastal waters increased from no reports during the 1950s and 1960s, to 10 in the 1970s, 25 in the 1980s and more than 100 in the 1990s (Yan and others, 2002). From 1980 to 2003, the area affected expanded to include the estuaries of the Pearl River and the Masinloc River and Manila Bay (Wang and others, 2008). Toxic species include potentially toxic *Noctiluca scintillans* (Pearl River estuary) and *Pyrodinium bahamense* (Philippine estuaries). *N. scintillans* has also been associated with hypoxia and the clogging of fish gills, and it may act as a vector of algal toxins to higher trophic levels (Escalera and others, 2007; Turkoglu, 2013).

4.8. Great Barrier Reef (Large Marine Ecosystem 40; 1,300,000 km²)

Since European settlement, annual riverine inputs of N and P to the Great Barrier Reef lagoon have increased from approximately 0.014 x 10^9 kg N per year to 0.080 x 10^9 kg N per year and from 1.8 x 10^6 kg P per year to 16 x 10^6 kg P per year (Brodie and others, 2011; Kroon and others, 2012). River-borne inputs of dissolved inorganic P (P-PO_4) can promote the growth of *Trichodesmium* spp. While limited broadscale monitoring of *Trichodesmium* spp. occurs across the Great Barrier Reef, long-term data at one site near the Yongala Wreck collected since 2010 indicate a gradual increase in its abundance (Robson and others, 2018; Great Barrier Reef Marine Park Authority (GBRMPA), 2019). The nitrogen-fixing ability of *Trichodesmium* spp. suggests that increasing levels of P-PO_4 alone may be driving increases in phytoplankton biomass, and there is some evidence that the trends are a significant factor in the decreasing condition of fringing reefs in the inner Great Barrier Reef lagoon. Long-term monitoring now shows that hard coral cover on the Great Barrier Reef has decreased by more than 70 per cent over the past century (Bell and others, 2014). The decline has been attributed mainly to storm damage, coral bleaching events, the widespread growth of *Acanthaster planci* (crown-of-thorns starfish) and coral skeletal diseases. Record levels of nanophytoplankton growth in river-affected regions of the lagoon appear to be promoting the growth of *A. planci* larvae and adult *A. planci* outbreaks (Bell, 1992). There is growing evidence that *A. planci* predation events and coral bleaching are promoted by eutrophication and that that is one of the reasons why the reefs have not recovered (Bell and others, 2014; GBRMPA, 2019).

4.9. East China Sea[18] (Large Marine Ecosystem 47; 1,008,100 km²)

The East China Sea is considered to be a highly productive system (> 300 g C/m²/year) and is in the "highest" risk category for eutrophication (Seitzinger and Mayorga, 2016). The flow of the Yangtze River (annual mean, 30,200 m³/s) accounts for more than 90 per cent of nutrient inputs to the Sea (Yuan and others, 2007; Tong and others, 2015). From 1968 to 1997, it is estimated that the anthropogenic nutrient load (e.g., nitrates) exported from the Yangtze River into the Sea increased more than tenfold (Yan and others, 2003). A comparison of nutrient concentrations in the Yangtze River estuary and the receiving waters of the Sea before 2002 and after the 2006 impoundment of the Three Gorges Dam (Chai and others, 2009) showed increases in the concentrations of total N (41.8 to 82.2 micrometres (μm), dissolved inorganic N (24.4 to 37.5 μm) and soluble reactive P (0.9 to 1.3 μm), and from 2006 to 2012,

[18] See http://lme.edc.uri.edu/LME/images/Content/LME_Briefs/lme_47.pdf.

total N load increased from 1,350 x 10^6 kg per year to 2,040 x 10^6 kg per year, while total P load increased from 122 x 10^6 kg per year to 240 x 10^6 kg per year (Tong and others, 2015). Atmospheric deposition of N was estimated to be about 1,750 x 10^6 kg per year, which is in the range of riverine inputs during that period (Tong and others, 2015).

While the atmospheric input is generally distributed over the entire East China Sea, during the summer monsoons, the impact of river-borne nutrients is largely focused in coastal waters. Thus, the concentration of sea surface chlorophyll a in the Sea is highest near the shore within the plume (> 10 mg/m^3) and decreases rapidly with distance to low concentrations (< 0.5 mg/m^3) in open waters beyond the continental shelf (Yuan and others, 2007). Inter-annual increase in nutrient loading has also led to increases in phytoplankton biomass over the years (Zhou and others, 2019).

Sinking organic matter produced by phytoplankton in the lower estuary and coastal plume fuel oxygen consumption and summer development of bottom-water hypoxia. The occurrence, frequency and spatial extent of hypoxia have been increasing since the late 1990s (Li and others, 2011; Wei and others, 2015). Today, the area of the Sea that is influenced by the coastal plume of the Yangtze River is regarded as one of the largest coastal hypoxic zones (> 12,000 km^2) in the world (Chen and others, 2007; Wang and others, 2016; Zhu and others, 2017).

As nutrient input from the Yangtze River increased, reported toxic algal bloom events along the coast of the Sea increased from zero in the 1950s and 1960s to 10 in the 1970s, 25 in the 1980s and more than 100 in the 1990s (Yan and others, 2002). In particular, large-scale blooms (covering an area of more than 1,000 km^2) have been recorded every year since 1998, and *Prorocentrum donghaiense* has been the recurrent bloom species for more than 10 years (Li and others, 2009; Lu and others, 2014). Blooms of potentially toxic *Karlodinium veneficum, Karenia mikimotoi, Alexandrium tamarense, Alexandrium catenella* and *Heterosigma akashiwo* have also been observed (Lu and others, 2014; Zhou and others, 2015; Wang and others, 2018).

5. Outlook

It is projected that anthropogenic N production will increase by nearly a factor of two during the first half of the twenty-first century and, based on projected increases of 40–45 per cent in dissolved inorganic N loading by 2050, the risk of coastal eutrophication will increase in 21 per cent of large marine ecosystems, most of which are in Africa, South America, South Asia and Oceania. The impacts of continued increases in N loading are likely to be exacerbated by climate-driven increases in ocean temperatures, vertical stratification, rainfall and the flux of atmospheric CO_2 into the ocean (Guinder and Molinero, 2013). Thus, it is likely that the severity and extent of coastal hypoxia, acidification and toxic algal events will also continue to increase in the absence of aggressive actions to reduce anthropogenic inputs of N and P (Townhill and others, 2018).

Important gaps in the current understanding of the impacts of anthropogenic nutrient inputs on the coastal ocean fall into two broad categories: (a) the lack of data on coastal ecosystems in the southern hemisphere (Altieri and others, 2019; Diaz and others, 2019); and (b) the need to understand synergies between the impacts of nutrient loading and climate-driven changes in coastal ecosystems (Paerl and others, 2014).

References

Akhil, V.P., and others (2016). Assessment of seasonal and year-to-year surface salinity signals retrieved from SMOS and Aquarius missions in the Bay of Bengal. *International Journal of Remote Sensing*, vol. 37, No.5, pp. 1089–1114.

Altieri, Andrew H., and Robert J. Diaz (2019). Dead zones: oxygen depletion in coastal ecosystems. In *World Seas: An Environmental Evaluation*, Elsevier, pp. 453–473.

Bachmann, R.W., and others (2006). Eutrophication in freshwater and marine systems. *Limnology and Oceanography*, vol. 51, pp. 351–800.

Baltic Marine Environment Protection Commission (HELCOM) (2018). *State of the Baltic Sea: Second HELCOM Holistic Assessment 2011–2016*. Baltic Sea Environment Proceedings 155. Baltic Marine Environment Protection Commission.

Barbier, Edward B., and others (2011). The value of estuarine and coastal ecosystem services. *Ecological Monographs*, vol. 81, No. 2, pp. 169–193.

Bargu, Sibel, and others (2016). Influence of the Mississippi River on *Pseudo-nitzschia* spp.: abundance and toxicity in Louisiana coastal waters. *Estuaries and Coasts*, vol. 39, No. 5, pp. 1345–1356.

Bell, Peter R.F., and others (2014). Evidence of large-scale chronic eutrophication in the Great Barrier Reef: quantification of chlorophyll *a* thresholds for sustaining coral reef communities. *Ambio*, vol. 43, No. 3, pp. 361–376.

Bell, Peter R.F. (1992). Eutrophication and coral reefs: some examples in the Great Barrier Reef lagoon. *Water Research*, vol. 26, No. 5, pp. 553–568.

Beusen, Arthur H.W., and others (2016). Global riverine N and P transport to ocean increased during the 20th century despite increased retention along the aquatic continuum. *Biogeosciences*, vol. 13, No. 8, pp. 2441–2451.

Boesch, Donald F. (2019). Barriers and bridges in abating coastal eutrophication. *Frontiers in Marine Science*, vol. 6, art. 123.

Bouwman, Lex, and others (2013). Exploring global changes in nitrogen and phosphorus cycles in agriculture induced by livestock production over the 1900–2050 period. *Proceedings of the National Academy of Sciences*, vol. 110, No. 52, pp. 20882–20887.

Boyer, Elizabeth W., and Robert W. Howarth (2008). Nitrogen fluxes from rivers to the coastal oceans. In *Nitrogen in the Marine Environment*, Elsevier Inc., pp. 1565–1587.

Breitburg, Denise, and others (2018). Declining oxygen in the global ocean and coastal waters. *Science*, vol. 359, No. 6371.

Bristow, Laura A., and others (2017). N_2 production rates limited by nitrite availability in the Bay of Bengal oxygen minimum zone. *Nature Geoscience*, vol. 10, No. 1, pp. 24–29.

Brodie, J.E., and others (2011). Assessment of the eutrophication status of the Great Barrier Reef lagoon (Australia). *Biogeochemistry*, vol. 106, No. 2, pp. 281–302.

Carstensen, Jacob, and others (2014). Deoxygenation of the Baltic Sea during the last century. *Proceedings of the National Academy of Sciences*, vol. 111, No. 15, pp. 5628–5633.

Chai, Chao, and others (2009). Nutrient characteristics in the Yangtze River Estuary and the adjacent East China Sea before and after impoundment of the Three Gorges Dam. *Science of the Total Environment*, vol. 407, no. 16, pp. 4687–4695.

Chen, Bingzhang, and others (2009). Estuarine nutrient loading affects phytoplankton growth and microzooplankton grazing at two contrasting sites in Hong Kong, China, coastal waters. *Marine Ecology Progress Series*, vol. 379, pp. 77–90.

Chen, Chung-Chi, and others (2007). Hypoxia in the East China Sea: one of the largest coastal low-oxygen areas in the world. *Marine Environmental Research*, vol. 64, No. 4, pp. 399–408.

Chen, Zhaoyun, and others (2017). Far-reaching transport of Pearl River plume water by upwelling jet in the northeastern South China Sea. *Journal of Marine Systems*, vol. 173, pp. 60–69.

Coles, Victoria J., and others (2013). The pathways and properties of the Amazon River Plume in the tropical North Atlantic Ocean. *Journal of Geophysical Research: Oceans*, vol. 118, No. 12, pp. 6894–6913.

Costanza, Robert, and others (2017). Twenty years of ecosystem services: how far have we come and how far do we still need to go? *Ecosystem Services*, vol. 28, pp. 1–16.

Dagg, Michael, and others (2004). Transformation of dissolved and particulate materials on continental shelves influenced by large rivers: plume processes. *Continental Shelf Research*, vol. 24, Nos. 7 and 8, pp. 833–858.

Demaster, David J., and Robert H. Pope (1996). Nutrient dynamics in Amazon shelf waters: results from AMASSEDS. *Continental Shelf Research*, vol. 16, No. 3., pp. 263–289.

Dentener, Frank, and others (2006). Nitrogen and sulfur deposition on regional and global scales: a multi-model evaluation. *Global Biogeochemical Cycles*, vol. 20, No. 4,

Diaz, Robert J., and Rutger Rosenberg (2008). Spreading dead zones and consequences for marine ecosystems. *Science*, vol. 321, No. 5891, pp. 926–929.

Diaz, Robert J., and others (2019). Hypoxia in estuaries and semi-enclosed seas. In *Ocean Deoxygenation–Everyone's Problem: Causes, Impacts, Consequences and Solutions*, D. Laffoley and J. M. Baxter, eds. Gland, Switzerland: IUCN.

Dortch, Quay, and others (1997). Abundance and vertical flux of *Pseudo-nitzschia* in the northern Gulf of Mexico. *Marine Ecology Progress Series*, vol. 146, pp. 249–264.

Duce, R.A., and others (2008). Impacts of atmospheric anthropogenic nitrogen on the open ocean. *Science*, vol. 320, No. 5878, pp. 893–897.

Elser, James J., and others (2007). Global analysis of nitrogen and phosphorus limitation of primary producers in freshwater, marine and terrestrial ecosystems. *Ecology Letters*, vol. 10, No. 12, pp. 1135–1142.

Escalera, Laura, and others (2007). *Noctiluca scintillans* may act as a vector of toxigenic microalgae. *Harmful Algae*, vol. 6, No. 3, pp. 317–320.

Fabricius, Katharina E. (2011). Factors determining the resilience of coral reefs to eutrophication: a review and conceptual model. In *Coral Reefs: An Ecosystem in Transition*, Zvy Dubinsky and Noga Stambler, eds. New York: Springer, pp. 493–505.

Fourqurean, James W., and others (2012). Seagrass ecosystems as a globally significant carbon stock. *Nature Geoscience*, vol. 5, No. 7, pp. 505–509.

Funkey, Carolina P., and others (2014). Hypoxia sustains cyanobacteria blooms in the Baltic Sea. *Environmental Science & Technology*, vol. 48, No. 5, pp. 2598–2602.

Galloway, James N., and others (2004). Nitrogen cycles: past, present, and future. *Biogeochemistry*, vol. 70, No. 2, pp. 153–226.

Glibert, Patricia, and Lex Bouwman (2012). Land-based nutrient pollution and the relationship to harmful algal blooms in coastal marine systems. *Loicz Newsletter Inprint*, vol. 2, pp. 5–7.

Glibert, Patricia M., and others (2005). The global, complex phenomena of harmful algal blooms. *Oceanography*, vol. 18, No. 2.

Glibert, Patricia M., and others (2014). Vulnerability of coastal ecosystems to changes in harmful algal bloom distribution in response to climate change: projections based on model analysis. *Global Change Biology*, vol. 20, No. 12, pp. 3845–3858.

Glibert, Patricia M., and others (2018). Key questions and recent research advances on harmful algal blooms in relation to nutrients and eutrophication. In *Global Ecology and Oceanography of Harmful Algal Blooms*. Springer, pp. 229–259.

Gomes, Helga Rosario, and others (2018). The influence of riverine nutrients in niche partitioning of phytoplankton communities – a contrast between the Amazon River Plume and the Chang Jiang (Yangtze) River diluted water of the East China Sea. *Frontiers in Marine Science*, vol. 5, art. 343.

Great Barrier Reef Marine Park Authority (GBRMPA) 2019. Great Barrier Reef Outlook Report 2019, GBRMPA, Townsville, Australia.

Green, Pamela A., and others (2004). Pre-industrial and contemporary fluxes of nitrogen through rivers: a global assessment based on typology. *Biogeochemistry*, vol. 68, No. 1, pp. 71–105.

Greenwood, N., and others (2010). Detection of low bottom water oxygen concentrations in the North Sea; implications for monitoring and assessment of ecosystem health. *Biogeosciences*, vol. 7, No. 4, pp. 1357–1373.

Guinder, Valeria, and Juan Carlos Molinero (2013). Climate change effects on marine phytoplankton. *Marine Ecology in a Changing World*, Andrés H. Arias and María C. Menendez, eds. Boca Raton, Florida, CRC Press, pp. 68–90.

Hargrave, Barry T., ed. (2005). *Environmental Effects of Marine Finfish Aquaculture*. Berlin: Springer.

Harrison, John A., and others (2005). Dissolved inorganic phosphorus export to the coastal zone: results from a spatially explicit, global model. *Global Biogeochemical Cycles*, vol. 19, No. 4.

Harrison, Paul J., and others (2008). Physical-biological coupling in the Pearl River Estuary. *Continental Shelf Research*, vol. 28, No. 12, pp. 1405–1415.

Heileman, S. (2008). Guinea Current LME. In *The UNEP Large Marine Ecosystem Report: A Perspective on Changing Conditions in LMEs of the World's Regional Seas*, K. Sherman and G. Hempel, eds. Nairobi: UNEP, 1.

Heisler, John, and others (2008). Eutrophication and harmful algal blooms: a scientific consensus. *Harmful Algae*, vol. 8, No.,1, pp. 3–13.

Heron, Scott Fraser, and others (2017). *Impacts of Climate Change on World Heritage Coral Reefs: A First Global Scientific Assessment*. Paris: UNESCO World Heritage Centre.

Hopkins, Jo, and others (2013). Detection and variability of the Congo River plume from satellite derived sea surface temperature, salinity, ocean colour and sea level. *Remote Sensing of Environment*, vol. 139, pp. 365–385.

Howarth, R.W., and others (1996). Riverine inputs of nitrogen to the North Atlantic Ocean: fluxes and human influences. *Biogeochemistry*, vol. 35, pp. 75–139.

Howarth, R.W., and others (2011). Coupled biogeochemical cycles: eutrophication and hypoxia in temperate estuaries and coastal marine ecosystems. *Frontiers in Ecology and the Environment*, vol. 9, No. 1, pp. 18–26.

Howarth, R.W., and others (2012). Nitrogen fluxes from the landscape are controlled by net anthropogenic nitrogen inputs and by climate. *Frontiers in Ecology and the Environment*, vol. 10, No. 1, pp. 37–43.

Howarth, Robert W. (2008). Coastal nitrogen pollution: a review of sources and trends globally and regionally. *Harmful Algae*, vol. 8, No. 1, pp. 14–20.

Howarth, Robert W., and others (2000). Nutrient pollution of coastal rivers, bays, and seas. *Issues in Ecology*, No. 7.

Howarth, Robert W., and Roxanne Marino (2006). Nitrogen as the limiting nutrient for eutrophication in coastal marine ecosystems: evolving views over three decades. *Limnology and Oceanography*, vol. 51, No. 1, part 2, pp. 364–376.

Hughes, Terry P., and others (2018). Spatial and temporal patterns of mass bleaching of corals in the Anthropocene. *Science*, vol. 359, No. 6371, pp. 80–83.

Intergovernmental Panel on Climate Change (IPCC) (2014). *Climate Change 2014: Impacts, Adaptation, and Vulnerability. Part A: Global and Sectoral Aspects. Contribution of Working Group II to the Fifth Assessment Report of the Intergovernmental Panel on Climate Change*. C. B. Field and others, eds. Cambridge: Cambridge University Press.

Jickells, T.D., and others (2017). A reevaluation of the magnitude and impacts of anthropogenic atmospheric nitrogen inputs on the ocean. *Global Biogeochemical Cycles*, vol. 31, No. 2, pp. 289–305.

Jilan, Su (2004). Overview of the South China Sea circulation and its influence on the coastal physical oceanography outside the Pearl River Estuary. *Continental Shelf Research*, vol. 24, No. 16, pp. 1745–1760.

Kay, Susan, and others (2018). Marine dynamics and productivity in the Bay of Bengal. In *Ecosystem Services for Well-Being in Deltas: Integrated Assessment for Policy Analysis*. Robert J. Nicholls and others, eds. Cham, Switzerland: Springer International Publishing, pp. 263–275. https://doi.org/10.1007/978-3-319-71093-8_14.

Kemp, W.M., and others (2009). Temporal responses of coastal hypoxia to nutrient loading and physical controls. *Biogeosciences*, vol. 6, No. 12, pp. 2985–3008.

Kroon, F.J., and others (2012). River loads of suspended solids, nitrogen, phosphorus and herbicides delivered to the Great Barrier Reef lagoon. *Marine Pollution Bulletin*, vol. 65, Nos. 4–9, pp. 167–181.

Lamsal, L.N., and others (2011). Application of satellite observations for timely updates to global anthropogenic NO_x emission inventories. *Geophysical Research Letters*, vol. 38, No. 5.

Lancelot, Christiane, and others (1987). *Phaeocystis* blooms and nutrient enrichment in the continental coastal zones of the North Sea. *Ambio*, No. 1.

Lancelot, Christiane, and others (1995). The mucilage phenomenon in the continental coastal waters of the North Sea. *Science of the Total Environment*, vol. 165, Nos.1–3, pp. 83–102.

Lee, Rosalynn Y., and others (2016). Land-based nutrient loading to LMEs: a global watershed perspective on magnitudes and sources. *Environmental Development*, vol. 17, pp. 220–229.

Li, Ji, and others (2009). Relationships between nitrogen and phosphorus forms and ratios and the development of dinoflagellate blooms in the East China Sea. *Marine Ecology Progress Series*, vol. 383, pp. 11–26.

Li, Xinxin, and others (2011). Historical trends of hypoxia in Changjiang River estuary: applications of chemical biomarkers and microfossils. *Journal of Marine Systems*, vol. 86, Nos. 3 and 4, pp. 57–68.

Liblik, T., and Lips, U. (2019). Stratification has strengthened in the Baltic Sea: an analysis of 35 years of observational data. *Frontiers In Earth Science*, vol. 7, art. 174. http://doi.org/10.3389/feart.2019.00174.

Limburg, Karin E., and Michele Casini (2018). Effect of marine hypoxia on Baltic Sea Cod *Gadus morhua*: evidence from otolith chemical proxies. *Frontiers in Marine Science*, vol. 5, art. 482.

Lu, Chaoqun, and Hanqin Tian (2017). Global nitrogen and phosphorus fertilizer use for agriculture production in the past half century: shifted hot spots and nutrient imbalance. *Earth System Science Data*, vol. 9, pp. 181–192.

Lu, Douding, and others (2014). Causative species of harmful algal blooms in Chinese coastal waters. *Algological Studies*, vol. 145, No. 1, pp. 145–168.

Luo, X.S., and others (2014). Chinese coastal seas are facing heavy atmospheric nitrogen deposition. *Environmental Research Letters*, vol. 9, No. 9, 095007.

Martínez, Maria Luiza, and others (2007). The coasts of our world: ecological, economic and social importance. *Ecological Economics*, vol. 63, Nos. 2 and 3, pp. 254–272.

Matli, V.R.R., and others (2018). Space-time geostatistical assessment of hypoxia in the Northern Gulf of Mexico. *Environmental Science and Technology*, vol. 52, No. 21, pp. 12484–12493. http://doi.org/10.1021/acs.est.8b03474.

Mohanty Ajit K., and others (2007). Red tide of *Noctiluca scintillans* and its impact on the coastal water quality of the near-shore waters, off the Rushikulya River, Bay of Bengal. *Current Science,* vol. 93, No. 5, pp. 616.–618.

Monolisha, S., and others (2018). Optical classification of the coastal waters of the Northern Indian Ocean. *Frontiers in Marine Science*, vol. 5, art. 87.

Mosier, Arvin R., and others (2004). Nitrogen fertilizer: an essential component of increased food, feed, and fiber production. *Agriculture and the Nitrogen Cycle: Assessing the Impacts of Fertilizer Use on Food Production and the Environment*, vol. 65, pp. 3–15.

Murray, C.J., and others (2019). Past, present and future eutrophication status of the Baltic Sea. *Frontiers in Marine Science*, vol. 6, art. 2.

Mvungi, Esther F., and Deena Pillay (2019). Eutrophication overrides warming as a stressor for a temperate African seagrass (*Zostera capensis*). *PloS One*, vol. 14, No. 4. e0215129.

Ngatia, Lucy, and others (2019). Nitrogen and phosphorus eutrophication in marine ecosystems. In *Monitoring of Marine Pollution*. London: IntechOpen.

Nixon, Scott W. (1995). Coastal marine eutrophication: a definition, social causes, and future concerns. *Ophelia*, vol. 41, No. 1, pp. 199–219.

Paerl, H.W., and others (2002). Atmospheric deposition of nitrogen: implications for nutrient over-enrichment of coastal waters. *Estuaries*, vol. 25, No. 4, pp. 677–693.

Paerl, H.W., and others (2014). Evolving paradigms and challenges in estuarine and coastal eutrophication dynamics in a culturally and climatically stressed world. *Estuaries and Coasts*, vol. 37, No. 2, pp. 243–258.

Paramor, O.A.L., and others (2009). *MEFEPO North Sea Atlas*. University of Liverpool.

Pedde, Simona, and others (2017). Modeling sources of nutrients in rivers draining into the Bay of Bengal: a scenario analysis. *Regional Environmental Change*, vol. 17, No. 8, pp. 2495–2506.

Peñuelas, Josep, and others (2013). Human-induced nitrogen-phosphorus imbalances alter natural and managed ecosystems across the globe. *Nature Communications*, vol. 4, art. 2934.

Pliński, Marcin, and others (2007). The potential causes of cyanobacterial blooms in Baltic Sea estuaries. *Oceanological and Hydrobiological Studies*, vol. 36, No. 1, pp. 134–137.

Qian, Wei, and others (2018). Current status of emerging hypoxia in a eutrophic estuary: the lower reach of the Pearl River Estuary, China. *Estuarine, Coastal and Shelf Science*, vol. 205, pp. 58–67.

Rabalais, Nancy N., and others (2007). Hypoxia in the northern Gulf of Mexico: does the science support the plan to reduce, mitigate, and control hypoxia? *Estuaries and Coasts*, vol. 30, No. 5, pp. 753–772.

Rabalais, Nancy N., and others (2009a). Dynamics and distribution of natural and human-caused coastal hypoxia. *Biogeosciences Discussions*, vol. 6, No. 5.

Rabalais, Nancy N., and others (2009b). Global change and eutrophication of coastal waters. *ICES Journal of Marine Science*, vol. 66, No. 7, pp. 1528–1537.

Radach, Günther, and Johannes Pätsch (2007). Variability of continental riverine freshwater and nutrient inputs into the North Sea for the years 1977–2000 and its consequences for the assessment of eutrophication. *Estuaries and Coasts*, vol. 30, No. 1, pp. 66–81.

Riebesell, Ulf, and others (2018). Toxic algal bloom induced by ocean acidification disrupts the pelagic food web. *Nature Climate Change*, vol. 8, No. 12, pp. 1082–1086.

Robson, B.J., and others (2018). *Trichodesmium* timeseries from the Yongala: IMOS National Reference Station, Integrated Marine Observing System, Tasmania.

Sahu, Gouri, and others (2014). Seasonality in the distribution of dinoflagellates with special reference to harmful algal species in tropical coastal environment, Bay of Bengal. *Environmental Monitoring and Assessment*, vol. 186, No. 10, pp. 6627–6644.

Santos, Maria L.S., and others (2008). Nutrient and phytoplankton biomass in the Amazon River shelf waters. *Anais da Academia Brasileira de Ciências*, vol. 80, No. 4, pp. 703–717.

Sattar, Md. Abdus, and others (2014). The increasing impact of food production on nutrient export by rivers to the Bay of Bengal 1970–2050. *Marine Pollution Bulletin*, vol. 80, Nos. 1 and 2, pp. 168–178.

Scheren, P.A., and others (2002). Environmental pollution in the Gulf of Guinea: a regional approach. *Marine Pollution Bulletin,* vol. 44, No. 7, pp. 633–641.

Schmale, Oliver, and others (2016). Dense bottom gravity currents and their impact on pelagic methanotrophy at oxic/anoxic transition zones. *Geophysical Research Letters*, vol. 43, No. 10, pp. 5225–5232.

Seitzinger, S.P., and others (2005). Sources and delivery of carbon, nitrogen, and phosphorus to the coastal zone: an overview of Global Nutrient Export from Watersheds (NEWS) models and their application. *Global Biogeochemical Cycles*, vol. 19, No. 4.

Seitzinger, S.P., and others (2010). Global river nutrient export: a scenario analysis of past and future trends. *Global Biogeochemical Cycles*, vol. 24, No. 4.

Seitzinger, S.P., and others (2010). Global river nutrient export: a scenario analysis of past and future trends. *Global Biogeochemical Cycles*, vol. 24, No. 4. https://doi.org/10.1029/2009GB003587.

Seitzinger, S.P., and E. Mayorga (2016). Chapter 7.3: Nutrients inputs from river systems to coastal waters. In *Large Marine Ecosystems: Status and Trends*, Nairobi: UNEP, pp.179–195.

Selman, Mindy, and others (2010). Eutrophication: sources and drivers of nutrient pollution. *Renewable Resources Journal*, vol. 26, No. 4, pp. 19–26.

Sengupta, Debasis, and others (2006). Surface freshwater from Bay of Bengal runoff and Indonesian throughflow in the tropical Indian Ocean. *Geophysical Research Letters*, vol. 33, No. 22.

Sinha, E., and others (2017). Eutrophication will increase during the 21st century as a result of precipitation changes. *Science*, vol. 357, No. 6349, pp. 405–408.

Smith, Val H., and others (2006). Eutrophication of freshwater and marine ecosystems. *Limnology and Oceanography*, vol. 51, No. 1, part 2, pp. 351–355.

Sonesten, Lars, and others (2018). *Sources and Pathways of Nutrients to the Baltic Sea: HELCOM PLC-6*. Baltic Sea Environment Proceedings 153.

Spencer, Robert G.M., and others (2012). An initial investigation into the organic matter biogeochemistry of the Congo River. *Geochimica et Cosmochimica Acta*, vol. 84, pp. 614–627.

Spokes, L.J., and Jickells (2005). Is the atmosphere really an important source of reactive nitrogen to coastal waters? *Continental Shelf Research*, vol. 25, No. 16, pp. 2022–2035.

Srinivas, Bikkina, and M.M. Sarin (2013). Atmospheric deposition of N, P and Fe to the Northern Indian Ocean: implications to C- and N-fixation. *Science of the Total Environment*, vol. 456, pp. 104–114.

Subramaniam, Ajit, and others (2008). Amazon River enhances diazotrophy and carbon sequestration in the tropical North Atlantic Ocean. *Proceedings of the National Academy of Sciences*, vol. 105, No. 30, pp. 10460–10465.

Sun, Che (2017). Riverine influence on ocean color in the equatorial South China Sea. *Continental Shelf Research*, vol. 143, pp. 151–158.

Szymczycha, B., and others (2019). Chapter 4: The Baltic Sea. *In World Seas: An Environmental Evaluation*, 2nd ed., vol. I: *Europe, the Americas and West Africa*. C. Sheppard, ed. London: Academic Press, pp. 85–111.

Tong, Yindong, and others (2015). Nutrient loads flowing into coastal waters from the main rivers of China (2006–2012). *Scientific Reports*, vol. 5, art. 16678.

Townhill, Bryony L., and others (2018). Harmful algal blooms and climate change: exploring future distribution changes. *ICES Journal of Marine Science*, vol. 75, No. 6, pp. 1882–1893.

Turkoglu, Muhammet (2013). Red tides of the dinoflagellate *Noctiluca scintillans* associated with eutrophication in the Sea of Marmara (the Dardanelles, Turkey). *Oceanologia*, vol. 55, No. 3, pp. 709–732.

United Nations (2017). Chapter 20: Coastal, riverine and atmospheric inputs from land. In *The First Global Integrated Marine Assessment: World Ocean Assessment I*. Cambridge: Cambridge University Press.

United Nations Environment Programme (UNEP) and others (2005). *South China Sea; GIWA Regional Assessment 54*. Kalmar, Sweden: University of Kalmar.

Valigura, Richard A., and others, eds. (2001). *Nitrogen Loading in Coastal Water Bodies: An Atmospheric Perspective*. vol. 57. Washington D.C.: American Geophysical Union.

Van Drecht, G., and others (2009). Global nitrogen and phosphate in urban wastewater for the period 1970 to 2050. *Global Biogeochemical Cycles*, vol. 23, No. 4.

Vermaat, Jan E., and others (2008). Past, present and future nutrient loads of the North Sea: causes and consequences. *Estuarine, Coastal and Shelf Science*, vol. 80, No. 1, pp. 53–59.

Vitousek, Peter M., and others (1997). Human alteration of the global nitrogen cycle: sources and consequences. *Ecological Applications*, vol. 7, No. 3, pp. 737–750.

Wallace, Ryan B., and others (2014). Coastal ocean acidification: the other eutrophication problem. *Estuarine, Coastal and Shelf Science*, vol. 148, pp. 1–13.

Wang, Hongjie, and others (2016). Eutrophication-driven hypoxia in the East China Sea off the Changjiang Estuary. *Environmental Science & Technology*, vol. 50, No. 5, pp. 2255–2263.

Wang, Mengqiu, and others (2019). The great Atlantic Sargassum belt. *Science*, vol. 365, No. 6448, pp. 83–87.

Wang, Sufen and others (2008). Occurrences of harmful algal blooms (HABs) associated with ocean environments in the South China Sea. *Hydrobiologia*, vol. 596, No. 1, pp. 79–93.

Wang, Yun-Feng and others (2018). Recurrent toxic blooms of *Alexandrium* spp. in the East China Sea: potential role of Taiwan warm current in bloom initiation. *Journal of Ecology and Toxicology*, vol. 2, No. 2.

Waycott, Michelle, and others (2009). Accelerating loss of seagrasses across the globe threatens coastal ecosystems. *Proceedings of the National Academy of Sciences*, vol. 106, No. 30, pp. 12377–12381.

Wear, Stephanie L., and R. Vega Thurber (2015). Sewage pollution: mitigation is key for coral reef stewardship. *Annals of the New York Academy of Sciences*, vol. 1355, No. 1, pp. 15–30.

Wei, Qinsheng, and others (2015). Recognition on the forming-vanishing process and underlying mechanisms of the hypoxia off the Yangtze River estuary. *Science China Earth Sciences*, vol. 58, No. 4, pp. 628–648.

World Water Assessment Programme (WWAP) (2017). *The United Nations World Water Development Report 2017: Wastewater – The Untapped Resource*. Paris: UNESCO.

Yan, Tian, and others (2002). A national report on harmful algal blooms in China. *Harmful Algal Blooms in the PICES Region of the North Pacific*, vol. 21. F.J.R. "Max" Taylor and Vera L. Trainer, eds. PICES Scientific Report, No. 23. Sidney, British Columbia, Canada: North Pacific Marine Science Organization (PICES).

Yan, Weijin, and others (2003). How do nitrogen inputs to the Changjiang basin impact the Changjiang River nitrate: a temporal analysis for 1968–1997. *Global Biogeochemical Cycles*, vol. 17, No. 4.

Yaremchuk, M., and others (2005). River discharge into the Bay of Bengal in an inverse ocean model. *Geophysical Research Letters*, vol. 32, No. 16.

Yeung, Laurence Y., and others (2012). Impact of diatom-diazotroph associations on carbon export in the Amazon River plume. *Geophysical Research Letters*, vol. 39, No. 18.

Yin, Kedong, and others (2001). Shift from P to N limitation of phytoplankton growth across the Pearl River estuarine plume during summer. *Marine Ecology Progress Series*, vol. 221, pp. 17–28.

Yin, Kedong, and Paul J. Harrison (2008). Nitrogen over enrichment in subtropical Pearl River estuarine coastal waters: possible causes and consequences. *Continental Shelf Research*, vol. 28, No. 12, pp. 1435–1442.

Zendong, Zita, and others (2016). Algal toxin profiles in Nigerian coastal waters (Gulf of Guinea) using passive sampling and liquid chromatography coupled to mass spectrometry. *Toxicon*, vol. 114, pp. 16–27.

Zhang, Bowen, and others (2017). Global manure nitrogen production and application in cropland during 1860–2014: a 5 arcmin gridded global dataset for Earth system modeling. *Earth System Science Data*, vol. 9, No. 2, p. 667.

Zhou, Chengxu, and others (2015). Interactions between *Karlodinium veneficum* and *Prorocentrum donghaiense* from the East China Sea. *Harmful Algae*, vol. 49, pp. 50–57.

Zhou, Mingjiang, and others (2019). Responses of a coastal phytoplankton community to increased nutrient input from the Changjiang River. In *Studies of the Biogeochemistry of Typical Estuaries and Bays in China*, Zhilian Shen, ed.

Zhu, Zhuo-Yi, and others (2017). Hypoxia off the Changjiang (Yangtze River) estuary and in the adjacent East China Sea: quantitative approaches to estimating the tidal impact and nutrient regeneration. *Marine Pollution Bulletin*, vol. 125, Nos. 1 and 2, pp. 103–114.

Chapter 11
Changes in liquid and atmospheric inputs to the marine environment from land (including through groundwater), ships and offshore installations

Contributors: Ralf Ebinghaus (convener: pharmaceuticals and personal care products), Bjørn Einar Grøsvik (convener: hydrocarbons), Ida-Maja Hassellöv (convener: ships), Colin F. Moffat (convener: persistent organic pollutants), Alan Simcock (convener: radioactivity), Lars Sonesten (convener: atmospheric inputs) and Penny Vlahos (convener: metals); and Eric P. Achterberg, Babajide Alo, Carlos Francisco Andrade, Maria João Bebianno (lead member), Miguel Caetano, Kissao Gnandi, Gi Hoon Hong, Suk Hyun Kim, Rainer Lohmann, Monika Stankiewicz and Juying Wang (co-lead member).

Keynote points

Persistent organic pollutants

- Persistent organic pollutants (POPs) continue to be a global issue, persisting at concentrations likely to cause biological effects.
- POPs are detected in remote locations far from their source of production, which includes the deepest parts of the ocean and the polar regions.
- The number of POPs continues to increase and thus the mixtures to which biota are exposed become more complex, making the determination of the likelihood of individual or population effects ever more challenging.

Metals

- There is a critical need to develop and expand coastal metal time series globally.
- Trends in metal concentrations vary regionally, although most show levelling of dissolved metals and a slight increase in higher trophic organisms.

Radioactivity

- There have been no significant nuclear accidents affecting the oceans since the first *World Ocean Assessment* (United Nations, 2017c).
- The generation of electricity from nuclear power plants continues to increase, with an increase of about 5 per cent globally between 2013 and 2018. Improved technology may be reducing discharges of many radionuclides, but those of tritium are probably increasing in line with electricity generation. Tritium is, however, only weakly radioactive.
- Published information on recent discharges of radioactive substances to the ocean from nuclear power plants and nuclear reprocessing plants is not available except for the North-East Atlantic and its adjacent seas. In that area, discharges to the ocean of radioactive substances from nuclear power plants and nuclear reprocessing plants continue to decline.
- On the basis of the information available, there is no reason to think that adverse impacts of radioactivity on the ocean have become significantly worse since the situation was reported in the first Assessment.

Pharmaceuticals and personal care products

- Hundreds of pharmaceuticals and personal care products (PPCPs) have been detected in the ocean, including in the Arctic and Antarctic.
- Novel analytical techniques have been developed for non-target analysis of PPCPs and their transformation products in the marine environment.
- A "watch list" of PPCPs should be formulated and incorporated into long-term international, national and regional monitoring programmes to serve as a scientific data basis for assessing the status of PPCPs in the ocean.

Shipping

- There is a globally decreasing trend regarding shipping accidents leading to oil spills (over 7 tons), and regionally improved surveillance and action capabilities indicate increased awareness leading to fewer spills.
- There is a general knowledge gap on the nature and impact of liquid input from ships, and the discharge of water from exhaust gas cleaning systems (scrubbers) is an emerging source of metals and polycyclic aromatic hydrocarbons.

Hydrocarbons

- Produced water from oil and gas exploration containing both hydrocarbons and metals is known to affect the marine environment, but knowledge gaps exist on the long-term impact of produced water discharges.
- There is a need for further studies at the community and population levels to advance the current knowledge on single species toxicity data.
- An increased rate of offshore platform decommissioning poses a challenge for the marine environment.

1. Introduction

Chemical production has continued to increase and change since 2003. The potential geographic impact of the chemical industry continued to change from the Atlantic Ocean to the Pacific Ocean, where almost 70 per cent of the industry is expected to operate by 2030, while new products are continually being developed, thus adding to the mixture of chemicals to which biota in the ocean is being exposed.

Different lists of hazardous substances have been identified by international organizations, although there is still no agreed single global list of substances that are of concern. The present chapter contains an assessment of the changes since the first Assessment in water and airborne inputs to the marine environment from land (including groundwater), ships and offshore installations. In addition, the information in the present chapter builds upon the assessment of the list of hazardous substances used in the first Assessment, namely, POPs, metals, hydrocarbons and radioactive substances. It includes new information on rare earth elements, PPCPs and airborne inputs of nitrogen oxides and sulfur oxides that were not included in the first Assessment.

2. Situation recorded in the first *World Ocean Assessment*

Chapter 20 of the first Assessment (United Nations, 2017b) contained the sources, main uses, production and related development, movements and impact of different hazardous substances included in the so-called black or grey lists of substances of concern that had been identified at the national level and by international organizations. Those lists evolved into a list of "priority substances" based on their toxicity, tendency to bioaccumulate and persistence in the ocean. Therefore, the hazardous substances included in the first Assessment were selected based on those for which action had been taken in all or some parts of the world ocean and included: metals (mercury, lead, cadmium), organometallic compounds (tributyltin), POPs (for example, halogenated hydrocarbons), polycyclic aromatic hydrocarbons and radioactive substances. Other substances, including pharmaceutical compounds (both human and veterinary) and cosmetic ingredients (e.g., musk xylene) that have been identified as emerging contaminants of concern are included in the present evaluation. Land-based point sources (wastewater treatment plants or industrial plants discharging into the ocean directly or through rivers), diffuse sources (run-off from land, seepage of groundwater directly to the ocean, accidental land-based or sea-based emissions of discharges) and atmospheric deposition (wet and dry deposition and emissions from sewage and from several industrial processes) that can reach and affect the ocean and their impact in several areas were identified.

The international commitment at the United Nations and the obligation at the regional level to take measures to reduce the impact of recognized emerging substances was also highlighted. From the data available at that time, it was difficult to make meaningful comparisons between areas and set priorities, not least because the data on hazardous substances in water, biota or sediments were expressed in different units. Methodological differences further complicated the picture, and the need to control sampling procedures and analytical methods was highlighted. For that reason, no detailed figures on concentrations of contaminants were included in the first Assessment.

The selected hazardous substances were found in all parts of the ocean, and those from waterborne origins were concentrated in coastal areas, whereas contaminants were transported much further out to the ocean. In the first Assessment, it was not possible to develop a general assessment of the relative impacts of those hazardous substances, but it was possible to identify the slow progress made to reduce their concentrations in some parts of the world ocean. It was also pointed out that there was increasing evidence of the significance of airborne inputs of metals and other hazardous substances to the ocean.

3. Persistent organic pollutants, including run-off from the use of agricultural pesticides

3.1. Introduction

Persistent organic pollutants represent a complex group of (often halogenated) substances that, as their name suggests, endure in the environment. Although the production of such compounds as polychlorinated biphenyls (PCBs) is no longer allowed under the Stockholm Convention on Persistent Organic Pollutants,[1] the Convention allows for equipment containing PCBs to continue to be used until 2025, thereby providing for a possible small, but new, source of PCBs. Movement through trophic levels and environmental recirculation of PCBs mean that they continue to be present in marine systems at concentrations likely to affect marine biota. As other halogenated hydrocarbons have been developed, they have added to the mixture of POPs to which marine biota is exposed. The mixtures, and their respective components, have very different physico-chemical characteristics. The consequence of that is that they exhibit different distributions in environmental compartments, distribution equilibria and analytical requirements.

Once in the environment, POPs recirculate and, through both atmospheric transport and transport by ocean currents, are translocated to locations far from their source. It is for that reason that POPs remain of concern in both the Arctic and Antarctic, as well as throughout the ocean.

3.2. Situation recorded in the first *World Ocean Assessment*

New substances are constantly being developed, and international organizations have prepared lists of chemicals presenting hazardous characteristics, including organohalogens and pesticides and/or biocides. Many of them are covered under the Stockholm Convention but others are not. Knowledge of the extent of the presence of those hazardous substances in the marine environment was patchy. The main observations in the first Assessment were:

1 United Nations, *Treaty Series*, vol. 2256, No. 40214.

(a) POPs are a global issue, however, concentrations in the open ocean were generally low, but detectable, with polybrominated diphenyl ethers (PBDEs) identified in tissues;

(b) Concentrations of POPs were often associated with urbanization and densely populated regions, such as densely populated coastal areas around the Mediterranean and in Africa, South America and the South Pacific, where there was also significant industrial activity;

(c) Some coastal areas were being affected by pesticides;

(d) POPs were found in the Arctic and concentrations, although decreasing, were likely to cause biological effects in some seabirds and polar bears;

(e) Biological effects of POPs were likely to be detected in coastal areas of the North-East Atlantic;

(f) Concentrations of POPs in the North-West Atlantic and the North-East Pacific were quite low, with a decreasing trend in concentration;

(g) Reductions in the concentrations of POPs were observed, but they tended to be localized;

(h) POPs were measurably present in most coastal areas of the East Asian seas;

(i) An area of concern was the exposure of the Great Barrier Reef to pesticides associated with intensive agriculture along the north-eastern coast of Australia;

(j) There was a dominance of comprehensive studies or time series in the northern Atlantic, Arctic, Baltic and northern Mediterranean areas.

3.3. Description of the environmental changes between 2010 and 2020

POPs continue to be a cause for concern in the marine environment, especially in top predators, such as cetaceans, which have been found to have mean blubber PCB concentrations likely to cause population declines and suppress population recovery (Jepson and others, 2016). In addition to the "legacy POPs", new POPs that represent a threat to the marine environment, including pesticides, industrial chemicals and by-products, have been regularly added to the Stockholm Convention (Stockholm Convention, 2018).[2]

Many studies continue to focus on the legacy chemicals, including PCBs and dichlorodiphenyltrichloroethane (DDT) (and its metabolites DDD and DDE). PBDEs were not, however, among the initial 12 POPs covered by the Stockholm Convention and are still grouped with the emerging contaminants, despite having been monitored in marine systems for many years. PBDEs are among the 16 "new" POPs to have been incorporated in the Convention since 2009. They include pentachlorobenzene, polychlorinated naphthalenes, short-chain chlorinated paraffins (SCCPs), perfluorooctane sulfonic acid (PFOS) and its salts, and perfluorooctane sulfonyl fluoride (PFOSF).[3] Chemicals recommended for listing include dicofol and pentadecafluorooctanoic acid (PFOA, perfluorooctanoic acid), its salts and PFOA-related compounds. Chemicals under review by the Persistent Organic Pollutants Review Committee[4] are perfluorohexane sulfonic acid (PFHxS), its salts and PFHxS-related compounds. The inclusion of additional chlorinated molecules, as well as

2 Twelve POPs, namely aldrin, chlordane, dichlorodiphenyltrichloroethane (DDT), dieldrin, endrin, heptachlor, hexachlorobenzene, mirex, toxaphene, polychlorinated biphenyls (PCBs) hexachlorobenzene, and polychlorinated dibenzo-p-dioxins and polychlorinated dibenzofurans (PCDD/PCDF) are recognized as causing adverse effects.

3 See http://chm.pops.int/TheConvention/ThePOPs/TheNewPOPs/tabid/2511/Default.aspx.

4 The Persistent Organic Pollutants Review Committee is a subsidiary body under the Stockholm Convention established for reviewing chemicals proposed for listing in the annexes to the Convention.

both brominated and fluorinated compounds, means that the breadth of contaminants covered by the term "POPs" has greatly increased, resulting in new challenges for environmental analytical laboratories. Short-chain chlorinated paraffins were detected in the Firth of Clyde, but the concentrations were method specific (Hussy and others, 2012), most likely owing to the presence of significant concentrations of medium- and long-chain chlorinated paraffins.

In the recent draft report on progress towards the elimination of PCBs (Stockholm Convention, 2018), it was highlighted that, for many countries, little, if any, relevant quantitative information was available. Extensive analytical work continues to be undertaken in some regions of the world, and it shows evidence of high concentrations of PCBs in some top predators, with the possibility of population consequences (Desforges and others, 2018) or altered adipose function in seal pups (Robinson and others, 2018). Both of those examples come from the North-East Atlantic. Recent data for the Arctic, based on long-term time series of PCBs in marine mammals and fish, show that concentrations are generally decreasing (Carlsson and others, 2018), although the rate of decrease has slowed in recent years (Arctic Monitoring and Assessment Programme (AMAP), 2016; Boitsov and others, 2019). Hexachlorobenzene (HCB) in fish liver decreased less with time compared with PCBs, DDT and its metabolites, trans-nonachlors and PBDEs (Boitsov and others, 2019). Exceptions exist, however, which are associated with changes in diet or a change in environmental processes that affect run-off and re-emissions (AMAP, 2016). For example, significant increasing trends for the concentration of a group of 10 PCBs have been observed in blue mussels from Iceland and juvenile polar bears from the east of Greenland and for two blue mussel time series from Iceland (AMAP, 2016).

There is some evidence that the presence of POPs, such as PCBs, peaked in ocean water in the 1970s and has been declining since (Wagner and others, 2019). In line with declining atmospheric concentrations, the Arctic Ocean has started to export those legacy POPs back into the atmosphere and through currents into the Atlantic Ocean (Ma and others, 2018).

The concentration of PCBs in fishes and shellfish in the North-East Atlantic has decreased, although local problems continue. Of the seven PCBs identified by the International Council for the Exploration of the Sea,[5] only PCB118 is found at a concentration in fishes and shellfish likely to cause biological effects (Commission for the Protection of the Marine Environment of the North-East Atlantic (OSPAR), 2017b). The other six PCBs are generally above background assessment concentrations, although in 4 of the 11 contaminant assessment areas defined by the Commission for the Protection of the Marine Environment of the North-East Atlantic (OSPAR Commission), PCB28 is at the background assessment concentration level. Furthermore, in 9 of 10 contaminant assessment areas where a temporal trend could be determined, the trend is downward. A similar state was described for PBDEs in fish, mussels and oysters in the majority of assessment areas of the Convention for the Protection of the Marine Environment of the North-East Atlantic (OSPAR Convention),[6] with declining concentrations being noted in all but the Skagerrak and Kattegat, where no change in concentration has been observed (OSPAR, 2017b).

PCBs were detected in fish from depths of between 600 and 1,800 m on the European continental slope to the west of Scotland, United Kingdom (Webster and others, 2014). Concentrations of the seven PCBs identified by the International Council for the Exploration of the Sea in the liver of three fish species were highly variable, ranging from 58.7 nanograms

5 PCB28, PCB52, PCB101, PCB118, PCB138, PCB153 and PCB180.

6 United Nations, *Treaty Series*, vol. 2354, No. 42279.

per gram (ng/g) lipid weight in black scabbard to 3,587 ng/g lipid weight in roundnose grenadier. Concentrations were mainly less than 500 ng/g lipid weight (or <1,250 ng/g lipid weight for the sum of 28 PCBs), a value used by some researchers as an indicator of concern. A total of 23 of the 95 fish livers collected between 2009 and 2012, inclusive, had PCB concentrations of more than 500 ng/g lipid weight for the seven PCBs identified by the Council. PCB118 was at a concentration at which biological effects are likely to be observed for all three fish species. Although there were species differences with respect to concentration, there were no temporal trends between 2006 and 2012, nor were there any differences detected with depth. Concentrations of PCBs were also examined in prey species (including lanternfish and Bean's bigscale) and were significantly lower compared with the concentrations found in predators. PBDEs were also detected in the predators, but at much lower concentrations than the PCBs.

Mean concentrations of PCBs in sediments in the Greater North Sea and the Celtic Seas are generally significantly above the congener's background assessment concentration, but below the environmental assessment criteria (OSPAR, 2017b). Sediments in both the northern North Sea and the Irish Sea were found to contain PBDEs, although most of the measured concentrations of PBDEs in sediments were low and often below detection levels. However, the lack of assessment criteria for PBDEs in sediments means that it is not possible to determine the environmental significance of the observed PBDE concentrations (OSPAR, 2017b).

Inputs of hazardous substances to the Baltic Sea are defined, on the basis of the Baltic Sea Impact Index (Baltic Marine Environment Protection Commission (HELCOM), 2018a), as the second most widely distributed pressure (HELCOM, 2018a, 2018b). In terms of POPs, PCBs, dioxins and furans do not appear to be a major driver of the integrated assessment status for the period 2011–2016. Atmospheric deposition of PCBs and polychlorinated dibenzo-p-dioxins and dibenzofurans (PCDD/PCDF) shows a steady decrease owing to the increased efficiency of various combustion and chlorination processes (HELCOM, 2018b). Hexachlorocyclohexane (γ-HCH, lindane), and DDT and its metabolites (DDD, DDE) are no longer considered of significant concern in the Baltic. The improved breeding success in the white-tailed sea eagle is attributed to such reductions (HELCOM, 2018c). However, elevated concentrations of PBDEs in fishes are a major contributor to the current impeded overall status of the Baltic Sea. Similarly, undue inputs of PCBs contaminating the food web in the Lagos Lagoon in Nigeria had been reported from activities on land (Alo and others, 2014).

Even if the deposition of PCDD/PCDF in the Baltic Sea is decreasing, atmospheric deposition has been found to be the major external source, and there is still noticeable elevated deposition in coastal areas of the North-East Atlantic and in the Baltic, the Mediterranean and the Caspian Sea (Wiberg and others, 2013). The atmospheric deposition of PCDD/PCDF and HCB is quite high in coastal areas of the North-East Atlantic and in the Baltic, the Mediterranean and the Caspian Sea, although there has been no intentional global production of HCB for decades (e.g., Wang and others, 2010), and emissions of PCDD/PCDF are supposed to cease in 2018 (Josefsson and Apler, 2019).

It is clear that various POPs continue to be present in the atmosphere (figure I), with a hotspot for PCB153 over Western Europe (figure I.B). High atmospheric concentrations of PCDD/PCDF have also been detected over Europe (figure I.A).

Figure I
Spatial distribution of global scale annual mean air concentrations, simulated for 2016

A. PCDD/PCDF (femtograms toxic equivalent /m³)

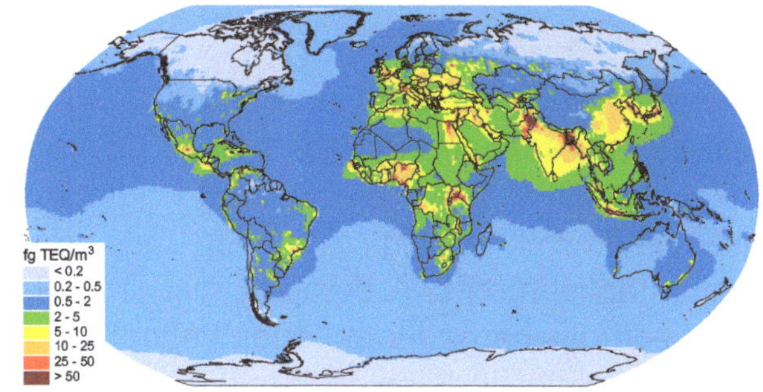

B. PCB153 (picograms/m³)

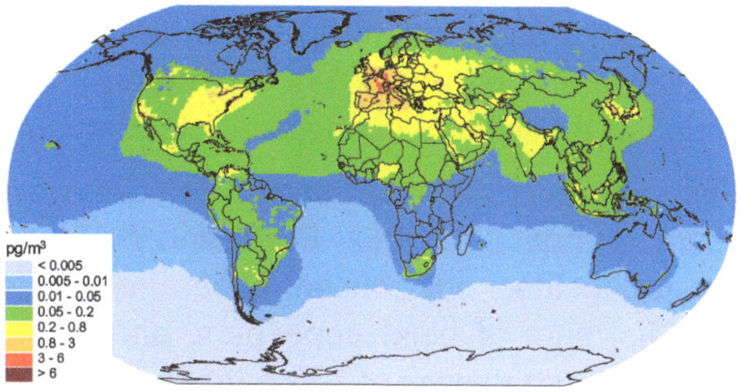

C. HCB (picograms/m³)

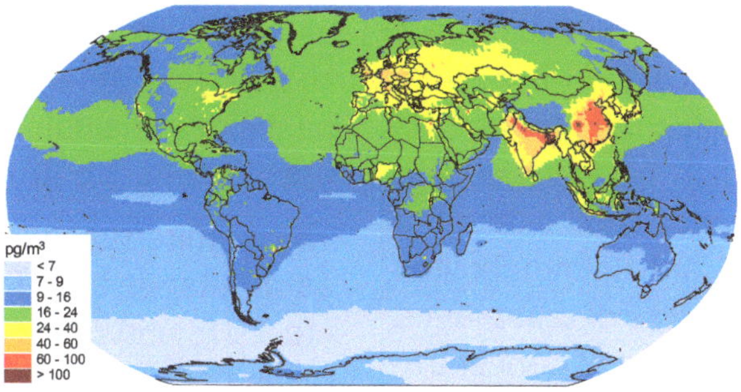

Source: Gusev, A., and others, 2018.

PBDEs have been used as flame retardants for many years and have become widespread across marine systems. As with other POP mixtures (e.g., PCBs), concentrations are based on a small number of the possible congeners. The lipophilic nature of PBDEs means that they, in the same way as PCBs, can be trapped in sediments. A review of PBDE concentrations on a worldwide basis for which the samples were collected prior to 2010 concluded that, in the majority of open ocean sediments, concentrations do not vary that much and are approximately 1 ng/g (Zhang and others, 2016). That contrasts with sediment concentrations close to the source of contamination, which were in excess of 7,000 ng/g. However, PBDEs were detected in amphipods from both the Mariana Trench and the Kermadec Trench, with the deepest sample collected at 10,250 m. The concentration for the sum of seven congeners ranged from 9.33 ng/g lipid weight to 318.71 ng/g lipid weight. PCBs were also detected in those samples, with concentrations, again for the sum of seven congeners, ranging from 62.02 ng/g lipid weight to 1,866.25 ng/g lipid weight (Jamieson and others, 2017). Although there is a scarcity of POPs data from the open ocean, the data available strongly indicate that those chemicals continue to be universally present in marine components far from their source. Concentrations of PBDE47 and PBDE99 in water to the west of Los Angeles, United States, were found to be in excess of 12,500 picograms per litre (pg/l) in 2012. In subsequent water samples, collected from progressively more westerly sites (towards Honolulu, United States), concentrations were very much lower (< 20 pg/l) but PBDEs were evident at all sites (Sun, 2015). Further studies show the presence of organophosphate flame retardants and PBDEs in the atmosphere, sediments, and surface and deeper waters of the Arctic Ocean and the North Atlantic Ocean (Li and others, 2017; Ma and others,

2017; McDonough and others, 2018) At present, atmospheric transport is presumed to be dominating over other modes of long-range transport for organophosphate flame retardants and PBDEs (Sühring and others, 2016; Vorkamp and others, 2019). Therefore, monitoring of those compounds needs to continue.

Fishes from around the South China Sea were found to contain PBDEs, PCBs and DDT and its metabolites, but concentrations in muscle (PBDEs, sum of eight congeners, and PCBs, sum of 19 congeners, < 200 ng/g lipid weight) were at the lower end of the global range and related to feeding habits among the various fish species (Sun and others, 2014). Staying in the South China Sea, more recent data from a range of species (xanthid crab, whiparm octopus, striated cone, Bower's parrotfish, bigeye scad and pike conger) from the Xuande Atoll illustrated that PCBs, PBDEs and DDT and its metabolites occur in the various components of that marine ecosystem; the PCB (17 congeners) concentrations ranged from 8.8 ng/g lipid weight in the whiparm octopus to 117.9 ng/g lipid weight in the pike conger (Sun and others, 2017).

Sediments carried from the Bering Sea through the Bering Strait, and from the Chukchi Sea, the Canada Basin and the Fram Basin to the Iceland Stations (central Arctic Ocean) contained organochlorine pesticides, PCBs and PBDEs. In depths below 500 m, the top 5 cm of sediments contained 286 ± 265 pg/g dry weight (d.w.) of PCBs (47 congeners), which was greater than concentrations from deeper sediments (149 ± 102 pg/g d.w.). There is also some evidence of increasing sediment concentrations of HCB, at least in the Baltic Sea (Josefsson, 2018), while in some environmental compartments in China, there is minimal change in the concentrations of HCB detected in the blubber of finless porpoises from the South China Sea. There were minimal differences between 1990, when the range of concentrations for HCB was 140–230 ng/g lipid weight, and 2000/2001, when the range was 87–250 ng/g lipid weight (Wang and others, 2010). The lack of reducing or even increasing HCB levels might be attributable to the unintentional production of HCB as a by-product in various combustion and chlorination processes (Josefsson and Apler, 2019).

There is no doubt that, in addition to widespread contamination of the marine environment by POPs, there are localized hotspots associated with urban proliferation and industrial establishments. A complex mix of POPs has been discharged into Lagos Lagoon on a daily basis. In addition to direct discharges, sawdust and other inland domestic wastes are ready sources of contaminants. POPs of interest were organochlorine pesticides since, in Nigeria and other developing countries, such pesticides, including DDT and lindane, are still used for pest control and as insecticides.

The Mediterranean has also been described as a hotspot area for POPs (Marsili and others, 2018 and references within table 7.1). The mean concentrations of PCBs in blubber from bottlenose dolphins in the Gulf of Ambrakia in 2013 were low (26,770 ng/g lipid weight; Gonzalvo and others, 2016) relative to the mean concentrations for the same species from the northern Adriatic Sea in 2011 (110,460 ng/g lipid weight; Jepson and others, 2016). However, the mean concentration for the northern Adriatic Sea was about 40,000 ng/g lipid weight higher than the mean obtained for bottlenose dolphins from Scotland, United Kingdom, that had been sampled over the period 2004–2012. Values for the Gulf of Mexico (Texas, United States); Hawaii, United States; and Reunion, France, were 47,700 (Balmer and others, 2015), 11,800 (Bachman and others, 2014) and 5,200 (Dirtu and others, 2016) ng/g lipid weight, respectively, with the animals all sampled around the period 2009–2012. Mean sperm whale blubber PCB concentrations in animals from the Corso-Ligurian Basin of the Mediterranean between 2006 and 2013 was 24,240 ng/g lipid weight and 16,880 g/g lipid weight for males and females, respectively (Marsili and others, 2018, table 7.2 and references within; Pinzone and others, 2015). That was not as high as in the Ligurian Sea and the Gulf of Lion (107,810 ng/g lipid weight; Praca and others, 2011), sampled between 2006 and 2009, but it was very much greater than the means obtained from the waters around the Galapagos Islands (1,320 ng/g lipid weight) and Papua New Guinea (1,140 ng/g lipid weight) in 2000 and 2001, respectively (Godard-Codding and others, 2011).

Although decreasing, the change in concentration of dieldrin in Arctic biota is slow, which is consistent with the air observations, where the change was very small over the period between 1993 and 2016. Chlordane compounds were also shown to be decreasing in concentration in Arctic biota (AMAP, 2016). The story for other "legacy" POPs (e.g., α-HCH, β-HCH and γ-HCH, PCBs) tends to be similar for Arctic biota.

As highlighted earlier in the present chapter, there are a range of fluorinated compounds that are of increasing interest. At coastal sites in the eastern North Sea, concentrations of 3.8 nanograms per litre (ng/l) were observed for PFOA, and 1.8 ng/l for PFOS. The concentrations decreased further, to 0.13 ng/l and 0.09 ng/l for PFOA and PFOS, respectively, towards the open sea (Theobald and others, 2011). Perfluorinated compounds have been found in seabirds in the Baltic (Rubarth and others, 2011), fish caught around Charleston, South Carolina, United States (Fair and others, 2019), a range of seafood in the Republic of Korea (Jeong, and others, 2019) and the marine food web of the Arctic (Butt, and others, 2010), as well as in biota from the Antarctic, which illustrates that those POPs are as ubiquitous in the global environment as the original 12 POPs detailed in the Stockholm Convention.

The presence of per- and polyfluorinated alkyl substances was documented in the Arctic and the global ocean over the past decade (Ahrens and others, 2010; Benskin and others, 2012; Yeung and others, 2017). The phase-out of PFOA and PFOS from production in the United States and Europe will result in declining concentrations in the surface ocean (Zhang and others, 2017), while replacement per- and polyfluorinated alkyl substances are likely to increase. Observed high concentrations of PFOS in the South Atlantic could be attributable to the use of a precursor chemical as a pesticide in Brazil (González-Gaya and others, 2014).

The ultimate challenge remains insofar as human ingenuity has resulted in the production of a wide range of halogenated hydrocarbons that have brought significant benefit to humankind but have been identified in the abiotic and biotic environment at a global scale. The full impact of those compounds on marine biota, especially when there is biomagnification, remains unclear, in particular as monitoring programmes tend to focus on a subset of compounds rather than the full spectrum of fluorinated, chlorinated and brominated compounds that are known to be present in the marine environment and that contribute to the total contaminant loading of individual animals. A detailed study of each subgroup is necessary owing to the toxicity and bioavailability of each compound.

3.4. Economic and social consequences and/or other economic or social changes

Highly toxic compounds, such as γ-HCH and p,p'-DDT, pose potentially unacceptable risks to aquatic organisms. More widely, there are risks to animals at the pinnacle of the food web, including humans. The pesticide residues γ-HCH and p,p'-DDE were shown to be the most persistent of all the POPs assessed and extrapolated for the Gulf of Guinea. In addition, γ-HCH was found to have high potential for long-range transport. The fact that such compounds can exert dioxin-like toxicity on lagoon biota is an indication of likely health risks to biota and to humans (Rose and others, 2017).

As the climate changes on a global basis, marine plants and animals will be subjected to additional stress from increasing temperatures and ocean deoxygenation. A reduction in pH has the potential to cause further stress. The marine plants and animals that are already experiencing some form of stress owing to their contaminant loading may be more vulnerable. Research is required to provide an understanding of the implications of multiple stressors, not only from the perspective of biodiversity, but also in the context of the shellfish and finfish industries, should there be population-level impacts.

POP concentrations alone could cause adverse biological effects that might have an impact beyond the level of the individual marine plant or animal. Localized population effects, or instances in which contaminant concentrations exceed compliance concentrations, have the potential to affect local industries. In 2018, the European Food Safety Authority Panel on Contaminants in the Food Chain reduced the tolerable weekly intake for dioxins and dioxin-like PCBs in food to 2 pg per kg of body weight, a figure that is seven times lower than the previous European Union tolerable intake.[7] That compares with the long-standing World Health Organization tolerable daily intake for dioxin-like PCBs of 1–4 pg toxic equivalent factor per kg of body weight. The United Nations Environment Programme (UNEP), which provides the secretariat for the PCB Elimination Network, has recently published a report (UNEP and United Nations Institute for Training and Research (UNITAR), 2018) detailing the progress made with respect to meeting the elimination deadline of 2028, as set out in the Stockholm Convention. Parties are not currently on track to achieve the 2028 goal. The consequence of that is that there is a need to continue to follow POP concentrations, both to understand the impact of an increasingly complex mixture of anthropogenic chemicals on marine systems and to assess the concentrations in seafood. Fishes and shellfish provide a valuable and nutritious source of protein, which must be safe to eat. That requires that emissions, discharges and losses of POPs are reduced and that concentrations in marine biota decline.

4. Metals

4.1. Introduction

Metals continue to be transported at elevated concentrations around the globe, with the potential to affect human life and the environment even in remote locations. Although metals occur naturally and are released into the environment from natural sources, anthropogenic emissions make important contributions to metal fluxes and even dominate fluxes for a number of metals. Highly toxic metals, such as mercury, cadmium and lead, along with tributyltin, which were assessed in the first Assessment, and rare earth elements are included in the present chapter.

4.2. Situation recorded in the first *World Ocean Assessment*

In the first Assessment, the sources, main uses, production and impact of metals (mercury, cadmium and lead) and tributyltin, an endocrine disruptor compound, were discussed; however, owing to the different analytical methods used and the fact that data were expressed in different units, the comparison was cumbersome.

The main sectors contributing to mercury emissions to the air were found to be combustion plants, mainly burning coal, and artisanal, small-scale gold mining. The share of those sources was estimated by UNEP to be approximately

7 See www.efsa.europa.eu/en/press/news/dioxins-and-related-pcbs-tolerable-intake-level-updated.

50 per cent of total anthropogenic mercury emissions, based on 2010 data (UNEP, 2019).

4.3. Description of the environmental changes between 2010 and 2020

Observations of metal concentrations in the global ocean have improved over the past 10 years, primarily owing to integrated efforts, such as the international GEOTRACES programme. Coastal observations and assessments of trends are lacking for most regions, with the exception of the Baltic Marine Environment Protection Commission, Convention for the Protection of the Marine Environment of the North-East Atlantic and Arctic Monitoring and Assessment Programme regions, and are thus focused on the European coasts and the North Atlantic and Arctic regions. The currently established trends vary across regions and for the different metals. Generally, there appears to be a levelling off in water column concentrations in the cases of lead and cadmium. However, mercury concentrations in fishes and other biota appear to be increasing in the Arctic regions. Efforts to address the lack of time series data in key regions, including the South Atlantic and South Pacific, should be prioritized, in particular in the midst of changing global temperatures and the increased projected mobility of metals. Those efforts are of particular importance in regions where decreasing permafrost will mobilize metals and increase exposure across food chains. Global fish catch[8] shows that all regions yield at least some higher trophic-level species that exceed recommended levels and, therefore, all ocean regions are affected. In summary, cadmium, mercury and lead can still be found at concentrations in biota above background levels, with both temporal and spatial differences. Top predators continue to be under pressure, with metal concentrations as a contributing factor.

According to the World Mineral Statistics archive (Brown and others, 2019), the annual world production of cadmium has been fairly constant at about 21,000–26,000 tons over the past decade, although production was at the higher end from 2014 to 2017. The mine production of lead has decreased almost 10 per cent since the peak production of 5,300,000 tons per year in the period 2013–2014. The production of refined lead has been fairly constant at around 11,000,000 tons during the same period. China alone is responsible for about half of the annual lead production. Annual mercury production doubled from 2010 to 2012 and reached 4,000,000 tons in 2017 (Brown and others, 2019). Also, during that period, the main producer, China, increased its share from about 75 per cent to almost 90 per cent.

Presently, based on 2015 data, UNEP estimates that stationary combustion of coal and artisanal gold mining are responsible for 60 per cent of total anthropogenic atmospheric mercury emissions (UNEP, 2019). However, it is not clear if the difference compared with 2010 is based on improved information or actual changes in emissions from those sectors. Overall, total anthropogenic emissions constitute about 30 per cent of the total mercury emissions to the air, whereas natural processes, such as the evaporation of mercury previously deposited to soils and water, are estimated to constitute 60 per cent, with the final 10 per cent coming from natural emissions from volcanoes (UNEP, 2019).

The global spatial distribution of mercury emissions to the air and atmospheric deposition reveals large hotspots in eastern and southern Asia, Central Africa and South America, as well as Central America and south-eastern North America (figure II). Subcontinental contributions to the global inventory in 2015 are very similar to those of 2010.

8 See www.fao.org/state-of-fisheries-aquaculture.

Figure II
Global distribution of model ensemble median mercury (Hg^0) concentrations in 2015

A. In surface air **B. Total (wet and dry) deposition flux**

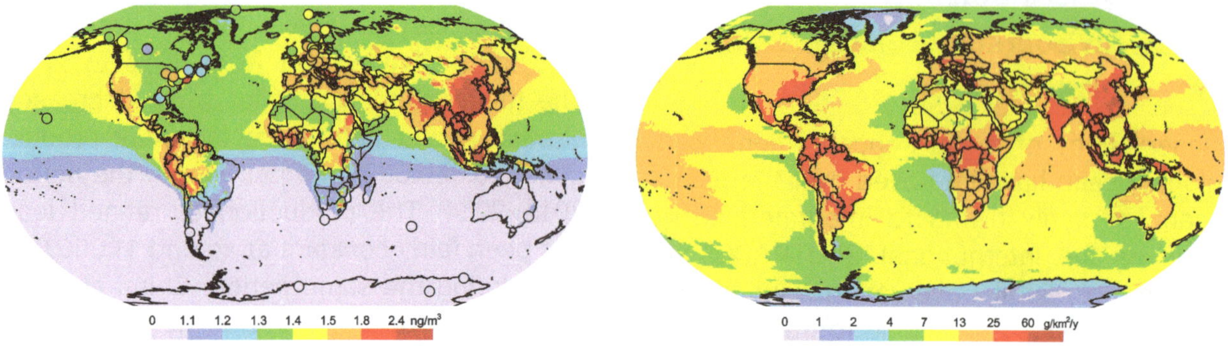

Source: Ilyin, I., and others, 2018.

4.4. Key region-specific changes and consequences

4.4.1. Arctic Ocean

The Arctic is changing rapidly and is the subject of increased research and monitoring efforts. Permafrost thawing is projected to increase the transfer of terrestrial mercury and other metals to Arctic coastal environments (Fisher and others, 2012). Metals do not disappear over time but can be trapped in sediments. However, data on metals in sediments in the Arctic are limited. The mean cadmium concentration in biota from the Barents Sea (north-west coast of Norway) was above the OSPAR background assessment concentrations but significantly below the European Commission maximum level for food (OSPAR, 2017d). The mean concentrations for both mercury and lead were at the background assessment concentration. None of the metals showed upward trends in concentrations in the water column.

A review of mercury in the marine environment of the Canadian Arctic has shown that the understanding of the biogeochemical cycling of that metal has improved but needs further characterization. Total mercury concentrations in sediments from the Hudson Bay are lower (8–58 ng/g d.w.) than in other marine regions of the circumpolar Arctic Ocean (e.g., up to approximately 290 ng/g d.w., Greenland Coast, 2000) (Fisher and others, 2012).

The mercury reservoir in permafrost is poorly quantified, and surface soils in the Arctic likely contain some portion of legacy mercury. Current estimates of riverine mercury export to the coastal Arctic stem from limited data and models and vary widely, ranging from 13 to 80 megagrams per year (Dastoor and Dunford, 2014), while mercury export through coastal erosion is estimated at 15–30 megagrams per year (Soerensen and others, 2016). Riverine mercury concentrations can increase up to sixfold in coastal areas following scenarios projecting up to 30 per cent increased terrestrial run-off (Jonsson and others, 2017). Riverine transport also exports a significant amount of toxic mercury, namely, methylmercury. Present flux estimates cannot close the mercury budget in the Arctic and, thus, hypothesized major mercury processing occurs in coastal zones, with evasion of gaseous mercury species to the atmosphere (Heimbürger and others, 2015).

There remains significant spatial variation in the total mercury concentration in Arctic biota, including with respect to marine mammals and birds. In the latter (thick-billed murres), the total mercury concentration increased in birds breeding at a higher latitude. There was an increase in total mercury concentrations in the eggs of seabirds (various species) over the period 1975–2012. The reasons for the increase remain unclear but are likely to be multifactorial. Greenland sharks have been found to contain high total mercury concentrations in their muscle (1.62 ± 0.52 µg/g wet weight (w.w.)), which is consistent with their high trophic position in the Arctic marine food web.

The fourth Global Mercury Assessment (2018) (UNEP, 2019), a joint venture between UNEP and the Arctic Monitoring and Assessment Programme, highlights the following:

(a) Loss of sea ice in the Arctic owing to climate change allows greater exchange of mercury between the ocean and the atmosphere;

(b) Coastal Arctic sites in Norway have slightly elevated levels of atmospheric mercury compared with those in Greenland, which is associated with direct transportation from continental Europe, especially during winter and spring;

(c) The Arctic is predominantly influenced by the long-range transport of atmospheric mercury;

(d) Dry deposition of mercury may be important in inland Arctic tundra;

(e) The deposition of mercury to the Arctic will not diminish by 2035 under current policies;

(f) The impacts of climate change on marine ecosystems in the Arctic are occurring rapidly, which amplifies its significance for a global understanding of mercury trends;

(g) Arctic birds tend to be at moderate or low risk with respect to mercury;

(h) Some Arctic marine mammals are in a high-risk category as a result of the uptake of methylmercury through their diet, with the mercury concentration in the muscle of pilot whales at the higher end of the concentration spectrum for toothed whales;

(i) Mercury in ringed seals from the North American Arctic has increased;

(j) Changes in mercury concentration in marine mammals and seabirds are a result of changes in feeding patterns and in environmental conditions and of climate change, which means that the reasons for the observed changes in mercury concentration in marine mammals and seabirds are not necessarily identifiable;

(k) The consumption of fish and marine mammals by Arctic people continues to put them at high risk as a result of mercury exposure; however, exposures have dropped over the past two decades.

In summary, cadmium, mercury and lead can still be found in biota at concentrations above background levels, with both temporal and spatial differences. Top predators continue to be under pressure, with heavy metal concentrations as a contributing factor.

4.4.2. North Atlantic Ocean, Baltic Sea, Black Sea, Mediterranean and North Sea

North Atlantic (including the OSPAR maritime area)

The Greater North Sea is the only OSPAR sea area that has sufficient waterborne metal input data to be used in an assessment. Mercury inputs via continental run-off have approximately halved between the period 1990–1995 and the period 2010–2014 (and atmospheric inputs have been reduced by approximately one third). Cadmium inputs through the atmosphere and run-off have both been reduced by two thirds. Advances in analytical methods resulting in improved (lowered) detection limits and higher precision mean that, while there is a downward trend in riverine inputs, the change is likely overestimated. However, it will require longer-term observation to establish the significance of the change (OSPAR, 2017a). Lead inputs through continental run-off have more than halved, while atmospheric lead deposition is less than a third of the level it was at in 1990. Secondary atmospheric pollution from resuspended material and from sources outside the OSPAR maritime area are now the major sources of airborne pollution.

Cooperation is needed beyond the OSPAR area to manage those sources, in addition to the waterborne inputs. Analyses of lead isotopes in the tropical North Atlantic show that up to 30–50 per cent of natural lead detected came from North African mineral dust, which indicates successful global efforts to reduce anthropogenic lead emissions (Bridgestock and others, 2016). Concentrations of dissolved lead in surface waters of the Celtic Sea in the North-East Atlantic decreased fourfold over the past four decades to 8 ng/l (Rusiecka and others, 2018), which is still one or two orders of magnitude higher than background concentrations. Atmospheric lead inputs have been reduced, and benthic dissolved lead fluxes (5.6–8.5 µg lead/(m^2/day) now exceed the atmospheric lead fluxes (0.006–2.5 µg lead/(m^2/day) in the Celtic Sea, indicating the significance of sediments as a contemporary lead source (Rusiecka and others, 2018).

Mean concentrations of mercury, cadmium and lead in marine sediments are either decreasing or show no significant change in the majority of areas assessed. Nevertheless, concentrations in all areas are above natural background levels, and four of the six areas assessed are above levels where adverse ecological effects cannot be ruled out (OSPAR, 2017c). Following bans on tributyltin in antifouling paints, there has been a marked improvement in the reproductive condition of marine snails in the North-East Atlantic over the assessment period 2010–2015. Compared with an assessment in 2010, levels of imposex have markedly improved. In most assessment areas, imposex induced by tributyltin is at or below the level at which harmful effects are expected to occur, and there is also evidence of downward temporal trends in the severity of imposex in all areas assessed. Nevertheless, some areas are still subject to high imposex levels. Although levels of imposex are reducing, imposex remains above background levels in all of the areas assessed (OSPAR, 2017d).

Following the ban on tributyltin, mean concentrations in sediments have measurably reduced in the southern part of the Greater North Sea and are very low or undetectable elsewhere in the North-East Atlantic. Most countries in the area have stopped monitoring organotins in sediments, especially at offshore locations, because concentrations are now often so low that they are below the limit of detection. That means that a reliable assessment of organotins in sediments could be carried out only in the southern North Sea (OSPAR, 2017e).

In most areas assessed in the first Assessment, concentrations of mercury, cadmium and lead in mussels and fishes are higher than the estimated background assessment concentration levels (figure III). Nevertheless, all concentrations are below European Commission limits for foodstuff. Concentrations are decreasing or show no significant change in all areas assessed except for cadmium in a few Greater North Sea and Irish Sea locations (OSPAR, 2017b). European Commission maximum levels for metal concentrations in fish and shellfish are at least five times greater than background concentrations. In all OSPAR regions assessed since 2009, the average metal concentrations are below European Commission maximum levels.

Figure III
Mean concentrations of each heavy metal in fishes and shellfish in each OSPAR contaminants assessment area relative to background assessment concentration (with 95 per cent upper confidence limits)

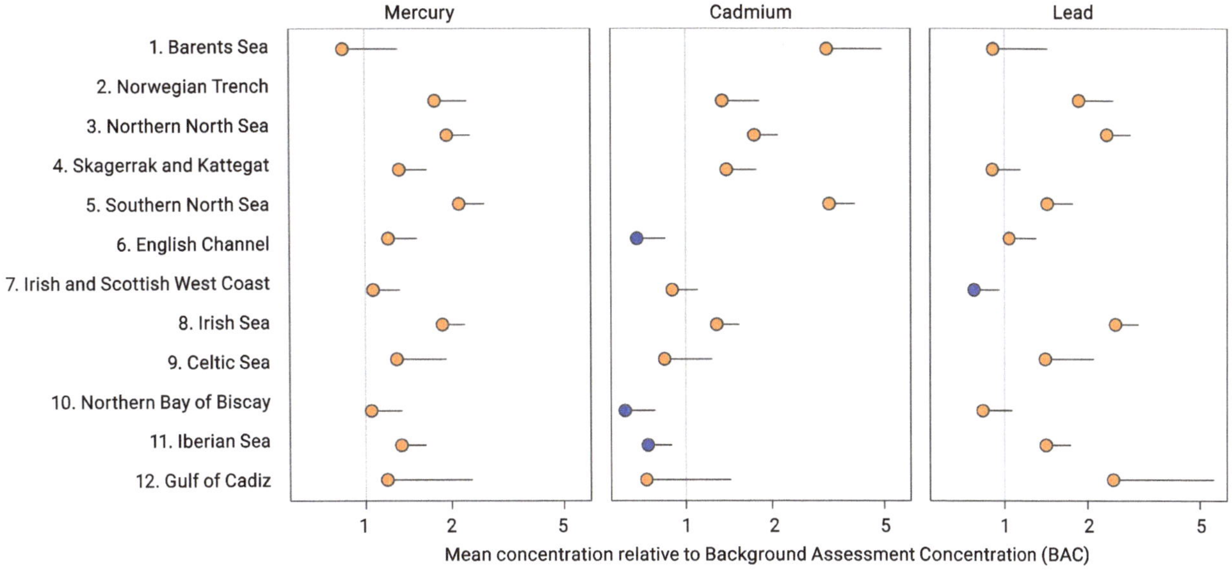

Source: OSPAR, 2017d.

Note: A value of 1 means that the mean concentration equals the background assessment concentration. Blue: the mean concentration is statistically significant below the background assessment concentration and European Commission maximum levels for food (*p* <0.05); orange: the mean concentration is at (if confidence limit crosses 1), or above the background assessment concentration, but significantly below the European Commission maximum levels for food. The European Commission maximum levels are more than five times higher than the background assessment concentration and hence are not shown. The geographical designations in the figure are those used by OSPAR.

Baltic Sea

There are large differences in the estimated total amounts of metals that enter the Baltic Sea every year, and their main route of entry is variable (Baltic Marine Environment Protection Commission (HELCOM), 2018a). It is estimated that the inputs of cadmium, mercury and lead to the Baltic Sea between 2012 and 2014 were in the range of 23–45, 4.8–5.6 and 443–565 tons per year, respectively (HELCOM, 2018a).

Mercury entering the Baltic Sea through atmospheric deposition constitutes about 70 per cent of the total, but levels decreased by 15 per cent from the 1990s to 2014.

Mercury concentrations in fish muscle (the most common species measured are herring and cod in open sea areas and flounder and perch in coastal areas) exceeded the established threshold level (20 µg/kg w.w.) in almost all monitored open sea sub-basins, indicating "not good" environmental status during the period 2011–2016 (HELCOM, 2018a). The threshold was also exceeded in some coastal areas and "good" status was achieved only in the Arkona Basin and in Danish and Swedish areas. There is no general trend for mercury in fish muscle for the investigated time series.

Riverine inputs of cadmium are dominant and make up 79 per cent of cadmium inputs to the Baltic Sea. Inputs through rivers with existing time series show large inter-annual variability that makes it hard to reveal any trend. Atmospheric cadmium deposition decreased by 60 per cent from the 1990s up to 2014.

For cadmium concentrations in seawater, biota (mussels) and sediments assessed by applying the "one-out-all-out" method, "good" status was achieved in only 35 per cent of open sea sub-basins assessed (HELCOM, 2018a) but no significant trends were observed in 89 per cent of the 38 trends evaluated, while there was a decreasing trend in 4 of 33 trends and only 1 showed an increasing trend. Threshold concentrations were 0.2 µg/l in water, 960 µg/kg d.w. (137.3 µg/kg w.w.) in mussel tissues and 2.3 mg/kg d.w. in sediments.

Riverine inputs of lead make up 64 per cent of the total input of lead to the Baltic Sea. The lead inputs of the existing time series show large inter-annual variability that makes it hard to reveal any tendencies. Atmospheric lead deposition has decreased by 80 per cent since the 1990s up to 2014.

Lead concentrations in biota (fishes and mussels) and sediments using the "one-out-all-out" approach indicate that "good" status was achieved only in four open sea sub-basins and in some coastal areas (HELCOM, 2018a). Furthermore, lead generally fails the established threshold value in biota (26 µg/kg w.w. in fish liver, and 1,300 µg/kg d.w. and 185.9 µg/kg w.w. in mussels). No consistent trend was observed.

In most areas, tributyltin is still a problem in water, sediments and biota (HELCOM, 2018b). For sediments, most of the sites failed the threshold level (1.6 µg/kg w.w.) and, even after two to three years of monitoring, no temporal trends could be assessed.

Levels of imposex measured for six or more years were found to be below the threshold value in the southern Kattegat and Skagerrak. In eight other sites, declining effects were observed, which is consistent with the findings in the North Sea area, where 48 per cent of the imposex sites showed decreasing trends.[9]

While the tributyltin situation is improving, levels of tributyltin in sediments and causal effects in marine gastropods indicate that historic pollution continues to affect the Baltic Sea. Uses of organotins other than in antifouling paints and their release from previously contaminated sediments should be investigated to ensure that decreasing trends continue.

Mediterranean

Metal contamination in the Mediterranean is the result of human activities (drivers and pressures) that take place all around the coastal and marine areas of the Mediterranean and cause imbalance to ecosystems from their natural steady-state conditions. Harmful contaminants enter the marine ecosystem through different routes, such as atmospheric deposition or inputs from land- and sea-based sources. Along the Mediterranean coast, small recreational marinas up to major commercial ports have created a number of different pressures in terms of chemical pollution. At present, there are still old threats and new pressures, although the trends and levels of metals have significantly decreased in most affected areas following the implementation of environmental measures (e.g., bans on leaded fuels and antifouling paints, mercury regulations), as observed in the western Mediterranean (UNEP/Mediterranean Action Plan (MAP)/Coordinated Mediterranean Pollution Monitoring and Research Programme (MED POL), 2011a), but Mar Menor is still highly affected by metals.

The latest available data sets of contaminants reported to the Coordinated Mediterranean Pollution Monitoring and Research Programme database continue to indicate lower levels of legacy pollutants and contaminants in the biota (mainly bivalves), despite known hotspots, as did the previous assessment reports (UNEP/MAP, 2009; UNEP/MAP/MED POL, 2011a; UNEP/MAP, 2012a, 2012b) and temporal

9 See https://oap.ospar.org/en/ospar-assessments/intermediate-assessment-2017/pressures-human-activities/contaminants/imposex-gastropods.

trends reports (UNEP/MAP/MED POL, 2011b, 2016b), while also indicating the accumulation and persistence of chemicals in coastal sediments. The monitored chemical contaminants in bivalves (e.g., mussels, clams), fishes and sediments and their assessment against background assessment concentrations, environmental concentrations and effects range-low criteria also point to that conclusion. For biota (bivalves and fishes), the percentage of sites with acceptable environmental conditions (below the European Commission threshold criteria), range from 92 to 100 per cent for cadmium, lead and total mercury. Only 8 per cent of sites assessed for lead in mussels were above environmental concentrations. Therefore, all the assessed sites for biota in the database show acceptable marine environmental conditions, except 8 per cent of them for lead, according those criteria. On the contrary, levels in the coastal sediments above the assessment criteria (greater than effects range-low criteria), that is, non-acceptable environmental conditions, are 4 per cent, 53 per cent and 15 per cent for cadmium, total mercury and lead, respectively. The level of 53 per cent for mercury indicates the need for revised subregional assessment criteria; a mixture of natural and anthropogenic known sources might influence the assessment, especially in the Adriatic Sea, the Aegean Sea and Levantine Basin. In that regard, a revision of the current assessment criteria is under consideration (UNEP/MAP/MED POL, 2016a) and should result in a further refinement of the findings in future assessments.

Based on the values of environmental assessment criteria recommended for indicative purposes by decision IG. 22/7 of the Contracting Parties to the Convention for the Protection of the Marine Environment and the Coastal Region of the Mediterranean at their nineteenth ordinary meeting, held in Athens from 9 to 12 February 2016, overall, assessments reflect non-acceptable environmental conditions, in particular for lead in mussels in some locations and for lead and total mercury (53 per cent of sites are greater than effects range-low criteria) in coastal sediments, although some are known Mediterranean hotspots and natural input areas. To guarantee the control and achievement of targets to maintain acceptable conditions for cadmium and total mercury in biota, there is a need for continuous monitoring and assessment.

4.4.3. South Atlantic Ocean and wider Caribbean

GEOTRACES cruises in the South Atlantic are providing new assessments of dissolved lead inputs. A major flux (0.9 to 1.5 x 10^6 kg/year) to the South Atlantic is from the Indian Ocean through the Agulhas Leakage, which supplies waters with elevated lead concentrations (annual mean concentration 5.8 μg/kg) that are equivalent to those provided by global atmospheric mineral dust deposition (1.6 x 10^9 g/year, assuming 8 per cent of lead released from dust to seawater) (Paul and others, 2015). Currently, dissolved lead concentrations in the South Atlantic remain higher than pre-industrial levels, with 58 per cent of the dissolved lead in those waters originating from anthropogenic sources (Schlosser and others, 2019). It is expected that GEOTRACES data will continue to develop and contribute to the next Assessment.

Significant concentrations of aluminium, mercury and copper were found in sediments and fishes in the Caribbean, mainly in Sea Lots and Point Lisas harbours, Trinidad and Tobago, (Mohammed and others, 2012). Tributyltin also remains a concern in the Caribbean.

Phosphorus mining

Phosphate deposits are found across the world, in both sedimentary and igneous minerals. Currently, China mines the largest volume of phosphate, but Morocco is the largest exporter; however, most of the phosphate extraction and processing takes place far from the sea. Phosphorite mining and processing is a major source of inputs of mercury, cadmium and lead, as well as chromium, nickel, copper, arsenic, thorium and uranium, to coastal waters (Gnandi and others, 2011). For example, in Togo, severe sediment, water and biota impact from metals has been documented, although other mining regions likely exhibit similar impacts. The phosphorite deposits of Togo, extracted since 1960 in the phosphate mines at Hahatoé and Kpogamé in southern Togo, are naturally enriched with metals and rare earth elements (Tanouayi and others, 2016). The ore processing allows the separation of the phosphorus-rich industrial fraction, leading to concentrations greater than 1 mm in seawater once phosphorite tailings are dumped into the ocean. Coastal sediments are highly enriched in trace metals and the calculated enrichment factors relative to the Earth's crust are high. Such high loads of trace metals were also found in biota (fishes and mussels). The ratio of measured trace metal concentrations in biota to threshold limits set by the World Health Organization, herein defined as the relative health factor, was high in fishes, listed here from the highest concentrations to the lowest: selenium, arsenic, silver, nickel, manganese, iron, lead, cadmium, chromium, copper and zinc. Cadmium and aluminium were not accumulated. In mussels, the relative health factor was highest for iron, followed by arsenic, lead, selenium, manganese, nickel, silver, cadmium and copper (Gnandi and others, 2011).

4.4.4. Indian Ocean, Arabian Sea, Bay of Bengal, Red Sea, Gulf of Aden and Persian Gulf

Fish continues to be an important food product and the potential for fishes to be contaminated with a range of metals remains. In the Persian Gulf, most metals regularly exceeded maximum allowable levels in fish muscle, but cadmium and mercury concentrations exceeded the levels only by 10 per cent (Cunningham and others, 2019).

A recent study of a fish (*Lethrinus nebulosus*) off Qatar (Al-Ansari and others, 2017) in the Persian Gulf showed that mercury levels had improved in the region. Total mercury was highest in the liver (602 ± 192 µg/kg w.w.) and lowest in the gonad (71 ± 31 µg/kg w.w.), with muscle falling between the two. The study found an increasing trend compared with the levels detected 20 years earlier, but the levels were more in line with those reported in 2007. Concentration of mercury in sediments was in the range of 8–34.3 µg/kg for total mercury (Hassan and others, 2019).

Stable isotope studies showed that, in the Indian Ocean and the Arabian Sea, lead concentrations have been greatly affected by anthropogenic inputs (Lee and others, 2015). Those data serve as a baseline but will require future sampling to establish trends. In the western Indian Ocean, lead and cadmium levels were below levels of concern, although mercury in higher trophic species (swordfish, wahoo and blue marlin) often surpassed 1 mg/kg w.w. (Bodin and others, 2017). Over 13 per cent of swordfishes sampled in the Indian Ocean had mercury levels that surpassed 1 mg/kg w.w. and, in a global catch for comparison mercury levels, Indian Ocean swordfish levels had the most frequent and highest average mercury concentrations (Esposito and others, 2018).

4.4.5. North Pacific Ocean

Inputs from the Asian continent to the East China Sea and the North Pacific exhibit large episodic and seasonal pulses related to biomass burning and fossil fuel combustion (Qin and others, 2016). Total mercury levels in the deep waters of the North Pacific are elevated relative to surface and intermediate waters, but comparisons with historical data suggest that concentrations have not increased over the past 20 years (Munson and others, 2015).

4.4.6. South Pacific Ocean

Detailed mercury distribution in the South Pacific showed elevated concentrations in the Peruvian upwelling region and significant methylmercury, as high as 20 per cent of total mercury (Bowman and others, 2016). Data in the region are not adequate to ascertain trends since the first Assessment, but values appear stable. The tropical South Pacific is a net source of mercury to the atmosphere but the exchange flux is lower than that of the North Atlantic (Mason and others, 2017).

4.4.7. Southern Ocean

Total mercury concentrations in the Southern Ocean are comparable to those in the South Pacific Ocean and the Atlantic Ocean. However, there are distinct regional features that include net mercury deposition along the ice edge of Antarctic sea ice, mercury enrichment in brine during sea ice formation and methylmercury formation south of the southern polar front (Cossa and others, 2011). Lead concentrations in water (6.2 µg/l) are comparable to those measured in more industrialized regions, such as the Baltic Sea, despite its remote location (Schlosser and others, 2016). Metal data in the region are too sparse to allow any trends since the first Assessment to be detected.

Rare earth elements

Contamination owing to "technology-critical elements" that are widely used in cost-effective low-carbon technologies, such as nuclear, solar, wind and bioenergy, as well as in carbon capture and storage technologies and electricity grids, and in medical products, has been observed since the beginning of the millennium (Bau and Dulski, 1996). Rare earth elements have been considered critical for the development and establishment of high-technology products. As a result of their application, an unavoidable release of such elements into the environment has been observed recently, thus increasing the number of trace elements acting as contaminants in the ocean. One of those elements, gadolinium, is used as a tracer of anthropogenic input in the study of positive anomalies (increased values relative to natural concentrations). The input of rare earth elements to the marine environment has been identified mainly through domestic sewage systems. In the past decade, positive anthropogenic gadolinium anomalies were found in marine waters globally as a result of drainage from densely populated areas, such as the North Sea (North-East Atlantic) (Kulaksiz and Bau, 2007), the San Francisco Bay and adjacent Pacific waters (Hatje and others, 2014), the Indian Ocean (Zhu and others, 2004; Ogata and Terakado, 2006; Akagi and Edanami, 2017) and the South Atlantic Ocean (Pedreira and others, 2018). In addition to gadolinium, other rare earth elements have been detected in raw phosphorite and mine tailings from phosphate mining at Hahatoé and Kpogamé (southern Togo) (Gnandi and others, 2011). However, scarce information exists on the environmental behaviour of those elements and on their impact on biota in marine systems. Although concentrations of anthropogenic gadolinium are rather low in marine waters, potential concerns regarding the effects of continuous exposure to low levels of gadolinium on aquatic organisms and human health have been arising (Hatje and others, 2018). The anthropogenic gadolinium complexes, originally considered to be safe for humans, have been shown to accumulate in humans and aquatic organisms.

4.5. Economic and social consequences and/or other economic or social changes

Metals of concern are non-essential trace elements that transfer through the trophic chain and ultimately bioaccumulate in the upper trophic levels of the oceans. The main social impact is that, despite some decreases in emissions, there are observed increases in concentrations of metals in higher trophic-level fish species, which have a direct impact on ecosystems, leading to apparent changes in food chains and, subsequently, human health risks (see chap. 8B) through ingestion. The risks are of particular concern to indigenous communities that rely on specific food sources. A second impact is the potential decrease in fish stocks and the subsequent hardship for fishers who are constrained to go further from the coast, often with poor equipment, to catch fish. In certain regions, inputs and mining activities lead to regional deterioration that affects tourism and local economies.

5. Radioactive substances

5.1. Introduction

The waters, biota and sediments of the ocean all contain radioactivity. Much of it is from natural sources. Since the 1940s, however, there have been significant inputs from human activities. It is important to distinguish between the occurrence of ionizing radiation, emitted through the decay of radionuclides, and the impact of such radiation on biota, which varies according to the nature of the radiation (in particular, whether the radiation is of α (alpha) or β (beta) particles) and the part of the biota concerned. Studies of radioactive impacts on biota have concentrated on humans, but in the period since 2000, the International Commission on Radiological Protection, the international body of experts that agrees standards of radiation protection, has developed approaches for considering how to protect non-human biota.

5.2. Situation recorded in the first *World Ocean Assessment*

In the first Assessment, the levels of naturally occurring radioactivity in the ocean, ranging from the lowest levels in the South-West Atlantic to the highest levels in the North-East Atlantic, and levels of a typical anthropogenic radionuclide, ranging from the lowest in the Southern Ocean to the highest in, again, the North-East Atlantic, were noted. The most significant anthropogenic input has been from the testing of nuclear weapons, but that is now purely historical. Nuclear reprocessing plants were the second most significant anthropogenic source: such plants existed in 2014 in China, France, India, Japan and the Russian Federation, and further plants were under construction or planned in China, India, Japan and the Russian Federation. The nuclear accidents at Chernobyl and Fukushima resulted in large inputs of radioactive material to the ocean but were of limited concern by the time the first Assessment was written; immediately after the accident at Fukushima, increments to the input were limited. At the end of 2013, there were 434 nuclear power reactors in 30 countries, resulting in radioactive discharges to the ocean in orders of magnitude less than those from weapons testing, reprocessing plants and major accidents, and such discharges tend to decrease over time with improved technology, except for discharges of tritium, which have low radiotoxicity. Also noted was an anthropogenic concentration of naturally occurring radionuclides, in particular from scale cleaned from offshore oil and gas pipelines and phosphogypsum.

5.3. Description of the environmental changes between 2010 and 2020

5.3.1. General

The assessment of global levels of natural and anthropogenic radioactivity in the ocean in the first Assessment was based on studies carried out by the International Atomic Energy Agency (IAEA) in 1995 and 2005 (IAEA, 1995, 2005). No similar studies have since been undertaken, and the picture presented in the first Assessment thus remains the best available. However, IAEA is planning new studies of that kind in the early 2020s (personal communication from IAEA, 5 July 2019).

For radioisotopes with long half-lives, carriage by ocean currents can be significant, unlike terrestrial radioactive contamination. As with airborne transport of radionuclides, ocean currents can transport radioactive substances introduced into the marine environment to areas thousands of km away from the point of introduction. For example, the ratio of plutonium-240 to plutonium-239 in the Kuroshio Current zone in the North-West Pacific provides evidence that those radionuclides are being transported to that zone from the former atomic-bomb and nuclear-bomb Pacific Proving Grounds in the Federated States of Micronesia (Hong and others, 2011; Wu and others, 2019).

Although there have been no global surveys of the level of radioactivity in the ocean, there have been major advances over the past decade in the ability to measure low levels of the long-lived radioisotope iodine-129 (half-life 15.7 million years), a product of nuclear weapons testing and nuclear fuel reprocessing plants. Studies have now revealed its global distribution throughout the ocean and its application as a circulation tracer (He and others, 2013).

In addition, the Scientific Committee on Oceanic Research, under the International Council for Science, has instituted the international GEOTRACES programme to determine the distribution of trace elements and their isotopes throughout the ocean. The programme also includes anthropogenic radionuclides. As part of the programme, intercalibration efforts have demonstrated the ability to identify plutonium-239, plutonium-240 and caesium-137 from relatively small samples (Kenna and others, 2012). Radioisotope data collected through the GEOTRACES programme have also contributed substantially to the understanding of movements of material in the ocean (Malakoff, 2014).

In 2015, the Scientific Committee on Oceanic Research also set up Working Group 146, "Radioactivity in the Ocean, 5 decades later (RiO5)", reverting to the theme of the first Working Group of the Committee in 1959. Working Group 146 has been tasked, among other things, with improving online resources for data on natural and anthropogenic radioisotopes in the ocean within the framework of the IAEA Marine Radioactivity Information System (MARIS) database, which contains measurements of radioactivity data in the marine environment found in seawater, biota, sediment and suspended matter (Scientific Committee on Oceanic Research (SCOR)-WG146, 2020).

5.3.2. Sources of radioactivity in the ocean

Developments with regard to the main sources of radioactive inputs to the ocean since 2014 (the base date for the relevant section of the first Assessment – chapter 20, section 10) have been as follows.

5.3.3. Nuclear weapons testing

The absence of atmospheric tests of nuclear weapons since 1980 has continued, and that source of inputs of radioactivity to the ocean therefore remains purely historical.

5.3.4. Nuclear reprocessing plants

The nuclear reprocessing plants mentioned in the first Assessment as functioning in 2014 (Gansu, China; Cap de la Hague, France; Kālpākkam, Tārāpur and Trombay, India; Tokai, Japan; Mayak, Russian Federation; and Sellafield, United Kingdom) remain in operation, but the Tokai plant is being decommissioned.

The nuclear reprocessing plants at Cap de la Hague and Sellafield continue to represent the dominant source of anthropogenic radioactive inputs to the North-East Atlantic, and they contributed approximately 90 per cent of the total alpha discharges and approximately 80 per cent of the total beta (excluding tritium) discharges over the period 2007–2013. Nevertheless, there had been substantial reductions by 2016 in average discharges from the reprocessing plants in that period over the average levels in the period 1995–2001 – a reduction of about 40 per cent in total alpha discharges and about 85 per cent in total beta discharges (OSPAR, 2017b).

In China, the planning of a further nuclear reprocessing plant in Gansu is continuing. In India, work started on a nuclear reprocessing plant at Kālpākkam in 2017. In Japan, the nuclear reprocessing plant at Rokkasho is expected to reach completion by October 2022 (Japan Nuclear Fuel Limited (JNFL), 2020). In the Russian Federation, a new nuclear reprocessing plant at Zheleznogorsk is expected to be operational as of 2022 (World Nuclear Association (WNA), 2020).

5.3.5. Nuclear power plants

There were 450 commercial nuclear power reactors in 30 countries in operation at the end of 2018 (as compared with 434 in the same 30 countries at the end of 2013). The plants containing them have a total capacity of over 395,000 megawatts (MW). A little over 300,000 MW of that capacity is in countries of the Organization for Economic Cooperation and Development (OECD). About 55 more reactors are under construction. The plants produce over 15 per cent of the world's electricity: the proportion ranges from about 70 per cent of the national supply in France to 2 per cent in the Islamic Republic of Iran (see table 1). That is a global average increase since 2013 of about 5 per cent. Other States that do not have nuclear power plants, such as Denmark and Italy, import substantial amounts of their electricity from neighbouring States that rely substantially on nuclear power (IAEA, 2019a).

For the nuclear power plants in the catchments of the Baltic and North-East Atlantic, the latest assessments show continuing reductions in the discharges of the various radionuclides that are monitored (other than tritium) (HELCOM, 2013; OSPAR, 2017b).

Detailed figures are not available for discharges in other global regions: the IAEA database on discharges of radionuclides to the atmosphere and the aquatic environment (information provided by national authorities on a voluntary basis) has not been updated since 2012, and much of the data in it are substantially older than that. As recorded in the first Assessment, tritium discharges from nuclear power plants are generally related to the level of electricity generation, and there is no accepted abatement technology.

Table 1
Proportion of electricity generated from nuclear power, 2018

State	Percetage of electricity from nuclear power	State	Percentage of electricity from nuclear power	State	Percentage of electricity from nuclear power
France	71.7 (73.3)	Bulgaria	34.7 (30.7)	Pakistan	6.8 (4.4)
Slovakia	55.0 (51.7)	Armenia	25.6 (29.2)	Japan	6.2 (1.7)
Ukraine	53.0 (43.6)	Republic of Korea	23.7 (27.6)	Mexico	5.3 (4.6)
Hungary	50.6 (50.7)	Spain	20.4 (19.7)	South Africa	4.7 (5.7)
Sweden	40.3 (42.7)	United States	19.3 (19.4)	Argentina	4.7 (4.4)
Belgium	39.0 (52.1)	Russian Federation	17.9 (17.5)	China	4.2 (2.1)
Switzerland	37.8 (36.4)	United Kingdom	17.8 (18.3)	Netherlands	3.1 (2.8)
Slovenia	35.9 (33.6)	Romania	17.2 (19.8)	India	3.1 (3.5)
Czechia	34.5 (35.9)	Canada	14.5 (16.0)	Brazil	2.7 (2.8)
Finland	32.5 (33.3)	Germany	11.8 (15.4)	Islamic Republic of Iran	2.1 (1.5)

Source: IAEA, 2019a.
Note: Figures for 2013 are provided in brackets, for comparison.

5.3.6. Non-nuclear sources of radioactive discharges to the ocean

A number of human activities other than nuclear installations result in discharges to the ocean both of naturally occurring radioactive material and of artificial radionuclides produced other than for nuclear energy purposes. The main activities of that kind are offshore hydrocarbon installations and pipelines, nuclear medicine and the production of agricultural fertilizer from phosphate rock. Published data on such discharges are not available except for the North-East Atlantic and its adjacent seas.

The collection of information on discharges of naturally occurring radioactive material and other non-nuclear discharges to the North-East Atlantic and its adjacent seas started in 2005. For the oil and gas industry, there are enough data to set a baseline (2005–2011), but it is not yet possible to identify trends in such discharges to the marine environment (OSPAR, 2017b). Recent studies by the OSPAR Commission conclude that the major source of naturally occurring radioactive material reaching the North-East Atlantic is the offshore oil and gas industry, where produced water (water coming from the reservoir with the oil and gas) and the scale that it deposits in pipelines (which has to be cleared periodically) contain low levels of radionuclides (mainly lead-210, polonium-210, radium-226 and radium-228). The total alpha and total beta discharges from the oil and gas sector are 97 per cent and 10 per cent of the discharges from all sectors, respectively (OSPAR, 2017b, 2018c). Of the total non-nuclear beta discharges, the largest contribution is iodine-131 from the medical subsector. Tritium discharges from the non-nuclear sector are insignificant compared with the nuclear sector (OSPAR, 2018c).

The production of agricultural fertilizers from phosphate rock results in the production of phosphogypsum (which is mainly a compound of calcium, but also contains naturally occurring radioactive material). It has often been discharged as slurry to the sea, but that now seems to have been widely phased out. Such discharge continues in Morocco (where there are new regulations and a review), Tunisia and elsewhere (Hermann and others, 2018; El Kateb and others, 2018). Morocco has, however, set up a system of improved management of phosphogypsum discharges (an investment of $120 million) so that discharges comply with international standards, in particular through marine outfalls equipped with diffusion systems along their ends (communication from the Government of Morocco).

5.3.7. Nuclear incidents

There have been no significant major nuclear incidents since 2011.

In relation to the 2011 incident in Fukushima, Japan, the United Nations Scientific Committee on the Effects of Atomic Radiation has reviewed the scientific work carried out on the maritime transport of radionuclides from the Fukushima Daiichi nuclear power plant since its 2013 report (which had concluded that effects on marine biota would be only local), and concluded that there were no reasons to change its conclusions.[10]

Activities to track the plume of low-level radioactivity in the North Pacific resulting from the Fukushima incident are ongoing (Men and others, 2015; Buesseler and others, 2017), and the plume has now been tracked into North American continental waters (Smith and others, 2015). Most notably, measurements of the long-lived iodine-129 (Hou and others, 2013; Otosaka and others, 2018; Suzuki and others, 2018) have provided critical information about ocean circulation and iodine biogeochemistry in the waters receiving radionuclides from Fukushima. Five years after the Fukushima accident, measurements of caesium-137 found highest activities in brackish groundwater underneath sand

10 See A/72/46, chap. II, sect. B.1.

beaches (Sanial and others, 2017), suggesting a previously undocumented submarine groundwater pathway for the storage and release of radionuclides to the ocean. However, the levels measured by Japan in the marine environment are low and relatively stable (IAEA, 2019b).

A study of Pacific bluefin tuna (*Thunnus orientalis*) caught off the coast of California, United States, around four months after the Fukushima accident showed a tenfold increase in radio-caesium concentrations (derived from Fukushima) compared with pre-Fukushima specimens. However, such radioactivity was approximately thirty times less than that emanating from concentrations of the naturally occurring radionuclide potassium-40 in both pre- and post-Fukushima fish samples (Madigan and others, 2012).

IAEA maintains databases on the dumping of radioactive waste at sea (which occurred between 1947 and 1993) and inputs from accidents and losses at sea. The last compilation of an inventory from the databases was published in 2015 (IAEA, 2015). The only incident that it records since 2010 is the entry into the ocean in 2015 of a Russian satellite with a small nuclear power pack.

5.4. Economic and social consequences and/or other economic or social changes

The pressures to increase the proportion of the world's supply of electricity that is not derived from fossil fuels means that there continues to be significant interest in the generation of electricity from nuclear power plants. As noted above, there has been a 5 per cent increase in the generation of such electricity in the period 2013–2018.

A new development is the construction of the world's first floating nuclear power plant by the Russian Federation. The Akademik Lomonosov completed initial testing in April 2019, to be ready to enter into service in December 2019 in the sea off the Russian port of Pevek, to replace an existing nuclear power plant and a combined heat and power plant (Power Engineering International (PEI), 2019). The Russian nuclear industry has also suggested collaboration with India over the development of floating nuclear power plants (Singh, 2019).

5.5. Regional aspects

There have been no significant studies of the global distribution of natural or anthropogenic radionuclides since the first Assessment, but, as noted above, IAEA is proposing to carry out some new assessments. As recorded in the first Assessment, both naturally occurring radioactivity in the ocean and the nuclear sources of anthropogenic inputs of radioactive material are significantly concentrated in the northern hemisphere. In the southern hemisphere, only Argentina, Brazil and South Africa have nuclear power plants.

5.6. Outlook

As noted in section 5.4, there may well be an increase in the number and scale of nuclear power plants. Linked with such increases is a likely increase in the scale of reprocessing of nuclear fuel. However, experience over recent decades suggests that there will be some offsetting reductions in the levels of radioactivity in discharges from such plants. As recorded in the first Assessment, the estimated highest current levels of committed effective doses to humans of radioactivity from food from the sea are less than a quarter of the IAEA recommended annual limit for the exposure of the general public to ionizing radiation. There is no evidence to suggest any recent significant change. Provided that adequate monitoring is maintained, therefore, such developments are not likely to be of concern.

6. Pharmaceuticals and personal care products

6.1. Introduction

As the population in the coastal regions grows, the size and number of cities grows with it. In particular, as megacities grow near the coast, river mouths and deltas, the anthropogenic pressure on coastal and marine ecosystems is increasing. The urbanization of coasts has direct implications for the input of PPCPs. An increasing number of people will need an increasing amount and number of pharmaceuticals and will apply an increasing amount and number of personal care products. At the same time, food production, such as aquaculture, will be of increased importance and will also lead to the input of pharmaceuticals for veterinary purposes. The picture is even more complicated when looking at demographic change and ageing populations, in particular in the western world. They will lead to an increasing application of certain pharmaceuticals per capita.

PPCPs include all chemicals used for health care, cosmetics and medical purposes. More than 3,000 PPCPs are currently marketed and new compounds enter the market yearly (Arpin-Pont and others, 2016). It is clear that the development of pharmaceuticals and their use in medicine is of considerable value to human society. Nevertheless, their fate is an environmental issue. PPCPs are often analysed together because their input pathways to the environment are similar. PPCPs reach the environment mainly indirectly through wastewater from households or agriculture (livestock farming). They are mostly washed off or excreted unchanged and released directly in the wastewater systems. As processes to remove PPCPs from wastewater are not efficient and most of the compounds are not degraded or are only slowly degraded, the products reach the aquatic environment through the wastewater effluents (Heberer, 2002; Verlicchi and others, 2012; Caldwell, 2016). Some PPCPs, such as ultraviolet filters in sunscreens, can also enter the ocean directly during recreational activities. They are often considered to be "pseudo-persistent" as their degradation is slow in relation to the large quantities that are input or discharged into the environment (Rivera-Utrilla and others, 2013; Bu and others, 2016).

However, it has been shown that several PPCPs may also be degraded to transformation products that could be more toxic (Kallenborn and others, 2018). Until now, most studies on PPCPs have been conducted in relation to the occurrence of PPCPs in influents and effluents of wastewater treatment plants (Fang and others, 2012; Rodil and others, 2012; Tamura and others, 2017), lakes and rivers (Sköld, 2000; Loos and others, 2010; Gothwal and Shashidar, 2015; Molins-Delgado and others, 2017). Many PPCPs have been detected in freshwater systems and, consequently, may end up in marine ecosystems. However, the available data are very limited. Consequently, PPCPs were not discussed or evaluated in the first Assessment.

The broad range of medicinal products available for human or veterinary use that can reach the marine environment may lead to a global environmental problem (Klatte and others, 2017). Owing to the continuous presence of pharmaceuticals in the aquatic environment entering through different entry pathways, they are regarded as a class of pseudo-persistent contaminants (Bu and others, 2016). Pharmaceuticals reach production volumes of up to 100,000 tons per year (Aus der Beek and others, 2016), representing nearly $1.5 trillion in the global pharmaceutical market by 2021, with further expansion predicted. The main drivers for the development are market expansion and demographic changes, including an ageing population (International Federation of Pharmaceutical Manufactures & Associations (IFPMA), 2017; Roig, 2010; Arnold and others, 2014). Pharmaceuticals go through a strict approval procedure in order to ensure effectiveness and patient safety (Taylor, 2016). However, long-term ecotoxicological studies for risk assessment to prevent undesirable environmental effects have only rarely been considered (Sanderson and others, 2003; Fent and others, 2006; Boxall and others, 2012). Since only limited data on the occurrence of a variety of pharmaceuticals in the coastal environment are available, pharmaceuticals with environmental relevance need to be monitored (Gaw and others, 2014; Richardson and Ternes, 2014; Arpin-Pont and others, 2016; Pazdro and others, 2016).

6.2. Situation recorded in the first *World Ocean Assessment*

PPCPs were included in section 2 of chapter 20 on hazardous substances (United Nations, 2017b), alongside classical POPs and heavy metals. They were not considered or evaluated in their own right.

6.3. Description of the environmental changes between 2010 and 2020

To date, there are few studies on the occurrence of PPCPs in marine ecosystems. However, there is increased interest in the occurrence of PPCPs in the ocean, not least because marine ecosystems are assumed to be affected by contamination by PPCPs and increasingly sensitive analytical capabilities are available (Picot-Groz and others, 2014). Available data based on the occurrence of PPCPs in seawater, sediment and marine organisms have recently been collected and published by Bebianno and Gonzalez-Rey (2015) and Arpin-Pont and others (2016). The most frequently investigated and detected compounds were antibiotics (erythromycin, sulfamethoxazole and trimethoprim; see figure IV), anti-epileptics (carbamazepine), caffeine, non-steroidal anti-inflammatories (ibuprofen, ketoprofen) and analgesics (acetaminophen). Among cardiovascular drugs, atenolol and gemfibrozil were most frequently detected or exhibited the highest relative concentrations (Arpin-Pont and others, 2016).

Limited amounts of data were available for personal care products (Bebianno and Gonzalez-Rey, 2015; Arpin-Pont and others, 2016). Available data cover musk fragrances, disinfectants (triclosan) and some ultraviolet filters, the most relevant of which are benzophenone-3 and octocrylene. Triclosan was detected at concentrations of up to 99.3 ng/l in water in Victoria Harbour, China (Wu and others, 2007). Concentrations of benzophenone-3 up to 2,013 ng/l were detected in water at Folly Beach, South Carolina, United States (Bratkovics and Sapozhnikova, 2011). Octocrylene that is used not only in sunscreens but also in food additives enters coastal areas either directly or indirectly through wastewater. Concentrations of octocrylene were up to 1,409 ng/l in water and up to 3,992 ng/g d.w. in mussel tissues (Arpin-Pont and others, 2016; Picot-Groz and others, 2014).

The majority of the measurements of PPCPs in marine waters have been conducted in the North Atlantic Ocean, the North Sea, the Baltic Sea, the Mediterranean and the Asian Pacific Ocean (table 2). In Asia, in particular in China, a number of different PPCPs were measured in seawater, sediments and biota in estuaries and in the Chinese marginal seas (Xu and others, 2013; Zhang and others, 2013b; Na and others, 2013; Nödler and others, 2014; Kallenborn and others, 2018; Kötke and others, 2019). The studies showed that PPCPs are present in all areas of the ocean, with higher levels in areas that are directly affected by anthropogenic activities. Recently, a number of studies have been carried out at coastal sites in the Arctic and Antarctic. In contrast, however, very few PPCP measurements have been taken in the marine environment of the southern hemisphere and very little information exists for PPCP levels in sediments (Arpin-Pont and others, 2016).

Figure IV
Geographical distribution of antibiotics in the world's oceans (ng/L)

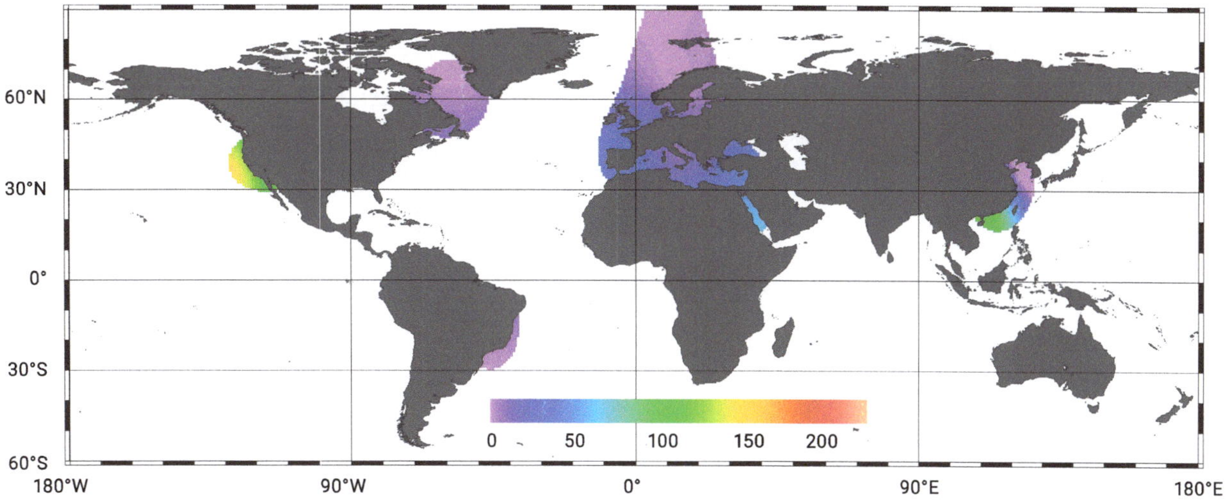

Source: Schlitzer, 2020.

In addition to the occurrence of antibiotics and their transformation products in the marine environment, antibiotic-resistant genes have also been found in bacteria and soil in the Pacific Ocean and the Arctic Ocean (McCann and others, 2019; Hatosy and Martiny, 2015). The occurrence of antibiotic-resistant genes in the marine environment can be linked to the coastal run-off of antibiotic-resistant bacteria from terrestrial sources, anthropogenic antibiotic run-off and selection for resistance in response to antibiotics introduced in the marine environment (Allen and others, 2010; Hatosy and Martiny, 2015).

The availability of data on PPCPs in the Arctic environment has been even more limited than for temperate marine systems. Nevertheless, Kallenborn and others (2018) concluded that the group of compounds are relevant pollutants, even in remote regions, including the Arctic. Based on recent studies, the character of local PPCP sources, such as sewage treatment, in combination with the low-temperature Arctic climate and limited technological standards for waste treatment facilities in Arctic settlements all contribute to extending the environmental stability of the residues compared with conditions found in lower-latitude regions (Kallenborn and others, 2018). More than 100 PPCP-related compounds have been identified in virtually all Arctic environmental matrices, from coastal seawater to high trophic-level biota. Some 22 of a total of 110 compounds were identified in seawater (Kallenborn and others, 2018), with the highest concentrations registered for citalopram (antidepressant), carbamazepine (anti-epileptic) and caffeine (stimulant). Relatively high levels of certain PPCPs in the Arctic environment are not necessarily linked to higher consumption rates but may more likely be explained by higher environmental stability in the low-temperature Arctic climate. That is considered to be of critical relevance when significant amounts of antimicrobial agents are released, thus enhancing the potential for the development of resistance (Gullberg and others, 2011; Kallenborn and others, 2018).

Although PPCPs have been suggested for inclusion in the list of hazardous substances for decision-making on control measures and there is clear evidence that PPCPs are present in all ocean areas and in marine organisms, the data are still insufficient for most PPCPs detected to assess the trend levels in water and the exposure effects on marine organisms.

Table 2
Concentrations of major pharmaceuticals and personal care products measured in coastal waters (ng/l)

Location	Erythromycin	Clarithromycin	Sulfamethoxazole	Sulfamethazine	Roxithromycin	Iomeprol	Iopromide	Diclofenac	Carbamazepine	Bezafibrate	Ibuprofen	Reference
Arctic, Tromsø (Norway)	n.a.	n.a.	n.a.	n.a.	n.a.	n.a.	n.a.	n.a.	n.a.	n.a.	n.a.	Kallenborn and others, 2018
Arctic, Longyearbyen (Norway)	n.a.	n.a.	n.d.	n.a.	n.a.	n.a.	n.a.	1.0–4.0	n.a.	n.a.	0.4–1	Kallenborn and others, 2018
Arctic, Tromsø (Norway)	n.a.	n.a.	n.a.	n.a.	n.a.	n.a.	n.a.	n.a.	n.a.	n.a.	n.d.–0.7	Weigel and others, 2004
Baltic Sea	n.d.–0.14	0.03–0.42	0.74–3.29	n.d.	n.d.–0.48	1.05–34.5	0.42–3.34	n.d.–0.84	1.98–10.6	n.d.–0.64	n.a.	Kötke and others, 2019
North Sea	0.13–0.94	0.4–1.66	1.78–13.0	n.d.	n.d.–2.86	7.66–207	7.27–34.1	n.d.–4.82	4.78–29.7	n.d–2.06	n.a.	Kötke and others, 2019
Himmerfjärden (Sweden)	n.a.	n.a.	n.a.	n.a.	n.a.	n.a.	n.a.	n.a.	4.0–12.0	n.a.	n.a.	Magnér and others, 2010
Baltic Sea	n.d.	14	21	n.a.	n.a.	98	45	9.2	22	n.a.	n.a.	Nödler and others, 2014
Oslofjord	n.a.	n.a.	n.d.	n.a.	n.a.	n.a.	n.a.	n.d.–48.0	n.a.	n.a.	n.d.–52.0	Kallenborn and others, 2018
Aegean Sea	n.d.	16	3.8	n.a.	n.a.	83	109	4.6	2.9	3.5	n.a.	Nödler and others, 2014
Adriatic Sea	5.8	n.d.	3.6	n.a.	n.a.	29	n.a.	n.d.	3.1	n.a.	n.a.	Nödler and others, 2014
Adriatic Sea	n.a.	n.a.	0.02–1.02	n.a.	n.a.	n.a.	n.a.	n.a.	0.11–0.36	0.02–0.14	n.a.	Loos and others, 2013
Mediterranean	9	5	14	n.a.	n.a.	n.a.	n.a.	n.a.	n.d.	n.a.	n.a.	Moreno-González and others, 2015
Santos Bay	n.a.	n.a.	n.a.	n.a.	n.a.	n.a.	n.a.	n.d.	n.a.	n.a.	326.1–2094	Pereira and others, 2016
Red Sea	n.a.	n.a.	63	n.a.	n.a.	n.a.	n.a.	14020	110	n.a.	508	Ali and others, 2017
Bohai Sea and Yellow Sea	0.69	0.07	1	0.01	n.a.	n.a.	n.a.	n.a.	n.a.	n.a.	n.a.	Zhang and others, 2013b
Jiaozhou Bay	4.5	0.58	9.6	0.04	n.a.	n.a.	n.a.	n.a.	n.a.	n.a.	n.a.	Zhang and others, 2013a
Yantai Bay	0.82	0.03	1.4	0.02	n.a.	n.a.	n.a.	n.a.	n.a.	n.a.	n.a.	Zhang and others, 2013a
Southern Yellow Sea	0.5	3	7.7	n.a.	n.a.	n.a.	n.a.	n.a.	n.a.	n.a.	n.a.	Du and others, 2017
East China Sea	n.a.	n.a.	0.5–3.5	n.a.	n.a.	n.a.	n.a.	n.a.	n.a.	n.a.	n.a.	Fisch and others, 2017
Pearl River delta	n.d.–126	n.a.	n.d.–40.6	n.a.	n.d.–12.0	n.a.	n.a.	n.a.	n.a.	n.a.	n.a.	Xu and others, 2013
South China Sea	21	n.a.	11.4	7.03	n.a.	n.a.	n.a.	n.a.	n.a.	n.a.	n.a.	Liang and others, 2013
Sydney (Australia) estuary	n.a.	n.a.	n.a.	n.a.	n.a.	n.a.	3.0–12.5	n.a.	n.d–2.7	n.a.	n.a.	Birch and others, 2015
Antarctic	n.d.	n.d.	n.d.	n.d.	n.d.	n.d.	n.d.	n.d.	n.d.	n.d.	n.d.	Hernández and others, 2019

Abbreviations: n.a., not available; n.d., not detected.

7. Atmospheric pollutants (nitrogen oxides, sulfur oxides)

7.1. Introduction

Combustion is a major source of nitrogen oxides (NO_x) and sulfur oxides (SO_x) in air emissions. Of particular interest for the marine environment are the emissions from shipping that contribute to air pollution. The local and regional environmental issues connected to shipping emissions are, to a large degree, coupled with shipping intensity, but such emissions can also contribute to global pollution.

7.2. Situation recorded in the first *World Ocean Assessment*

In chapter 17 of the first Assessment (United Nations, 2017a), emissions of NO_x and SO_x in areas of heavy traffic were discussed, as were the contribution of those compounds to acid rain and to human health.

7.3. Description of the environmental changes between 2010 and 2020

Total annual NO_x emissions from shipping have been estimated at about 19,000 kilotons (2013–2015), of which about 91 per cent derives from international shipping, with the rest deriving from domestic shipping and fishing vessels (6 per cent and 3 per cent, respectively) (Olmer and others, 2017). Total annual nitrogen emissions from international shipping on the Baltic Sea amount to approximately 80 tons, which is about 5 per cent of the total NO_x emissions in the Baltic Sea countries (Gauss and others, 2018).

The adverse effects of air pollution caused by shipping are an issue of interest to the International Maritime Organization (IMO), which, on the basis of annex VI to the International Convention for the Prevention of Pollution from Ships, 1973, as modified by the Protocol of 1978 relating thereto,[11] endeavours to reduce emissions of, for example, SO_x (and, indirectly, particulate matter) and NO_x from ships through international agreements. There are also IMO-designated emission control areas, in which the restrictions with respect to emissions of SO_x and/or NO_x are more stringent. As at 1 January 2020, the global limit for sulfur content in the fuel oils used by shipping was reduced from 3.5 per cent by mass to 0.5 per cent, while since 2015, the limit has been reduced to 0.1 per cent in the emission control areas. There are four emission control areas: the Baltic Sea area and the North Sea area (presently only for SO_x, but will include NO_x from 2021), the North American area and the United States Caribbean Sea area. The implementation of the North Sea and Baltic Sea SO_x emission control areas led to a significant reduction of sulfur dioxide concentrations in bordering port cities and coastal regions, which benefited the health of coastal citizens (European Union, 2018). The requirement was also set to reduce acidification resulting from SO_x deposition at sea (European Environment Agency (EEA), 2013). It is estimated that the implementation of the Baltic Sea NO_x area will reduce nitrogen deposition at sea by about 40 per cent by 2040 (Karl and others, 2019). Despite those improvements, a modelling study of the longer-term perspective shows that, without additional measures, the current IMO and European Union regulations will cut SO_2 emissions of international shipping up to 2030, but that after that, emissions will grow again. The pattern is even more pronounced for NO_x emissions; it is expected that, after 2030, the emissions from international shipping will exceed those from land-based sources in the European Union, if no further control is applied (International Institute for Applied Systems Analysis (IIASA), 2018).

11 United Nations, *Treaty Series*, vol. 1340, No. 22484.

To meet the stricter sulfur regulations without switching to more expensive fuel of lower sulfur content, an increasing number of ships (7 ships in 2010, 256 in 2015 and more than 4,400 in 2020) have been equipped with an exhaust gas cleaning system, also known as scrubbers, which allows for continued use of heavy fuel oil. In the scrubber, the exhausts are washed in a fine spray of water, and in the simplest and most common form, open-loop scrubbers, the wash water is directly discharged back to the sea. In addition to sulfur oxides, other substances, such as metals and organic pollutants, are also washed out of the exhausts, and there is increasing concern that wide-scale discharge of scrubber wash water may affect the marine environment negatively (Koski and others, 2017; Ytreberg and others, 2019; Teuchies and others, 2020). For that reason, some ports, regions and countries have taken a precautionary approach and prohibited such discharges in their waters (Turner and others, 2017). They include many European ports, such as Rotterdam, the Netherlands, and ports in California, United States, and Singapore and, recently, China and Egypt also proposed such a ban in Chinese waters and the Suez Canal, respectively.

Further efforts to reduce the environmental impact of shipping include the IMO International Code for Ships Operating in Polar Waters,[12] which promotes the identification of hazardous substances on the basis of routine operations and navigational and shipping accident reports. As a consequence of the stricter global sulfur rules and the encouragement to not use heavy fuel oil in the Arctic, more alternative fuel blends have entered the market. More research is needed to determine the potential toxicity of the new fuels.

8. Hydrocarbons from terrestrial sources, ships and offshore installations, including arrangements for response to spills and discharges

8.1. Situation recorded in the first *World Ocean Assessment*

As described in the first Assessment, the impact of hydrocarbons, for example, from oil spills, can affect the marine ecosystem both physically, through the oiling of birds, mammals and beaches, and chemically, through toxic components, such as polycyclic aromatic hydrocarbons. Depending on concentration and exposure, the effects may be acute or chronic (Lindgren and others, 2012). Hydrocarbons enter the marine environment through many pathways. Land-based sources include urban run-off and coastal refineries, while shipping-related sources include operational discharges and accidents and, for offshore oil and gas facilities, operational discharges, accidents and blow-outs. In addition, atmospheric fallout and natural seeps are substantial sources. It was posited in 2003 that the total range from all sources may have reached 470,000 tons to 8.4 million tons a year (National Research Council and Transportation Research Board, 2003), which can be compared with world crude oil production, for example in 1999, which was about 3.5 billion tons. The levels of polycyclic aromatic hydrocarbons are expected to decrease owing to tighter regulations of combustion plants, vehicles and so forth. In 2017, crude oil production increased by almost

[12] International Maritime Organization, document MEPC 68/21/Add.1, annex 10.

25 per cent and was approaching 4.4 billion tons (Global Energy Statistics Yearbook, 2018).

8.2. Description of the environmental changes (between 2010 and 2020)

Based on global models of long-range atmospheric deposition of benzo(a)pyrene (B[a]P), one of the polycyclic aromatic hydrocarbon compounds, it is notably higher in the Adriatic Sea and the Aegean Sea in the Mediterranean, in coastal areas of the North Sea, in the North-East Atlantic and in the south-eastern part of the Baltic Sea, as well as in the northern Caspian Sea (figure V.A). However, on a global scale, the major emissions and deposition of B[a]P are to be found in the eastern and southern parts of Asia, where the atmospheric deposition is a magnitude higher or even more, compared with the levels illustrated in figure V (Gusev and others, 2018). The deposition of B[a]P in the Baltic Sea increased up to 2000, after which time the deposition rate seems to have levelled off.

Other important sources of hydrocarbons entering the ocean are shipping accidents, operational losses and illegal discharges from shipping. The global trend regarding shipping accidents leading to oil spills above 7 tons is, however, decreasing. According to the International Tanker Owners Pollution Federation (2019), the annual average number of spills in the period 2009–2018 was 6.4, compared with 35.8 for the period 1990–1999. The decrease in tanker spills is likely the result of improved safety measures in terms of the phaseout of single-hull tankers, which came into effect in 2003 (IMO, 2019), through an accelerated process following the disastrous accident involving the *Erika* tanker in 1999. The *Erika* and *Prestige* (2003) accidents also marked the starting point for maritime vetting inspections as a possible measure for cargo owners to demand higher safety standards, primarily for chemical and oil tankers (Powers, 2008). The declining trend in the number of tanker spills is even more pronounced taking into account the steady growth – close to an 80 per cent increase from 1990 to 2017 – in loaded crude, petroleum and gas shipping (United Nations Conference on Trade and Development (UNCTAD), 2018).

During the past 10 years, offshore oil production has remained at the same level, about 26 million–27 million barrels per day (International Energy Agency (IEA), 2018a), but its market share has shrunk as global oil production increased to approximately 95 million barrels per day in 2017 (IEA, 2018b). Aside from oil spills, the main impact from the offshore production of oil and gas is associated with the discharge of produced water, with global volume estimated to be up to 39.5 million m^3 per day (Jiménez and others, 2018), and the disposal of drilling waste (Bakke and others, 2013). Although several studies (e.g., Moodley and others, 2018) indicate sublethal effects from produced water on marine species, there is a general understanding that there is a low risk of long-term, widespread impact from produced water and the disposal of drilling waste, but it cannot be verified from published literature (Bakke and others, 2013). However, the observed levels of DNA adducts in the livers of wild-caught fishes from regions with oil production in the North Sea above environmental assessment criteria raise concern regarding the effects of oil compounds on early life stages (Balk and others, 2011; Pampanin and others, 2017). There is a need for further studies on community and population levels to advance the current knowledge based on single-species toxicity data (Camus and others, 2015). The need is also relevant for environmental risk assessment prior to new offshore exploration. If a risk assessment is based on worst-case scenarios that are limited in their holistic validity, it may be biased in the handling of associated uncertainties (Hauge and others, 2014). From the marine environment perspective, an increasing area of concern is the decommissioning of offshore platforms.

The International Energy Agency (2018a) estimated that 2,500–3,000 offshore projects will probably need to be decommissioned, while today the annual average decommissioning rate is 120 platforms per year. The costliest part of platform decommissioning is plugging and abandoning wells. In the North Sea, the removal of all topsides and substructures has been required since 1998, under the OSPAR Convention. However, the "rig to reef" approach has been adopted in the United States and South-East Asia, allowing for parts of the subsea structures to be left and converted to artificial reefs. In the Gulf of Mexico, there are already more than 500 such permanently converted decommissioned rigs (IEA, 2018a).

Figure V.A
Spatial distribution of global scale annual mean modelled air concentrations (ng/m³) of B[a]P for 2016

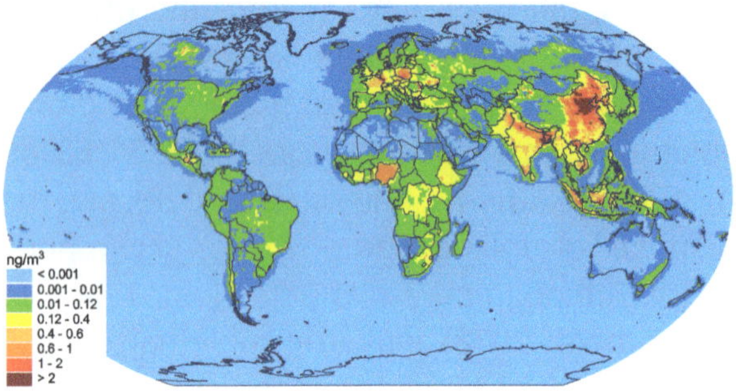

Figure V.B
Spatial distribution of global scale deposition fluxes (g/km²/year) of B[a]P for 2016

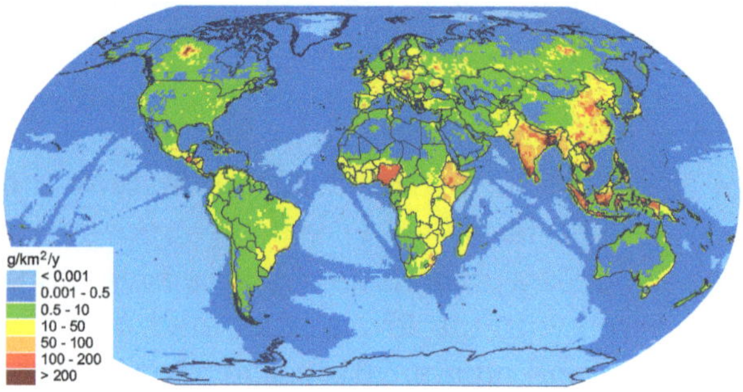

Source: Gusev, A., and others, 2018.

9. Other substances used on, and discharged from, offshore installations

Beyond the environmental impact caused by its hydrocarbon content, produced water also contains elevated concentrations of metals, such as arsenic, cadmium, chromium, copper, lead, mercury, nickel, silver and zinc, some in the range of 10^2–10^5 times higher than background concentrations (Jiménez and others, 2018).[13] Naturally occurring radioactive material originating from geological formations may also be present as dissolved solids in produced water. The most common such compounds are radium-226, radium-228 and barium (Bou-Rabee and others, 2009).[14] To minimize the negative environmental impact of produced water, efforts are being made to: (a) use a small volume of water for the oil extraction process; (b) reuse the water; and (c) dispose of it at sea (Jiménez and others, 2018).

As concluded in the first Assessment, there are still knowledge gaps with respect to an assessment of the large-scale impact of produced water (OSPAR, 2018a). In the North Sea region, the OSPAR Commission has worked hard to achieve phaseout of the most toxic chemicals used in the offshore production industry until 2017. Although the target was not entirely reached, at least the chemicals on the OSPAR List of Chemicals for Priority

13 The potentially negative effects of metals are described in section 4 of the present chapter.

14 The potentially negative effects of naturally occurring radioactive material are described in section 5 of the present chapter.

Action were not used at all on the Norwegian continental shelf from 2014 to 2016. The total use and discharge quantity of chemicals on the Norwegian continental shelf peaked in 2013, and there was a similar trend regarding discharge on the United Kingdom continental shelf (OSPAR, 2018b). The total quantity of chemicals used offshore was 398,158 tons in 2016. A total of 71 per cent (weight) of the used chemicals were on the OSPAR List of Substances Used and Discharged Offshore which are Considered to Pose Little or No Risk to the Environment, 28 per cent (weight) were other non-substitution chemicals and 1 per cent comprised substitution chemicals (i.e., chemicals that contain one or more substances that are candidates for substitution). In addition to the work done on phasing out toxic chemicals, new technologies, for example, advanced oxidation processes for the remediation of produced water, have also been proposed (Jiménez and others, 2018).

10. Relationship to the Sustainable Development Goals

The atmospheric deposition of various pollutants on water (or land) is directly related to Goal 14 but is also relevant to most, if not all, Sustainable Development Goals,[15] for example, Goals 2 and 6 or Goals that might have an impact on air emissions, including Goals 1 and 8, as one of the prerequisites for life on earth is the supply of clean and healthy water.

The presence of POPs at concentrations likely to cause deleterious effects means that it is unlikely that Sustainable Development Goal target 14.1 will have been achieved by 2025. For many of the legacy POPs, such as PCBs, emissions, discharges and losses are very low; the issue is the re-emergence from sediments owing to the resistance of POPs to biodegradation. There also remains a clear need to increase scientific knowledge (Sustainable Development Goal target 14.a and other Goals) around the cumulative impacts of the growing mixture of chemicals to which marine biota are being exposed.

Sustainable Development Goal target 3.9 will be hard to achieve with respect to POPs, metals, PPCPs and hydrocarbons, specifically with respect to achieving a substantial reduction in water pollution. The impacts of POPs, metals, PPCPs and hydrocarbons on human health have not been evaluated in the present chapter but it has been recognized that marine mammals are being affected by POPs, with concentrations for some POPs and metals decreasing only slowly and with increasing concentrations affecting top predators.

The achievement of Sustainable Development Goal target 2.1 will require more concerted monitoring programmes covering the edible portion of marine plants and animals to ensure the quality of marine food sources.

The available information on the impact of ionizing radiation from anthropogenic sources on the marine environment suggests that it probably does not pose a significant problem for the achievement of Sustainable Development Goal target 14.1. However, there are significant gaps in the information available on discharges of radionuclides in much of the world.

Relevant PPCPs should be included in already established long-term international, national and regional monitoring programmes to serve as a scientific basis for region-specific "watch lists" for PPCPs, in particular in coastal waters. There should be no segregation of environmental regulations and legislation between terrestrial and marine ecosystems at the national and international levels, with coastal areas treated as a transition zone in

15 See General Assembly resolution 70/1.

the "catchment-to-sea continuum" and as the link between Goals 6 and 14.

As the impacts of increased anthropogenically produced carbon dioxide become more significant in the ocean, it becomes more evident that the marine biota is being exposed to yet another stressor – ocean acidification. The pH decreases (see chap. 9), along with the increase in temperature and decrease in dissolved oxygen, pose a risk that biota already made vulnerable by their contaminant loading will succumb to the multiple stressors (see also chap. 25) they are experiencing. It would be desirable to reduce the presence of multiple stressors in the ocean along with climate action.

11. Key remaining knowledge gaps

In the first Assessment, the need to work through a number of different organizations was highlighted as limiting the possibility of making clear comparisons between the environmental quality of different ocean areas because of the use of different measuring techniques and the very different ranges of the varieties of chemicals being observed. That situation remains.

Information on the atmospheric deposition of various pollutants is heavily dependent on the modelling approaches used to increase the spatial coverage. To be able to model the deposition, there is a significant need for high quality data on emissions and deposition. The data need to be collected and used in regional and/or global modelling to facilitate the production of high-resolution spatial and temporal deposition estimates. However, the availability of that kind of fundamental data is limited, especially for some ocean areas, which is quite evident from the present Assessment, for which there is a lack of information for a large part of the ocean.

Changes in industrial production will result in changes in compartmental patterns as well as the point sources and substance mixtures. With the broadening of the Stockholm Convention, there is a need for information on the concentrations of the compounds detailed in the Convention that are found in the environment to permit the consideration of cumulative impacts (see chap. 25) and the effectiveness of the processes aimed at eliminating the emissions and use of those compounds.

Critically, the biological effects and cumulative impacts of the chemicals detailed in the Stockholm Convention require considerable research to allow appropriate status assessments to be prepared, especially in cases where changes are attributable to the impact of increased atmospheric greenhouse gas concentrations (e.g., ocean warming, ocean deoxygenation, ocean acidification and changing rates of respiration).

Current efforts and ongoing time series under the GEOTRACES programme will improve both global and regional resolution. However, significantly higher resolution is needed to improve estimates of trends with respect to trace elements and their isotopes. Time series in the South Atlantic and across the South Pacific are currently lacking for hazardous substances, as are data for the Southern Ocean. The extent of transboundary marine pollution is yet to be properly investigated. The mapping of contamination of coastal waters and sediments requires a more integrated effort, together with more globally targeted studies of biota such that effects can be determined on a larger (oceanic) basis.

It is necessary to coordinate spatial and temporal sampling for metals such that the data reflect a global strategy. That will require integrated efforts, possibly including the UNEP regional seas conventions and action plans, with both coastal and open ocean sampling. As sampling resolution is optimized, such that changes in concentration can be detected with a known confidence, quality control and quality assurance guidelines, including intercalibrations, will be required.

Very limited detailed information is published of the levels of discharges of radioactive substances to the marine environment, outside the North-East Atlantic and its adjacent seas. It is known that substantial monitoring is carried out. There is, therefore, a case for restarting and extending the IAEA database on discharges of radionuclides to the atmosphere and the aquatic environment as a means to provide much wider publication of the information.

Likewise, the intention of IAEA to repeat the studies carried out by the Agency in 1995 and 2005 (IAEA, 1995, 2005) on the levels of natural and anthropogenic radioactivity in fishes and seawater in the different major fishing areas is welcomed. It would be an appropriate contribution to the United Nations Decade of Ocean Science for Sustainable Development (2021–2030).

A review of the studies of the impact of ionizing radiation on crustaceans concludes that there is poor coverage of data, in particular in the field, on the subject and suggests that similar problems may exist with other phyla (Fuller and others, 2019), which implies that there is a need for further research on the subject.

The quite large number of PPCPs identified in marine ecosystems is primarily indicative of the capability of today's analytical method for the identification and quantification of those substances and their metabolites. It does not necessarily reflect the full range of PPCPs present in the marine environment. The ultratrace concentrations of PPCPs in seawater, sediments and biota are still a significant challenge for existing analytical methods. However, technological developments and novel applications will further decrease limits of quantification and, in addition, will lead to the identification of new and presently unidentified PPCPs (Kallenborn and others, 2018).

Both active and passive sampling strategies and analytical methodologies for the analysis of PPCPs and their metabolites in the marine environment need to be harmonized. That will ensure common data quality and allow more effective data comparison between laboratories and geographic regions (Arpin-Pont and others, 2016).

Because PPCPs are mostly excreted unchanged or as metabolites, it is not appropriate to target only the parent compounds; the major transformation products must be included in both the analytical procedures and the risk assessments (Rivera-Utrilla and others, 2013).

To date, there is no comprehensive data set available covering the worldwide occurrence of PPCPs in the coastal regions and the open ocean, which means that it has not been possible to conduct any potential assessment of the impacts of PPCPs on marine organisms. It would be desirable to create a database to support the risk assessment and modelling and provide information for the international management of PPCPs. Owing to the lack of adequate data, especially for the different trophic levels in marine webs, a safety factor of 10,000 needs to be applied, which results in a high uncertainty of the risk characterization of the compounds (European Medicines Agency (EMEA), 2018).

To further evaluate the ecotoxicity of the investigated PPCPs and to estimate whether the observed concentrations may have an effect on marine ecosystems, it will be important to improve the data on marine test organisms. Such efforts should focus on the impacts of chronic toxicity characterized by low-dose exposure in long-term studies, which should include the behaviour of mixtures of chemicals (Deruytter and others, 2017).

12. Key remaining capacity-building gaps

The complex nature of the mixtures that comprise POPs and PPCPs, coupled with the fact that, even at very low concentrations, those compounds can be toxic, means that there is a need to develop the necessary analytical capabilities on a global scale.

Sampling and subsequent analyses in the open ocean and in coastal and shelf seas need to be undertaken in a systematic, quality-assured manner on a global basis, covering both the original and new POPs, as detailed in the Stockholm Convention, as well as metals, PPCPs, radioactive substances, NO_x, SO_x and hydrocarbons. Although significant analytical challenges are expected, such an approach will permit precise spatial and temporal assessments to be made, which will ultimately inform better management decisions with respect to the utilization of POPs, PPCPs and other materials that may be deleterious to the marine environment.

POPs continue to accumulate in the polar regions and in top predators but neither present straightforward sampling opportunities. Therefore, greater effort must be put into more harmonized monitoring plans such that the collection of samples for the determination of POP concentrations is integral to as many programmes as practicable, especially in regions known to be affected by POPs. Furthermore, there needs to be a greater awareness and understanding of the movement of POPs through food webs. The development of trophic magnification factors should allow concentrations across food webs to be modelled, providing an indication of the probable concentrations of POPs in species that are difficult to sample.

Re-emergence is a significant source of POPs that is contributing to the sustained elevated concentrations of, for example, PCBs. However, establishing a clear understanding of the routes and pathways through which contaminants enter seas will enable better evaluation and targeting of measures, provide information on issues of potential re-emergence and potentially offer the possibility of predicting recovery times. In addition, a major consideration for future assessments should be the determination of the environmental realities attributable to multiple mixed effects, in particular, the impact on the environment not just of single substances or substance groups but the complex and potentially magnifying effects of numerous contemporary hazardous substances.

Over the many decades of analyses, the instrumentation has improved, as have sampling methodology and sample preservation. However, in determining temporal trends, it is often the determined concentration that is given the most attention, with less consideration given to the relevant limit of detection of the instrument for that sample. In that context, there is a need to consider the more technical and specific aspects of the analysis (Mangano and others, 2017). In addition, to support future assessments, it will be necessary to review and harmonize the threshold values utilized in the individual indicators, to ensure their relevance and application. Furthermore, gaining a comprehensive overview of novel sources of contaminants, in particular those emerging from offshore activities, such as wind farms, will also be beneficial.

There is a need to develop laboratory facilities that can improve knowledge of the toxicity of POPs and PPCPs in marine systems. Furthermore, it is essential that an infrastructure be put in place that will permit assessments of the contribution of POPs and PPCPs to the wider cumulative impacts of the multiple stressors to which marine species and habitats are being exposed, especially a changing climate and ocean acidification.

As with other monitoring of hazardous substances, there are major gaps in the capacities of most developing countries to monitor concentrations of POPs, metals, PPCPs and radionuclides in the marine environment.

The Minamata Convention on Mercury[16] entered into force on 16 August 2017 and includes articles to support parties thereto, including with respect to capacity-building and technical assistance, as well as health aspects, public awareness, education and monitoring. There are 113 parties to the Convention (as of July 2020).

Moreover, efforts should be made to reduce all the sources of inputs of those hazardous substances to the ocean.

References

Ahrens, Lutz and others (2010). Distribution of polyfluoroalkyl compounds in water, suspended particulate matter and sediment from Tokyo Bay, Japan. *Chemosphere*, vol. 79, No. 3, pp. 266–272.

Akagi, Tasuku, and Keisuke Edanami (2017). Sources of rare earth elements in shells and soft-tissues of bivalves from Tokyo Bay. *Marine Chemistry*, vol. 194, pp. 55–62.

Al-Ansari, Ebrahim M.A.S., and others (2017). Mercury accumulation in Lethrinus nebulosus from the marine waters of the Qatar EEZ. *Marine Pollution Bulletin*, vol. 121, Nos. 1–2, pp. 143–153.

Ali, Aasim M., and others (2017). Occurrence of pharmaceuticals and personal care products in effluent-dominated Saudi Arabian coastal waters of the Red Sea. *Chemosphere*, vol. 175, pp. 505–513.

Allen, Heather K., and others (2010). Call of the wild: antibiotic resistance genes in natural environments. *Nature Reviews Microbiology*, vol. 8, No. 4, pp. 251–259.

Arctic Monitoring and Assessment Programme (AMAP) (2015). Temporal trends in Persistent Organic Pollutants in the Artic. ISBN – 978-82-7971-100-1.

________ (2016). www.amap.no.

Alo, B., others (2014) Studies and transactions on pollution assessment of the Lagos Lagoon system, Nigeria. In *The Land/Ocean Interactions in the Coastal Zone of West and Central Africa*. S. Diop and others, eds. Springer International Publishing, Switzerland, pp. 65–76. https://doi.org/10.1007/978-3-319-06388-1.

Arnold, Kathryn E., and others (2014). *Medicating the Environment: Assessing Risks of Pharmaceuticals to Wildlife and Ecosystems*. The Royal Society.

Arpin-Pont, Lauren, and others (2016). Occurrence of PPCPs in the marine environment: a review. *Environmental Science and Pollution Research*, vol. 23, No. 6, pp. 4978–4991.

Aus der Beek, Tim, and others (2016). *Pharmaceuticals in the Environment: Global Occurrence and Potential Cooperative Action under the Strategic Approach to International Chemicals Management (SAICM)*. Dessau-Roßlau: German Environment Agency.

Bachman, Melannie J., and others (2014). Persistent organic pollutant concentrations in blubber of 16 species of cetaceans stranded in the Pacific Islands from 1997 through 2011. *Science of the Total Environment*, vol. 488, pp. 115–123.

Bakke, Torgeir, and others (2013). Environmental impacts of produced water and drilling waste discharges from the Norwegian offshore petroleum industry. *Marine Environmental Research*, vol. 92, pp. 154–169.

Balk, Lennart, and others (2011). Biomarkers in natural fish populations indicate adverse biological effects of offshore oil production. *PLoS One*, vol. 6, No. 5.

16 UNEP(DTIE)/Hg/CONF/4, annex II.

Balmer, Brian C. and others (2015). Persistent organic pollutants (POPs) in blubber of common bottlenose dolphins (*Tursiops truncatus*) along the northern Gulf of Mexico coast, USA. *Science of the Total Environment* vol. 527, pp. 306–312.

Baltic Marine Environment Protection Commission (HELCOM) (2013). *Thematic Assessment of Long-Term Changes in Radioactivity in the Baltic Sea, 2007-2010*. Baltic Sea Environmental Proceedings 135. Helsinki, Finland: HELCOM. http://stateofthebalticsea.helcom.fi/wp-content/uploads/2018/07/HELCOM_Thematic-assessment-of-hazardous-substances-2011-2016_pre-publication.pdf.

________ (2018a). *HELCOM Thematic Assessment of Hazardous Substances 2011-2016: Supplementary Report to the 'State of the Baltic Sea' Report*. http://stateofthebalticsea.helcom.fi/wp-content/uploads/2018/07/HELCOM_Thematic-assessment-of-hazardous-substances-2011-2016_pre-publication.pdf.

________ (2018b). *Inputs of Hazardous Substances to the Baltic Sea*. Baltic Sea Environment Proceedings 161. www.helcom.fi/Lists/Publications/BSEP162.pdf.

________ (2018c). *Metals HELCOM Core Indicator 2018. HELCOM Core Indicator Report. ISSN: 2343-2543,*. HELCOM Core Indicator Report. www.helcom.fi/baltic-sea-trends/indicators/metals.

Bau, Michael, and Peter Dulski (1996). Anthropogenic origin of positive gadolinium anomalies in river waters. *Earth and Planetary Science Letters*, vol. 143, Nos. 1–4, pp. 245–255.

Bebianno, M.J., and M. Gonzalez-Rey (2015). Ecotoxicological risk of personal care products and pharmaceuticals. In *Aquatic Ecotoxicology*, pp. 383–416. Elsevier.

Benskin, Jonathan P., and others (2012). Perfluoroalkyl acids in the Atlantic and Canadian Arctic oceans. *Environmental Science & Technology*, vol. 46, No. 11, pp. 5815–5823.

Birch, G.F., and others (2015). Emerging contaminants (pharmaceuticals, personal care products, a food additive and pesticides) in waters of Sydney estuary, Australia. *Marine Pollution Bulletin*, vol. 97, Nos. 1–2, pp. 56–66.

Bodin, Nathalie, and others (2017). Trace elements in oceanic pelagic communities in the western Indian ocean. *Chemosphere*, vol. 174, pp. 354–362.

Boitsov, Stepan, and others (2019). Levels and temporal trends of persistent organic pollutants (POPs) in Atlantic cod (*Gadus morhua*) and haddock (*Melanogrammus aeglefinus*) from the southern Barents Sea. *Environmental Research*, vol. 172, pp. 89–97.

Bou-Rabee, F., and others (2009). Technologically enhanced naturally occurring radioactive materials in the oil industry (TENORM). A review. *Nukleonika*, vol. 54, No. 1, pp. 3–9.

Bowman, Katlin L., and others (2016). Distribution of mercury species across a zonal section of the Eastern Tropical South Pacific. *Marine Chemistry*, vol. 186, pp. 156–166.

Boxall, Alistair B.A., and others (2012). Pharmaceuticals and personal care products in the environment: what are the big questions? *Environmental Health Perspectives*, vol. 120, No. 9, pp. 1221–1229.

Bratkovics, Stephanie, and Yelena Sapozhnikova (2011). Determination of seven commonly used organic UV filters in fresh and saline waters by liquid chromatography-tandem mass spectrometry. *Analytical Methods*, vol. 3, pp. 2943–2950.

Bridgestock, Luke, and others (2016). Return of naturally sourced Pb to Atlantic surface waters. *Nature Communications*, vol. 7, art. 12921.

Brown, T.J., and others (2019). *World Mineral Production 2013–17*. British Geological Survey.

Bu, Qingwei, and others (2016). Assessing the persistence of pharmaceuticals in the aquatic environment: challenges and needs. *Emerging Contaminants*, vol. 2, No. 3, pp. 145–147.

Buesseler, Ken, and others (2017). Fukushima Daiichi-derived radionuclides in the ocean: transport, fate, and impacts. *Annual Review of Marine Science*, vol. 9, pp. 173–203.

Butt, Craig M., and others (2010). Levels and trends of poly-and perfluorinated compounds in the Arctic environment. *Science of the Total Environment*, vol. 408, No. 15, pp. 2936–2965.

Caldwell, Daniel J. (2016). Sources of pharmaceutical residues in the environment and their control. In *Pharmaceuticals in the Environment. Issues in Environmental Science and Technology*, No. 41, R.E. Hester and R.M. Harrison, eds., pp. 92–119.

Camus, L., and others (2015). Comparison of produced water toxicity to Arctic and temperate species. *Ecotoxicology and Environmental Safety*, vol. 113, pp. 248–258.

Carlsson, Pernilla, and others (2018). Polychlorinated biphenyls (PCBs) as sentinels for the elucidation of Arctic environmental change processes: a comprehensive review combined with ArcRisk project results. *Environmental Science and Pollution Research*, vol. 25, No. 23, pp. 22499–22528.

Commission for the Protection of the Marine Environment of the North-East Atlantic (OSPAR) (2017a). *Inputs of Mercury, Cadmium and Lead via Water and Air to the Greater North Sea*. OSPAR Intermediate Assessment 2017. https://oap.ospar.org/en/ospar-assessments/intermediate-assessment-2017/pressures-human-activities/contaminants/heavy-metal-inputs.

________ (2017b). *Intermediate Assessment 2017*. https://oap.ospar.org/en/ospar-assessments/intermediate-assessment-2017.

________ (2017c). *Status and Trend for Heavy Metals (Cadmium, Mercury and Lead) in Sediment*. OSPAR Intermediate Assessment 2017. https://oap.ospar.org/en/ospar-assessments/intermediate-assessment-2017/pressures-human-activities/contaminants/metals-sediment.

________ (2017d). *Status and Trend for Heavy Metals (Mercury, Cadmium, and Lead) in Fish and Shellfish*. OSPAR Intermediate Assessment 2017. https://oap.ospar.org/en/ospar-assessments/intermediate-assessment-2017/pressures-human-activities/contaminants/metals-fish-shellfish.

________ (2017e). *Status and Trends in the Levels of Imposex in Marine Gastropods (TBT in Shellfish)*. OSPAR Intermediate Assessment 2017. https://oap.ospar.org/en/ospar-assessments/intermediate-assessment-2017/pressures-human-activities/contaminants/imposex-gastropods.

________ (2018a). *Assessment of the Discharges, Spills and Emissions from Offshore Installations on the Norwegian Continental Shelf in 2012–2016*.

________ (2018b). *Assessment of the Discharges, Spills and Emissions from Offshore Installations on the United Kingdom Continental Shelf in 2012–2016*.

________ (2018c). *Annual Report on Discharges of Radioactive Substances from the Non-Nuclear Sector in 2016*. www.ospar.org/documents?v=38960.

Cossa, Daniel, and others (2011). Mercury in the Southern Ocean. *Geochimica et Cosmochimica Acta*, vol. 75, No. 14, pp. 4037–4052.

Cunningham, Patricia A., and others (2019). Assessment of metal contamination in Arabian/Persian Gulf fish: a review. *Marine Pollution Bulletin*, vol. 143, pp. 264–283.

Dastoor, Ashu P., and Dorothy A. Durnford (2013). Arctic Ocean: is it a sink or a source of atmospheric mercury? *Environmental Science & Technology*, vol. 48, No. 3, pp. 1707–1717.

Deruytter, David, and others (2017). Mixture toxicity in the marine environment: model development and evidence for synergism at environmental concentrations. *Environmental Toxicology and Chemistry*, vol. 36, No. 12, pp. 3471–3479.

Desforges, Jean-Pierre, and others (2018). Predicting global killer whale population collapse from PCB pollution. *Science*, vol. 361, No. 6409, pp. 1373–1376.

Dirtu, Alin C., and others (2016). Contrasted accumulation patterns of persistent organic pollutants and mercury in sympatric tropical dolphins from the south-western Indian Ocean. *Environmental Research*, vol. 146, pp. 263–273.

Du, Juan, and others (2017). Antibiotics in the coastal water of the South Yellow Sea in china: occurrence, distribution and ecological risks. *Science of the Total Environment*, vol. 595, pp. 521–527.

El Kateb, Akram, and others, 2020. Impact of industrial phosphate waste discharge on the marine environment in the Gulf of Gabes (Tunisia), *PloS One*, 17 May 2018.

Esposito, Mauro, and others (2018). Total mercury content in commercial swordfish (*Xiphias gladius*) from different FAO fishing areas. *Chemosphere*, vol. 197, pp. 14–19.

European Environment Agency (EEA) (2013). The impact of international shipping on European air quality and climate forcing. EEA Technical report No. 4/2013. www.eea.europa.eu/publications/the-impact-of-international-shipping.

European Medicines Agency (EMEA) (2018). *Draft Guideline on the Environmental Risk Assessment of Medicinal Products for Human Use*. London.

European Union (2018). Report on the implementation and compliance with Directive (EU) 2016/802 which is transposing MARPOL Annex VI requirements into EU law. https://eur-lex.europa.eu/legal-content/EN/TXT/?uri=CELEX:52018DC0188.

Fair, P.A., and others (2019). Perfluoroalkyl substances (PFASs) in edible fish species from Charleston Harbor and tributaries, South Carolina, United States: exposure and risk assessment. *Environmental Research*, vol. 171, pp. 266–277.

Fang, Tien-Hsi, and others (2012). The occurrence and distribution of pharmaceutical compounds in the effluents of a major sewage treatment plant in Northern Taiwan and the receiving coastal waters. *Marine Pollution Bulletin*, vol. 64, No. 7, pp. 1435–1444.

Fent, Karl, and others (2006). Ecotoxicology of human pharmaceuticals. *Aquatic Toxicology*, vol. 76, No. 2, pp. 122–159.

Fisch, Kathrin, and others (2017). Occurrence of pharmaceuticals and UV-filters in riverine run-offs and waters of the German Baltic Sea. *Marine Pollution Bulletin*, vol. 124, No. 1, pp. 388–399. https://doi.org/10.1016/j.marpolbul.2017.07.057.

Fisher, David, and others (2012). Recent melt rates of Canadian Arctic ice caps are the highest in four millennia. *Global and Planetary Change*, vol. 84, pp. 3–7.

Fuller, Neil, and others (2019). Impacts of ionising radiation on sperm quality, DNA integrity and post-fertilisation development in marine and freshwater crustaceans. *Ecotoxicology and Environmental Safety*, vol. 186, 109764.

Gauss, Michael, and others (2018). Atmospheric supply of nitrogen, cadmium, mercury, benzo(a)pyrene and PVB-153 to the Baltic Sea in 2016. *EMEP/MSC-W Technical Report 1/2018*.

Gaw, Sally, and others (2014). Sources, impacts and trends of pharmaceuticals in the marine and coastal environment. *Philosophical Transactions of the Royal Society B: Biological Sciences*, vol. 369, No. 1656, 20130572.

Global Energy Statistics Yearbook (2018). https://yearbook.enerdata.net/crude-oil/world-production-statitistics.html.

Gnandi, Kissao, and others (2011). Increased bioavailability of mercury in the lagoons of Lomé, Togo: the possible role of dredging. *Ambio*, vol. 40, No. 1, pp. 26–42.

Godard-Codding, Céline A.J., and others (2011). Pacific Ocean-wide profile of CYP1A1 expression, stable carbon and nitrogen isotope ratios, and organic contaminant burden in sperm whale skin biopsies. *Environmental Health Perspectives*, vol. 119, No. 3, p. 337.

González-Gaya, Belén, and others (2014). Perfluoroalkylated substances in the global tropical and subtropical surface oceans. *Environmental Science & Technology*, vol. 48, No. 22, pp. 13076–13084. https://doi.org/10.1021/es503490z.

Gonzalvo, J., and others (2016.) The Gulf of Ambracia's common bottlenose dolphins, *Tursiops truncatus*: a highly dense and yet threatened population. *Advances in Marine Biology*, vol. 75, pp. 259–296.

Gothwal, Ritu, and Thhatikkonda Shashidhar (2015). Antibiotic pollution in the environment: a review. *Clean–Soil, Air, Water*, vol. 43, No. 4, pp. 479–489.

Gullberg, Erik, and others (2011). Selection of resistant bacteria at very low antibiotic concentrations. *PLoS Pathogens*, vol. 7, No. 7, e1002158.

Gusev, A. (2018). *Atmospheric Deposition of Benzo(a)Pyrene on the Baltic Sea. HELCOM Baltic Sea Environment Fact Sheets.*

Gusev, A., and others (2018). Persistent Organic Pollutants: assessment of transboundary pollution on global, regional, and national scales. *EMEP Status Report 3/2018*, July 2018. http://en.msceast.org/reports/3_2018.pdf.

Hassan, Hassan, and others (2019). Baseline concentrations of mercury species within sediments from Qatar's coastal marine zone. *Marine Pollution Bulletin*, vol. 142, pp. 595–602.

Hatje, Vanessa, and others (2014). Determination of rare earth elements after pre-concentration using NOBIAS-chelate PA-1® resin: method development and application in the San Francisco Bay plume. *Marine Chemistry*, vol. 160, pp. 34–41.

Hatje, Vanessa, and others (2018). Trace-metal contaminants: human footprint on the ocean. *Elements: An International Magazine of Mineralogy, Geochemistry, and Petrology*, vol. 14, No. 6, pp. 403–408.

Hatosy, Stephen M., and Adam C. Martiny (2015). The ocean as a global reservoir of antibiotic resistance genes. *Applied and Environmental Microbiology*, vol. 81, No. 21, pp. 7593–7599.

Hauge, K.H., and others (2014). Inadequate risk assessments – a study on worst-case scenarios related to petroleum exploitation in the Lofoten area. *Marine Policy*, vol. 44, pp. 82–89.

He, P., and others (2013). A summary of global ^{129}I in marine waters. *Nuclear Instruments and Methods in Physics Research Section B: Beam Interactions with Materials and Atoms*, vol. 294, pp. 537–541.

Heberer, T. (2002). Occurrence, fate, and removal of pharmaceutical residues in the aquatic environment: a review of recent research data. *Toxicology Letters*, vol. 131, Nos. 1–2, pp. 5–17.

Heimbürger, Lars-Eric, and others (2015). Shallow methylmercury production in the marginal sea ice zone of the central Arctic Ocean. *Scientific Reports*, vol. 5, art. 10318.

Hermann, L., and others (2018). Phosphorus processing – potentials for higher efficiency. *Sustainability*, vol. 10, No. 5, art. 1482.

Hernández, F., and others (2019). Occurrence of antibiotics and bacterial resistance in wastewater and sea water from the Antarctic. *Journal of Hazardous Materials*, vol. 363, pp. 447–456.

Hong, G.-H., and others (2011). Applications of anthropogenic radionuclides as tracers to investigate marine environmental processes. In *Handbook of Environmental Isotope Geochemistry*, pp. 367–394. Springer.

Hou, X., and others (2013). Iodine-129 in seawater offshore Fukushima: distribution, inorganic speciation, sources, and budget. *Environmental Science & Technology*, vol. 47, pp. 3091–3098.

Hussy, Ines, and others (2012). Determination of chlorinated paraffins in sediments from the Firth of Clyde by gas chromatography with electron capture negative ionisation mass spectrometry and carbon skeleton analysis by gas chromatography with flame ionisation detection. *Chemosphere*, vol. 88, No. 3, pp. 292–299.

Ilyin, I., and others (2018). Assessment of heavy metal transboundary pollution on global, regional and national scales. *EMEP Status Report 2/2018*. http://en.msceast.org/reports/2_2018.pdf.

International Atomic Energy Agency (IAEA) (1995). Sources of radioactivity in the marine environment and their relative contributions to overall dose assessment from marine radioactivity (MARDOS), IAEA-TECDOC-838, IAEA, Vienna.

________ (2005). Worldwide marine radioactivity studies (WOMARS): radionuclide levels in oceans and seas, IAEA-TECDOC-1429, IAEA, Vienna.

________ (2015). *Inventory of Radioactive Material Resulting from Historical Dumping, Accidents and Losses at Sea*. TECDOC Series 1776. Vienna: IAEA. www.iaea.org/publications/10925/inventory-of-radioactive-material-resulting-from-historical-dumping-accidents-and-losses-at-sea.

________(2019a). Power reactor information system (PRIS) database. www.iaea.org/resources/databases/power-reactor-information-system-pris.

________ (2019b). Events and highlights on the progress related to recovery operations at Fukushima Daiichi Nuclear Power Station. www.iaea.org/sites/default/files/19/09/events-and-highlights-july-2019.pdf.

International Energy Agency (IEA) (2018a). *Offshore Energy Outlook*. World Energy Outlook Series. https://doi.org/10.1787/weo-2018-en.

________(2018b). *Oil Information: Overview.* https://www.iea.org/reports/oil-information-overview.

International Federation of Pharmaceutical Manufacturers & Associations (IFPMA) (2017). *The Pharmaceutical Industry and Global Health: Facts and Figures 2017.* www.ifpma.org/wp-content/uploads/2017/02/IFPMA-Facts-And-Figures-2017.pdf.

International Institute for Applied Systems Analysis (IIASA) (2018). The potential for cost-effective air emission reductions from international shipping through designation of further Emission Control Areas in EU waters with focus on the Mediterranean Sea. Final Report. https://iiasa.ac.at/web/home/research/researchPrograms/air/Shipping_emissions_reductions_main.pdf.

International Maritime Organization (2019). Construction requirements for oil tankers – double hulls. www.imo.org/en/OurWork/Environment/Pages/constructionrequirements.aspx.

International Tanker Owners Pollution Federation (ITOPF) (2019). *Oil Tanker Spill Statistics 2018*. www.itopf.org/fileadmin/data/Documents/Company_Lit/Oil_Spill_Stats_2019.pdf.

Jamieson, Alan J., and others (2017). Bioaccumulation of persistent organic pollutants in the deepest ocean fauna. *Nature Ecology & Evolution*, vol. 1, No. 3, art. 0051.

Japan Nuclear Fuel Limited (JNFL) (2020). Reprocessing. www.jnfl.co.jp/en/business/reprocessing.

Jeong, Yu-Jin, and others (2019). Comparing levels of perfluorinated compounds in processed marine products. *Food and Chemical Toxicology.*

Jepson, Paul D., and Robin J. Law (2016) Persistent pollutants, persistent threats. *Science*, vol. 352, No. 6292, pp.1388–1389.

Jepson, Paul D., and others (2016). PCB pollution continues to impact populations of orcas and other dolphins in European waters. *Scientific Reports*, vol. 6, 18573.

Jiménez, S., and others (2018). State of the art of produced water treatment. *Chemosphere*, vol. 192, pp. 186–208.

Jonsson, Sofi, and others (2017). Terrestrial discharges mediate trophic shifts and enhance methylmercury accumulation in estuarine biota. *Science Advances*, vol. 3, No. 1, e1601239.

Josefsson, Sarah (2018). *Hexaklorbensen i Svenska Sediment 1986–2015.*

Josefsson, Sarah, and Anna Apler (2019). *Miljöföroreningar i Utsjösediment–Geografiska Mönster Och Tidstrender.*

Kallenborn, Roland, and others (2018). Pharmaceuticals and personal care products (PPCPs) in Arctic environments: indicator contaminants for assessing local and remote anthropogenic sources in a pristine ecosystem in change. *Environmental Science and Pollution Research*, vol. 25, No. 33, pp. 33001–33013.

Karl, M., and others (2019). Impact of a nitrogen emission control area (NECA) on the future air quality and nitrogen deposition to seawater in the Baltic Sea region. *Atmospheric Chemistry and Physics*, vol. 19, No. 3, pp. 1721–1752. https://doi.org/10.5194/acp-19-1721-2019.

Kenna, Timothy C., and others (2012). Intercalibration of selected anthropogenic radionuclides for the GEOTRACES program. *Limnology and Oceanography: Methods*, vol. 10, No. 8, pp. 590–607.

Klatte, Stephanie, and others (2017). Pharmaceuticals in the environment – a short review on options to minimize the exposure of humans, animals and ecosystems. *Sustainable Chemistry and Pharmacy*, vol. 5, pp. 61–66.

Koski, M., and others (2017). Ecological effects of scrubber water discharge on coastal plankton: potential synergistic effects of contaminants reduce survival and feeding of the copepod *Acartia tonsa*. *Marine Environmental Research*, vol. 129, pp. 374–385.

Kötke, Danijela, and others (2019). Prioritised pharmaceuticals in German estuaries and coastal waters: occurrence and environmental risk assessment. *Environmental Pollution*, vol. 255, part I, 113161.

Kulaksiz, Serkan, and Michael Bau (2007). Contrasting behaviour of anthropogenic gadolinium and natural rare earth elements in estuaries and the gadolinium input into the North Sea. *Earth and Planetary Science Letters*, vol. 260, Nos. 1–2, pp. 361–371.

Lee, Jong-Mi, and others (2015). Impact of anthropogenic Pb and ocean circulation on the recent distribution of Pb isotopes in the Indian ocean. *Geochimica et Cosmochimica Acta*, vol. 170, pp. 126–144.

Li, Jing, and others (2017). Organophosphate esters in air, snow, and seawater in the North Atlantic and the Arctic. *Environmental Science & Technology*, vol. 51, No. 12, pp. 6887–6896.

Liang, Ximei, and others (2013). The distribution and partitioning of common antibiotics in water and sediment of the Pearl River Estuary, South China. *Chemosphere*, vol. 92, No. 11, pp. 1410–1416.

Lindgren, J. Fredrik, and others (2012). Meiofaunal and bacterial community response to diesel additions in a microcosm study. *Marine Pollution Bulletin*, vol. 64, No. 3, pp. 595–601.

Loos, Robert, and others (2010). Pan-European survey on the occurrence of selected polar organic persistent pollutants in ground water. *Water Research*, vol. 44, No. 14, pp. 4115–4126.

Loos, Robert, and others (2013). Analysis of polar organic contaminants in surface water of the northern Adriatic Sea by solid-phase extraction followed by ultrahigh-pressure liquid chromatography–QTRAP® MS using a hybrid triple-quadrupole linear ion trap instrument. *Analytical and Bioanalytical Chemistry*, vol. 405, No. 18, pp. 5875–5885.

Ma, Yuxin, and others (2017). Organophosphate ester flame retardants and plasticizers in ocean sediments from the North Pacific to the Arctic Ocean. *Environmental Science & Technology*, vol. 51, No. 7, pp. 3809–3815.

Ma, Yuxin, and others (2018). Concentrations and water mass transport of legacy pops in the Arctic Ocean. *Geophysical Research Letters*, vol. 45, No. 23, pp. 12972–12981.

Madigan, Daniel J., and others (2012). Pacific bluefin tuna transport Fukushima-derived radionuclides from Japan to California. *Proceedings of the National Academy of Sciences*, vol. 109, No. 24, pp. 9483–9486.

Magnér, Jörgen, and others (2010). Application of a novel solid-phase-extraction sampler and ultra-performance liquid chromatography quadrupole-time-of-flight mass spectrometry for determination of pharmaceutical residues in surface sea water. *Chemosphere*, vol. 80, No. 11, pp. 1255–1260.

Malakoff, David (2014). *Chemical Atlas Shows Where Seas Are Tainted – And Where They Can Bloom*. American Association for the Advancement of Science.

Mangano, Maria Cristina, and others (2017). Monitoring of persistent organic pollutants in the polar regions: knowledge gaps & gluts through evidence mapping. *Chemosphere*, vol. 172, pp. 37–45.

Marsili, Letizia, and others (2018). Persistent organic pollutants in cetaceans living in a hotspot area: the Mediterranean Sea. In *Marine Mammal Ecotoxicology*, pp. 185–212. Elsevier.

Mason, Robert P., and others (2017). The air-sea exchange of mercury in the low latitude Pacific and Atlantic Oceans. *Deep Sea Research Part I: Oceanographic Research Papers*, vol. 122, pp. 17–28.

McCann, Clare M., and others (2019). Understanding drivers of antibiotic resistance genes in High Arctic soil ecosystems. *Environment International*, vol. 125, pp. 497–504.

McDonough, Carrie A., and others (2018). Dissolved organophosphate esters and polybrominated diphenyl ethers in remote marine environments: Arctic surface water distributions and net transport through Fram Strait. *Environmental Science & Technology*, vol. 52, No. 11, pp. 6208–6216.

Men, Wu, and others (2015). Radioactive status of seawater in the northwest Pacific more than one year after the Fukushima nuclear accident. *Scientific Reports*, vol. 5, art. 7757.

Mohammed, Azad, and others, 2012. Metals in sediments and fish from Sea Lots and Point Lisas Harbors, Trinidad and Tobago. *Marine Pollution Bulletin* vol. 64, No. 1, pp. 169–173.

Molins-Delgado, Daniel, and others (2017). UV filters and benzotriazoles in urban aquatic ecosystems: the footprint of daily use products. *Science of the Total Environment*, vol. 601, pp. 975–986.

Moodley, Leon, and others (2018). Effects of low crude oil chronic exposure on the northern krill (*Meganyctiphanes norvegica*). *Journal of Experimental Marine Biology and Ecology*, vol. 500, pp. 120–131.

Moreno-González, R., and others (2015). Seasonal distribution of pharmaceuticals in marine water and sediment from a Mediterranean coastal lagoon (SE Spain). *Environmental Research*, vol. 138, pp. 326–344.

Munson, Kathleen M., and others (2015). Mercury species concentrations and fluxes in the central tropical Pacific Ocean. *Global Biogeochemical Cycles*, vol. 29, No. 5, pp. 656–676.

Na, Guangshui, and others (2013). Occurrence, distribution, and bioaccumulation of antibiotics in coastal environment of Dalian, China. *Marine Pollution Bulletin*, vol. 69, pp. 233–240.

National Research Council and Transportation Research Board (2003). *Oil in the Sea III: Inputs, Fates, and Effects*. Washington, D.C.: The National Academies Press. https://doi.org/10.17226/10388.

Nödler, Karsten, and others (2014). Polar organic micropollutants in the coastal environment of different marine systems. *Marine Pollution Bulletin*, vol. 85, No. 1, pp. 50–59.

Ogata, Tomoya, and Yasutaka Terakado (2006). Rare earth element abundances in some seawaters and related river waters from the Osaka Bay area, Japan: significance of anthropogenic Gd. *Geochemical Journal*, vol. 40, No. 5, pp. 463–474.

Olmer, Naya, and others (2017). Greenhouse gas emissions from global shipping, 2013–2015. *The International Council on Clean Transportation*.

Otosaka, S., and others (2018). Distribution and fate of ^{129}I in the seabed sediment off Fukushima. *Journal of Environmental Radioactivity*, vol. 192, pp. 208–218.

Pampanin, Daniela M., and others (2017). DNA adducts in marine fish as biological marker of genotoxicity in environmental monitoring: the way forward. *Marine Environmental Research*, vol. 125, pp. 49–62.

Paul, Maxence, and others (2015). Tracing the Agulhas leakage with lead isotopes. *Geophysical Research Letters*, vol. 42, No. 20, pp. 8515–8521.

Pazdro, Ksenia, and others (2016). Analysis of the residues of pharmaceuticals in marine environment: state-of-the-art, analytical problems and challenges. *Current Analytical Chemistry*, vol. 12, No. 3, pp. 202–226.

Pedreira, Rodrigo M.A., and others (2018). Tracking hospital effluent-derived gadolinium in Atlantic coastal waters off Brazil. *Water Research*, vol. 145, pp. 62–72.

Pereira, Camilo D. Seabra, and others (2016). Occurrence of pharmaceuticals and cocaine in a Brazilian coastal zone. *Science of the Total Environment*, vol. 548, pp. 148–154.

Picot-Groz, M., and others (2014). Detection of emerging contaminants (UV filters, UV stabilizers and musks) in marine mussels from Portuguese coast by QuEChERS extraction and GC–MS/MS. *Science of The Total Environment*, vol. 493, pp. 162–69. https://doi.org/10.1016/j.scitotenv.2014.05.062.

Pinzone, Marianna, and others (2015). POPs in free-ranging pilot whales, sperm whales and fin whales from the Mediterranean Sea: influence of biological and ecological factors. *Environmental Research*, vol. 142, pp. 185–196.

Power Engineering International (PEI) (2019). *World's First Floating Nuclear Power Unit Set to Start Operations*. www.powerengineeringint.com/nuclear/reactors/world-s-first-floating-nuclear-power-unit-set-to-start-operatiing.

Powers, Maria (2008). Vetting – selected legal aspects of the vessel selection process: with special focus on seaworthiness, duty of care and charter party vetting clauses. PhD Thesis, Faculty of Law, Lund University.

Praca, Emilie, and others (2011). Toothed whales in the northwestern Mediterranean: insight into their feeding ecology using chemical tracers. *Marine Pollution Bulletin* vol. 62, No. 5, pp. 1058–1065.

Qin, Xiaofei, and others (2016). Seasonal variation of atmospheric particulate mercury over the East China Sea, an outflow region of anthropogenic pollutants to the open Pacific Ocean. *Atmospheric Pollution Research*, vol. 7, No. 5, pp. 876–883.

Richardson, Susan D., and Thomas A. Ternes (2011). Water analysis: emerging contaminants and current issues. *Analytical Chemistry*, vol. 83, No. 12, pp. 4614–4648.

Rivera-Utrilla, José, and others (2013). Pharmaceuticals as emerging contaminants and their removal from water. A review. *Chemosphere*, vol. 93, No. 7, pp. 1268–1287.

Robinson, Kelly J., and others (2018). Persistent organic pollutant burden, experimental POP exposure, and tissue properties affect metabolic profiles of blubber from gray seal pups. *Environmental Science & Technology*, vol. 52, No. 22, pp. 13523–13534.

Rodil, Rosario, and others (2012). Transformation of phenazone-type drugs during chlorination. *Water Research*, vol. 46, No. 7, pp. 2457–2468.

Roig, Benoit (2010). *Pharmaceuticals in the Environment*. IWA publishing.

Rose, Alani, and others (2017). Modeling and Risk Assessment of Persistent, Bioaccumulative and Toxic (PBT) Organic Micropollutants in the Lagos Lagoon. *International Journal of Environmental Analytical Chemistry*, vol. 2, No. 2, pp. 22–26.

Rubarth, Janne, and others (2011). Perfluorinated compounds in red-throated divers from the German Baltic Sea: new findings from their distribution in 10 different tissues. *Environmental Chemistry*, vol. 8, No. 4, pp. 419–428.

Rusiecka, D., and others (2018). Anthropogenic signatures of lead in the Northeast Atlantic. *Geophysical Research Letters*, vol. 45, No. 6, pp. 2734–2743. https://doi.org/10.1002/2017GL076825.

Sanderson, Hans, and others (2003). Probabilistic hazard assessment of environmentally occurring pharmaceuticals toxicity to fish, daphnids and algae by ECOSAR screening. *Toxicology Letters*, vol. 144, No. 3, pp. 383–395.

Sanial, Virginie, and others (2017). Unexpected source of Fukushima-derived radiocesium to the coastal ocean of Japan. *Proceedings of the National Academy of Sciences*, vol. 114, No. 42, pp. 11092–11096.

Schlitzer, Reiner (2020). Ocean Data View. https://odv.awi.de.

Schlosser, Christian, and others (2016). Distribution and cycling of lead in the high and low latitudinal Atlantic Ocean. American Geophysical Union, Ocean Sciences Meeting 2016, abstract No. CT14B–0130.

Schlosser, Christian, and others (2019). Distribution of dissolved and leachable particulate Pb in the water column along the GEOTRACES section GA10 in the South Atlantic. *Deep Sea Research Part I: Oceanographic Research Papers*, vol. 148, pp. 132–142. https://doi.org/10.1016/j.dsr.2019.05.001.

Scientific Committee on Oceanic Research, Working Group 146 (SCOR-WG146) (2020). *Radioactivity in the Ocean, 5 Decades Later (RiO5). First Report of SCOR Working Group #146, September 2015*. https://scor-int.org/group/146.

Shamsudheen, S.V., and others (2015). Atmospheric Supply of nitrogen, lead, cadmium, mercury and PCBs to the Baltic Sea in 2013. *EMEP/MSC-W Technical Report*, vol. 2.

Singh, Surendra (2019). Russia wants to jointly develop small, medium-sized N-plants, including floating N-station, with India. *Times of India*.

Sköld, Ola (2000). Sulfonamide resistance: mechanisms and trends. *Drug Resistance Updates*, vol. 3, No. 3, pp. 155–160.

Smith, John N., and others (2015). Arrival of the Fukushima radioactivity plume in North American continental waters. *Proceedings of the National Academy of Sciences*, vol. 112, No. 5, pp. 1310–1315.

Soerensen, Anne L., and others (2016). A mass budget for mercury and methylmercury in the Arctic Ocean. *Global Biogeochemical Cycles*, vol. 30, No. 4, pp. 560–575.

Stockholm Convention (2018). *Draft Report on Progress towards the Elimination of Polychlorinated Biphenyls, Stockholm Convention on Persistent Organic Pollutants, Small Intersessional Working Group on Polychlorinated Biphenyls, Fourth Meeting (First Face-to-Face Meeting), 12–14 December 2018.*

Sühring, Roxana, and others (2016). Organophosphate esters in Canadian Arctic air: Occurrence, levels and trends. *Environmental Science & Technology*, vol. 50, No. 14, pp. 7409–7415.

Sun, Caoxin (2015). Persistent organic pollutants in the Arctic, Atlantic and Pacific Oceans. PhD Thesis, University of Rhode Island.

Sun, Yu-Xin, and others (2014). Persistent organic pollutants in marine fish from Yongxing Island, South China Sea: levels, composition profiles and human dietary exposure assessment. *Chemosphere*, vol. 98, pp. 84–90.

Sun, Yu-Xin, and others (2017). Halogenated organic pollutants in marine biota from the Xuande Atoll, South China Sea: levels, biomagnification and dietary exposure. *Marine Pollution Bulletin*, vol. 118, Nos. 1–2, pp. 413–419.

Suzuki, T., and others (2018). Vertical distribution of ^{129}I released from the Fukushima Daiichi Nuclear Power Plant in the Kuroshio and Oyashio current areas. *Marine Chemistry*, vol. 204, pp. 163–171.

Tamura, Ikumi, and others (2017). Contribution of pharmaceuticals and personal care products (PPCPs) to whole toxicity of water samples collected in effluent-dominated urban streams. *Ecotoxicology and Environmental Safety*, vol. 144, pp. 338–350.

Tanouayi, Gnon, and others (2016). Distribution of Fluoride in the Phosphorite Mining Area of Hahotoe–Kpogame (Togo). *Journal of Health and Pollution*, vol. 6, No. 10, pp. 84–94.

Taylor, David (2016). The pharmaceutical industry and the future of drug development. In *Pharmaceuticals in the Environment. Issues in Environmental Science and Technology*, vol. 41, pp. 1–33, R.E. Hester and R.M. Harrison, eds.

Teuchies, J., and others (2020). The impact of scrubber discharge on the water quality in estuaries and ports. *Environmental Sciences Europe*, vol. 32, No. 1, art. 103.

Theobald, Norbert, and others (2011). Occurrence of perfluorinated organic acids in the North and Baltic seas. Part 1: distribution in sea water. *Environmental Science and Pollution Research*, vol. 18, No. 7, pp. 1057–1069.

Turner, David R., and others (2017). Shipping and the environment: smokestack emissions, scrubbers and unregulated oceanic consequences. *Elementa-Science of the Anthropocene*, vol. 5.

United Nations (2017a). Chapter 17: Shipping. In *The First Global Integrated Marine Assessment: World Ocean Assessment I*. Cambridge: Cambridge University Press.

________ (2017b). Chapter 20: Coastal, riverine and atmospheric inputs from land. In *The First Global Integrated Marine Assessment: World Ocean Assessment I*. Cambridge: Cambridge University Press.

________ (2017c). *The First Global Integrated Marine Assessment: World Ocean Assessment I*. Cambridge: Cambridge University Press.

United Nations Conference on Trade and Development (UNCTAD) (2018). *Review of Maritime Transport*. United Nations.

United Nations Environment Programme (UNEP) (2019). *Global Mercury Assessment 2018*. UNEP.

United Nations Environment Programme and United Nations Institute for Training and Research (UNEP and UNITAR) (2018). PCB: A Forgotten Legacy. 2028: Final Elimination of PCB. UNEP.

United Nations Environment Programme/Mediterranean Action Plan (UNEP/MAP) (2012a). Initial integrated assessment of the Mediterranean Sea: Fulfilling step 3 of the ecosystem approach process. United Nations Environment Programme, Mediterranean Action Plan, Athens.

________ (2012b). State of the Mediterranean Marine and Coastal Environment. United Nations Environment Programme, Mediterranean Action Plan, Athens.

UNEP/MAP/Coordinated Mediterranean Pollution Monitoring and Research Programme (MED POL) (2011a). Hazardous substances in the Mediterranean: a spatial and temporal assessment. United Nations Environment Programme, Mediterranean Action Plan, Athens.

________ (2011b). Analysis of trend monitoring activities and data for the MED POL Phase III and IV (1999–2010). United Nations Environment Programme, Mediterranean Action Plan, Athens.

Vanderford, Brett J., and others (2003). Analysis of endocrine disruptors, pharmaceuticals, and personal care products in water using liquid chromatography/tandem mass spectrometry. *Analytical Chemistry*, vol. 75, No. 22, pp. 6265–6274.

Verlicchi, Paola, and others (2012). Occurrence of pharmaceutical compounds in urban wastewater: removal, mass load and environmental risk after a secondary treatment – a review. *Science of the Total Environment*, vol. 429, pp. 123–155.

Vorkamp, Katrin, and others (2019). Current-use halogenated and organophosphorous flame retardants: a review of their presence in Arctic ecosystems. *Emerging Contaminants*, vol. 5, pp. 179–200.

Wagner, Charlotte C., and others (2019). A global 3-D ocean model for PCBs: benchmark compounds for understanding the impacts of global change on neutral persistent organic pollutants. *Global Biogeochemical Cycles*, vol. 33, No. 3, pp. 469–481.

Wang, Guang, and others (2010). Hexachlorobenzene sources, levels and human exposure in the environment of China. *Environment International*, vol. 36, No. 1, pp. 122–130.

Webster, Lynda, and others (2014). Halogenated persistent organic pollutants in relation to trophic level in deep sea fish. *Marine Pollution Bulletin*, vol. 88, Nos. 1–2, pp. 14–27.

Weigel, Stefan, and others (2004). Determination of selected pharmaceuticals and caffeine in sewage and seawater from Tromsø/Norway with emphasis on ibuprofen and its metabolites. *Chemosphere*, vol. 56, No. 6, pp. 583–592.

Wiberg, K., and others (2013). *Managing the dioxin problem in the Baltic region with focus on sources to air and fish*. Swedish Environmental Protection Agency Report 6566.

World Nuclear Association (WNA) (2020). Country profiles. 2020. www.world-nuclear.org/information-library/country-profiles.aspx.

Wu, Jian-Lin, and others (2007) Triclosan determination in water related to wastewater treatment. *Talanta*, vol. 72, pp. 1650–1654.

Wu, Junwen, and others (2019). Plutonium in the western North Pacific: transport along the Kuroshio and implication for the impact of Fukushima Daiichi Nuclear Power Plant accident. *Chemical Geology*, vol. 511, pp. 256–264.

Xu, Weihai, and others (2013). Antibiotics in riverine runoff of the Pearl River Delta and Pearl River Estuary, China: concentrations, mass loading and ecological risks. *Environmental Pollution*, vol. 182, pp. 402–407.

Yeung, Leo W.Y., and others (2017). Vertical profiles, sources, and transport of PFASs in the Arctic Ocean. *Environmental Science & Technology*, vol. 51, No. 12, pp. 6735–6744.

Ytreberg, E., and others (2019). Effects of scrubber washwater discharge on microplankton in the Baltic Sea. *Marine Pollution Bulletin*, vol. 145, pp. 316–324.

Zhang, Ruijie, and others (2013a). Antibiotics in the offshore waters of the Bohai Sea and the Yellow Sea in China: occurrence, distribution and ecological risks. *Environmental Pollution*, vol. 174, pp. 71–77.

Zhang, Ruijie, and others (2013b). Occurrence and risks of antibiotics in the coastal aquatic environment of the Yellow Sea, North China. *Science of the Total Environment*, vol. 450, pp. 197–204.

Zhang, Xianming, and others (2017). North Atlantic Deep Water formation inhibits high Arctic contamination by continental perfluorooctane sulfonate discharges. *Global Biogeochemical Cycles*, vol. 31, No. 8, pp. 1332–1343.

Zhang, Ying, and others (2016). Environmental characteristics of polybrominated diphenyl ethers in marine system, with emphasis on marine organisms and sediments. *BioMed Research International*, vol. 2016, art. 1317232.

Zhu, Yanbei, and others (2004). Gadolinium anomaly in the distributions of rare earth elements observed for coastal seawater and river waters around Nagoya City. *Bulletin of the Chemical Society of Japan*, vol. 77, No. 10, pp. 1835–1842.

Chapter 12
Changes in inputs and distribution of solid waste, other than dredged material, in the marine environment

Contributors: François Galgani (convener: marine debris) and Aleke Stöfen-O'Brien (convener: dumping), Archis Ambulkar, Maurizio Azzaro, Maria João Bebianno (lead member), Joan Bondareff, Huw Griffiths, Martin Hassellov, Christos Ioakeimidis, Jenna Jambeck, Paula Keener, Fernanda de Oliveira Lana, Iryna Makarenko, Chelsea Rochman, Qamar Schuyler, Paula Sobral, Ca Thanh Vu (co-lead member), Konstantinos Topouzelis, Dick Vethaak, Penny Vlahos, Juying Wang (co-lead member) and Judith Weis.

Keynote points

- Plastics now represent the major share of marine litter or marine debris.
- Most marine litter is from land-based sources, resulting from poor waste management practices, especially in some rural and developing regions.
- Marine litter is present in all marine habitats, affecting the environment and marine organisms through entanglement, ingestion and the rafting of invasive species.
- Amounts of marine litter are increasing in remote and unpopulated areas.
- Time series data are needed to assess and monitor impacts of marine litter, including microplastics and nanoplastics.
- Although a decreasing trend is observed, there is a need to harmonize reporting on dumping at sea.

1. Activities resulting in marine debris, including plastics, abandoned fishing gear, microparticles and nanoparticles, and estimates of sources from land, ships and offshore installations

1.1. Introduction

The term "marine litter" refers to any persistent, manufactured or processed solid material discarded, disposed of or abandoned in marine and coastal environments (Joint Group of Experts on the Scientific Aspects of Marine Environmental Protection (GESAMP), 2019) and covers an extremely wide variety of materials, ranging in size from mega-litter (> 1 m), to macro-litter (> 25 mm), meso-litter (> 5 mm), micro-litter (> 1 μm) and nano-litter (< 1 μm). It is classified by the nature of the material, such as plastic, metal, glass, rubber or wood, or by sources or uses, such as fishing gear, industrial pellets, sanitary items and single-use plastics. Plastic, defined as polymers synthesized from hydrocarbon molecules or biomass with thermoplastic or thermoset properties, comprises the main component of marine litter and exhibits a wide range of properties, shapes and compositions (GESAMP, 2016). In 2018, approximately 348 million tons of plastic waste had been generated worldwide (PlasticsEurope, 2019), with annual amounts entering the ocean in the range of 4.8 to 12.7 million tons, based on data from 2010 (Jambeck and others, 2015).

Marine litter is most obvious on shorelines, where it accumulates from water currents, wave and wind action and river outflows. However, marine litter, mainly plastic, is also found on the ocean surface in convergent zones (ocean gyres), in the water column, on the sea floor and in association with marine biota, where it can cause harm (Barnes and others, 2009).

The present section provides a robust description of changes in the state of marine litter, including key region-specific features, and describes the consequences of those changes for human communities, economies and well-being.

1.2. Situation recorded in the first *World Ocean Assessment*

The first *World Ocean Assessment* (United Nations, 2017a) contained only limited understanding of the sources, fate, transport, degradation and impacts of marine litter. Economic impacts and reduction measures were not considered in depth, owing to a lack of information and knowledge about marine litter, including its spatial and temporal extent. A consideration of remote or ultra-deep areas, specific sources and fluxes for specific types of marine litter (e.g., riverine inputs, wastewater and atmospheric inputs of microplastics) was not included, and impacts were not discussed. More recently, however, discussions have begun in earnest, as a result of the increased number of surveys and extensive studies highlighting, for example, that more than 1,400 species had been affected by marine litter by 2019 (Claro and others, 2019).

Similarly, there was little discussion of microplastics, which are polymer particles of less than 5 mm (upper limit) and larger than 1 micron, as defined by the Joint Group of Experts on the Scientific Aspects of Marine Environmental Protection (GESAMP, 2019), with reference only to primary microplastics that were crafted to be microplastics, and the fact that larger pieces of plastic break up into smaller pieces (secondary microplastics).

1.3. Description of environmental changes between 2010 and 2020

Marine litter is present in all marine habitats, from densely populated areas to remote regions (Barnes and others, 2009), from beaches and shallow waters to deep-ocean trenches (Pierdomenico and others, 2019). Most of it originates from land-based sources (GESAMP, 2016; 2019), wastewater, combined sewer overflows, onshore recreational uses, solid waste disposal, inappropriate or illegal discharges and dumping, mismanaged waste dumps and runoff (see figure I). It is estimated that more than 1 million tons of plastic waste enter the ocean every year from rivers, with the top 20 polluting rivers, mostly located in Asia, accounting for a large percentage of the global total (Lebreton and others, 2017; Van Emmerick and others, 2018; Schmidt and others, 2017). Plastic pollution also enters the marine environment as a result of deficiencies in waste management infrastructures, with microplastics from wastewater treatment plants potentially reaching up to 10 million particles/m^3 (Science Advice for Policy by European Academies (SAPEA), 2019). Inputs resulting from extreme events and natural disasters, such as hurricanes, floods, earthquakes and tsunamis, along with accidents, may reach millions of tons every year and match the magnitude of regular inputs from land (Murray and others, 2018).

Single-use plastic items are the biggest contributors to marine litter (Addamo and others, 2017). It is estimated that 1 to 5 trillion plastic bags are consumed worldwide each year (United Nations Environment Programme (UNEP), 2018). The remaining sources of marine litter can be attributed to maritime transport, industrial exploration and offshore oil platforms, fishing and aquaculture (GESAMP, 2016; 2019), as well as the loss and purposeful disposal of, for example, containers, ballast weights and cargoes. In commonly used fishing grounds, large marine litter is entirely composed of abandoned, lost or otherwise discarded fishing gear (Pham and others, 2014). The amount of such litter is not well known, although some estimates are available (e.g., 640,000 tons per year, according to Macfadyen and others (2009)), and about 70 per cent (by weight) of floating macroplastics in the open ocean is related to fishing (Eriksen and others, 2014). It is also estimated that 5.7 per cent of all fishing nets, 8.6 per cent of all traps and 29 per cent of all lines are lost around the world each year (Richardson and others, 2019).

Figure I
Plastics: production, use by sectors, end use by citizens and flows back into the economy or the environment

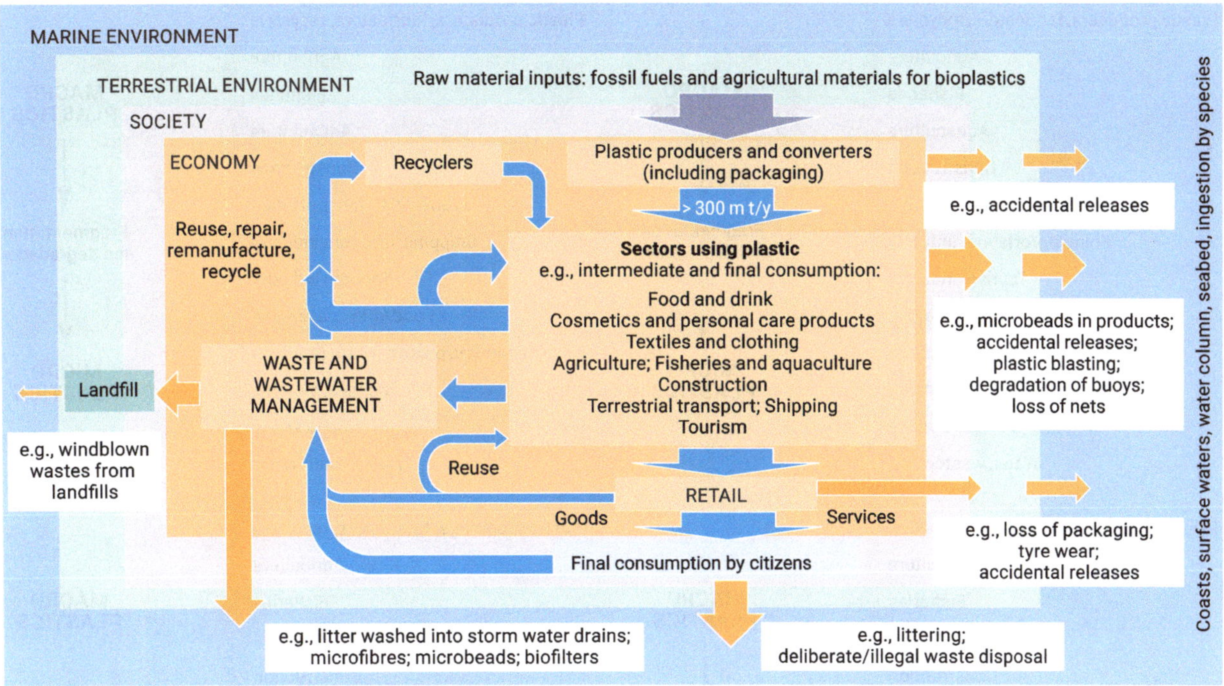

Source: United Nations Environment Programme (2017).

Primary microplastics, such as microbeads or industrial granulated pellets, enter the marine environment directly, while secondary microplastics result from the weathering, abrasion and fragmentation of single-use plastics (e.g., cutlery, trays, straws, cigarette butts, caps and lids, plastic bottles and shopping bags), synthetic textiles and clothing, coatings and paints, and tyres (see figure II). Recent studies suggest that the atmospheric transport and deposition of microplastics may also be an important pathway (Rochman, 2018).

The most common impacts of marine litter on marine life include the entanglement and ingestion of plastic marine litter (GESAMP, 2016; 2019). Entanglement poses a threat mainly to larger marine animals, such as top predators. Ingestion is common in a wider range of marine organisms, including marine mammals, turtles, sea birds, fish and invertebrate species, given that plastics occur in various sizes. Other impacts of plastic marine litter include changes to marine communities, with structures acting as new habitats (Reisser and others, 2014), across several levels of biological organization (Rochman and others, 2018) or by infestation of the marine environment by non-indigenous species, harmful algal blooms and pathogens dispersed on anthropogenic flotsam (Carlton and others, 2017; Viršek and others, 2017). As a result, it can increase the genetic exchange of bacteria and the spread of antibiotic resistance (Arias-Andrés and others, 2018).

Figure II
Sources of plastic entering the marine environment via rivers (green), coastlines (orange), direct inputs (blue) and the atmosphere (red)

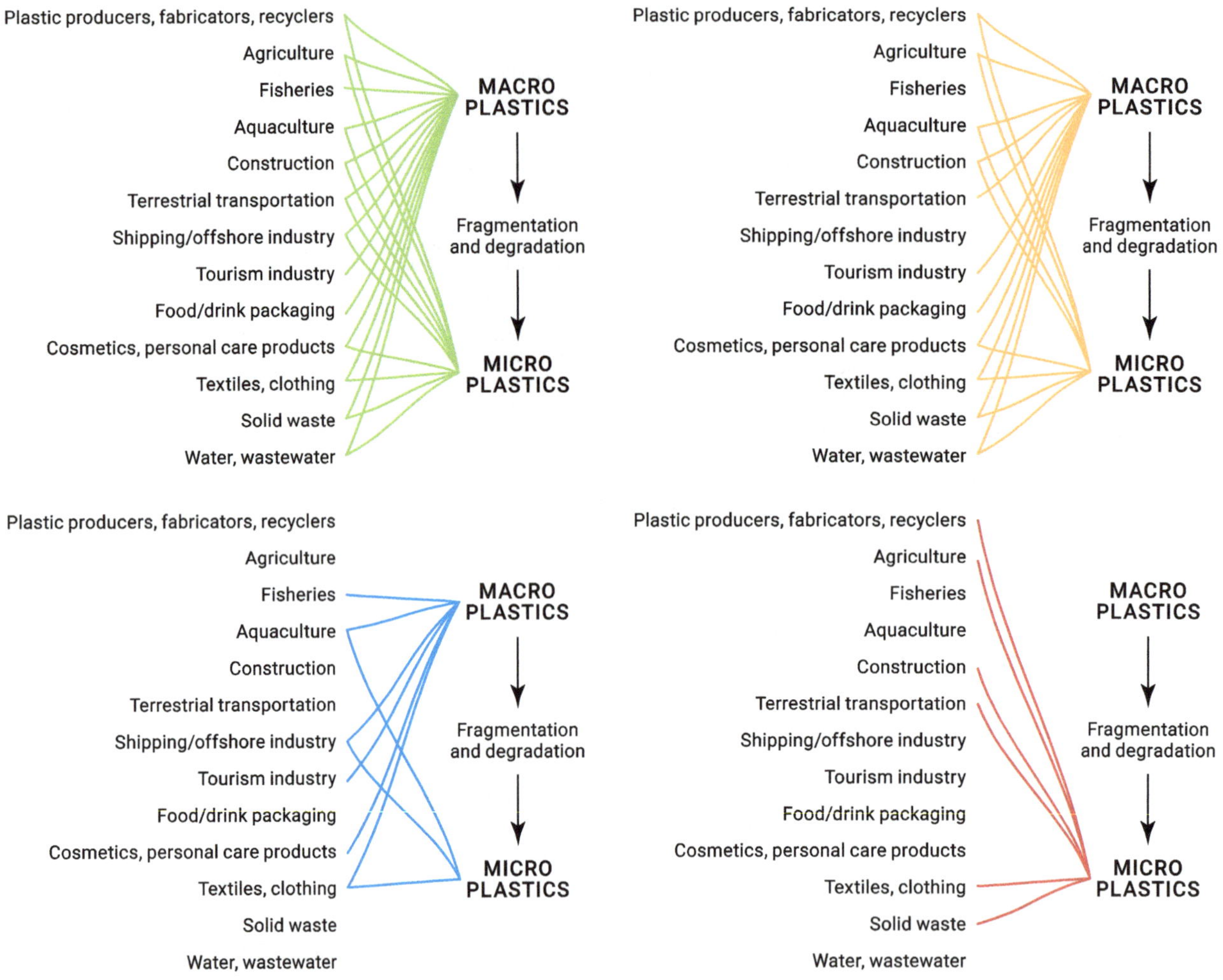

Source: adapted from GESAMP (2016).

Plastic marine litter also smothers and damages benthic organisms. The potential impact is not only at the level of organisms, but also at the population and ecosystem levels (Rochman and others, 2016). The Intergovernmental Science-Policy Platform on Biodiversity and Ecosystem Services confirmed the negative impact of plastics on biodiversity, with possible imbalances and disruptions in ecosystem diversity (Intergovernmental Science-Policy Platform on Biodiversity and Ecosystem Services, 2019). After the tsunami in Japan in 2011, 289 species of macrofauna and macroflora were rafted to North America in just six years (Carlton and others, 2017), a very uncommon scheme, with potential long-term consequences (Murray and others, 2018).

Aside from being a physical contaminant, plastics and microplastics often contain chemical additives, such as phthalates and brominated flame retardants (see chap. 11) and capture other contaminants. Laboratory studies demonstrate that microplastics can harm organisms and populations at higher concentrations than those found in nature. However, the best available evidence suggests that microplastics do not yet pose a widespread ecological risk (as opposed to a risk to individual organisms), except in some coastal waters and sediments (SAPEA, 2019).

Human health is a primary concern, despite a rather limited knowledge of impacts such as injuries and accidents or through possible contamination after a potential release of chemicals (SAPEA, 2019) or owing to the presence of microplastics in seafood, and there are few appropriate risk assessment studies. Such concerns may cause people to change their behaviour (e.g., tourism habits or reduction in the consumption of seafood).

More data have become available since the first Assessment and, as a consequence, modelling studies, assessments of riverine inputs, new technologies such as automated sensors, including aerial sensors and satellites, and new ecosystem approaches such as risk assessments for marine species and communities (Everaert and others, 2018) are improving understanding of how marine litter and plastics, in particular nanoplastics and microplastics, can cause harm.

To better support assessments and monitoring, new technical approaches, using tools such as drones, remote systems and automated sensors (Maximenko and others, 2019), and new indicators may support implementation of the harmonized monitoring of marine litter trends and improve the efficiency of global approaches and measures (GESAMP, 2019). Remote-sensing technology is the only approach that can be used to monitor large coastal or open-sea areas in several spatial resolutions and thus help in meeting the requirements of indicator 14.1.1 of the Sustainable Development Goals.[1] Space agencies are considering both optical and remote-sensing methods for testing and possible application in regular monitoring (Topouzelis and others, 2019; Martínez-Vicente and others, 2019). In terms of understanding the effects of plastics on wildlife and the environment, risk assessment is also a promising tool, by helping to model interactions between animal species and plastic. That approach is becoming more widely used, though further work is needed on quantifying the effect of the interaction, in particular in terms of the lethality and sub-lethality (e.g., changes to feeding, reproduction and growth) of ingested plastics (Schuyler and others, 2016; Wilcox and others, 2018).

1.4. Key region-specific changes and consequences

Many regional seas programmes have developed thematic strategies or plans for marine litter. The Regional Seas Indicators Working Group, established under the United Nations Environment Programme (UNEP) regional seas conventions, protocols and action plans, has developed a core set of 22 regional seas indicators on marine litter. Work is under way to develop common methodologies for the indicators, building upon monitoring programmes in each region (GESAMP, 2019). Some regional seas conventions, instruments or bodies (e.g., the Coordinating Body on the Seas of East Asia, the Convention for the Prevention of Marine Pollution by Dumping from Ships and Aircraft, the Convention for the Prevention of Marine Pollution from Land-based Sources and the Action Plan for the Protection of the Marine Environment and the Sustainable Development of the Coastal Areas of the Mediterranean) have updated or are considering updates to action plans to include port reception facilities to better manage administrative and legal matters and to enforce, control and monitor systems, infrastructure and alternatives for collecting and treating ship-generated waste. Table 1 provides an overview of the state of knowledge in the various basins of the world ocean.

1 See General Assembly resolutions 70/1 and 71/313, annex.

1.5. Trends

Understanding the factors associated with changes in the quantities and impact of marine litter and the magnitude of such changes remains difficult because of the lack of standardization of methods for collection and analysis. It is therefore difficult to accurately compare counts or levels from different locations and over time. Moreover, reports often address a specific component of the marine environment, such as types of litter and impacts, without giving attention to natural environmental variability (GESAMP, 2019), which impairs complete understanding of the state of and possible changes in marine litter densities and impacts.

Table 2 summarizes the available information on beach, sea floor, floating and ingested marine litter worldwide. Additional information can be obtained from the online portal for marine litter.[2] While several modelling studies predict increasing trends (Kako and others, 2014; Everaert and others, 2018; Lebreton and others, 2018), they may potentially be balanced by reduction measures. Most of the work based on regular surveys did not demonstrate any trend, other than in specific cases such as remote islands in the Antarctic (Barnes and others, 2009), ingested plastic in the Atlantic petrel (Petry and Benemann, 2017) or specific features such as converging currents above the Arctic Circle (Tekman and others, 2017). The increase in remote areas could be interpreted as a long-term transfer from affected areas to regions where human activity is either extremely reduced or non-existent. Decreasing trends were demonstrated in certain cases, such as the ingestion of debris, especially in respect of industrial granules. Brandon and others (2019) and Wilcox and others (2019) also suggested an increase in sediment microplastics in California and floating microplastics in the North Atlantic in relation to plastic production worldwide. The challenge now is to better understand how plastic is cycled through marine ecosystems, where it goes and how it degrades.

2 See https://litterbase.awi.de/litter.

Table 1
Overview of the state of knowledge of marine litter in the various basins of the world ocean

Basin	Sources/distribution	Importance	Circulation	Impacts
Arctic Ocean	Plastic and microplastics are in sea ice, surface and deep waters, deep-sea sediments and biota (Kanhai and others, 2018; Peeken and others, 2018).	There are low quantities of marine debris; microplastics are several orders of magnitude higher in sea ice (Cózar and others, 2017; Barrows and others, 2018); there is a high prevalence of ghost fishing gear and an impact in fishing grounds.	Debris is transported to the north via the surface branch of the thermohaline circulation.	There are low concentrations of microplastics in the polar cod (*Boreogadus saida*), the bigeye sculpin (*Triglops nybelini*) (Kühn and others, 2018; Morgana and others, 2018) and 11 species of benthic invertebrates (Fang and others, 2018); plastic is accumulated by the Greenland shark (*Somniosus microcephalus*) (Leclerc and others, 2012; Nielsen and others, 2014).
North Atlantic Ocean, Baltic Sea and North Sea	Litter and microparticles have been found in all components of the marine environment; there are monitoring data since 1988 in the North-East Atlantic (Commission for the Protection of the Marine Environment of the North-East Atlantic, 2017); and since 2005, along the coast of the United States of America.	Beaches in the maritime area under the Convention for the Protection of the Marine Environment of the North-East Atlantic[a] have litter in the range of hundreds of items per 100 m (maximum: 6,090); litter is widespread on the sea floor (Maes and others, 2018); abandoned and lost fishing gear is the most important type of litter in the Baltic sea.	Surface litter from populated areas of the North-East Atlantic is transported to the Arctic; litter from the South-East Atlantic travels through the equatorial current to the Western Atlantic, and from the North-West Atlantic to the North Atlantic gyre (Van Sebille and others, 2015).	There are many species with ingested litter or microplastics; 94 per cent of birds in the North Sea have pieces of plastic in their stomach; entanglement (affecting seals, sea turtles, birds and invertebrates) is a common pattern in the North Atlantic.
Mediterranean Sea and Black Sea	The amount of municipal solid waste ranges from 208 to 760 kg/capita/year; 250 billion particles are afloat (Collignon and others, 2012); the highest concentration worldwide is for floating microplastics (64 million items/km^2) (Van der Hall and others, 2017) and sea floor debris (1.3 million items/km^2) (Pierdomenico and others, 2019); the Black Sea beaches and seabed are largely affected by abandoned or lost fishing gear.	The Mediterranean Sea is one of the most affected areas worldwide (Ioakeimidis and others, 2017); five types of single-use plastics (cutlery/trays/straws, cigarette butts, caps/lids, plastic bottles and shopping bags) account for more than 60 per cent of all types of marine litter.	The Mediterranean Sea and the Black Sea are closed basins, with important large rivers (Nile, Po, Danube) (Lechner and others, 2014; Lebreton and others, 2017); they are tourist destinations, with a high volume of maritime traffic.	All types of impacts are described in the Mediterranean Sea, including ingestion by many species, entanglement, release of chemicals and rafting of various species.

Basin	Sources/distribution	Importance	Circulation	Impacts
South Atlantic Ocean	All types of litter in the South Atlantic are due to highly populated areas and large rivers; pelagic plastics are restricted to the tropical Atlantic Ocean (Eriksen and others, 2014); in all islands (Ivar do Sul and others, 2014), there are hard plastic fragments, plastic films, paint chips, fibres and strands; there is deep-sea floor litter with high densities in the south-east (Woodall and others, 2015), dominated by single-use items and microplastics.	Litter is at very high concentrations locally, but the basin is not the most affected area; there are higher densities of macroplastics in the islands of the Caribbean Sea compared with other islands in the Atlantic Basin; sources are more directly related to human occupation than to fisheries (Ivar do Sul and others, 2014).	Besides the general circulation scheme linked to geostrophic currents and the presence of the South Atlantic gyre, transport to remote islands is an important driver (Monteiro and others, 2018).	Despite a lack of data from the eastern part, all types of impacts are described in the South Atlantic, including ingestion by many species, entanglement, release of chemicals and rafting of various species.
Indian Ocean	South-East Asia and India are the main sources of marine debris (Jambeck and others, 2015; Lebreton and others, 2017); available data from South Africa and India are very recent.	The Indian Ocean has a greater surface particle count and weight of plastic, a large part of which is in the Gulf of Bengal and the central part of the basin, than the South Atlantic and the South Pacific combined (Eriksen and others, 2014); deep-sea floor litter is at high densities far from coasts (Woodall and others, 2015), dominated by fishing gear, but with patchy distribution in the south-eastern part (Woodall and others, 2014); plastic and microplastics are also in adjacent seas of the Indian Ocean, including the Red Sea (Arossa and others, 2019); in the Persian Gulf, low-density polyethylene and polypropylene are in seawater and sediments (Abayomi and others, 2017).	Because of the nature of currents, marine litter dumped anywhere is transported to the southern Indian Ocean gyre (Van Sebille and others, 2015), as well as to the western part by the residual circulation (Veerasingam and others 2016), thus reaching remote and unpopulated islands; the western Indian Ocean and the Arabian Sea are heavily trafficked by commercial shipping and fishing vessels, and the loss of fishing gear and the dumping of garbage are prevalent (Woodall and others, 2015).	Data are limited; the impacts described include ingestion by many species (e.g., fish, invertebrates and sea turtles), entanglement (sea turtles and birds), release of chemicals and rafting of various species.
North Pacific Ocean	Besides the Mediterranean Sea, the North-West Pacific is the most affected region (Chiba and others, 2018); shores of the Pacific Ocean and the marginal seas of East Asia are surrounded by countries undergoing rapid economic expansion; there are high inputs from countries such as China, Indonesia, the Philippines, and Viet Nam (UNEP and GRID-Arendal, 2016).	The North Pacific is disproportionately affected by plastic (Eriksen and others, 2014) from land-based sources and often sea-based sources in highly populated islands (Filho and others, 2019); abandoned, lost or otherwise discarded fishing gear represents 46 per cent of the mass of debris larger than 5 cm, which accounts for one third of the total mass of floating litter (Lebreton and others, 2018); densities of marine debris reach millions of items/km^2 (Eriksen and others, 2014; Van Sebille and others, 2015), with plastic as the predominant material (90 per cent of small pieces).	Besides the general circulation scheme linked to geostrophic currents and the presence of North Atlantic gyres, natural disasters such as tsunamis and earthquakes act as drivers in the generation of litter.	All types of impacts, including entanglement and ingestion by marine organisms, including birds, sea turtles and mammals, are detected in the deepest invertebrates of the Mariana Trench (Jamieson and others, 2019); in some regions, because of fisheries (Alaska) or drifting litter (Hawaii), entanglement seriously affects marine ecosystems, such as coral reefs and animal forests, or untargeted populations, such as pinnipeds (Claro and others, 2019).

Basin	Sources/distribution	Importance	Circulation	Impacts
South Pacific Ocean	Compared with other ocean basins, there is relatively little new information on plastic concentrations; data are mainly from Australia and Chile.	The highest concentrations of debris on beaches (239.4 ± 347.3 items/m^2, maximum: 671.6 items per m^2) are on Henderson Island (Lavers and Bond, 2017); in Isla Salas y Gómez, close to the centre of the South Pacific subtropical gyre, debris levels are significantly lower (< 1 item/km^2) (Miranda-Urbina and others, 2015); the highest recorded floating plastics are in the South Pacific subtropical gyre – more than 390,000 items/km^2 (maximum: 50,000 items/km^2) (Miranda-Urbina and others, 2015; Eriksen and others, 2018).	Different oceanographic models and empirical data sets suggest that marine debris counts and concentrations in the South Pacific subtropical gyre are lower than in subtropical gyres in the northern hemisphere (Van Sebille and others, 2015); locally, rivers may also play an important role in the distribution of marine litter (Gaibor and others, 2020).	A total of 97 different species of animals, including turtles, fish, seabirds, mammals and corallimorphs, either ingested or became entangled in plastics (Thiel and others, 2018; Markic and others, 2018); there is evidence of ingestion closer to subtropical gyres (Thiel and others, 2018); microplastics are ingested in ultra-deep amphipods (Jamieson and others, 2019).
Southern Ocean	The Southern Ocean has the lowest densities of plastic litter in the world, owing to small-scale human activity; marine debris is on a very local scale; the potential input is of approximately 44–500 kg of microplastics per decade (Waller and others, 2017); microplastics are generated from macroplastic degradation or transferred across the limit of the Polar region (polar front).	There are microplastics in intertidal sediments from a sub-Antarctic island (Barnes and others, 2009) in deep-sea sediments in the Weddell Sea (Van Cauwenberghe and others, 2013), in surface waters of the Pacific sector (Waller and others 2017; Isobe and others, 2015; 2017) and in shallow sediments and macroalgae at sites on King George Island near scientific research stations (Waller and others 2017); concentrations of 0.100–0.514 g/km^2 are found in the south polar front, ranging from 46,000 to 99,000 particles/km^2 south of latitude 60° south, with higher concentrations in coastal regions of the Ross Sea (Cincinelli and others, 2017; Cózar and others, 2014; Isobe and others, 2017); there are plastics in sediments from Terra Nova Bay, with a total of 1,661 items (3.14 g), fibres being the most frequent type (Munari and others, 2017); in surface trawls in the Antarctic peninsula, there is debris estimated at 1,794 items/km^2, an average weight of 27.8 g/km^2, not originating from latitudes lower than 58° south; paint fragments are 30 times more abundant than plastics (Lacerda and others, 2019).	The transfer of litter from northern waters to Antarctica is common.	There has been macroplastic and fishing debris on beaches and in seabird colonies at Bird Island Research Station since the austral summer of 1992/93 (Barnes and others, 2009); plastic particles are ingested by 12 species of seabirds, the majority in association with the wandering albatross and the grey-headed albatross, and recently in penguins (Bessa and others, 2019); there have been encounters between marine mammals and marine debris – mainly Antarctic fur seals entangled in plastic packaging bands, synthetic line and fishing nets; the number of incidents has declined significantly since the introduction of legislation in the late 1980s to prohibit the disposal of plastics overboard and improvements in the disposal of packaging bands (Barnes and others, 2009).

[a] United Nations, *Treaty Series*, vol. 2354, No. 42279.

Table 2
Marine litter trends in the marine environment (locations, compartments) (compilation of data from reports and scientific literature)

Location	Compartment/ species	Period (duration)	Methods	Trends	Observation	Reference
East Greenland	Ingested microplastics in little auk (*Alle alle*)	2005 and 2014	Collected from live birds in nests	No evident temporal trend		Amélineau and others, 2016
East Greenland	Subsurface microplastics	2005 and 2014	WP-2 net; vertical tows -50 m to surface	Significant increase		Amélineau and others, 2016
North Atlantic/ Arctic Circle, Fram Strait	Deep-sea floor; two stations at 2,500 m, 79–79°35′ north	2002–2014	Towed camera	Clear increase in litter densities and abundance of small-sized plastics	Possible spreading from Europe to North Atlantic and Arctic Basin	Tekman and others, 2017
North-East Atlantic	78 beaches	2001–2011	Convention for the Protection of the Marine Environment of the North-East Atlantic; Marine Strategy Framework Directive protocol	No large-scale trends	Hydrodynamics; climate-related drivers for local short-term changes	Schulz and others, 2013
North-East Atlantic (Rockall Trough)	Microplastics ingestion in deep-sea benthic invertebrates (> 2,000 m)	1976–2015	Epibenthic sled; Agassiz trawl	No trends between overall abundance or polymer types	Two species	Courtene-Jones and others, 2019
North Atlantic	Floating/subsurface	1957–2016	Debris trapped in towed continuous plankton recorders (16,725 tows)	Increase since 1957; no trend since 2000; no change in Arctic waters	6.5 million nautical miles	Ostle and others, 2019
North Sea, waters of the United Kingdom of Great Britain and Northern Ireland	Seabed; 17–150 stations/year	1992–2017	Marine Strategy Framework Directive classification system	No detectable trend	Unit: presence of plastic	Maes and others, 2018
North Sea/ Netherlands	Birds (fulmars, 973 samples stranded)	1979–2012	Regular protocol to the Convention for the Protection of the Marine Environment of the North-East Atlantic (mass and number)	Increase to mid-1990s; stable in the past decade; significant decrease in pellets		Van Franeker and Lavender Law, 2015

Location	Compartment/ species	Period (duration)	Methods	Trends	Observation	Reference
Waters of Ireland	Cetaceans (stranded and by-catches)	1990–2015	Stomach content	No trend for ingestion of litter and entanglement		Lusher, 2015
Baltic Sea	2,377 hauls; 53 cruises	2012–2017	Marine Strategy Framework Directive; Baltic International Trawl Surveys	Increase in plastics in past two years; no trend for litter from fishing	Plastic (35 per cent of litter)	Zablotski and Kraak, 2019
Baltic Sea	245 stations; floating/ingested microplastics; Atlantic herring and sprat (814 samples)	1987–2015	Plankton samples; trawling; stomach content	No change in floating or ingested microplastics		Beer and others, 2018
North Atlantic subtropical gyre	Floating microplastics	1986–2008	6,136 surface Neuston nets, 335-μm mesh	No trend	Sea Education Association archived plankton samples	Lavender Law and others, 2010
North Atlantic subtropical gyre	Floating plastics (2,624 tows)	1987–2012	Surface Neuston nets, 335-μm mesh	No significant change in user plastics; highly significant decrease in industrial plastics	Extension of work from Lavender Law and others (2010)	Van Franeker and Lavender Law, 2015
North-East Adriatic Sea	Seabed (67 stations)	2011–2016	Otter trawl	Decrease in total litter; no trend for plastic	50 per cent of plastic is from fishing/aquaculture	Strafella and others, 2019
France, Mediterranean	Sea floor; shelves and canyons	1994–2017	Trawling (1,902 hauls); Marine Strategy Framework Directive classification system	No regular increase but higher levels in 1999–2001, and since 2012	Plastic (up to 62 per cent)	Gerigny and others, 2019
Spain, Mediterranean	Seabed shelves (1,323 hauls)	2007–2017	Trawling, Marine Strategy Framework Directive classification system	No temporal trend; decrease in Alboran Sea	MEDITS project	García-Rivera and others, 2018
Western Mediterranean	Ingested debris; sea turtles	1995–2016	Marine Strategy Framework Directive classification system	Slight decrease	195 samples	Domènech and others, 2019
Balearic Islands	Floating	2005–2015	Onshore/offshore cleaning boats	No trend (all types of debris); increase in summer	Cleaning operations	Compa and others, 2019

Location	Compartment/ species	Period (duration)	Methods	Trends	Observation	Reference
Southern Brazil	Birds (white-chinned petrel, 122 samples, stranded)	1990–2014	Stomach content	Increase in fragments and pieces; decrease in virgin pellets		Petry and Benemann, 2017
North Pacific subtropical gyre	Floating microplastics	2001–2012	2,500 surface Neuston nets, 335-µm mesh	No evident temporal trend	Confounded spatial and temporal variability	Lavender Law and others, 2014
Taiwan Province of China	Beach litter (541 clean-up events)	2004–2016	Clean-up events	No temporal trend	Data from ocean coastal clean-up events	Walther and others, 2018
China	National monitoring; beaches, surface and sea floor	2011–2018	State Oceanic Adminsitration protocols	No trend		Ministry of Ecology and Environment, China, 2019
China	23 sites (beaches and adjacent waters; floating and seabed)	2007–2014	North-West Pacific action plan; State Oceanic Administration protocols	No clear trend	Percentage of plastic increasing in seabed litter	Zhou and others, 2016
Chile	Beaches (all coasts); 3 surveys, 69 beaches	2006–2016	Participative science; main categories	No trend	Three sampling years	Hidalgo-Ruz and others, 2018
Ecuador	Beaches (26 sites)	2018–2020	Participative science (400 volunteers)	No trend	One sampling year	Gaibor and others, 2020

1.6. Consequences of changes for human communities, economies and well-being

The most significant impact of the use of plastic in products and packaging is marine pollution (UNEP, 2014), but it is important to emphasize that it is difficult to quantify the economic impact of marine litter. Based on figures from 2011, the economic costs of marine plastic, in relation to marine natural capital, are conservatively conjectured to be between $3,300 and $33,000 per ton per year (Beaumont and others, 2019). While the input of plastic into the ocean is limited in European coastal areas (Jambeck and others, 2015), the estimated costs of cleaning up marine litter in coastal areas can amount to up to €630 million per year (Crippa and others, 2019). More recently (McIlgorm and others, 2020), a ninefold increase in the direct economic costs of marine litter was found from 2009 to 2015, reaching $10.8 billion.

In addition to indirect impacts (i.e., on biodiversity and ecosystems), beach litter is perhaps the most visible direct impact and affects the patrimonial value of coastal areas that can be translated as the financial expenditure of cleaning up (UNEP, 2019). Damage and costs to marine ecosystems and services must be considered in the future despite an actual limited understanding of the detrimental impacts on the structure and functioning of the marine ecosystem.

Marine litter can also result in increased costs for the shipping sector and recreational activities, including yachting (e.g., fouled motors, entangled propellers, lost output and repair costs) (Hong and others, 2017), but the damage and associated social costs extend to other sectors such as aquaculture and fisheries. The removal of 10 per cent of derelict fishing pots alone would provide estimated additional revenues of $831 million annually for the global crustacean fishery industry (Scheld and others, 2016).

Most microplastics in marine organisms are found in their digestive system, which people do not ordinarily consume, with the exception of shellfish and small fish that are eaten whole. Besides accidents and injuries, there is no evidence that microplastics concentrations have a negative impact on fish and shellfish health or commercial stocks (Barboza and others, 2018). Links to human health are not sufficiently addressed, and gaps in knowledge are even greater in relation to nanoplastics (< 1 micron), in particular their absorption and behaviour (GESAMP, 2016; see also chap. 8) and how they may pass through biological barriers via different mechanisms (Wright and Kelly, 2017). As relevant toxicity data are absent, the European Food Safety Authority concluded that it was currently not possible to evaluate the human health risk of nanoplastics and microplastics (European Food Safety Authority Panel on Contaminants in the Food Chain, 2016). Moreover, there are indications that microplastic fibre ingestion by humans through the consumption of contaminated seafood is only a minimal contribution to the microplastic contamination of the total food basket (Catarino and others, 2018).

The socioeconomic impact of marine litter and the potential cost for key sectors and activities in or depending on the marine and coastal environment have not been well assessed, resulting in the mispricing of ecosystem values and the externalization of pollution costs. Approaches for giving value to marine litter are also not well known. Efforts need to be focused on assessing the environmental and socioeconomic costs of the damage caused by marine litter and a cost-benefit analysis of marine litter prevention and reduction measures (see table 3).

Table 3
Marine litter in relation to economic sectors, sources, amounts and impacts

Economic sector	Non-construction zone	Natural risks	Extreme events	Climate change	Agriculture and forestry run-off	Coastal urbanization	Damming (demand on water)	Wastewater discharges	Industry	Tourism frequentation	Yachting	Marine mining	Dredging	Desalinization	Coastal artificialization	Port operations
Quantities of marine litter in relation to the economic sector																
Riverine inputs	No	Significant	Significant	Limited	Limited	Significant	No	Significant	Significant	Limited	No	Limited	No	No	Limited	No
Beach litter	No	Limited	High	No	Significant	High	Limited	Significant	Significant	High	Limited	No	No	No	Significant	Significant
Surface litter	No	Limited	High	No	Limited	Significant	Limited	Significant	Significant	Significant	Limited	Limited	No	No	Limited	Limited
Sea floor litter	No	Limited	Limited	No	No	Limited	No	Limited	Limited	Limited	Limited	Limited	Limited	No	Significant	Limited
Microplastics	No	No	Limited	No	Limited	High	No	Significant	Limited	Significant	Limited	No	No	No	Significant	Limited
Impacts from marine litter in relation to the economic sector																
Wildlife	No	Significant	High	No	No	Significant	No	Significant	Limited	Significant	Limited	No	No	No	Significant	Limited
Human health	No	Limited	Limited	No	No	Significant	No	Limited	Limited	Significant	Limited	No	No	No	Limited	Limited
Tourism	No	Limited	Limited	No	Limited	High	No	Significant	Limited	High	Limited	No	No	No	Significant	Limited
Fisheries	No	Limited	Limited	No	Limited	High	No	Significant	Limited	Significant	Significant	No	No	No	Significant	Limited
Navigation	No	Limited	Limited	No	No	Significant	No	Limited	No	High	Limited	No	No	No	Significant	Significant
Municipalities	No	Limited	Limited	No	Limited	Significant	No	Significant	Limited	High	Limited	No	No	No	Significant	Significant
Clean-up	No	Limited	Limited	No	No	Significant	No	Significant	No	High	Limited	No	No	No	Limited	Limited

Economic sector	Offshore structures	cables and pipelines	Shipping	Oil and gas extraction	Renewable energy	Fishing (including recreationnal)	Sea-based food harvesting	Extraction of genetic resources	Aquaculture	Solid waste disposal	Storage of gases	Research and education	Defence operations	Dumping of munitions
Quantities of marine litter in relation to the economic sector														
Riverine inputs	No	No	No	No	No	No	No	No	No	High	No	No	No	No
Beach litter	Limited	Limited	Limited	No	No	Significant	No	No	No	High	No	No	No	No
Surface litter	Limited	Limited	High	Limited	No	High	No	No	Significant	Significant	No	No	Limited	Limited
Sea floor litter	Limited	Limited	High	Limited	No	High	No	No	Significant	Significant	No	No	Significant	Significant
Microplastics	Limited	Limited	Limited	Limited	No	Limited	Limited	No	Limited	High	No	No	No	No
Impacts from marine litter in relation to the economic sector														
Wildlife	Limited	Significant	Significant	Limited	No	High	Limited	No	High	High	No	No	Limited	Limited
Human health	No	No	Limited	No	No	Significant	No	No	Limited	Significant	No	No	Limited	Limited
Tourism	No	No	Limited	Limited	No	Significant	No	No	Limited	High	No	No	Limited	Limited
Fisheries	Limited	Limited	Limited	Limited	Limited	Significant	No	No	Limited	High	No	No	Limited	Limited
Navigation	Limited	Limited	Significant	Limited	No	Significant	No	No	Limited	Significant	No	No	Limited	Limited
Municipalities	No	No	Significant	No	No	Significant	No	No	Limited	High	No	No	No	No
Clean-up	No	No	No	No	No	Significant	No	No	Limited	No	No	Limited	No	No

High contribution | Significant contribution | Limited contribution | No contribution

Source: UNEP (2019).

1.7. Relevance to the Sustainable Development Goals and other frameworks

Global commitments on marine litter have been made in the context of the General Assembly and the United Nations Environment Assembly, as well as the Convention on Biological Diversity,[3] and in recent declarations by the Group of Seven (action plan to combat marine litter) and the Group of 20 (action plan on marine litter) (United Nations Environment Assembly (UNEA), 2019). In 2016, the United Nations Environment Assembly adopted resolution 2/11 on combating marine plastic litter and microplastics[4] and, in 2019, it published guidelines for the monitoring and assessment of plastic litter in the ocean.[5]

Marine debris is directly linked to Sustainable Development Goal 14 on conserving and sustainably using the oceans, seas and marine resources for sustainable development. Target 14.1 is currently classified as a tier III indicator, for which no internationally established methodology or standards are available (UNEA, 2019). To advance measurements of indicator 14.1.1,[6] more harmonized methods have been proposed to encourage the development and implementation of regional or global monitoring programmes and facilitate the exchange of results. The methods will help to move indicator 14.1.1 from tier III to tier II (for which conceptually clear, established methodology and standards exist, but data are not regularly produced).

Microplastics and nanoplastics also relate to Goal 12 on ensuring sustainable consumption and production patterns. Goal 11 should be mentioned as well since plastic marine litter also originates from the mismanaged waste of urban settlements. Solid waste ending in the ocean is directly related to Goal 6, as plastic litter and microplastics are carried by mismanaged wastewater and storm water.

In 2019, the Group of Seven reviewed the ongoing activities within the regional seas conventions and set priorities for further action, ensuring effective coordination through United Nations bodies to address monitoring and the socioeconomic impacts and consequences on human health and biota, as well as the involvement of industry in developing and implementing responses relating to the management of waste and prevention. Also, in the framework of the Basel Convention on the Control of Transboundary Movements of Hazardous Wastes and Their Disposal,[7] the parties adopted amendments to the annexes thereto to bring certain plastic wastes within the scope of the Convention in order to, inter alia, address the impacts of plastic on the marine environment.[8]

In addition to many national plans, interregional policies, such as the European Union plastics strategy of 2018 and its various legally binding directives (the Marine Strategy Framework Directive (2008/56/EC), the directive on port reception facilities (2019/883/EU) and the directive on single-use plastics (2019/904/EU)),[9] constitute a good example of an approach to address marine litter, taking into account circular economy principles, with many measures that are now being implemented (e.g., new materials, wastewater treatment, bans and extended producer responsibility).

---

3 United Nations, *Treaty Series*, vol. 1760, No. 30619.
4 See United Nations Environment Assembly of the United Nations Environment Programme, document UNEP/EA.2/Res.11.
5 Joint Group of Experts on the Scientific Aspects of Marine Environmental Protection, Report and Studies No. 99.
6 See General Assembly resolution 71/313, annex.
7 United Nations, *Treaty Series*, vol. 1673, No. 28911.
8 See United Nations Environment Programme, document UNEP-CHW.14-28. See also www.basel.int/TheConvention/ConferenceoftheParties/Meetings/COP14/tabid/7520/Default.aspx.
9 See https://eur-lex.europa.eu/legal-content/en/TXT/?uri=CELEX:32008L0056, https://eur-lex.europa.eu/eli/dir/2019/883/oj, and https://eur-lex.europa.eu/legal-content/EN/TXT/PDF/?uri=CELEX:32019L0904.

Many initiatives have been launched to include scientific, political, social and economic action in projects, from both the individual and the global system perspective. As an example, the free massive open online course on marine litter[10] is aimed at forming a global network of actors actively involved in addressing marine litter challenges. New tools such as mobile applications also enable citizens to log data on the location and type of debris found on coastlines and waterways in scientific databases. Other effective tools, such as the publicly available European Marine Observation and Data Network (EMODnet),[11] include digital maps of litter, thus providing a comprehensive tool for marine policy and society as a whole.

More than 60 countries have introduced bans and levies to curb single-use plastic waste (UNEP, 2018), often without data, metrics or monitoring in place to evaluate the effectiveness and consequences of such actions. Measures include a ban on certain items (such as plastic bags), the introduction of levies or deposit systems and voluntary agreements at the industry level.

A variety of existing measures have already been implemented (Food and Agriculture Organization of the United Nations, 2016), including gear marking; port state measures; onshore collection; payment for retrieved gear; better location and reporting of lost gear; disposal and recycling; alternatives to single-use plastics, especially polystyrene fish boxes; and awareness-raising schemes.

Pursuant to the International Convention for the Prevention of Pollution from Ships, 1973, as modified by the Protocol of 1978 relating thereto,[12] efforts to address marine litter include the development of a port reception facility database as a module of the International Maritime Organization (IMO) Global Integrated Shipping Information System.

1.8. Outlook

The management of marine litter pollution is exceptionally complex and requires an integrated approach, encompassing science, legislation, economics, ocean literacy, education, social participation and international cooperation on capacity-building, and technology transfer, as well as technical and financial support at multiple levels, from the global to the regional and local levels, owing to the diversity of the actors, sources, materials, socioeconomic aspects and regulatory frameworks involved. Without improved international policies and mobilization, plastic pollution will only worsen (Jambeck and others, 2015). It is estimated that, if current consumption patterns and waste management practices do not improve, there will be about 12 billion tons of plastic litter in landfills and the natural environment by 2050 (Geyer and others, 2017). The consequences will not be purely economic, and the environmental impact will be huge.

A variety of options exist to deal with critical levels of marine litter, some of which include approaches to address the issue, while understanding that not all are applicable to or supported by every country and some do not consider adverse impacts: the reduction of plastic consumption; support for eco-design and innovation (especially research into end-of-life plastic issues and alternatives); resource efficiency and better management of waste and water; long-term, efficient and viable recycling targets for municipal waste, packaging and plastic waste; greater use of policy instruments and control measures, including incentives, taxes and other regulatory measures, such as bans or extended producer responsibility schemes; and the adoption of remanufacturing initiatives and the coordination of policy investments in the waste sector (Ten Brink and others, 2018). There is also a

10 See https://sustainablehighereducation.com/2019/03/22/mooc2019.
11 See www.emodnet-bathymetry.eu/approach.
12 United Nations, *Treaty Series*, vol. 1340, No. 22484.

need for tight regulation and supervision of global waste trading, especially scrap plastic.

Plastic pollution is also a gateway to effective environmental education. The challenge is to change people's perceptions and understanding of the issue, so that they can see plastic pollution as a vector of education, awareness and literacy, as well as to find potential strategies to overcome political, economic and cultural barriers. Within the context of marine litter science, the objectives may be related to policy-relevant goals and thus increase the stimulus to citizens (GESAMP, 2019).

1.9. Key remaining knowledge and capacity-building gaps

For microplastics, major gaps in knowledge include the quantification of microplastics in the marine environment using standardized methods and information on how plastic degrades in various components of the marine environment and on the presence and impact of nanoplastics. More research is needed into the role of plastic debris as a transport vector of pathogens, antibiotic resistance, chemicals and biotoxins and the potential for dispersing diseases among marine life and human populations. Finally, in many countries, the lack of adequate national and regional monitoring of the quantities and impacts of marine litter, including plastics, is a major bottleneck for addressing the issue and for assessing the effectiveness of measures already taken.

Recent programmes (under the Commonwealth Scientific and Industrial Research Organisation, the University of Baltimore and the Marine Strategy Framework Directive) have been designed to answer some of the scientific questions surrounding the factors that govern the distribution of land-based litter and the quantity of litter that flows from the land to the sea. Outputs of the initiatives are expected to include data-driven estimates of leakage rates to the sea and help countries to understand where best to target effective interventions to stop debris from entering the ocean. With different methodologies developed to measure plastic leakage into waterways and oceans, from either mismanaged waste or in the form of microplastics, there is a need to harmonize the different approaches.

Most significantly, there are insufficient infrastructures and policies for recycling, or for wastewater and solid waste management (UNEP, 2017). In addition, and although illegal stakeholders may be active in solid waste collection and recovery, legislation is weak and there are huge disparities between countries with informal sectors, illegal manufacturing and black markets, which are limiting the implementation of reduction measures in relation to use, waste management and prevention (UNEP, 2019). There is, however, general agreement and a raft of initiatives from all stakeholders on implementing more sustainable patterns of production and consumption, including the circular economy, which is aimed at eliminating waste and the continual use of resources, by promoting reuse, sharing, repair, remanufacturing and recycling to create a closed-loop system. The recent actions taken by the United Nations Environmental Assembly at its fourth session (UNEA, 2019) largely support that approach, with resolutions on sustainable consumption, production and business practices, waste management and single-use plastic products.

Additional gaps include weak enforcement, separate collection, strong regional disparities between urban and rural areas and poor storm water management. Essential measures include those aimed at securing landfills, developing port waste management, promoting best practices for the fishing industry and improving maritime transport to limit container losses or primary microplastics spills.

Understanding that the reduction in plastic consumption should lead to a reduction in plastic waste generation, barriers in tackling marine litter and microplastics may be related to unsustainable patterns of consumption and production. Collaboration with the private sector and industry is needed to promote a shift towards sustainable solutions. Insufficient economic incentives may be an underlying reason for challenges related to changing behaviours. Finally, the design of chemical products and processes that reduce or eliminate the use or generation of hazardous substances is of particular interest for both producers and users of plastic (see table 4).

Table 4
Summary of knowledge and capacity-building gaps

Knowledge gaps
Incomplete knowledge about root causes of marine litter. Research is not being focused on the sources and fate of plastic pollution.
Accurate accounting methods and analytical tools for microplastics and nanoplastics related to throughput, detection limits, precision and quality are limited.
Polymer identification is complex and time-consuming for μm range particles.
Fragmented scientific knowledge (about scale of plastic pollution, microplastics, scientific and technical basis for monitoring, coordination on data, toxicity of plastics, risk assessment and fate).
Unknown effects on human health of ingestion of plastic-contaminated seafood.
Little knowledge of the contribution and impacts of abandoned, lost or otherwise discarded fishing gear and aquaculture-related marine litter.
Knowledge of the degradation of plastics and the leaching of additives or other chemical classes in different environments remains limited.
Scope and granularity of computational models are insufficiently developed.
Knowledge gaps regarding economic impacts of plastics on fisheries, tourism and maritime transportation. Links between marine litter fluxes and regional economy poorly understood.
Impact of marine plastic litter on climate change through extreme events and possible release of emissions or by limiting the ability of the ocean to act as a carbon sink needs to be studied further.
Extended producer responsibility is difficult to implement in some countries, especially archipelagic countries.
Lack of public awareness, behavioural change and circular economy models, with differences in educational levels depending on the countries.
Capacity-building gaps
Monitoring not in place in many parts of the world.
Technical difficulties in locating accumulation areas and specific types of litter (abandoned, lost or otherwise discarded fishing gear).
Technological shortcomings (i.e., deficiencies in waste management infrastructures). Solid policies must relate to environmentally sustainable and efficient waste management, recycling capacity and materials substitution.
Economic evaluation methodologies need to incorporate costs of plastic in the environment.
Lack of integrated decision-making at different levels and coordination in the establishment and implementation of programmes, including measures that target regional priorities.
Weak enforcement of measures.
Insufficient or inefficient waste treatment infrastructures and policies; inexistent waste management in many parts of the world.
Strong regional disparities between urban and rural areas.
Poor storm water management.
Inadequate infrastructures for waste collection, management, recycling and port reception.
Recyclability must be improved.
Collaboration and coordination with the private sector and industry are needed to reduce and transform the production, demand and consumption of plastic.
Awareness, information and education must be improved.

2. Dumping at sea, including garbage from ships and sewage sludge

2.1. Introduction

Dumping is any deliberate disposal of waste or other matter from vessels, aircraft, platforms or other human-made structures at sea, according to article 1, paragraph 5 (a) (i), of the United Nations Convention on the Law of the Sea,[13] the Convention on the Prevention of Marine Pollution by Dumping of Wastes and Other Matter, 1972 (the London Convention), and the 1996 London Protocol thereto.

The dumping of substances such as dredged material, sewage sludge, industrial waste, fish waste, discharges from vessels and human-made structures, organic and inorganic chemicals, radioactive material, war explosives and military chemicals has had an impact on marine ecosystems and created environmental challenges (Commission for the Protection of the Marine Environment of the North-East Atlantic (OSPAR), 2010b; IMO, 2018). In addition to the United Nations Convention on the Law of the Sea, to counter the environmental challenges resulting from waste dumping, the London Convention and the London Protocol contain provisions for controlling the unregulated dumping and incineration of waste at sea. Those regulatory requirements have been amended on several occasions (IMO, 2018). In addition, many countries have developed regional initiatives and approaches to control and assess waste dumping activities. Initiatives have also been taken in the framework of the Stockholm Convention on Persistent Organic Pollutants[14] and the Basel Convention[15] to address the control of transboundary movements of hazardous wastes and their disposal, as well as to protect human health and the environment from persistent organic pollutants.

2.2. Situation recorded in the first *World Ocean Assessment*

Chapter 24 of the first Assessment, on solid waste disposal (United Nations, 2017b), outlined the regulatory system relating to dumping and important international milestones, such as the adoption of the London Convention and the London Protocol. An overview of the regulatory techniques and waste streams covered under both instruments was provided, as well as efforts to understand the quantity and nature of waste and other matter being dumped. The Assessment also identified concerns about underreporting by many contracting parties to the London Convention and the London Protocol, leading to difficulties in deriving a clear picture for assessing regime implementation and understanding waste dumping status.

2.3. Changes in the state of dumping at sea

The London Protocol prohibited all dumping of waste, except for a limited number of categories such as: (a) dredged material; (b) sewage sludge; (c) fish waste, or material resulting from industrial fish processing operations; (d) vessels and platforms or other human-made structures at sea; (e) inert, inorganic geological material; (f) organic material of natural origin; (g) bulky items primarily comprising iron, steel, concrete and similar non-harmful materials, for which the concern is physical impact; and (h) sub-seabed sequestration of CO_2 streams in sub-seabed geological formations (IMO, 2018).

---

13 United Nations, *Treaty Series*, vol. 1833, No. 31363.
14 Ibid., vol. 2256, No. 40214.
15 Ibid., vol. 1673, No. 28911.

Changes in the overall waste dumping status can be understood by reviewing published waste dumping data and permits issued under the London Convention and the London Protocol (IMO, 2019). The sections below provide an overview of the different solid waste dumping categories.

2.3.1. Sewage sludge dumping

Sewage sludge dumping has an impact on sediment quality, benthic assemblages, aquatic flora and fauna and, in general, the whole marine ecosystem. Excessive nutrient loads from sewage discharges can lead to a reduction in oxygen content in water, cause marine life mortality and destroy entire habitats and ecosystems (see chap. 10). A total of 13 contracting parties reported the disposal of sewage sludge for the period from 1976 to 2016, with a total amount of 393 x 10^6 tons (IMO, 2019). Figure III shows that dumping has declined dramatically to the point that many contracting parties prohibit the activity and very few report disposal operations. In 2011, a total of 0.6 million tons was dumped while, in 2016, the quantity fell to only 0.00041 million tons.

Figure III
Amount of sewage sludge dumped

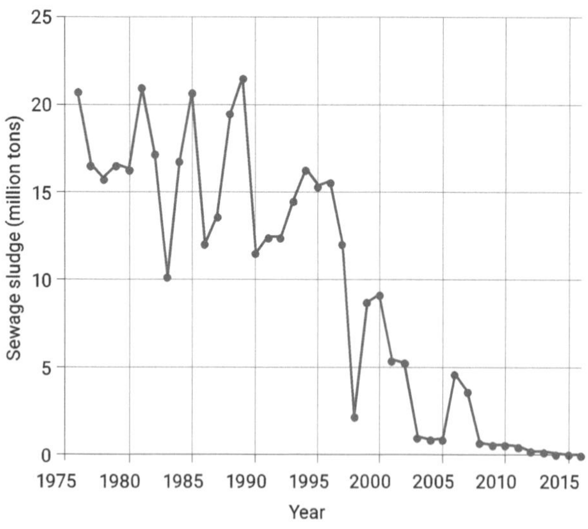

Source: IMO (2019).

In 2016, IMO published a report on the current state of knowledge regarding marine litter in waste dumped at sea, under the London Convention and the London Protocol. It sought to review whether sewage sludge or dredged material contained marine litter, depending on litter types, properties and quantities. It concluded that such an assessment was difficult at the time owing to an overall shortage of data, differences in methodology and reporting and the lack of systematic sampling in space and time (IMO, 2016a).

2.3.2. Disposal of vessels at sea

A total of 22 contracting parties to the London Convention and the London Protocol reported disposing of 758 vessels from 1976 to 2010 (IMO, 2016a). Some of the vessels were disposed of to create reefs (Hess and others, 2001), but in other cases the contracting parties only permitted the dumping of vessels when no land-based disposal options existed and the vessels were dumped in deeper waters and not for the purpose of creating reefs. Other disposal vectors from vessels include material for scientific experiments (IMO, 2016b).

2.3.3. Dumping of organic and inorganic wastes

Organic and inorganic wastes have long been disposed of at sea, primarily loaded from land and transported offshore for disposal from vessels and platforms. Many nations continue to use the ocean as an ongoing depository for certain wastes generated within their borders. A total of 15 contracting parties to the London Convention and the London Protocol reported disposing of inert, inorganic geological material at sea for the period from 1983 to 2010, with a total amount of 315,227 x 10^6 tons (IMO, 2016a). In 2011, 3.82248 million tons were dumped; in 2013, 1.453725 million tons; and in 2016, 1.229620 million tons (see figure IV).

Figure IV
Amount of inert, inorganic geological material licensed

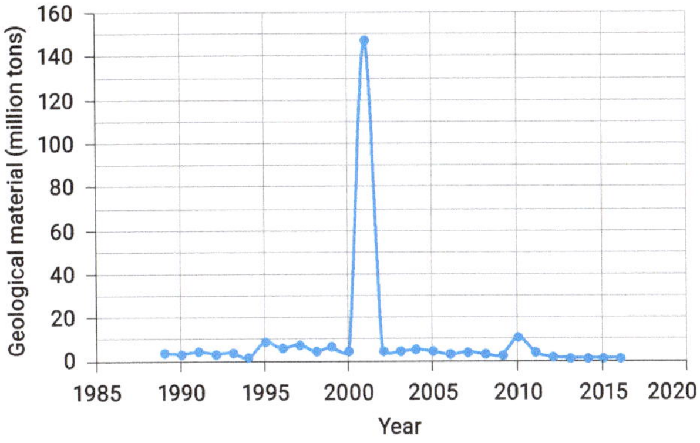

Source: IMO (2019).

Figure V
Global distribution of documented marine sites with munitions dumped at sea

Source: Global distribution of documented marine sites with munitions dumped at sea, produced by the James Martin Center for Nonproliferation Studies, Middlebury Institute of International Studies at Monterey (www.nonproliferation.org/chemical-weapon-munitions-dumped-at-sea).

Also, a total of 17 contracting parties to the London Convention and the London Protocol reported disposing of organic material of natural origin at sea during the period from 1977 to 2010, with a total amount, including spoilt cargo (consisting of organic material of natural origin), of 37,628 x 10^6 tons (IMO, 2016a). Spoilt cargo was reported as disposed of at sea (totalling 31,833 x 10^6 tons) by seven contracting parties from 2003 to 2010 (IMO, 2016a).

2.3.4. Dumping of industrial wastes and war chemicals

A total of 23 contracting parties reported the disposal of industrial wastes at sea from 1976 to 1995 (232 x 10^6 tons), including scrapped vessels, waste explosives in concrete, sludge, waste acids or alkali, cattle industry waste, glass dust, industrial dust, ceramics, ammunition, concrete pipes, demolition rubble, sodium hydrosulphite, sludge containing heavy metals and fluorides, titanium oxide production waste, chlorophenol production waste, chromate production waste, gunpowder, fly ash, fermentation waste and potassium mining waste (IMO, 2019).

War explosives and military chemicals dumped at sea since the First World War still present significant risks to the marine ecosystem and for various users of the sea (see figure V). Environmental samples typically show low concentrations of munitions compounds in water and sediments (in the order of ng/L and µg/kg, respectively), and the ecological risk appears to be generally low (Helsinki Commission, 2013; OSPAR, 2010a). Nonetheless, recent work demonstrates the possibility of sublethal genetic and metabolic effects in aquatic organisms (Beck and others, 2018).

Moreover, the catching of munitions in fisher nets, the interaction of explosives with submarine infrastructure or offshore installations, as well as related material floating to the surface, can lead to accidental burning or explosions (OSPAR, 2010a).

2.3.5. Incineration at sea

Incineration at sea is the disposal of waste at sea, using specially designed incinerator ships, by burning organochlorine compounds and other toxic wastes that are difficult to dispose of. Amendments to the London Convention that entered into force in 1994 banned the incineration at sea of industrial wastes, but incineration did not end until 2000 (IMO, 2016a).

2.4. Factors associated with changes

The present section covers various factors that led to changes in dumping practices, namely: (a) factors that triggered increased dumping activities in the past; and (b) sustained actions taken to mitigate such a grave environmental issue. For centuries, communities have disposed of waste in the ocean and seas, assuming that they were convenient and safe dumping grounds for diverting land-based pollution. Factors such as ignorance, negligence and a lack of proper waste disposal systems played an important role in harmful waste dumping practices, as did the lack of strict regulations and monitoring.

Improved scientific understanding, awareness in scientific communities and increased government participation, along with growing global concern, increased the need for international instruments to regulate the dumping of waste in the ocean (IMO, 2018). In addition to the United Nations Convention on the Law of the Sea, the regulatory measures taken in the framework of the London Convention and the London Protocol were important factors in improving the dumping situation.

Generic and comprehensive guidelines have been developed for all wastes that may be considered for dumping at sea (OSPAR, 2016; IMO, 2018). Also, guidance on national implementation of the London Protocol has been developed and updated and provides an outline of the types of action that States should consider at the national level. Based on the underreporting issues highlighted in the first *Assessment*, contracting parties to the London Protocol and the London Convention took further action to address the situation, including the adoption a strategic plan (IMO, 2019).

2.5. Impacts of changes on and interactions with other components of the marine system

The impacts of discharged materials on the marine ecosystem are the crux of the global solid waste dumping issue. Owing to the dynamic nature of the ocean, determining the fate of various dumped materials is a challenging task. Furthermore, the existence of different pollution sources and the complexity associated with tracking down specific contaminants make it difficult to establish the extent to which ocean dumping contributes to observed ecological effects and impacts. In general, dumping effects depend upon the type, quantity and quality of waste materials as well as the characteristics of affected areas of the ocean. In addition, the duration of the extended time period of dumping practices contributes to the ecological effects. To comprehend the dynamics, it is necessary to understand the possible impacts of major waste categories on marine components, as well as how changes in dumping practices are alleviating the problems (IMO, 2018).

Solid waste dumping in the ocean and seas can have varied impacts on the marine ecosystem, flora and fauna, as well as human beings who rely on saline water sources. These may include chemical pollution (see chap. 11), nutrient pollution and eutrophication (see chap. 10), water quality degradation, depletion of water oxygen levels, suffocation of marine creatures, decreased submerged vegetation, poisoning and death of oceanic plants and animals, and human health hazards. While different pollution pathways and associated sources exist, solid waste dumping actions carry their share of responsibility for the burden on the ocean and seas (IMO, 2018).

2.6. Ecosystem and socioeconomic consequences of continued changes in the system

In the ocean, undesired shifts between ecosystem states are caused by the combination of external forces that have an impact on the system and the internal resilience of the system. As resilience declines, the ecosystem becomes vulnerable and, as a consequence, progressively smaller external events can cause shifts. Thus, anthropogenic actions leading to perturbation increase the likelihood of undesired regime shifts (Scheffer and others, 2001).

Just as there is limited knowledge of socioeconomic consequences, the same holds true when assessing the consequences of continued change in the system. Ecosystem state shifts can cause large losses in terms of ecological and economic resources. Restoring a desired state may depend on the degradation affecting the system and require drastic and expensive intervention. One estimate suggests that removing litter from the wastewater streams of South Africa would cost about $279 million per year (Lane and others, 2007). With regard to other dumping activities, there are significant gaps in knowledge of socioeconomic consequences and market-based instruments.

2.7. Relevance to the Sustainable Development Goals

The issue of waste dumping is closely connected to Sustainable Development Goal 14, in particular targets 14.1 and 14.c. In the context of the present chapter, the relevant objectives of Goal 14 are also linked to Goal 12 on ensuring sustainable consumption and production patterns, as well as Goal 11 on making cities and human settlements inclusive, safe, resilient and sustainable. A considerable amount of work has been undertaken to further support the integration of the Goals among sectors, which may have a spillover effect on dumping at sea. Most notably, the Global Partnership on Waste Management (UNEP, 2010) is an important convergence and integration nexus, in particular as its six thematic areas include integrated waste management, marine litter and waste minimization. Pursuant to the London Convention, efforts to address marine litter include the development of a port reception facility database as a module of the IMO Global Integrated Shipping Information System.

2.8. Outlook

Drivers of change in reference to dumping are associated with modifications to the production and consumption patterns of materials that are currently dumped in the ocean. Whereas different and distinct waste streams are covered under the London Convention and the London Protocol, each stream is associated with separate industries and drivers that may lead to change. Therefore, changing production and consumption patterns need to include stakeholders from a diverse set of industries.

The strategic plan, adopted in 2016 at the thirty-eighth Consultative Meeting of Contracting Parties to the London Convention and the eleventh Meeting of Contracting Parties to the London Protocol, provides some indication of near- to medium-term development with regard to dumping (IMO, 2018). The plan outlines four strategic directions. Strategic direction 1 is aimed at promoting ratification of or accession to the London Protocol and outlines a target substantially to increase the rate per year of new ratifications or accessions thereto. Strategic direction 2 is aimed at enhancing the effective implementation of the London Protocol and the London Convention through the provision of technical assistance and support to the contracting parties and the development of guidance and measures to support implementation by addressing regulatory, scientific and technical barriers, as well as encouraging and facilitating improved compliance, including reporting, and the participation of the contracting parties in the work of both instruments. Strategic direction 3 is aimed at promoting the work of the London Protocol and the London Convention externally; and strategic direction 4 is aimed at identifying and addressing emerging issues in the marine environment within the scope of both instruments. To that end, several graded targets have been formulated, stating that, by 2030, 100 per cent of the contracting parties should be meeting their reporting obligations and have a national authority in place and appropriate legislative or regulatory authority to implement the London Convention and the London Protocol.

Future goals under both the London Convention and the London Protocol are the regulation of ocean fertilization and geoengineering and a review of the impacts of new marine "geoengineering" technologies. Further work is envisaged on the basis of collaboration between IMO (under the London Protocol), the United Nations and the Joint Group of Experts on the Scientific Aspects of Marine Environmental Protection on mine tailings, habitat destruction or restoration and marine litter, in order to address gaps in the international legal framework. Furthermore, easy online reporting will be introduced, a database established and monitoring activities reviewed. Finally, the environmental effects of the legacy of chemical munitions dumped at sea in the past will be addressed.

2.9. Key remaining knowledge and capacity-building gaps

Since the adoption of the United Nations Convention on the Law of the Sea, the London Convention and the London Protocol, regulations for solid waste disposal at sea were introduced by coastal States and significant progress has been made (IMO, 2018). However, owing to substantial underreporting by many contracting parties and a lack of published data, it is difficult to track implementation and understand the current extent of the challenge that exists.

Knowledge gaps include:

- Scale of the impacts of dumped fibre-reinforced plastic vessels
- Socioeconomic impacts of all waste streams allowed to be dumped, including the legacy of dumping activities
- Understanding of the impacts of relevant policies on dumping activities and marine environmental impacts (such as waste policies)
- Understanding of the extent and impact of marine litter
- Cumulative impacts of current and previous dumping activities and prevailing pollution from other sources

Capacity-building gaps include:

- Monitoring (and reporting) of dumping activities
- Understanding of the impacts of land-based activities on the amount of waste streams dumped in the ocean
- New techniques to manage the risks associated with dumped munitions, the development of guidelines on encounters with munitions (such as individuals working in the fishing industry, techniques for the safe removal and monitoring of possible effects of dumped munitions
- Development of sustainable alternatives to ocean-based dumping or prevention of the need for dumping by changing production patterns

While the dumping of most allowable waste streams has been significantly reduced, other waste streams may increase. Distant areas of the world are also increasingly connected as consumption, production and governance decisions influence materials, waste, energy and information flows in other countries, which can generate aggregate economic gains while shifting economic and environmental costs. With over 60 per cent of the urban infrastructure expected to exist by 2050 yet to be built, understanding the role of dumping activities from urban construction and development activities is crucial. The land-ocean impacts of those activities on the marine environment need to be considered.

References

Abayomi, Oyebamiji, and others (2017). Microplastics in coastal environments of the Arabian Gulf. *Marine Pollution Bulletin*, vol. 124, No. 1, pp. 181–188. http://doi.org/10.1016/j.marpolbul.2017.07.011.

Addamo, A.M., and others (2017). *Top Marine Beach Litter Items in Europe*. Luxembourg: Publications Office of the European Union. ISBN 978-92-79-87711-7. http://doi.org/10.2760/496717.

Amélineau, Françoise, and others (2016). Microplastic pollution in the Greenland Sea: background levels and selective contamination of planktivorous diving seabirds. *Environmental Pollution*, vol. 219, pp. 1131–1139.

Arias-Andrés, María, and others (2018). Microplastic pollution increases gene exchange in aquatic ecosystems. *Environmental Pollution*, vol. 237, pp. 253–261.

Arossa, Silvia, and others (2019). Microplastic removal by Red Sea giant clam (Tridacna maxima). *Environmental Pollution*, vol. 252, part. B.

Barboza, Luís Gabriel Antão, and others (2018). Marine microplastic debris: an emerging issue for food security, food safety and human health. *Marine Pollution Bulletin*, vol. 133, pp. 336–348.

Barnes, David K.A., and others (2009). Accumulation and fragmentation of plastic debris in global environments. *Philosophical Transactions of the Royal Society B: Biological Sciences*, vol. 364, No. 1526, pp. 1985–1998.

Barrows, Abigail, and others (2018). Marine environment microfiber contamination: global patterns and the diversity of microparticle origins. *Environmental Pollution*, vol. 237, pp. 275–284.

Beaumont, Nicola J., and others (2019). Global ecological, social and economic impacts of marine plastic. *Marine Pollution Bulletin*, vol. 142, pp. 189–195.

Beck, Aaron J., and others (2018). Spread, behavior, and ecosystem consequences of conventional munitions compounds in coastal marine waters. *Frontiers in Marine Science*, vol. 5, art. 141.

Beer, Sabrina, and others (2018). No increase in marine microplastic concentration over the last three decades: a case study from the Baltic Sea. *Science of the Total Environment*, vol. 621, pp. 1272–1279.

Bessa, Filipa, and others (2019). Microplastics in gentoo penguins from the Antarctic region. *Scientific Reports*, vol. 9, No. 1, art. 14191. https://doi.org/10.1038/s41598-019-50621-2.

Brandon, Jennifer A., and others (2019). Multidecadal increase in plastic particles in coastal ocean sediments. *Science Advances*, vol. 5, No. 9. https://doi.org/10.1126/sciadv.aax0587.

Carlton, James T., and others (2017). Tsunami-driven rafting: transoceanic species dispersal and implications for marine biogeography. *Science*, vol. 357, No. 6358, pp. 1402–1406.

Catarino, A.I., and others (2018). Low levels of microplastics (MP) in wild mussels indicate that MP ingestion by humans is minimal compared to exposure via household fibres fallout during a meal. *Environmental Pollution*, vol. 237, pp. 675–684

Chiba, Sanae and others (2018). Human footprint in the abyss: 30-year records of deep-sea plastic debris. *Marine Policy*, vol. 96, pp. 204–212.

Cincinelli, Alessandra, and others (2017). Microplastic in the surface waters of the Ross Sea (Antarctica): occurrence, distribution and characterization by FTIR. *Chemosphere*, vol. 175, pp. 391–400.

Claro, Francoise, and others (2019). Tools and constraints in monitoring interactions between marine litter and megafauna: Insights from case studies around the world. *Marine Pollution Bulletin*, vol. 141, pp. 147–160.

Collignon, Amandine, and others (2012). Neustonic microplastic and zooplankton in the North Western Mediterranean Sea. *Marine Pollution Bulletin*, vol. 64, No. 4, pp. 861–864. https://doi.org/10.1016/j.marpolbul.2012.01.011.

Commission for the Protection of the Marine Environment of the North-East Atlantic (OSPAR) (2010a). *Overview of Past Dumping at Sea of Chemical Weapons and Munitions in the OSPAR Maritime Area: 2010 Update*. London.

________ (2010b). Quantities of dredged material dumped. In *Quality Status Report 2010*. https://qsr2010.ospar.org/media/assessments/p00433_supplements/p00433_suppl_3_total_annual_amounts.pdf.

________ (2016). OSPAR annual report on dumping and placement of wastes or other matter at sea in 2014. Environmental Impacts of Human Activities Series.

________ (2017). *Marine Litter chapter*. OSPAR Intermediate Assessment 2017. https://oap.ospar.org/en/ospar-assessments/intermediate-assessment-2017/pressures-human-activities/marine-litter.

Compa, Montserrat, and others (2019). Spatio-temporal monitoring of coastal floating marine debris in the Balearic Islands from sea-cleaning boats. *Marine Pollution Bulletin*, vol. 141, pp. 205–214.

Courtene-Jones, Winnie, and others (2019). Consistent microplastic ingestion by deep-sea invertebrates over the last four decades (1976–2015), a study from the North East Atlantic. *Environmental Pollution*, vol. 244, pp. 503–512.

Cózar A., and others (2014). Plastic debris in the open ocean. *Proceedings of the National Academy of Sciences of the United States of America*, vol. 111, No. 28, pp. 10239–10244.

Cózar, A., and others (2017). The Arctic Ocean as a dead end for floating plastics in the North Atlantic branch of the Thermohaline Circulation. *Science Advances*, vol. 3, No. 4, p. e1600582.

Crippa, Maurizio, and others (2019). *A Circular Economy for Plastics: Insights from Research and Innovation to Inform Policy and Funding Decisions*. M.D. Smet and M. Linder, eds. Brussels: European Commission.

Domènech, F., and others (2019). Two decades of monitoring in marine debris ingestion in loggerhead sea turtle, *Caretta caretta*, from the western Mediterranean. *Environmental Pollution*, vol. 244, pp. 367–378.

Eriksen, Marcus, and others (2014). Plastic pollution in the world's oceans: more than 5 trillion plastic pieces weighing over 250,000 tons afloat at sea. *PLOS One*, vol. 9, No. 12, p. e111913.

Eriksen, Marcus, and others (2018). Microplastic sampling with the AVANI trawl compared to two neuston trawls in the Bay of Bengal and South Pacific. *Environmental Pollution*, vol. 232, pp. 430–439.

European Food Safety Authority Panel on Contaminants in the Food Chain (2016). Presence of microplastics and nanoplastics in food, with particular focus on seafood. *EFSA Journal*, vol. 14, No. 6, e04501.

Everaert, Gert, and others (2018). Risk assessment of microplastics in the ocean: modelling approach and first conclusions. *Environmental Pollution*, vol. 242, pp. 1930–1938.

Fang, Chao, and others (2018). Microplastic contamination in benthic organisms from the Arctic and sub-Arctic regions. *Chemosphere*, vol. 209, pp. 298–306.

Filho W., and others (2019). Plastic debris on Pacific islands: ecological and health implications. *Science of the Total Environment*, vol. 670, pp. 181–187, https://doi.org/10.1016/j.scitotenv.2019.03.181.

Food and Agriculture Organization of the United Nations (2016). *The State of World Fisheries and Aquaculture: Contributing to Food Security and Nutrition for All*. Rome.

Gaibor, Nikita, and others (2020). Composition, abundance and sources of anthropogenic marine debris on the beaches from Ecuador: a volunteer-supported study. *Marine Pollution Bulletin*, vol. 154, art. 111068. http://doi.org/10.1016/j.marpolbul.2020.111068.

García-Rivera, Santiago, and others (2018). Spatial and temporal trends of marine litter in the Spanish Mediterranean seafloor. *Marine Pollution Bulletin*, vol. 137, pp. 252–261.

Gerigny, O., and others (2019). Seafloor litter from the continental shelf and canyons in French Mediterranean water: distribution, typologies and trends. *Marine Pollution Bulletin*, vol. 146, pp. 653–666.

Geyer, Roland, and others (2017). Production, use, and fate of all plastics ever made. *Science Advances*, vol. 3, No. 7, p. e1700782.

Helsinki Commission (2013). *Chemical Munitions Dumped in the Baltic Sea. Report of the Ad Hoc Expert Group to Update and Review the Existing Information on Dumped Chemical Munitions in the Baltic Sea*. Baltic Sea Environment Proceeding (BSEP), No. 142.

Hess, Ronald W., and others (2001). *Disposal Options for Ships*. Santa Monica, California: RAND Corporation. www.rand.org/pubs/monograph_reports/MR1377.html.

Hidalgo-Ruz, Valeria, and others (2018). Spatio-temporal variation of anthropogenic marine debris on Chilean beaches. *Marine Pollution Bulletin*, vol. 126, pp. 516–524.

Hong, Sunwook, and others (2017). Navigational threats by derelict fishing gear to navy ships in the Korean seas. *Marine Pollution Bulletin*, vol. 119, No. 2, pp. 100–105. https://doi.org/10.1016/j.marpolbul.2017.04.006.

Intergovernmental Science-Policy Platform on Biodiversity and Ecosystem Services (2019). *Global Assessment Report on Biodiversity and Ecosystem Services: Summary for Policymakers*. J. Brondizio, S. Settele and H.T.N. Díaz, eds. Bonn: IPBES secretariat. www.ipbes.net/global-assessment-biodiversity-ecosystem-services.

International Maritime Organization (IMO) (2016a). *Overview of Statistics of Dumping Permits for the Period from 1972–2010 for the Twentieth Anniversary of the Adoption of the London Protocol*. Final Report on Permits Issued in 2010, LC 38-7-1.

________ (2016b). *Review of the Current State of Knowledge Regarding Marine Litter in Wastes Dumped at Sea under the London Convention and Protocol*. Final Report.

________ (2018). *Report of the Forty-First Meeting of the Scientific Group of the London Convention and the Twelfth Meeting of the Scientific Group of the London Protocol*, LC/SG 41/16.

________ (2019). *London Convention and Protocol Overview of Statistics of Dumping Permits for the Period 1976 to 2016 (2019)*. Direct Communication from the Secretariat for London Convention/Protocol and Ocean Affairs.

Ioakeimidis C., and others (2017). Occurrence of marine litter in the marine environment: a world panorama of floating and seafloor plastics. In *Hazardous Chemicals Associated with Plastics in Environment*. H. Takada and H.K. Karapanagioti, eds. Handbook of Environmental Chemistry Series, vol. 78. Cham, Switzerland: Springer, pp. 93–120. https://link.springer.com/chapter/10.1007/698_2017_22.

Isobe, Atsuhiko, and others (2015). East-Asian seas: a hot spot of pelagic microplastics. *Marine Pollution Bulletin*, vol. 101, No. 2, pp. 618–623.

Isobe, Atsuhiko, and others (2017). Microplastics in the Southern Ocean. *Marine Pollution Bulletin*, vol. 114, No. 1, pp. 623–626.

Ivar do Sul, Juliana A., and others (2014). Microplastics in the pelagic environment around oceanic islands of the Western Tropical Atlantic Ocean. *Water, Air, and Soil Pollution*, vol. 225, No. 7, art. 2004.

Jambeck, Jenna R., and others (2015). Plastic waste inputs from land intothe ocean. Science, vol. 347, No. 622, pp. 768–771.

Jamieson, Alan J., and others (2019). Microplastics and synthetic particles ingested by deep-sea amphipods in six of the deepest marine ecosystems on Earth. *Royal Society Open Science*, vol. 6, No. 2, art. 180667.

Joint Group of Experts on the Scientific Aspects of Marine Environmental Protection (GESAMP) (2016). *Sources, Fate and Effects of Microplastics in the Marine Environment: Part 2 of a Global Assessment*. P.J. Kershaw and C.M. Rochman, eds. GESAMP Report and Studies Series, No. 93. IMO/FAO/UNESCO-IOC/UNIDO/WMO/IAEA/UN/UNEP/UNDP Joint Group of Experts on the Scientific Aspects of Marine Environmental Protection.

________ (2019). *Guidelines for the Monitoring and Assessment of Plastic Litter and Microplastics in the Ocean*. P.J. Kershaw and F. Galgani, eds. GESAMP Report and Studies Series, No. 99.

Kako, Shin'ichiro, and others (2014). A decadal prediction of the quantity of plastic marine debris littered on beaches of the East Asian marginal seas. *Marine Pollution Bulletin*, vol. 81, No. 1, pp. 174–184.

Kanhai, La Daana K., and others (2018). Microplastics in sub-surface waters of the Arctic Central Basin. *Marine Pollution Bulletin*, vol. 130, pp. 8–18. https://doi.org/10.1016/j.marpolbul.2018.03.011.

Kühn, Fabienne, and others (2018). Plastic ingestion by juvenile polar cod (*Boreogadus saida*) in the Arctic Ocean. *Polar Biology*, vol. 41, pp. 1269–1278. https://doi.org/10.1007/s00300-018-2283-8.

Lacerda, Ana L.D.F., and others (2019). Plastics in sea surface waters around the Antarctic Peninsula. *Scientific Reports*, vol. 9, No. 1, art. 3977.

Lane, S.B., and others (2007). Regional overview and assessment of marine litter related activities in the West Indian Ocean region. *Report to the United Nations Program*.

Lavender Law, Kara, and others (2010). Plastic accumulation in the North Atlantic subtropical gyre. *Science*, vol. 329, No. 5996, pp. 1185–1188. http://doi.org/10.1126/science.1192321.

Lavender Law, Kara, and others (2014). Distribution of surface plastic debris in the eastern Pacific Ocean from an 11-year data set. *Environmental Science and Technology*, vol. 48, No. 9, pp. 4732–4738.

Lavender Law, Kara, and others (2017). Plastics in the marine environment. *Annual Review of Marine Science*, vol. 9, pp. 205–229.

Lavers, Jennifer L., and Alexander L. Bond (2017). Exceptional and rapid accumulation of anthropogenic debris on one of the world's most remote and pristine islands. *Proceedings of the National Academy of Sciences*, vol. 114, No. 23, pp. 6052–6055.

Lebreton, L., and others (2017). River plastic emissions to the world's oceans. *Nature Communications*, vol. 8, art. 15611.

Lechner, Aaron, and others (2014). The Danube so colourful: a potpourri of plastic litter outnumbers fish larvae in Europe's second largest river. *Environmental Pollution*, vol. 188, pp. 177–181.

Leclerc, Lisa-Marie E., and others (2012). A missing piece in the Arctic food web puzzle? Stomach contents of Greenland sharks sampled in Svalbard, Norway. *Polar Biology*, vol. 35, No. 8, pp. 1197–1208. https://doi.org/10.1007/s00300-012-1166-7.

Lusher, Amy (2015). Microplastics in the Marine Environment: Distribution, Interactions and Effects. In *Marine Anthropogenic Litter*. Springer, pp. 245–307.

Macfadyen, Graeme, and others (2009). *Abandoned, Lost or Otherwise Discarded Fishing Gear*. UNEP Regional Seas Reports and Studies, No. 185 and FAO Fisheries and Aquaculture Technical Paper, No. 523. Rome: FAO.

Maes, Thomas, and others (2018). Below the surface: twenty-five years of seafloor litter monitoring in coastal seas of North West Europe (1992–2017). *Science of the Total Environment*, vol. 630, pp. 790–798. https://doi.org/10.1016/j.scitotenv.2018.02.245.

Markic, Ana, and others (2018). Double trouble in the South Pacific subtropical gyre: increased plastic ingestion by fish in the oceanic accumulation zone. *Marine Pollution Bulletin*, vol. 136, pp. 547–564.

Martínez-Vicente, Víctor, and others (2019). Measuring marine plastic debris from space: initial assessment of observation requirements. *Remote Sensing*, vol. 11, No. 20.

Maximenko, Nikolai, and others (2019). Toward the Integrated Marine Debris Observing System. *Frontiers in Marine Science*, vol. 6, art. 447. https://doi.org/10.3389/fmars.2019.00447.

Ministry of Ecology and Environment, China (2019). *Bulletin of Marine Ecological Environmental Status of China in 2018*. http://hys.mee.gov.cn/dtxx/201905/P020190529532197736567.pdf.

McIlgorm, A., and others (2020). *Update of 2009 APEC report on Economic Costs of Marine Debris to APEC Economies*. A report to the APEC Ocean and Fisheries Working Group by the Australian National Centre for Ocean Resources and Security (ANCORS). Australia: University of Wollongong.

Miranda-Urbina, Diego, and others (2015). Litter and seabirds found across a longitudinal gradient in the South Pacific Ocean. *Marine Pollution Bulletin*, vol. 96, Nos. 1–2, pp. 235–244.

Monteiro, Raqueline C.P., and others (2018). Plastic pollution in islands of the Atlantic Ocean. *Environmental Pollution*, vol. 238, pp. 103–110.

Morgana, Silvia, and others (2018). Microplastics in the Arctic: a case study with sub-surface water and fish samples off Northeast Greenland. *Environmental Pollution*, vol. 242, pp. 1078–1086.

Munari, Cristina, and others (2017). Microplastics in the sediments of Terra Nova Bay (Ross Sea, Antarctica). *Marine Pollution Bulletin*, vol. 122, Nos. 1–2, pp. 161–165.

Murray, Cathryn Clarke, and others (2018). The influx of marine debris from the Great Japan Tsunami of 2011 to North American shorelines. *Marine Pollution Bulletin*, vol. 132, pp. 26–32.

Nielsen, Julius, and others (2014). Distribution and feeding ecology of the Greenland shark (*Somniosus microcephalus*) in Greenland waters. *Polar Biology*, vol. 37, No. 1, pp. 37–46. https://doi.org/10.1007/s00300-013-1408-3.

Ostle, Clare, and others (2019). The rise in ocean plastics evidenced from a 60-year time series. *Nature Communications*, vol. 10, No. 1, art. 1622.

Peeken, Ilka, and others (2018). Microplastics in the marine realms of the Arctic with special emphasis on sea ice. *Arctic Report Card*, vol. 2018, pp. 89–99.

Petry, Maria V., and Victória R.F. Benemann (2017). Ingestion of marine debris by the white-chinned petrel (*Procellaria aequinoctialis*): is it increasing over time off southern Brazil? *Marine Pollution Bulletin*, vol. 117, Nos. 1–2, pp. 131–135.

Pham, Christopher, and others (2014). Marine litter distribution and density in European seas, from the shelves to deep basins. *PLOS One*, vol. 9, No. 4, p. e95839.

Pierdomenico, Martina, and others (2019). Massive benthic litter funnelled to deep sea by flash-flood generated hyperpycnal flows. *Scientific Reports*, vol. 9, art. 5330. www.nature.com/articles/s41598-019-41816-8.

PlasticsEurope (2019). *Plastics: The Facts 2018 – An Analysis of European Plastics Production, Demand and Waste Data*. Brussels.

Reisser, Julia, and others (2014). Millimeter-sized marine plastics: a new pelagic habitat for microorganisms and invertebrates. *PLOS One*, vol. 9, No. 6, p. e100289.

Richardson, Kelsey, and others (2019). Estimates of fishing gear loss rates at a global scale: a literature review and meta-analysis. *Fish and Fisheries*, vol. 20, No. 6, pp. 1218–1231. https://doi.org/10.1111/faf.12407.

Rochman, Chelsea M. (2018). Microplastics research: from sink to source. *Science*, vol. 360, No. 6384, pp. 28–29.

Rochman, Chelsea M., and others (2016). The ecological impacts of marine debris: unraveling the demonstrated evidence from what is perceived. *Ecology*, vol. 97, No. 2, pp. 302–312. https://doi.org/10.1890/14-2070.1.

Scheffer, Marten, and others (2001). Catastrophic shifts in ecosystems. *Nature*, vol. 413, No. 6856, pp. 591–596.

Scheld, Andrew, and others (2016). The dilemma of derelict gear. *Scientific Reports*, vol. 6, art. 19671. https://doi.org/10.1038/srep19671.

Schmidt, Christian, and others (2017). Export of plastic debris by rivers into the sea. *Environmental Science and Technology*, vol. 51, No. 21, pp. 12246–12253. https://pubs.acs.org/doi/full/10.1021/acs.est.7b02368.

Schulz, Marcus, and others (2013). A multi-criteria evaluation system for marine litter pollution based on statistical analyses of OSPAR beach litter monitoring time series. *Marine Environmental Research*, vol. 92, pp. 61–70.

Schuyler, Qamar A., and others (2016). Risk analysis reveals global hotspots for marine debris ingestion by sea turtles. *Global Change Biology*, vol. 22, No. 2, pp. 567–576.

Science Advice for Policy by European Academies (SAPEA) (2019). *A Scientific Perspective on Microplastics in Nature and Society*. Berlin. https://doi.org/10.26356/microplastics.

Strafella, P., and others (2019). Assessment of seabed litter in the Northern and Central Adriatic Sea (Mediterranean) over six years. *Marine Pollution Bulletin*, vol. 141, pp. 24–35.

Tekman, Mine B., and others (2017). Marine litter on deep Arctic seafloor continues to increase and spreads to the North at the HAUSGARTEN observatory. *Deep Sea Research Part I: Oceanographic Research Papers*, vol. 120, pp. 88–99.

Ten Brink, Patrick, and others (2018). Circular economy measures to keep plastics and their value in the economy, avoid waste and reduce marine litter. *Economics*. Discussion paper, No. 2018-3, pp. 1–15. www.economics-ejournal.org/economics/discussionpapers/2018-3.

Thiel, Martin, and others (2018). Impacts of marine plastic pollution from continental coasts to subtropical gyres: fish, seabirds, and other vertebrates in the SE Pacific. *Frontiers in Marine Science*, vol. 5, art. 238.

Topouzelis, Konstantinos, and others (2019). Detection of floating plastics from satellite and unmanned aerial systems (Plastic Litter Project 2018). *International Journal of Applied Earth Observation and Geoinformation*, vol. 79, pp. 175–183. https://doi.org/10.1016/j.jag.2019.03.011.

United Nations (2017a).*The First Global Integrated Marine Assessment: World Ocean Assessment I*. Cambridge: Cambridge University Press.

________ (2017b). Chapter 24: Solid waste disposal. In *The First Global Integrated Marine Assessment: World Ocean Assessment I*. Cambridge: Cambridge University Press.

United Nations Environment Assembly (UNEA) (2019). *Resolutions and Decisions Adopted by the Committee of the Whole of the United Nations Environment Assembly at Its Fourth Session on 11 – 15 March 2019. Ministerial Declaration, Resolutions and Decisions for UNEA 4*. https://environmentassembly.unenvironment.org/proceedings-report-ministerial-declaration-resolutions-and-decisions-unea-4.

United Nations Environment Programme (2010). *Global Partnership on Waste Management*.

________(2014) *Valuing Plastics: The Business Case for Measuring, Managing and Disclosing Plastic Use in the Consumer Goods Industry*. Nairobi.

________(2017). *Marine Litter Socio Economic Study*. Nairobi.

________(2018). *Single-Use Plastics: A Roadmap for Sustainability*. Nairobi.

________(2019). *State of the Environment and Development in the Mediterranean. UNEP MAP.* Information document of the 21st Meeting of the Contracting Parties to the Barcelona Convention. Naples, Italy, 2–5 December 2019. UNEP/MED IG.24/Inf.11. Nairobi.

United Nations Environment Programme (UNEP) and GRID-Arendal (2016). *Marine Litter Vital Graphics.* Nairobi.

Van Cauwenberghe, Lisbeth, and others (2013). Microplastic pollution in deep-sea sediments. *Environmental Pollution*, vol. 182, pp. 495–499.

Van der Hall, N., and others (2017). Exceptionally high abundances of microplastics in the oligotrophic Israeli Mediterranean coastal waters. *Marine Pollution Bulletin*, vol. 116, pp. 151–155.

Van Emmerick, Tim, and others (2018). Methodology to characterize riverine macroplastic emission into the ocean. *Frontiers in Marine Science*, vol. 5, art. 372. https://doi.org/10.3389/fmars.2018.00372.

Van Franeker, Jan A., and Kara Lavender Law (2015). Seabirds, gyres and global trends in plastic pollution. *Environmental Pollution*, vol. 203, pp. 89–96.

Van Sebille, Erik, and others (2015). A global inventory of small floating plastic debris. *Environmental Research Letters*, vol. 10, No. 12, p. 124006.

Veerasingam, S., and others (2016). Characteristics, seasonal distribution and surface degradation features of microplastic pellets along the Goa coast, India. *Chemosphere*, vol. 159, pp. 496–505.

Viršek, Manca Kovač, and others (2017). Microplastics as a vector for the transport of the bacterial fish pathogen species *Aeromonas salmonicida*. *Marine Pollution Bulletin*, vol. 125, Nos. 1–2, pp. 301–309.

Waller, Catherine, and others (2017). Microplastics in the Antarctic marine system: an emerging areaof research. *Science of the Total Environment*. vol. 598, pp. 220–227. https://doi.org/10.1016/j.scitotenv.2017.03.283.

Walther, Bruno A., and others (2018). Type and quantity of coastal debris pollution in Taiwan. *Marine Pollution Bulletin*, vol. 135, pp. 862–872.

Wilcox, Chris, and others (2018). A quantitative analysis linking sea turtle mortality and plastic debris ingestion. *Scientific Reports*, vol. 8, No. 1, p. 12536. https://doi.org/10.1038/s41598-018-30038-z.

Wilcox, Chris, and others (2019). Abundance of Floating Plastic Particles Is Increasing in the Western North Atlantic Ocean. *Environmental Science and Technology*, vol. 54, No. 2, pp. 790–796. https://doi.org/10.1021/acs.est.9b04812.

Woodall, Lucy C., and others (2014). The deep sea is a major sink for microplastic debris. *Royal Society Open Science*, vol. 1, No. 4, art. 140317.

Woodall, Lucy C., and others (2015). Deep-sea litter: a comparison of seamounts, banks and a ridge in the Atlantic and Indian Oceans reveals both environmental and anthropogenic factors impact accumulation and composition. *Frontiers in Marine Science*, vol. 2, art. 3.

Wright, Stephanie, and Franck Kelly (2017). Plastic and human health: a micro issue? *Environmental Science. and Technology*, vol. 51, No. 12, pp. 6634–6647. https://doi.org/10.1021/acs.est.7b00423.

Zablotski, Yury, and Sarah B.M. Kraak (2019). Marine litter on the Baltic seafloor collected by the international fish-trawl survey. *Marine Pollution Bulletin*, vol. 141, pp. 448–461. https://doi.org/10.1016/j.marpolbul.2019.02.014.

Zhou, Changchun, and others (2016). Assessment of marine debris in beaches or seawaters around the China seas and coastal provinces. *Waste Management*, vol. 48, pp. 652–660.

Chapter 13
Changes in erosion and sedimentation

Contributors: Ca Thanh Vu (convener and lead member), Paulette Bynoe, Trang Minh Duong, Matt Eliot, Frank Hall, Roshanka Ranasinghe, Matthieu de Schipper and Joshua T. Tuhumwire (co-lead member).

Keynote points

- Coastal erosion can lead to coastal retreat, habitat destruction and loss of land, which result in significant negative ecological and socioeconomic impacts on the global coastal zones.
- Sediment budget and geology determine coastal morphology and dynamics, which influence the nature and health of coastal ecosystems. Human activities affecting the sediment dynamics, both on the coast and on land, modify the naturally occurring patterns of erosion and sedimentation.
- Globally, the abstraction or interruption of sediment supplies to and along the coast has been increasing, through upstream dams, coastal and river sand mining, and coastal infrastructures. Reduced sediment supply enhances shoreline retreat.
- Distinct from sand or muddy coasts, cliffs experience progressive erosion, which is largely caused by a combination of geotechnical instability, weathering on the upper cliff profile and wave action on the lower profile.
- The results of recent investigations reveal that, at approximately 15 per cent of all sandy beaches worldwide, the shoreline has been retreating, with an average trend of 1 m or more per year over the past 33 years, while almost half of the world's sandy beaches are currently stable.
- Many areas of the observed historical shoreline advance are related to reclamation and impoundment by coastal structures. Those human activities modify coastal dynamics, typically resulting in downdrift erosion.
- Climate change impacts, including sea level rise and potential increases in the frequency and intensity of severe tropical and extratropical storms, can accelerate coastal erosion. Human activities have the strongest impacts on deltas and adjacent coasts, with potentially severe impacts on other coastal systems, such as sand spits, barrier islands and wave-dominated estuaries.

1. Introduction

Coastal erosion and subsequent damage to coastal properties were briefly discussed in chapter 26 of the first *World Ocean Assessment* (United Nations, 2017a). However, there was limited discussion of the wider causes, geographical distribution and impacts of coastal erosion and sedimentation, the effects of the increased use of coastal protection structures, the impacts of coastal erosion on coastal ecological systems, and the capacity for modelling and forecasting coastal erosion and sedimentation.

The above-mentioned gaps are addressed in the present chapter, with a particular focus on the trends and changes to coastal erosion and sedimentation patterns over the period 2010–2020, following the baseline state described in the first Assessment (United Nations, 2017c). Aspects that have been considered include changes in river management that alter the sediment supply to the coasts; sand mining, dredging and the disposal of dredged materials; changes in coastal infrastructures affecting coastal sediment transport processes; coastal erosion and sedimentation with respect to coastal and ocean ecological systems and social economics (natural resources or capital, livelihoods and well-being); management practices for coastal erosion and sedimentation prevention; and advances in knowledge and capacity that have contributed to the evaluation of the changes in state.

2. Changes in state of coastal erosion and sedimentation

Factors that influence coastal erosion and sedimentation encompass characteristics of coastal sediment, exchanges between the land, the coast and the shelf, and geomorphic responses to oceanic forcing. Human activities may both substantially influence and be affected by coastal erosion and sedimentation (Hapke and others, 2013; Angamuthu and others, 2018; Mentaschi and others, 2018).

A modern evaluation of the change to deltaic sediment supply was undertaken using satellite imagery approaches with consideration of sediment trapping on the floodplain or estuaries (Nyberg and others, 2018); relative distribution between the shelf and the coast; and fluvial sediment mobility compared with in situ material on muddy, sandy or rocky coasts. Factors that influence geographically variable shoreline responses to sediment availability include underlying geological frameworks, wave action, tidal hydrodynamics, aeolian processes and ecomorphodynamic feedback, such as for dunes or mangroves (Moore and others, 2018).

Widespread impacts owing to human activities can occur if longshore sediment transport is disrupted by the installation of coastal structures or sand mining (Hapke and others, 2013; International Council for the Exploration of the Sea, 2016). Furthermore, low-lying coastal areas that are identified as being sensitive to the projected, rapid sea level rise include coastal wetlands, barrier coasts, deltas and small islands (Nicholls and others, 1999).

Until recently, there has been no reliable global-scale assessment of the occurrence of sandy beaches or their rates of shoreline morphological change. Exploiting the increased availability of satellite images, advanced image processing analysis techniques and computing resources, Luijendijk and others (2018a) presented an up-to-date global assessment of the occurrence and evolution of sandy shorelines using a fully automated analysis of 33 years (1984–2016) of satellite imagery. Their analysis showed that 31 per cent of the ice-free world shoreline is sandy, with the highest presence of sandy beaches reported in Africa (66 per cent), although the nature and characteristics of those beaches examined in the study vary substantially.

2.1. Changes in drivers

Human civilizations originated and thrived in the floodplains and the deltaic coastal zones of the world's large rivers, which are now inhabited by about 2.7 billion people (Best, 2019). The rapid increase in the demand for water, food, land and power has led to human interventions, such as the construction of large dams, deforestation, intensive agriculture expansion, urbanization, infrastructural construction and sand mining. Such human activities have placed those systems under immense stress, leading to large-scale and irreversible changes.

According to the International Commission on Large Dams (2018), globally, there are 59,071 dams with heights of more than 15 m and related reservoirs of more than 3 million m^3. The largest densities of hydropower dams are found in South America, South Asia and Northern Europe. The largest dams, including those have been built, are under construction or are planned, are located in the Mekong River basin, the Amazon River basin and the Congo River basin (Kondolf and others, 2014; Warner and others, 2019).

The construction of dams and reservoirs can reduce the sediment supply to the coast by different degrees (Slagel and Griggs, 2008), sometimes by more than 50 per cent (Besset and others, 2019), leading to the erosion of deltas and adjacent coasts. The reduction in sediment supply to the coasts is expected to increase greatly in the twenty-first century (Dunn and others, 2018), by 50 to 100 per cent

(Kondolf and others, 2014; Besset and others, 2019). For example, in the Pearl River, China, the construction of two mega dams (Yangtan and Longtan) has reduced the fluvial sediment supply to the coast by 70 per cent over the period 1992–2013 (Ranasinghe and others, 2019). Kondolf and others (2014) found that 140 dams had been built, were under construction or were planned for the Mekong River or its tributaries. Under a "definite future", if 38 dams that are planned or are under construction are actually completed, the cumulative sediment reduction to the Mekong Delta would be 51 per cent; and if all dams that are planned and under construction are completed, there would be a cumulative sediment reduction to the Mekong Delta of 96 per cent. That would lead to a serious decay of mangrove systems and, as a consequence, the erosion of the coast and irreversible changes in the surrounding ecosystem. On the other hand, there are substantial efforts in States to remove large dams, such as the Elwha Dam in Washington State, United States (Warrick and others, 2015).

Sand mined from rivers, beaches and coastal seabeds is used for land reclamation, beach nourishment and industry (Bendixen and others, 2019). That removes significant amounts of sand that would otherwise contribute to littoral transport, consequently resulting in a coastal sediment deficit (Montoi and others, 2017) and affecting the coastal morphology (International Council for the Exploration of the Sea, 2016; Abam and Oba, 2018). Presently, coastal beach and seabed sand mining is common practice in many countries, although it is sometimes illegal. Sand mining, in general, is known to take place in 73 countries on five continents, although there is no reliable figure on the practice worldwide (Peduzzi, 2014; Jayappa and Deepika, 2018).

2.2. Changes in pressure

Economics and population growth commonly drive human occupation of the coastal zone, which is offset by the socioeconomic costs of coastal management and adverse effects upon coastal ecosystem services. The balance between those pressures is commonly challenged by jurisdictional or economic divisions, with benefits and impacts often separated geographically (e.g., updrift accretion and downdrift erosion affect different communities) or occurring over different time scales (e.g., building a sea wall may defer the erosion pressure by a generation, but may effectively commit a community to subsequent construction of additional or larger works).

Secular changes to erosion and sedimentation may exceed the tolerance of coastal systems to adjust. For natural systems, such changes can lead to a loss of ecosystem services (Xu and others, 2019). Human activities may be intolerant of coastal dynamics, such as infrastructure that may be damaged or lose function owing to changing shoreline or seabed position. The perceived need to respond to erosion or sedimentation generally depends on the nature of human activities in the coastal zone, as follows:

(a) Port facilities, including harbour basins and navigable access channels, typically extend across the bulk of the active coastal zone, and the retention of port functions frequently requires coastal sediment management using breakwaters and dredging (see also chap. 14);

(b) Substantial urban growth has occurred along the coasts since the 1950s, with the number of coastal cities with more than 100,000 inhabitants increasing from 472 in 1950 to 2,129 in 2012 (Barragán and Andrés, 2015; see also chap. 14);

(c) Coastal management responses vary substantially, depending upon economics, legislation and social values, and are broadly classified into strategies of protection, accommodation, managed retreat and sacrifice (Williams and others, 2018);

(d) Rural sensitivity to erosion and sedimentation is typically determined by the impacts to the drainage and flood mitigation structures (Hou and others, 2016); as they are commonly located in the supratidal zone, their sensitivity to coastal change is not always apparent.

2.3. Changes in state

Besset and others (2019) examined the coastal area changes of 54 selected deltas around the world over 30 years, based on literature and the analysis of satellite imagery. They found that 29 deltas are in overall retreat, 18 shorelines are advancing and 7 do not show any significant change. Luijendijk and others (2018a), using Landsat images and supervised classification algorithms for shoreline detection, found that, over the period 1984–2016, 24 per cent of the world's sandy beaches were retreating at a rate greater than 0.5 m per year, while 28 per cent were advancing and 48 per cent were stable. They also found that about 4 per cent of the world's sandy beaches were retreating at rates exceeding 5 m per year, while about 2 per cent of global sandy shorelines were retreating at rates exceeding 10 m per year (see figure below). Continental Australia and Africa are experiencing net erosion (0.20 m/year and 0.07 m/year, respectively), while all other continents appear to be experiencing net accretion. Globally, 8 per cent, 6 per cent and 3 per cent of sandy beaches have accreted at rates of 3 m per year, 5 m per year and 10 m per year, respectively, over the period 1984–2016. Asia is the continent with the largest advancing rate (1.27 m/year), which is probably attributable to large land reclamation projects in the past few decades. Relatively high erosion rates are also seen at latitudes just south of the equator and are associated with large-scale land losses adjacent to the mouth of the Amazon River.

Global hotspots of beach erosion and accretion

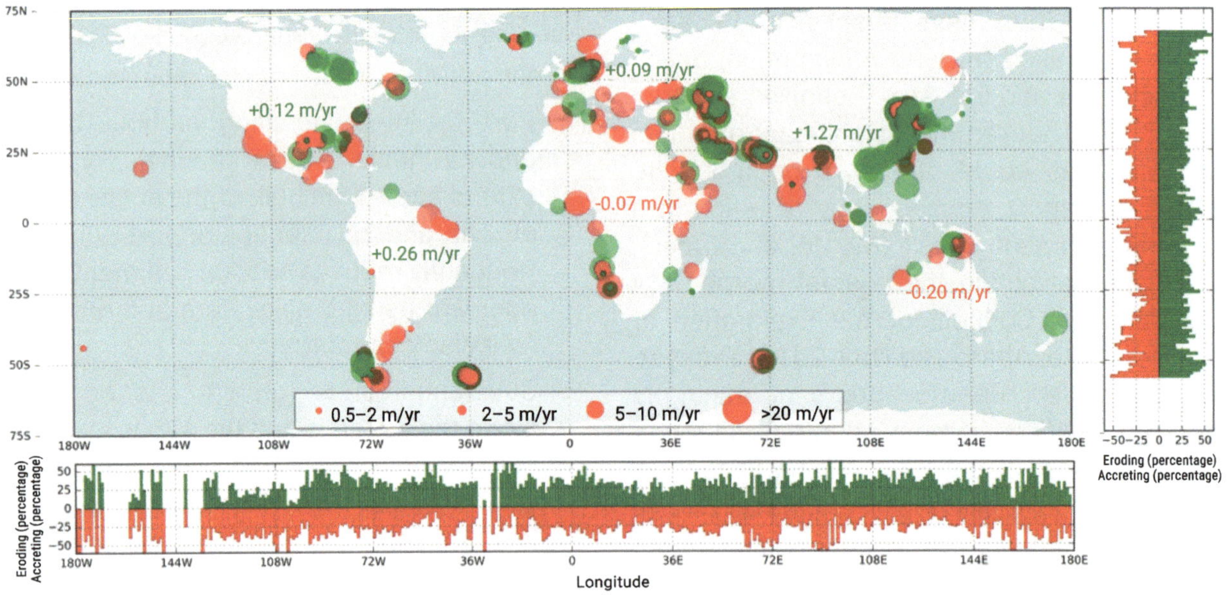

Source: Luijendijk and others, 2018a, reprinted under the Creative Commons licence (https://creativecommons.org/licenses/by/4.0).

Notes: The red (green) circles indicate erosion (accretion) for the four relevant shoreline dynamic classifications (see legend). The bar plots to the right and at the bottom present the relative occurrence of eroding (accreting) sandy shorelines per degree latitude and longitude, respectively. The numbers presented in the main plot represent the average change rate for all sandy shorelines per continent.

As a result of climate change, especially with the projected rise in sea level and increased frequency and severity of extreme waves, changes in coastal erosion and sedimentation patterns are likely to occur globally, as shown by several modelling efforts on projecting the future evolution of shorelines at the local, regional and global scales (Anderson and others, 2015; Antolínez and others, 2019; Castelle and others, 2014; Long and Plant, 2012; Ranasinghe and others, 2012; Splinter and others, 2014; Vitousek and others, 2017, Dastgheib and others, 2019; Bamunawala and others, 2020; Athanasiou and others, 2020; Vousdoukas and others, 2020). Recent observations have also indicated an acceleration in coastal cliff erosion (Hurst and others, 2016; Sunamura, 2015; Castedo and others, 2017).

2.4. Changes in impact

Coastal erosion and changes in sedimentation pose severe risks to coastal infrastructure, property, economic activities and ecological systems, and adaptation calls for significant investment. There is a tendency towards increasing damage from coastal erosion in specific locations that severely affects coastal socioeconomic activities and properties (Gopalakrishnan and others, 2016; Nguyen and others, 2018 ; Stronkhorst and others, 2018). The projection for risk and damage associated with coastal erosion and changes in sedimentation indicates that they are likely to increase in the future (Dunn and others, 2019).

Ecosystem impacts from coastal erosion and changes in sedimentation can be substantial, in particular if there is a transformation from long-term accretion to erosion. Coastal wetlands are at significant risk, as many of them were developed during the relative mean sea level standstill of the late Holocene (Jones and others, 2019) and may not keep up with the rising seas in the future (Myers and others, 2019). Other geomorphic features sensitive to changing patterns of erosion and sedimentation include mangrove coasts, barrier coasts and small islands. There is a high risk of ecological disturbance for organisms that exclusively use the coastal zone for nesting or nurseries, with increased proliferation of human-occupied and modified shorelines also reducing the overall bioproductivity of the coastal zone (Rangel-Buitrago and others, 2018b).

Major socioeconomic impacts will occur at locations where erosion coincides with high population density. Existing problems have been identified adjacent to the Ganges, Mekong, Yellow, Yangtze, Volta and Mississippi river deltas. For other parts of the coast, the management of erosion hazards through the use of engineering interventions requires long-term commitments to maintenance, including the cost of upgrading coastal defensive works, with potential risk to human safety and livelihoods if defences are subject to decline.

Local sea level rise and storminess vary significantly between regions. Based on long-term satellite data, wave height shows an overall global increase (Young and Ribal, 2019), but large regional differences are reported, from large changes in the Southern Ocean to negligible effects in the North Sea (De Winter and others, 2012). Such spatial variations are likely to result in regional variations in erosion and sedimentation (Brown and others, 2016).

2.5. Changes in response

Coastal erosion and sediment management practices have progressively matured from being almost wholly responsive to external changes to arriving at recognition of the need for coastal resilience, using adaptive management and assessing the coast from a more holistic, longer-term perspective (Rangel-Buitrago and others, 2018b).

The increase in larger, coastal-scale studies, an initial change from local-scale stabilization to regional assessment of erosion and accretion, has been followed by the recognition that conditions may be variable, with the potential for complex interactions between sedimentary coastal features (French and others, 2016; Psuty and others, 2018). Interconnections between coastal sediment supply and transport have been demonstrated over large scales, may occur over hundreds of kms and are likely to be further complicated by the potential impacts of projected sea level rise and other climate change-driven variations (Hapke and others, 2013). Therefore, changing modal conditions may introduce substantial uncertainty into future coastal change, leading to an increased need for coastal resilience planning using adaptive design (Wright and Thom, 2019).

A major outcome of understanding large-scale coastal systems has been demonstrated by changes in the applied scales of beach nourishment, for example, with the Sand Engine concept, which involves both onshore and nearshore placement of sediment, allowing natural hydrodynamics to redistribute sediment along the shore over a sustained period of time (Stive and others, 2013; De Schipper and others, 2016; Luijendijk and others, 2018b).

Recent developments in coastal protection strategies have involved supplementing structural engineering approaches with "softer" or "greener" forms of coastal stabilization, which aim both to increase ecological co-benefits and to utilize the resilient attributes of natural systems, such as the adaptive capacity displayed by coastal dunes or the disturbance-recovery behaviour demonstrated by coastal wetlands and mangrove forests (Narayan and others, 2016; Reguero and others, 2018).

There is also an emerging trend towards the use of probabilistic analysis frameworks, instead of the traditional deterministic approach, which take into account the uncertainties associated with climate change impacts to facilitate risk-informed decision-making (Wainwright and others, 2014; Jongejan and others, 2016).

3. Consequences of the changes for human communities, economies and well-being

Coastal erosion and changes in sedimentation continue to pose severe threats to the livelihood and well-being of households that are dependent on coastal resources, as well as damaging ecosystems and causing environmental stress. The closeness of human and ecological systems, and the risks created by accelerated erosion and sedimentation changes are evident in many areas all over the world (Jones and others, 2019). Furthermore, erosion and changes in sedimentation have physical and chemical consequences for water quality and the health of fragile aquatic ecological systems (Prosser and others, 2018).

Coastal erosion and changes in sedimentation can have serious implications for the achievement of the integrated set of global priorities and objectives set out in the 2030 Agenda for Sustainable Development, especially Sustainable Development Goals 14 and 15.[1] Those processes may damage coastal infrastructure and habitats and increase risks to coastal communities, forcing them to adapt and/or reallocate resources.

1 See General Assembly resolution 70/1.

4. Key region-specific changes and consequences

4.1. North Atlantic Ocean, Baltic Sea, Black Sea, Mediterranean and North Sea

The coasts of the North Atlantic Ocean, the Mediterranean and adjacent seas are densely populated and highly developed (Collet and Engelbert, 2013; Zhang and Leatherman, 2011; European Union, 2013; Neumann and others, 2015). Areas that are highly sensitive to coastal change include the extensively reclaimed Netherlands shore, the subsiding Venetian coast and the barrier islands along the eastern seaboard of the United States, and the coasts of the Gulf of Mexico. The high economic value of the hinterland and coastal zone results in a low tolerance for erosion, and human interventions are common. Beach nourishment is the most common intervention along the coastlines of the eastern seaboard and the Gulf of Mexico. Widespread erosion has been observed along the Gulf coast, associated with substantially reduced sediment load from the Mississippi River (Blum, 2009; Thorne and others, 2008). An extensive decline in fluvial sediment inputs has also been identified for major European river systems draining into the Mediterranean that support productive wetland areas.

4.2. South Atlantic Ocean and wider Caribbean

The South Atlantic Ocean and wider Caribbean regions have densely populated coastal cities, such as the city of João Pessoa, Brazil,[2] and important coastal ecological systems, such as the Amazon mangrove forest, as well as sparsely populated coastal areas, such as the coasts of many States in Southeast Africa and the southern coast of Argentina (Zhang and Leatherman, 2011; United Nations Educational, Scientific and Cultural Organization, 2009; Neumann and others, 2015). The input of sediments transported by rivers is limited to areas near large basins, such as the Amazon River and the Rio de la Plata. The reduction in sediment supply to the coasts caused by upstream dam construction and beach sand extraction has caused serious coastal erosion at various places, such as the coast of Ghana and many other places on the south-west coast of Africa and the east coast of South America. Locally, many coastal sectors used and still prefer to use hard structures for erosion control, which, in many cases, have exacerbated the problem, as in Colombia (Rangel-Buitrago and others, 2018b) and Brazil (Bonetti and others, 2018), for example.

4.3. Indian Ocean, Arabian Sea, Bay of Bengal, Red Sea, Gulf of Aden and Persian Gulf

Indian Ocean coasts include the east coast of Africa, southern coasts of the Middle East, South Asia, the Indonesian archipelago, the west coast of Australia and Indian Ocean islands, including Madagascar and Sri Lanka. Deltas of major rivers include the Ganges, the Indus, the Ayeyarwady, the Chao Phraya, the Shatt al-Arab, the Zambezi and the Limpopo, many of which are highly dynamic and adjacent to areas with high populations (Neumann and others, 2015). Africa, Australia and the Middle East have predominantly arid sandy coasts, with barrier lagoons, estuaries and, in some areas, extensive salt-flat coasts characteristic of the late Holocene sea level highstand that limit the transfer of fluvial sediments to the coast. Substantial coastal engineering projects, including the construction of artificial islands

2 See www.ibge.gov.br/en/cities-and-states/pb/joao-pessoa.html.

through dredging and reclamation, have been undertaken along the west and south coasts of the Persian Gulf, in particular along the coast of the United Arab Emirates (Peduzzi, 2014).

4.4. North Pacific Ocean

North Pacific coasts include the west coast of North America, the east coast of Asia and the northern Pacific islands, comprising the Philippines, Japan and Hawaii, United States. Areas of high population density are found on the east coast of Asia and the west coast of the United States and coincide with significant coastal interventions and declining sediment yield from the major river systems of the Pearl River, the Yellow River and the Red River, and rivers flowing to the west coast of the United States (Neumann and others, 2015). For example, on the west coast of the United States, coastal erosion is caused by the reduction in fluvial sediment supply, coastal structures, and climate change and variations, such as El Niño (Barnard and others, 2017; Hapke and others, 2009; Patsch and Griggs, 2007; Allan and Komar, 2006). Islands in the North Pacific are highly sensitive to potential coastal change and the impacts of severe events, including typhoons and tsunamis. In addition, deforestation is resulting in increased fluvial sediment delivery to the coast associated with the Fly River, Papua New Guinea.

4.5. South Pacific Ocean

South Pacific coasts include the east coast of Australia, the west coast of South America and the shores of Pacific islands, including New Zealand, New Caledonia and numerous island and archipelagic States with different population sizes (United Nations, 2017b). The continental coasts are characterized by their geological structure and relatively low volumes of fluvial sediment reaching the ocean, resulting in compartmentalized coasts, with intermittent exchange related to along-shelf sediment transport (Thom and others, 2018). Changes to relative sediment supply are therefore most apparent at regional coastal sediment sinks and sources, with the potential susceptibility of estuarine settings, barrier coasts and coastal wetlands to sea level rise. Coastal change impacts identified throughout the South Pacific are typically episodic, associated with extreme storms and tropical cyclones, with more widespread pressure during phases of elevated mean sea level.

Pacific island coasts include volcanic land masses, seamounts, uplifted limestone and coral atolls. Sediment productivity is low, resulting in limited capacity for coastal adjustment to projected sea level rise (Nunn and others, 2015), in particular for low-lying reclaimed areas.

4.6. Arctic Ocean and Southern Ocean

In the context of climate change, with rising air temperatures, declining sea ice extent and increasing wave action owing to the possibility of greater storm intensity, storm-induced tide and water surface area, the permafrost coasts of the Arctic Ocean are now experiencing severe erosion (Bull and others, 2019; Gibbs and Richmond, 2017; Tanski and others, 2016; Frederick and others, 2016; Fritz and others, 2015). The erosion rate of the Arctic coasts of the United States has doubled from the 1950s to the present, and it appears to be accelerating; in particular, the coastline of the Beaufort Sea in Alaska is retreating at a rate of more than 30 m per year (Frederick and others, 2016; Wobus and others, 2011). The release of organic carbon to the Arctic Ocean through coastal erosion can enhance global warming (Tanski and others, 2016). The ice sheet in Antarctica is also rapidly melting (Rignot and others, 2019; Gardner and others, 2018; Li and others, 2016).

5. Outlook

Human activities affecting the incidence of coastal erosion and sedimentation include the substantial growth in the number and scale of dams on major waterways, land-use changes leading to catchment deforestation and increased human occupation of the coastal zone, coincident with a proliferation of coastal structures (Rangel-Buitrago and others, 2018a, 2018c). The evaluation of global coastal change is not sufficiently mature to establish metrics for human-induced change to secular trends. However, identified hotspots of shoreline displacement, mostly associated with coastal erosion and accretion, are areas that are strongly linked to human activity, producing estimated 33-year trends exceeding 5 m per year for approximately 4 per cent of the world's coasts (Luijendijk and others, 2018a). Compared with the knowledge of preceding conditions, substantial coastal erosion has been observed for a majority of deltas owing to a significant reduction in riverine sediment loads from 1970 to 2014 (Besset and others, 2019). Overall decreases in riverine sediment supply to the coast are expected to reduce the stability of adjacent downdrift coasts and, for parts of the coast, will reverse long-term accretive trends, which will exacerbate the demand for coastal management works and reduce the effectiveness of existing works, in particular those that act to redistribute sediment supply. Furthermore, that situation will increase the proliferation of coastal works, which have historically been developed in response to increased coastal population levels and a corresponding low tolerance for coastal change. As demonstrated by shoreline monitoring, the increased manipulation of coastal dynamics and the strict regulation of sand mining permits provide opportunities for substantial secular change to coastal trends, including both accretion and erosion (Williams and others, 2018; Bergillos and others, 2019). With sea level rise and an increase in the frequency and intensity of extreme climate events owing to climate change, coastal erosion will be more serious for islands where riverine sediment does not exist.

6. Key remaining knowledge and capacity-building gaps

At present, significant knowledge has been accumulated on the interaction of coastal dynamic processes and sediment transport. However, the accuracy of models for sediment transport and coastal erosion and/or sedimentation is still limited; therefore, more research is needed. More information is also needed on the extent of coastal erosion for the identification of appropriate management strategies for coastal erosion and sedimentation, including the management of riverine sediment supply and other management strategies, such as protection, accommodation and retreat.

Although there have been substantial advances in data sets, in particular owing to the use of satellite imagery (Besset and others, 2019; Luijendijk and others, 2018a; Shirzaei and Bürgmann, 2018), in many regions, especially in developing States, the available data remain immature for local and regional decision-making, with many data sets requiring substantial further interpretation and better worldwide spatial resolution. A better understanding of how to attribute driving processes and determine responses, and of how those processes will change with sea level rise and climate change, is required. Furthermore, quantified erosion or sedimentation rates need to be placed in the context of thresholds for coastal ecosystems or morphological systems. The interpretation of impacts from both changing fluvial sediment supply and the application of coastal defence strategies requires an improved understanding of the spatial dimensions associated with the alongshore redistribution of available sediment supply, in particular in situations where it occurs across international boundaries.

References

Abam, T.K.S., and Tamunotonye Oba (2018). Recent case studies of sand mining, utilization and environmental impacts in the Niger delta. *Journal of Environmental Geology*, vol. 2, No. 2.

Allan, J.C., and P.D. Komar (2006). Climate Controls on US West Coast Erosion Processes. *Journal of Coastal Research*, vol. 22, No. 3, pp. 511–529. http://doi.org/10.2112/03-0108.1.

Anderson, Tiffany R., and others (2015). Doubling of coastal erosion under rising sea level by mid-century in Hawaii. *Natural Hazards*, vol. 78, No. 1, pp. 75–103.

Angamuthu, Balaji, and others (2018). Impacts of natural and human drivers on the multi-decadal morphological evolution of tidally-influenced deltas. *Proceedings of the Royal Society A*, vol. 474, No. 2219, 20180396.

Antolínez, José A.A., and others (2019). Predicting climate driven coastlines with a simple and efficient multiscale model. *Journal of Geophysical Research: Earth Surface*. https://doi.org/10.1029/2018JF004790.

Athanasiou, Panagiotis, and others (2020). Uncertainties in projections of sandy beach erosion due to sea level rise: an analysis at the European scale. *Scientific Reports*, vol. 10, art. 11895.

Bamunawala, Janaka, and others (2020). A Holistic Modeling Approach to Project the Evolution of Inlet-Interrupted Coastlines Over the 21st Century. *Frontiers in Marine Science*, https://doi.org/10.3389/fmars.2020.00542.

Barnard, Patrick L., and others (2017). Extreme oceanographic forcing and coastal response due to the 2015–2016 El Niño. *Nature Communications*, vol. 8, art. 14365.

Barragán, Juan Manuel, and María de Andrés (2015). Analysis and trends of the world's coastal cities and agglomerations. *Ocean & Coastal Management*, vol. 114, pp. 11–20.

Bendixen, Mette, and others (2019). *Time Is Running out for Sand*. Nature Publishing Group.

Bergillos, R., and others (2019). Management of Coastal Erosion Under Climate Change Through Wave Farms. In *Ocean Energy and Coastal Protection*, ed. 1, pp. 59–73. New York: Springer International Publishing.

Besset, Manon, and others (2019). Multi-decadal variations in delta shorelines and their relationship to river sediment supply: an assessment and review. *Earth-Science Reviews*.

Best, Jim (2019). Anthropogenic stresses on the world's big rivers. *Nature Geoscience*, vol. 12, No. 1, pp. 7–21.

Blum, M.D., and H. Roberts (2009). Drowning of the Mississippi Delta due to Insufficient Sediment Supply and Global Sea-Level Rise. *Nature Geoscience*, vol. 2, No. 7, pp. 488–491. http://doi.org/10.1038/ngeo553.

Bonetti, J., and others (2018). Geoindicator-based assessment of Santa Catarina (Brazil) sandy beaches susceptibility to erosion. *Ocean & Coastal Management*, vol. 156, pp. 198–208.

Brown, Sally, and others (2016). Spatial variations of sea-level rise and impacts: An application of DIVA. *Climatic Change*, vol. 134, No. 3, pp. 403–416.

Bull, D.L., and others (2019). Development of a Tightly Coupled Multi-Physics Numerical Model for an Event-Based Understanding of Arctic Coastal Erosion. AGUFM, 2019, pp. C12B-04.

Castedo, R., and others (2017). The Modelling of Coastal Cliffs and Future Trends. In *Hydro-Geomorphology Models and Trends*, Dericks P. Shukla, ed. Croatia: InTech, http://doi.org/10.5772/intechopen.68445. Available at www.intechopen.com/books/hydro-geomorphology-models-and-trends/the-modelling-of-coastal-cliffs-and-future-trends.

Castelle, Bruno, and others (2014). Equilibrium shoreline modelling of a high-energy meso-macrotidal multiple-barred beach. *Marine Geology*, vol. 347, pp. 85–94.

Collet, C., and A. Engelbert (2013). Coastal regions: people living along the coastline, integration of NUTS 2010 and latest population grid. Statistics in focus 30, ISSN:2314-9647, Catalogue number: KS-SF-13-030-EN-N.

Dastgheib, Ali, and others (2018). Regional Scale Risk-Informed Land-Use Planning Using Probabilistic Coastline Recession Modelling and Economical Optimisation: East Coast of Sri Lanka. *Journal of Marine Science Engineering*, vol. 6, No. 4, 120; https://doi.org/10.3390/jmse6040120.

De Schipper, Matthieu A., and others (2016). Initial spreading of a mega feeder nourishment: Observations of the Sand Engine pilot project. *Coastal Engineering*, vol. 111, pp. 23–38.

De Winter, Renske C., and others (2012). The effect of climate change on extreme waves in front of the Dutch coast. *Ocean Dynamics*, vol. 62, No. 8, pp. 1139–1152.

Dunn, Frances E., and others (2018). Projections of historical and 21st century fluvial sediment delivery to the Ganges-Brahmaputra-Meghna, Mahanadi, and Volta deltas. *Science of the Total Environment*, vol. 642, pp. 105–116.

________(2019). Projections of declining fluvial sediment delivery to major deltas worldwide in response to climate change and anthropogenic stress. *Environmental Research Letters*, vol. 14, No. 8, 084034.

European Union (2013). ESaTDOR European Seas and Territorial Development, Opportunities and Risks. ANNEX 4 to the Scientific Report: Baltic Sea Regional Profile.

Frederick, Jennifer M., and others (2016). The Arctic Coastal Erosion Problem. Sandia Report SAND2016-9762. Sandia National Laboratories, Albuquerque, New Mexico 87185 and Livermore, California 94550.

French, Jon, and others (2016). Appropriate complexity for the prediction of coastal and estuarine geomorphic behaviour at decadal to centennial scales. *Geomorphology*, vol. 256, pp. 3–16.

Fritz, M., and others (2015). Arctic coastal erosion and the transport of terrigenous material into the Arctic Ocean during the Holocene , XIX. INQUA-Congress, Nagoya, Japan, 26 July –2 August 2015.

Gardner, Alex S., and others (2018). Increased West Antarctic and unchanged East Antarctic ice discharge over the last 7 years. *The Cryosphere*, vol. 12, pp. 521–547, https://doi.org/10.5194/tc-12-521-2018.

Gibbs, A.E., and B.M. Richmond (2017). National assessment of shoreline change – Summary statistics for updated vector shorelines and associated shoreline change data for the north coast of Alaska, U.S.-Canadian border to Icy Cape: U.S. Geological Survey Open-File Report 2017–1107. https://doi.org/10.3133/ofr20171107.

Gopalakrishnan, Sathya, and others (2016). Economics of coastal erosion and adaptation to sea level rise. *Annual Review of Resource Economics*, vol. 8, pp. 119–139.

Hapke, Cheryl J., and others (2009). Rates and trends of coastal change in California and the regional behavior of the beach and cliff system. *Journal of Coastal Research*.

Hapke, Cheryl J., and others (2013). Geomorphic and human influence on large-scale coastal change. *Geomorphology*, vol. 199, pp. 160–170.

Hou, Xi Yong, and others (2016). Characteristics of coastline changes in mainland China since the early 1940s. *Science China Earth Sciences*, vol. 59, No. 9, pp. 1791–1802.

Hurst, M.D., and others (2016). *Proceedings of the National Academy of Sciences*, vol. 113, No. 47, pp. 13336–13341. www.pnas.org/cgi/doi/10.1073/pnas.1613044113.

International Council for the Exploration of the Sea (2016). Effects of extraction of marine sediments on the marine environment 2005–2011. Report (Scientific report). https://archimer.ifremer.fr/doc/00326/43700.

International Commission on Large Dams (2018). www.icold-cigb.org.

Jayappa, K.S., and B. Deepika (2018). Impacts of Coastal Erosion, Anthropogenic Activities and their Management on Tourism and Coastal Ecosystems: A Study with Reference to Karnataka Coast, India. In *Beach Management Tools – Concepts, Methodologies and Case Studies*, pp. 421–440. Springer.

Jones, Miriam C., and others (2019). Rapid inundation of southern Florida coastline despite low relative sea-level rise rates during the late-Holocene. *Nature Communications*, vol. 10, No. 1, pp. 1–13.

Jongejan, Ruben, and others (2016). Drawing the line on coastline recession risk. *Ocean & Coastal Management*, vol. 122, pp. 87–94.

Kondolf, G.M., and others (2014). Dams on the Mekong: cumulative sediment starvation. *Water Resources Research*, vol. 50, No. 6, pp. 5158–5169.

Li, Xin, and others (2016). Ice flow dynamics and mass loss of Totten Glacier, East Antarctica, from 1989 to 2015. *Geophysical Research Letters*, vol. 43, pp. 6366–6373. https://doi.org/10.1002/2016GL069173.

Long, Joseph W., and Nathaniel G. Plant (2012). Extended Kalman Filter framework for forecasting shoreline evolution. *Geophysical Research Letters*, vol. 39, No. 13.

Luijendijk, Arjen, and others (2018a). The state of the world's beaches. *Scientific Reports*, vol. 8, No. 1, p. 6641.

Luijendijk, Arjen, and others (2018b). The initial morphological response of the Sand Engine: A process-based modelling study. *Coastal Engineering*, vol. 119, pp. 1–14.

Mentaschi, Lorenzo, and others (2018). Global long-term observations of coastal erosion and accretion. *Scientific Reports*, vol. 8, No. 1, art. 12876.

Montoi, Jayawati, and others (2017). A Study on Tuaran River Channel Planform and the Effect of Sand Extraction on River Bed Sediments. *Transactions on Science and Technology*, vol. 4, No. 4, pp. 442–48.

Moore, Laura J., and others (2018). The role of ecomorphodynamic feedbacks and landscape couplings in influencing the response of barriers to changing climate. In *Barrier Dynamics and Response to Changing Climate*, pp. 305–336. Springer.

Myers, Monique R., and others (2019). A multidisciplinary coastal vulnerability assessment for local government focused on ecosystems, Santa Barbara area, California. *Ocean and Coastal Management*, vol. 182. https://doi.org/10.1016/j.ocecoaman.2019.104921.

Narayan, Siddharth, and others (2016). The effectiveness, costs and coastal protection benefits of natural and nature-based defences. *PloS One*, vol. 11, No. 5, e0154735.

Neumann, B., and others (2015). Future coastal population growth and exposure to sea-level rise and coastal flooding: a global assessment. *PloS One*, vol. 10, No. 6, e0131375. https://doi.org/10.1371/journal.pone.0131375.

Nguyen, A.T., and others (2018). Tourism and beach erosion: valuing the damage of beach erosion for tourism in the Hoi An World Heritage site, Vietnam. *Environment, Development and Sustainability* https://doi.org/10.1007/s10668-018-0126-y.

Nicholls, Robert J., and others (1999). Increasing flood risk and wetland losses due to global sea-level rise: regional and global analyses. *Global Environmental Change*, vol. 9, pp. S69–S87.

Nunn, Patrick, and others (2015). Regional coastal susceptibility assessment for the Pacific Islands: Technical Report. Australian Government and Australian Aid, Canberra.

Nyberg, Björn, and others (2018). The distribution of rivers to terrestrial sinks: implications for sediment routing systems. *Geomorphology*, vol. 316, pp. 1–23.

Patsch, K., and G. Griggs (2007). Development of Sand Budgets for California's Major Littoral Cells. Institute of Marine Sciences, University of California, Santa Cruz.

Peduzzi, Pascal (2014). Sand, rarer than one thinks. *Environmental Development*, vol. 11, pp. 208–218.

Prosser, Diann J., and others (2018). Impacts of coastal land use and shoreline armoring on estuarine ecosystems: an introduction to a special issue. *Estuaries and Coasts*, vol. 41, No. 1, pp. 2–18.

Psuty, Norbert P., and others (2018). Responding to coastal change: creation of a regional approach to monitoring and management, northeastern region, USA. *Ocean & Coastal Management*, vol. 156, pp. 170–182.

Ranasinghe, Roshanka, and others (2019). Disentangling the relative impacts of climate change and human activities on fluvial sediment supply to the coast by the world's large rivers: Pearl River Basin, China. *Scientific Reports*, vol. 9, No. 1, art. 9236.

Ranasinghe, Roshanka, and others (2012). Estimating coastal recession due to sea level rise: beyond the Bruun rule. *Climatic Change*, vol. 110, pp. 561–574.

Rangel-Buitrago, Nelson, and others (2018a). Preface to the special issue: Management strategies for coastal erosion processes. *Ocean & Coastal Management*, vol. 156, pp. 1–3. https://doi.org/10.1016/j.ocecoaman.2017.11.020.

Rangel-Buitrago, Nelson, and others (2018b). Risk Assessment to Extreme Wave Events: The Barranquilla–Cienaga, Caribbean of Colombia Case Study. In *Beach Management Tools – Concepts, Methodologies and Case Studies*, pp. 469–496. Springer.

Rangel-Buitrago, Nelson, and others (2018c). How to make integrated coastal erosion management a reality. *Ocean & Coastal Management*, vol. 156, pp. 290–299.

Reguero, Borja G., and others (2018). Comparing the cost effectiveness of nature-based and coastal adaptation: a case study from the Gulf Coast of the United States. *PloS One*, vol. 13, No. 4, e0192132.

Rignot, Eric, and others (2019). Four decades of Antarctic Ice Sheet mass balance from 1979–2017. *Proceedings of the National Academy of Sciences*, vol. 116, No. 4, pp. 1095–1103. www.pnas.org/cgi/doi/10.1073/pnas.1812883116.

Shirzaei, M., and R. Bürgmann (2018). Global climate change and local land subsidence exacerbate inundation risk to the San Francisco Bay Area. *Science Advances*, vol. 4, No. 3, eaap9234. https://doi.org/10.1126/sciadv.aap9234.

Slagel Matthew J., and Gary B. Griggs (2008) Cumulative Losses of Sand to the California Coast by Dam Impoundment. *Journal of Coastal Research,* No. 243, pp. 571–584. https://doi.org/10.2112/06-0640.1.

Splinter, Kristen D., and others (2014) A generalized equilibrium model for predicting daily to interannual shoreline response. *Journal of Geophysical Research: Earth Surface*, vol. 119, No. 9, pp. 1936–1958.

Stive, Marcel J.F., and others (2013). A new alternative to saving our beaches from sea-level rise: The Sand Engine. *Journal of Coastal Research*, vol. 29, No. 5, pp. 1001–1008.

Stronkhorst, J., and others (2018). Regional coastal erosion assessment based on global open access data: a case study for Colombia. *Journal of Coastal Conservation*, vol. 22, pp. 787–798. https://doi.org/10.1007/s11852-018-0609-x.

Sunamura, T. (2015). Rocky coast processes: with special reference to the recession of soft rock cliffs. *Proceedings of the Japan Academy, Series B: Physical and Biological Sciences*, vol. 91, No. 9, pp. 481–500. http://doi.org/10.2183/pjab.91.481.

Tanski, G., and others (2016). Erosion of ice-rich permafrost coasts and the release of dissolved organic carbon into the Arctic Ocean.

Thom, Bruce G., and others (2018). National sediment compartment framework for Australian coastal management. *Ocean & Coastal Management*, vol. 154, pp. 103–120.

Thorne, Colin R., and others (2008). Current and Historical Sediment Loads in the Lower Mississippi River. European Research Office of the U.S. Army, London, England under contract number 1106-EN-01, from School of Geography, University of Nottingham.

United Nations (2017a). Chapter 26: Land-sea physical interaction. In *The First Global Integrated Marine Assessment: World Ocean Assessment I*. Cambridge: Cambridge University Press.

________ (2017b). Chapter 36D: South Pacific Ocean. In *The First Global Integrated Marine Assessment: World Ocean Assessment I*. Cambridge: Cambridge University Press.

________ (2017c). *The First Global Integrated Marine Assessment: World Ocean Assessment I*. Cambridge: Cambridge University Press.

United Nations Educational, Scientific and Cultural Organization (2009). *African Oceans and Coasts*. M. Odido M. and S. Mazzilli S., eds. IOC Information Document 1255. UNESCO Office Nairobi and Regional Bureau for Science in Africa.

Vitousek, Sean, and others (2017). A model integrating longshore and cross-shore processes for predicting long-term shoreline response to climate change. *Journal of Geophysical Research: Earth Surface*, vol. 122, No. 4, pp. 782–806.

Vousdoukas, Michalis I., and others (2020). Sandy coastlines under threat of erosion. *Nature Climate Change*, vol. 10, pp. 260–263.

Wainwright, David J., and others (2014). An argument for probabilistic coastal hazard assessment: Retrospective examination of practice in New South Wales, Australia. *Ocean & Coastal Management*, vol. 95, pp. 147–155.

Warner, Jeroen, and others (2019). The Fantasy of the Grand Inga Hydroelectric Project on the River Congo. *Water*, vol. 11, No. 3, p. 407.

Warrick, Jonathan A., and others (2015). Large-scale dam removal on the Elwha River, Washington, USA: source-to-sink sediment budget and synthesis. *Geomorphology*, vol. 246. http://doi.org/10.1016/j.geomorph.2015.01.010.

Williams, A.T., and others (2018). The management of coastal erosion. *Ocean & Coastal Management*, vol. 156, pp. 4–20.

Wobus, Cameron, and others (2011). Thermal Erosion of a Permafrost Coastline: Improving Process-Based Models Using Time-Lapse Photography. *Arctic, Antarctic, and Alpine Research*, vol. 43, No. 3, pp. 474–484.

Wright, Lynn Donelson, and Bruce G. Thom (2019). Promoting Resilience of Tomorrow's Impermanent Coasts. In *Tomorrow's Coasts: Complex and Impermanent*, Lynn Donelson Wright and C. Reid Nichols, eds., pp. 341–353. Springer.

Xu, Weihua, and others (2019). Hidden loss of wetlands in China. *Current Biology*, vol. 29, No. 18, pp. 3065–3071.

Young, Ian R., and Agustinus Ribal (2019). Multiplatform evaluation of global trends in wind speed and wave height. *Science*, vol. 364, No. 6440, pp. 548–552.

Zhang, K., and S. Leatherman (2011). Barrier Island Population along the U.S. Atlantic and Gulf Coasts. *Journal of Coastal Research*, vol. 27, No. 2, pp. 356–363. https://doi.org/10.2112/JCOASTRES-D-10-00126.1.

Chapter 14
Changes in coastal and marine infrastructure

Contributors: Ca Thanh Vu (convener and lead member), Sam Bentley, Lionel Carter, Catherine Creese, Robert Dapa, Regina Folorunsho, Alan Simcock (co-lead member) and Alix Willemez.

Keynote points

- Coastal and marine infrastructures are necessary for the use, exploitation and protection of the coastal and marine natural resources and environment for socioeconomic development.
- In general, if well-designed and well-built, coastal infrastructure development can be ecologically as well as economically and socially sustainable, increase the resilience of coasts and lead to sustainable economic growth.
- Infrastructures can influence natural systems and their use and create pressures and conflicts or favourable conditions.
- Between 2010 and 2020, there was an upward trend in newly developed, renovated or upgraded marine and coastal infrastructure.
- The most significant changes are coastal and offshore land reclamation, especially in East Asian countries, for new coastal urban development, roads, coastal defence structures, port and harbour facilities and tourist facilities.
- Depending on the case, coastal and marine infrastructures may cause substantial damage or reduce damage to coastal and marine ecosystems.
- The new coastal infrastructure development approach, known as "blue infrastructure development", can harmonize coastal protection and development, as well as habitat and ecological protection, thereby reducing ecological damage.
- Coastal and marine infrastructure development in general has created new opportunities for coastal dwellers and supported sustainable socioeconomic coastal development.

1. Introduction

1.1. Scope

The present chapter covers changes in coastal and marine infrastructure during the period from 2010 to 2020 since the baseline described in the first *World Ocean Assessment* (United Nations, 2017).

1.2. First *World Ocean Assessment*

Chapters 17–19, 26–28 and 30 of the first *Assessment* cover coastal and marine infrastructures, including waste-receiving facilities at ports, their operation and impacts on the local marine environment worldwide and their contribution to economic activity; knowledge gaps and capacity-building at ports, including the improvement of operational skills; waste-reception facilities; the capacity to examine dredged material for safe redeposition in the sea; the history, development and present state of submarine communications and power cables; the impacts of submarine communications and power cables on the marine environment; the threat to cables from the marine environment; capacity-building on safe routing for submarine cables and resolving any conflicting demands with other parties; land reclamation, including the present state, trend and socioeconomic and environmental impacts; tourism and recreation and related infrastructures, such as roads, ports, harbours and airports, and other coastal infrastructures; sustainable tourism and capacity-building for tourism management; desalination and related coastal infrastructures; ocean and coastal scientific research facilities; and impacts of the coastal built environment on wildlife. There is increasing financial investment in adaptation and mitigation strategies across sectors, from insurance to coastal protection.

It is clear from the first *Assessment* that issues relating to marine and coastal infrastructures need to be more systematically addressed. The present chapter will fill in the gaps and provide additional data for evaluating the trends in marine and coastal infrastructures, in particular for the period from 2010 to 2020.

2. Documented changes in the state of marine and coastal infrastructures

2.1. Changes in land reclaimed from the sea

Coastal and marine land reclamation transforms ocean areas into land in many ways, including infilling with dredged material or waste from land and building dykes. That is often the case for coastal and island cities with dense urbanization and a need for more space. Sengupta and others (2018) reviewed land reclamation from the mid-1980s to the present day in 16 coastal megacities using Landsat Thematic Mapper satellite imagery. The total land reclamation area of the 16 megacities is 1,249.8 km^2, mostly in China, where the policy changed in 2018. However, on the basis of current trends, more land reclamation is to be expected in the near future globally.

2.2. Extent of new land defences against the sea, and extent of abandoned sea defences

Common strategies for adaptation to coastal erosion include hard or soft protection (to hold or advance the line), accommodation, managed retreat and sacrifice (i.e., no active intervention) (Williams and others, 2018).

The most rudimentary method for coastal protection, using hard structures, has evolved from line protection to surface protection and, in addition to preventing storm surges and high tides, is aimed at protecting sandy beaches where there is reduced sediment supply to the coasts. Typical coastal hard structure defences include sea walls, revetments, groynes of various types, onshore and detached breakwaters and headlands, which protect beaches from wave attack or modify nearshore wave field and related sediment transport processes to create a new sediment balance at coasts that favours sedimentation instead of erosion.

Improperly designed or ageing coastal defence structures cannot function properly and may be abandoned or repaired. Causes of coastal defence structure degradation include damage caused by corrosion; structure sinking owing to foundation liquidation under wave action; toe scouring; wave overtopping; wave forces on structures; and sea level rise owing to climate change. It is very difficult to make decisions about removing degraded coastal protection structures because they may already host endemic habitats with habitat value and the effects of removal are difficult to predict. Thus, in many cases, degraded coastal structures are simply abandoned.

Nature-based solutions for coastal protection, including artificial wetlands or salt marshes, beach nourishment, oyster reef creation and mangrove re-establishment and protection, have the advantage of being able to grow with sea level and increasing CO_2 storage capacity (Davis and others, 2015). However, at eroded coasts, owing to a reduction in sediment supply, hard coastal defence structures can effectively prevent natural hazards and protect the environment and natural habitats when used in combination with natural barriers or natural systems, such as mangroves or coral reefs. Such structures can be called "blue infrastructures" (Kazmierczak and Carter, 2010; Edwards and others, 2013).

2.3. Extent of coastal development, including for tourism: roads, town sites, tourism and recreation facilities, artificial beaches and other coastal development structures

Coastal developments along estuaries and coastlines have become hotspots for population explosion and magnets for various industries as well as non-industrial activities such as residential, tourism and recreational development. Many coastal cities are transformed into megacities within a few years as a result of socioeconomic activities (Blackburn and others, 2013).

The demand for seafront living globally makes such areas develop into densely populated cities with road networks and businesses. Wealthier populations with coastal and marine tourism and recreation demands also contribute to the rapid development of coastal tourism and recreation cities, such as on Asia-Pacific coasts and islands. Coastal tourism also demands the creation of many artificial beaches worldwide, such as Waikiki Beach in Hawaii, United States of America, and beaches in Singapore.

2.4. Adaptation strategies for coastal communities dealing with sea level rise

Climate change and sea level rise will increase the risks of natural hazards to coasts (Intergovernmental Panel on Climate Change, 2019). Adaptation strategies will have to determine risks and develop and implement management approaches to reduce to an acceptable level the risks to individuals, communities, societies and ecological systems at the coast and on the sea. Among the common adaptation strategies mentioned in section 2.2, accommodation and protection require the building or upgrading of present infrastructures, in many cases in combination with the restoration of coastal habitats or ecological systems.

The upgrading of coastal infrastructure is also influenced by economic factors. For example, 5 of the top 10 ports in the world that are most susceptible to sea level rise are located in the eastern and south-eastern United States. While the ports are working to rebuild infrastructure to higher standards, they must balance the requirements for predicted increases in international trade with the need to address both sea level rise and stronger and more frequent extreme weather events.

2.5. Changes in port, harbour and marina installations and their management, including dredging

According to the United Nations Conference on Trade and Development (UNCTAD), container transport is expanding rapidly: in 2017, it grew at a rate of 6 per cent, or 42.3 million twenty-foot equivalent units (UNCTAD, 2018). Port competition also heightened, thus providing opportunities for shipping lines to improve management skills and increase bargaining power and influence.

The highest regions in terms of container port volume are Asia (63 per cent) and the Americas (16 per cent). Measured by total tons of all cargo handled, of the world's 10 largest ports, 8 are in Asia, mainly China. Profit levels vary considerably between ports but averages across volumes suggest that only $4 are earned for each ton of cargo (UNCTAD, 2018). Employees are categorized across traditional lines that have yet to reflect the technological shift in working methods and skill sets. While few new large seaports are being planned or constructed, it was suggested that, beyond 2020, 80 per cent of world trade would be conducted through seaports that would require additional facilities. Also, there is a growing interest in offshore, deep-water ports such as the Louisiana Offshore Oil Port.

Owing to an increase in global fishing fleets (Rousseau and others, 2019), there was a rise in the number of fishing ports in the past, but as world ocean fisheries resources decrease, it is likely that the trend will not continue.

The global recreational boating market is also on the rise. In 2009, total revenue for the industry was $18.12 billion, increasing to $40 billion in 2017, with a 2 per cent growth rate from 2015 to 2017, the highest increases being in North America and the Asia-Pacific region (Value Market Research, 2017).

Dredging to maintain, create or increase navigation depth in existing ports (regular operations, renovation and expansion) or newly constructed ports, harbours and marinas is increasing in line with the global economic growth rate (International Association of Dredging Companies, 2018).

2.6. Changes in submarine cables and submarine pipelines

After a marked decline in production between 2006 and 2010, from 2010 to 2018 the length of communications cables installed in all oceans, measured in km, increased, at an average rate of over 70,000 km per year. As at early 2018, there were approximately 448 submarine cables with a length of over 1,000,000 km in service around the world. There has been a noticeable increase in Oceania and South-East Asia. Also, the growth of cables connecting countries in Africa, as well as from Africa to Asia, Europe and South America, continued. Before 2009, only 16 African countries were connected to a submarine cable system. Currently, only one coastal country – Eritrea – has yet to be connected. Over 50 submarine projects have been proposed so far for the period from 2019 to 2021, worth a total investment of around $7.2 billion dollars. About 30 per cent of the expected deployment will be in the Pacific Ocean, followed by the Atlantic Ocean and the Indian Ocean, which are projected to receive about 21 per cent and 17 per cent, respectively, of investment planned for the coming years.

A new industry has emerged for the recovery of old cables for their scrap value. In the past 10 years, some 62,000 km of cables have been recovered, with projections that over 100,000 km will be contracted for recovery by the end of 2020.

Transmission power cable installations have seen more modest growth. However, large numbers of power cables have been installed in association with marine wind farms.

There continue to be few cable faults in the deep ocean below a depth of approximately 2,000 m , as there is little human disturbance in that area. For example, in the vast areas beyond national jurisdiction, an average of four cable breaks are recorded annually, compared with approximately150–200 breaks worldwide. However, deep-ocean mining is a potential future threat and subject to ongoing discussions between the cable industry and the International Seabed Authority (International Seabed Authority, 2018).

As discussed in the first *Assessment*, seabed disturbance from cable installation is temporary, with natural restoration occurring over weeks to years, depending on the vigour of wave or current action and the supply of sediment (Kraus and Carter, 2018). Because submarine landslides and sediment flows can be triggered by storms, as well as earthquakes and, potentially, tsunamis, climate change may influence the hazard risk to telecommunications cables by affecting storm frequency and intensity (Gavey and others, 2017). New research (Gutscher and others, 2019) suggests that natural hazards, too weak to break a cable, may still deform glass fibres to produce a detectable signal, raising the possibility of using cables as environmental monitors and early warning systems for hazards.

3. Consequences of changes for human communities, economies and well-being

The development or improvement of coastal infrastructures, especially blue infrastructures, can bring huge benefits to coastal communities. Coastal and marine infrastructures are very important for disaster risk reduction, economic development and coastal and marine science development. Coastal infrastructures support intermodal connections to maritime connections and critical global supply chains; provide public access to coastal recreation, tourism and other uses; and support access for development. Coastal defence structures can help to minimize damage caused by, for example, coastal erosion, flooding, high waves and storm surge. Hotel and recreation infrastructures support tourism and recreation and generate employment. New cable connectivity brings the benefits of global communications, telemedicine and learning to otherwise isolated communities and supports economic development, ocean science development and management implementation.

Coastal infrastructures play critical roles in achieving the Sustainable Development Goals[1] (Economist, 2019). The improvement of coastal and marine infrastructures contributes, in particular, to the implementation of Goals 1, 2, 6, 8, 9, 10, 13 and 14. With regard to Goal 14 in particular, coastal and marine infrastructures may enable better observation, monitoring and surveys of coastal and ocean environments, ecological systems and biodiversity in order to provide better data for improved management. However, the development of coastal and marine infrastructure may damage habitats and ecological systems, including their extent, structures and functions. Careful planning, with the aid of evidence-based marine spatial planning and functional analysis and the use of blue infrastructures, can help to reduce negative effects. In the United States, for example, federally approved state coastal management programmes are required to consider all stakeholder interests related to the ocean-coast interface.

4. Key region-specific changes and consequences

4.1. North Atlantic Ocean, Baltic Sea, Black Sea, Mediterranean Sea and North Sea

4.1.1. North Atlantic Ocean

The coasts of the North Atlantic Ocean extend to eastern Canada, the eastern United States and Western European and West African countries. The levels of economic development of nations in those regions vary markedly, as do coastal defence and other infrastructure development. In Canada and the United States, coastal habitats are used as natural defence infrastructures (Elkin, 2017). In Western Europe, owing to the limited land area, coastal defence and other marine infrastructures are developed for coastal protection.

1 See General Assembly resolution 70/1. See also https://sustainabledevelopment.un.org/?menu=1300.

For North-West Africa, numerous interrelated issues exist, including severe coastal erosion, flooding, poverty and inadequate development of coastal infrastructures. In 2018, the Global Environment Facility and the World Bank Group funded the West Africa Coastal Areas Resilience Investment Project with a budget of $210 million to help to build the resilience of coastal communities in Benin, Côte d'Ivoire, Mauritania, Sao Tome and Principe, Senegal and Togo (World Bank, 2018).

4.1.2. Baltic Sea

The total length of coastline in the Baltic Sea is about 40,000 km, and large flood-prone areas exist in Denmark, Germany and Poland. Therefore, the coastal infrastructure, such as dykes, needs to be upgraded to better adapt to flooding. There is also considerable development of tourist-related infrastructure, ports, harbours, marinas, shipyards, wind farms, solar power stations and submarine power and communications cables.

4.1.3. Black Sea

The Black Sea coastline is 2,042 km long, with 1,228 beaches and an area of 224 km^2. Some coastal parts are densely populated and are also popular for tourism, with many facilities, such as hotels, resorts and marinas. There are also ports and harbours. Coastal defence structures have been built to ameliorate severe coastal erosion and flooding.

4.1.4. Mediterranean Sea

The Mediterranean Sea has 46,000 km of coastline bordering 22 countries. As one of the busiest maritime regions in the world, there are many important ports. The densely populated Mediterranean coast also faces risks of erosion and flooding – a situation that will become more serious in the future owing to climate change and sea level rise, and coastal infrastructure will need to be upgraded.

4.1.5. North Sea

The coastal and low-lying inland areas bordering the North Sea are at risk of flooding. As with other areas, coastal flooding risks will increase in the future owing to sea level rise and more intense or frequent storms. Thus, new as well as upgraded coastal defence structures are needed to meet the challenge.

4.2. South Atlantic Ocean and wider Caribbean

The South Atlantic Ocean and the wider Caribbean coasts encompass countries in South America and South-West Africa. South-West African coasts are generally in a natural condition. For example, some parts are protected by coastal ecological systems, such as mangrove forests. Coastal infrastructures in the South Atlantic Ocean and the wider Caribbean include coastal defence structures, tourism facilities, ports and harbours, but newly developed or upgraded structures are needed to adapt to climate change. Caribbean nations are also exposed to pronounced earthquake and volcanic activity. Natural infrastructures are used for coastal prevention and hazard protection, and research (e.g., Powell and others, 2018) found that investments in natural infrastructure in the coastal zone can have a measured value for coastal communities while increasing ecological persistence and resilience, but more research is needed for the development of best practices.

4.3. Indian Ocean, Bay of Bengal, Arabian Sea, Red Sea, Gulf of Aden and Persian Gulf

The Indian Ocean and the Bay of Bengal encompass many developing countries in Asia and Africa. Coastal natural hazards for countries bordering the North Indian Ocean include storm surge, sea level rise, earthquakes and tsunamis. However, environmental degradation and exploitation through unsustainable economic activity resulted in reduced adaptive capacity of coastal communities, requiring huge investments for adaptation infrastructures and sustainable economies. Perhaps the most feasible way forward for coastal States of the Indian Ocean is to restore degraded and damaged coastal habitats in order to create coastal blue infrastructures.

Coastal and marine infrastructures of the Arabian Sea, the Red Sea, the Gulf of Aden and the Persian Gulf are, in general, better developed than those in the Indian Ocean and the Bay of Bengal.

4.4. North Pacific Ocean

As with the coasts of the North Atlantic Ocean, developed countries such as Canada, Japan, the Republic of Korea and the United States have high-quality coastal and marine infrastructures that not only protect coasts and reduce hazard risks but in some cases also promote the protection and conservation of coastal and ocean environments, habitats and biodiversity (Gillies and others, 2019). For many Pacific coastal States, there are the ever-present risks of major earthquakes and volcanic eruptions. However, the coastal and marine infrastructures in developing countries in the region are not so advanced (Partnerships in Environmental Management for the Seas of East Asia, 2018; Connell, 2018). To remediate the situation of underdeveloped coastal infrastructures in developing countries in Asia, the Asian Development Bank launched an ambitious action plan with a proposed investment of $5 billion for healthy oceans, which includes developing or improving coastal infrastructure (Asian Development Bank, 2019).

4.5. South Pacific Ocean

South Pacific coasts include the east coast of Australia, the west coast of South America and shores of the Pacific Islands, including Papua New Guinea, New Zealand and New Caledonia. Coastal infrastructure in those nations is mainly aimed at supporting economic development, preventing damage due to natural hazards, especially extreme storms and rising sea level, and adapting to climate change. Major earthquakes, tsunamis and volcanic eruptions are also a consideration.

4.6. Arctic Ocean and Southern Ocean

The low population densities of these regions mean that coastal and ocean infrastructures are less developed than those of highly populated regions such as the circum-Pacific and Mediterranean.

4.6.1. Arctic Ocean

Coastal infrastructure development in the Arctic Ocean is faced with rapidly changing weather and ice conditions owing to climate change. Declining sea ice cover is leading to increased shipping and related infrastructure (United States Committee on the Marine Transportation System, 2018). Progress has been made with the installation of a 1,900 km fibre-optic communications cable off the coast of north Alaska, with branch lines to six coastal communities (Submarine Cable Networks, 2017) and extensions of national networks in Greenland and Norway, among others (Quintillion, 2020).

4.6.2. Southern Ocean

Much of the Southern Ocean comes under the aegis of the Antarctic Treaty System, including the Commission for the Conservation of Antarctic Marine Living Resources (Antarctic Treaty System, 2019; Commission for the Conservation of Antarctic Marine Living Resources, 2017; 2019). Nevertheless, there is a strong focus on scientific research into the roles played by Antarctica and the Southern Ocean in influencing the global climate and ocean. Such research is supported by permanently occupied stations along the Antarctic coast and on some sub-Antarctic islands.

5. Outlook

5.1. Outlook for the state of marine and coastal infrastructures over the near-to-medium term (approximately 10–20 years)

In the next 10 to 20 years, more upstream hydropower dams will be constructed and river sand mining will continue with increasing sediment deficit at coasts, leading to accelerated coastal erosion (see also chap. 13) and the need for more coastal protection structures. Coastal and offshore land reclamation, together with coastal erosion, will continue to damage or degrade important coastal and offshore shallow water marine habitats. There is also an increase in maritime tourism and associated infrastructure. At the same time and at many coastal places, socioeconomic development will lead to an increase in coastal populations and requirements for coastal and marine infrastructure. All of these factors, together with climate change, manifested by increasing ocean temperature, rising sea level and growth in the frequency and intensity of extreme weather events, increase risks to the coasts of maritime natural hazards. Thus, there is a need to develop new or upgrade existing infrastructure to mitigate risks and ensure the sustainable development of the coasts and maritime economy.

Progress in knowledge and capacity anticipated in the future will contribute to an evaluation of the change in the state of marine and coastal infrastructures and promote the development of more effective and environmentally friendly marine and coastal infrastructures; and there will be an increase in the use of blue infrastructure or natural barriers to harmonize coastal and environmental protections.

5.2. Socioeconomic consequences of continued change in ecosystems

In general, the development of marine and coastal infrastructures, especially coastal and offshore land reclamation, will damage coastal and marine habitats and ecological systems (Duan and others, 2016; McManus, 2017; Lin and Yu, 2018). The impacts of coastal structures on the ecology of coastal systems include obstructing animal access routes, destroying coastal habitats and ecological systems and changing the coastal environment (Hill, 2015). Coastal defence structures can modify the sediment budget at the coast and thus change coastal morphology, with corresponding changes in coastal biotic communities, but, in certain cases, coastal defence structures can protect coastal habitats that will otherwise be destroyed by coastal erosion (Schmitt and Albers, 2014). Coastal land reclamation may also help to create and restore coastal habitats for hazard prevention (Khalil and Raynie, 2015).

Researchers, such as Taormina and others (2018), failed to show conclusively any influence of cable-based electromagnetic fields on the abundance and biodiversity of organisms, and confirmed the generally low environmental footprint of telecommunications cables, especially in the deep ocean (depths > 2,000 m) (Burnett and others, 2013). Records of submarine cable breaks caused by landslides and sediment-laden currents are important observations of such processes that transfer heat, carbon and nutrients from land to the deep ocean and hence may influence marine ecosystems (Pope and others, 2017).

During the past 10 years, there has been a clear tendency towards alleviating or mitigating damage to coastal and ocean ecosystems from coastal and offshore development using a new development approach: blue economy development (Partnerships in Environmental

Management for the Seas of East Asia, 2018). Blue coastal infrastructures can harmonize coastal protection and habitat or ecological protection and promote carbon sequestration (Sutton-Grier and others, 2015; Wellman and others, 2017).

Coastal and marine infrastructures, in general, have a positive socioeconomic impact on coastal communities. Good infrastructure is the most important condition for coastal hazard risk mitigation, sustainable socioeconomic development and poverty eradication.

6. Key remaining knowledge and capacity-building gaps

In general, at the global level, not enough is known about the extent of coastal infrastructures, especially built coastal defence infrastructures, and their ecological and socioeconomic impacts. Also, scientific understanding of the interactions between coastal dynamics, sediment transport and the environment, and between ecological processes and marine and coastal infrastructures, is still lacking. The problems are especially serious for developing countries where little money is invested to undertake coastal and marine scientific research. A lack of proper knowledge and data also hinders correct design and construction and increases the environmental and ecological damage of coastal and marine infrastructures.

A science-policy interface is particularly important when considering decision-making related to the sustainable development of blue and nature-based marine and coastal infrastructures in order to optimize the use of and minimum damage to coastal and marine infrastructures.

Acknowledgement: The authors would like to thank John Manock for his contribution to section 2.6.

References

Antarctic Treaty System (2019). The Antarctic Treaty. www.ats.aq/index_e.html.

Asian Development Bank (2019). *Action Plan for Healthy Oceans: Investing in Sustainable Marine Economies for Poverty Alleviation in Asia and the Pacific.*

Blackburn, Sophie, and others (2013). Mega-urbanisation on the coast: global context and key trends in the twenty-first century. In *Megacities and the Coast: Risk, Resilience and Transformation*, M. Pelling and S. Blackburn, eds., pp. 1–21.

Burnett, Douglas R., and others (2013). *Submarine Cables: The Handbook of Law and Policy.* Martinus Nijhoff Publishers.

Carter, L., and others (2019). Chemical and physical stability of submarine fibre-optic cables in the Area Beyond National Jurisdiction (ABNJ). In *SubOptic 2019 Conference.*

Commission for the Conservation of Antarctic Marine Living Resources (2017). CCAMLR to create world's largest Marine Protected Area. www.ccamlr.org/en/news/2016/ccamlr-create-worlds-largest-marine-protected-area.

________ (2019). About the Commission for the Conservation of Antarctic Marine Living Resources. www.ccamlr.org/en/organisation.

Connell, John (2018). Effects of climate change on settlements and infrastructure relevant to the Pacific Islands. *Pacific Marine Climate Change Report Card Scientific Review.*

Davis, McKenna, and others (2015). Coastal protection and SUDS: nature-based solutions. RECREATE Project Policy Brief, No. 4.

Duan, Huabo, and others (2016). Characterization and environmental impact analysis of sea land reclamation activities in China. *Ocean and Coastal Management*, vol. 130, pp. 128–137.

Economist (2019). *The Critical Role of Infrastructure for the Sustainable Development Goals.* https://content.unops.org/publications/The-critical-role-of-infrastructure-for-the-SDGs_EN.pdf?mtime=20190314130614.

Edwards, P.E.T., and others (2013). Investing in nature: restoring coastal habitat blue infrastructure and green job creation. *Marine Policy*, vol. 38, pp. 65–71.

Elkin, R.S. (2017). Beyond Restoration: Planting Coastal Infrastructure. In *Climate Change Adaptation in North America*, pp. 119–135. Springer.

Gavey, R., and others (2017). Frequent sediment density flows during 2006 to 2015, triggered by competing seismic and weather events: observations from subsea cable breaks off southern Taiwan. *Marine Geology*, vol. 384, pp. 147–158.

Gillies, Chris, and others (2019). Coastal habitat squeeze: a review of adaptation solutions for saltmarsh, mangrove and beach habitats. *Ocean and Coastal Management*, vol. 175, pp. 180–190.

Gutscher, Marc-André, and others (2019). Fiber optic monitoring of active faults at the seafloor: I the FOCUS project. *Photoniques,* pp. 32–37.

Hill, Kristina (2015). Coastal infrastructure: a typology for the next century of adaptation to sea-level rise. *Frontiers in Ecology and the Environment*, vol. 13, No. 9, pp. 468–476.

Intergovernmental Panel on Climate Change (2019). Summary for Policymakers. In: *IPCC Special Report on the Ocean and Cryosphere in a Changing Climate.* H-O. Pörtner and others, eds. https://report.ipcc.ch/srocc/pdf/SROCC_FinalDraft_FullReport.pdf.

International Association of Dredging Companies (2018). *Dredging in Figures 2017.*

International Seabed Authority (2018). *Deep Seabed Mining and Submarine Cables.*

Kazmierczak, A., and J. Carter (2010). Adaptation to climate change using green and blue infrastructure. A database of case studies. University of Manchester.

Khalil, S.M., and R.C. Raynie (2015). Coastal restoration in Louisiana: an update. *Shore and Beach*, vol. 83, No. 4, p. 4.

Kraus, Christoph, and Lionel Carter (2018). Seabed recovery following protective burial of subsea cables: observations from the continental margin. *Ocean Engineering*, vol. 157, pp. 251–261.

Lin, Qiaoying, and Shen Yu (2018). Losses of natural coastal wetlands by land conversion and ecological degradation in the urbanizing Chinese coast. *Scientific Reports*, vol. 8, No. 1, p. 15046.

McManus, John W. (2017). Offshore coral reef damage, overfishing, and paths to peace in the South China Sea. *The International Journal of Marine and Coastal Law*, vol. 32, No. 2, pp. 199–237.

Partnerships in Environmental Management for the Seas of East Asia (2018). *State of Oceans and Coasts 2018: Blue Economy Growth in the East Asian Region.* http://pemsea.org/publications/reports/state-oceans-and-coasts-2018-blue-economy-growth-east-asian-region.

Pope, Ed. L., and others (2017). Damaging sediment density flows triggered by tropical cyclones. *Earth and Planetary Science Letters*, vol. 458, pp. 161–169.

Powell, E.J.,and others (2018). A review of coastal management approaches to support the integration of ecological and human community planning for climate change. *Journal of Coastal Conservation*, vol. 23, No. 1, pp.1–18.

Quintillion (2020). System Specifications. http://qexpressnet.com/system.

Rousseau, Yannick, and others (2019). Evolution of global marine fishing fleets and the response of fished resources. *Proceedings of the National Academy of Sciences of the United States of America*, vol. 116, No. 25, pp. 12238–12243. https://doi.org/10.1073/pnas.1820344116.

Schmitt, Klaus, and Thorsten Albers (2014). Area coastal protection and the use of bamboo breakwaters in the Mekong Delta. In *Coastal Disasters and Climate Change in Vietnam*, pp. 107–132. Elsevier.

Sengupta, Dhritiraj, and others (2018). Building beyond land: an overview of coastal land reclamation in 16 global megacities. *Applied Geography*, vol. 90, pp. 229–238.

Submarine Cable Network (2017). Quintillion activates Arctic subsea cable. www.submarinenetworks.com/en/systems/asia-europe-africa/arctic-fiber/quintillion-activates-arctic-subsea-cable.

Sutton-Grier, Ariana E., and others (2015). Future of our coasts: the potential for natural and hybrid infrastructure to enhance the resilience of our coastal communities, economies and ecosystems. *Environmental Science and Policy*, vol. 51, pp. 137–148.

Taormina, Bastien, and others (2018). A review of potential impacts of submarine power cables on the marine environment: knowledge gaps, recommendations and future directions. *Renewable and Sustainable Energy Reviews*, vol. 96, pp. 380–391.

United Nations (2017). *The First Global Integrated Marine Assessment: World Ocean Assessment I*. Cambridge: Cambridge University Press.

United Nations Conference on Trade and Development (UNCTAD) (2018). *Review of Maritime Transport*. Geneva.

United States Committee on the Marine Transportation System (2018). *Revisiting Near-Term Recommendations to Prioritize Infrastructure Needs in the U.S. Arctic*. Washington, D.C., p. 43.

Value Market Research (2017). Recreational Boating Industry Report: Trends, Forecast and Competitive Analysis. www.valuemarketresearch.com/report/recreational-boating-market.

Wellman, Emory, and others (2017). Catching a wave? A case study on incorporating storm protection benefits into habitat equivalency analysis. *Marine Policy*, vol. 83, pp. 118–125.

Williams, A.T., and others (2018). The management of coastal erosion. *Ocean and Coastal Management*, vol. 156, pp. 4–20.

World Bank (2018). World Bank Board Approves West Africa Coastal Areas (WACA) Resilience Investment Project. www.worldbank.org/en/news/press-release/2018/04/09/world-bank-board-approves-west-africa-coastal-areas-waca-resilience-investment-project.

Chapter 15
Changes in capture fisheries and harvesting of wild marine invertebrates

Contributors: Porter Hoagland (convener), Megan Bailey, Lena Bergstrom, Alida Bundy, Karen Evans (co-lead member), Manuel Hidalgo, Andrew Johnson, Melina Kourantidou, Fernanda de Oliveira Lana, Enrique Marschoff (lead member), Essam Mohammed (co-lead member), Henn Ojaveer (co-lead member), Franklin Ormaza-Gonzalez, Imants George Priede, Ylenia Randrianisoa (co-lead member), Jörn Schmidt (co-lead member), Zacharie Sohou, Burcu Bilgin Topçu and Chang Ik Zhang.

Keynote points

- Worldwide, from 2012 to 2017, estimated landings in marine capture fisheries increased by 3 per cent to 80.6 million tons, and estimated gross landed value increased by 1 per cent to $127 billion.
- Some of the world's capture fisheries continued to experience overexploitation, vessel subsidization, ineffective management, by-catch and discards, habitat degradation, abandoned, lost or otherwise discarded fishing gear, and illegal, unreported or unregulated fishing.
- In 2017, the World Bank estimated that annual net losses to global capture fisheries were $88.6 billion for the year 2012 (expressed in 2017 dollars) owing to overfishing. If allowed to continue for the foreseeable future, such losses would constitute a lost natural capital asset worth trillions of dollars.
- The large majority of small-scale, artisanal or subsistence fishery landings were destined for local human consumption, thus contributing in a vital way to food security and nutrition in developing States, but illegal, unreported or unregulated fishing continued to pose risks to many people who depended upon fisheries for protein, exacerbating poverty, augmenting food insecurity and potentially hindering efforts to achieve the targets of the Sustainable Development Goals.
- Promisingly, scientific stock assessments and management were shown to lead to more sustainable[1] outcomes, and management reforms were predicted to lead to rapid (decade-scale) rebuilding of stocks. These were important lessons as the world began to look to unexploited and as yet unregulated fisheries in the polar regions and the deep ocean (the mesopelagic zone).
- The adverse effects of climate change on the oceans are expected to hinder sustainable outcomes, and fishery-dependent developing States, in particular their small-scale fisheries, are highly vulnerable to climate-related changes.

1. Introduction

Global landings of marine capture fisheries expanded significantly from the 1950s (Food and Agriculture Organization of the United Nations (FAO), 2016d; 2018b; 2019a), but have levelled off since the late 1980s, with a growth rate of less than 1 per cent since 2010 (FAO, 2019a). Between 2012 and 2017, world marine capture fisheries production (mainly harvests constituting landings) remained flat, ranging from 78.4 million tons in 2012 to 80.6 million tons in 2017. From 2010 to 2017, capture fishery yields (inland and marine) increased slightly in both the developed world, from 24.1 million to 24.8 million tons (2.9 per cent), and the developing world, from 63.0 million to 67.6 million tons (7.3 per cent) (FAO, 2019a).

In 2017, the world price averaged over all fisheries was $1.57 per kg, which translated into an estimated gross landed value for the world's marine capture fisheries of $126.8 billion (FAO, 2019a). Estimated annual net benefits from the landings were only $3 billion (2012 data expressed in 2017 dollars) (World Bank and others, 2012; Tai and others, 2017; World Bank,

[1] In the present chapter, "sustainable", "biologically sustainable" and "maximally sustainable" are, according to the definitions given by the Food and Agriculture Organization of the United Nations, applied primarily to single stocks.

2017). Excessive fishing efforts, leading to lowered biomass, resulted in estimated annual lost net benefits of $88.9 billion. If allowed to continue, that would constitute natural capital asset losses (i.e., the discounted or "present" value of future losses occurring annually at the same level as the 2012 estimate) in the range of $1.3 trillion to $4.4 trillion, when applying social rates of time preference of 7 to 2 per cent.

In the past decade, fish markets exhibited fast-paced globalization, thus increasing the vulnerability of small-scale fisheries to the depletion of some locally important stocks (Crona and others, 2015; Kramer and others, 2017). In 2017, about 38 per cent of global fish production entered international trade, for either human consumption or fishmeal and fish oil (FAO, 2018b). In 2017, the export value of seafood was $156.5 billion, of which $84.6 billion (54 per cent) was attributed to exports from developing States.

Reported world landings and value suggested that little had changed since the publication of the first *World Ocean Assessment* (United Nations, 2017), which relied upon data up to 2012. Governance improved in some regions, however, including the rebuilding of some fisheries as a result of prudent management (FAO, 2018b; Hilborn and others, 2020). The ecosystem-based approach to management has been recommended in the scientific literature as a helpful tool in the longer term, bringing commercial fisheries closer to the ideals expressed in the Code of Conduct for Responsible Fisheries of 1995 (Long and others, 2015; Patrick and Link, 2015; FAO, 2018a; Marshall and others, 2018; see also chap. 27).

There was extensive evidence that some of the world's fisheries were not managed sustainably (Sustainable Development Solutions Network, 2019), meaning that the targets of the Sustainable Development Goals,[2] in particular the fisheries-related targets under Goal 14, as well as others relating to food security, had not yet been met. Some progress was noted, however (United Nations, 2019b). During the period from 2012 to 2017, the most salient issues were as follows:

- Of the world's marine capture fisheries for which data existed, about 60 per cent were "maximally sustainably fished", and that proportion has been increasing since 1990 (FAO, 2020b). The combined sum of the proportions of maximally sustainably fished and "underfished" stocks was reflected in indicator 14.4.1 (proportion of fish stocks within biologically sustainable levels) (see figure I). Equally, the indicator also revealed the growing proportion of "overfished" stocks since 1974 (Sustainable Development Solutions Network, 2019; FAO, 2020a; 2020b; World Bank, 2017; see also figure I). While 66 per cent of fish stocks currently are either maximally sustainably fished or underfished, figure I emphasizes the need to reverse a declining trend in the combined sum of the two categories by improving management approaches to the 34 per cent of fish stocks outside biologically sustainable levels.
- Illegal, unreported or unregulated fishing continued, thus weakening fisheries governance and contributing to illicit trade in seafood (Macfadyen and others, 2019; Sumaila and others, 2020).
- The subsidization of fishing vessels continued (Sumaila and others, 2019a), including subsidies that contributed to overcapacity, excessive fishing and stock depletion (Rousseau and others, 2019). Sala and others (2018) estimated that 54 per cent of high seas fishing grounds would be rendered unprofitable if subsidies were eliminated. Negotiations under the auspices of the World Trade Organization (WTO) on eliminating illegal, unreported or unregulated fishery subsidies and prohibiting certain other forms of subsidies continued at

2 See General Assembly resolution 70/1.

an accelerated pace, and agreement was expected during 2020 (WTO, 2020).

- The impacts of bottom trawling on marine habitats continued, but measures were implemented by individual States or regional fisheries management organizations or arrangements to mitigate the impacts on seabeds and seamounts, and progress was made on the development of indices of seabed integrity to assess levels of impact (Eigaard and others, 2017; Hiddink and others, 2017; Kroodsma and others, 2018).
- Fishing power in demersal and pelagic fisheries continued its steady, but often imperceptible, increase in efficiency (known as "technology creep"), at 0.2 per cent annually on average, necessitating compensating adjustments in management (Palomares and Pauly, 2019a).
- Abandoned, lost or otherwise discarded fishing gear continued to diminish ecosystem integrity, thus imposing costs for both industry and national authorities (FAO, 2018b).
- Some regional fisheries management organizations or arrangements covering the high seas were not effective enough in assessing stocks, enforcing catch limits or providing observer coverage to account for catches, by-catches or discards (Cullis-Suzuki and Pauly, 2010; Crespo and Dunn, 2017; International Council for the Exploration of the Sea (ICES), 2018a), but States were increasingly motivated to achieve sustainable outcomes by increasing the effectiveness of such organizations or arrangements, as exemplified by agreements on regional cooperative initiatives, such as the Jakarta Concord of 2017 for the Indian Ocean.[3]
- Significant gaps remained in establishing and reaching consensus on management practices for sustaining healthy fish stocks, including: disputed jurisdictions in the central Pacific Ocean and the south-western Atlantic Ocean (Harrison, 2019); less than fully effective management of high seas fisheries on deep ocean shelves and seamounts (ICES, 2018b); limited progress in the conservation of potential fish stocks in the central Arctic Ocean (with a temporary 16-year moratorium on unregulated fishing awaiting entry into force); and the absence of management of prospective fisheries in the mesopelagic zone, where regulation was either nascent or non-existent (Priede, 2017; Hidalgo and Browman, 2019; Remesan and others 2019).

Figure I
Proportion of fish stocks within biologically sustainable levels (Sustainable Development Goal indicator 14.4.1)

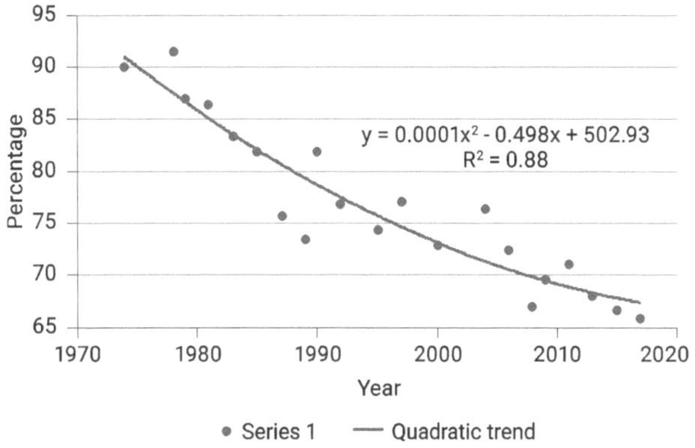

The data reflect the sum of the percentages of the world's marine capture fisheries that were considered to be either "maximally sustainably fished" (59.6 per cent in 2017) or "underfished" (6.2 per cent in 2017). The percentage of fisheries that were maximally sustainably fished increased from 1990 to 2017. Alternatively, if subtracted from 100 per cent in any year, the data reflect the growing percentage of the number of fisheries that were "overfished" (FAO, 2020a; 2020b).

Abbreviations: x, year; y, proportion of fish stocks within biologically sustainable levels; R^2, coefficient of determination for the relationship between y and both x and x^2.

3 The Indian Ocean Rim Association, Jakarta Concord: promoting regional cooperation for a peaceful, stable and prosperous Indian Ocean (adopted on 7 March 2017). See www.iora.int/media/23699/jakarta-concord-7-march-2017.pdf.

Notwithstanding these issues and gaps, recent scientific research has suggested that, with appropriate governance, the median time required to rebuild overfished stocks could be less than 10 years, and, if reforms were implemented, 98 per cent of overfished stocks could be considered healthy by the middle of the century (Sumaila and others, 2012; Neubauer and others, 2013; Costello and others, 2016; Hilborn and Costello, 2018; Garcia and others, 2018). Little consensus existed among scientists, however, on whether recovered ecosystems and populations could assume their original functions (Van Gemert and Andersen, 2018; Ingeman and others, 2019), and for some extremely depleted stocks, such as the Atlantic cod (*Gadus morhua*), potential recovery times were projected to be much longer (Neuenhoff and others, 2019).

The scientific assessment and management of fish stocks was shown to improve their sustainability (Hilborn and others, 2020). Scientists reasoned that management reforms, including rights-based approaches, had the potential to yield significant increases in annual catches (2 million–16 million tons) and profits ($31 billion–$53 billion) (Costello and others, 2016). Scientists also maintained that increases in biomass and biodiversity concomitant with fishery management reforms would facilitate the adaptation of ocean ecosystems to global climate change (Berkes and Ross, 2013; Armitage and others, 2017). Consequently, the rebuilding of fish stocks remained a high priority for States and international organizations (Delpeuch and Hutniczak, 2019).

Even with appropriate governance leading to stock rebuilding, the adverse effects of global climate change were expected to impede progress toward sustainability (Lam and others, 2016; Pentz and others, 2018; Intergovernmental Panel on Climate Change, 2019; Lotze and others, 2019; see also chap. 5). Despite a limited understanding of the extent to which changing conditions contributed to ecosystem shifts, scientists found that alterations in the structures and functions of marine ecosystems were more common than expected, and they contended that such changes could be hard to reverse (Selkoe and others, 2015; Samhouri and others, 2017).

The impacts of climate change were also expected to include increases in the intensity and frequency of natural hazards, thus affecting the local distribution and abundance of fish populations (Barange and others, 2014; Bryndum-Buchholz and others, 2018). Scientists predicted that fishery-dependent developing States would be affected the most severely, and, owing to expected changes in species distribution and consequent increases in the transboundary migration of stocks, such redistributions would need to be accounted for in future international governance (Pinsky and others, 2018; Sumaila and others, 2019b).

2. Catch-landing disparities, Sustainable Development Goals and small-scale fisheries

2.1. National jurisdictions

Between 2012 and 2017, global landings were stable, and the density of fishing efforts continued to be highly concentrated in coastal oceans (Tickler and others, 2018). Catches by incoming distant water fleets were found to grow faster than catches by home States, and 78 per cent of trackable industrial fishing in the exclusive economic zones of lower-income States was carried out by vessels flagged to high-income States (McCauley and others, 2018). In 2016, landings in tropical areas continued to grow strongly to 23.8 million tons, were flat in temperate areas at 38.9 million tons, and in upwelling areas exhibited high variability, declining to 14.5

million tons (FAO, 2018b). Tables 1 and 2 depict national and regional variations in average landings between 2005 and 2014 compared with the period from 2015 to 2016 (FAO, 2018b).

Table 1
Marine capture fisheries production by country

Country	Production (tons)			Percentage variation		Variation, 2015 to 2016 (tons)
	Average 2005–2014	2015	2016	2005–2014 (average) to 2016	2015 to 2016	
China	13 189 273	15 314 000	15 246 234	15.6	-0.4	-67 766
Indonesia	5 074 932	6 216 777	6 109 783	20.4	-1.7	-106 994
United States	4 757 179	5 019 399	4 897 322	2.9	-2.4	-122 077
Russian Federation	3 601 031	4 172 073	4 466 503	24	7.1	294 430
Peru Total	6 438 839	4 786 551	3 774 887	-41.4	-21.1	-1 011 664
Excluding anchoveta	989 918	1 016 631	919 847	-7.1	-9.5	-96 784
India	3 218 050	3 497 284	3 599 693	11.9	2.9	102 409
Japan[a]	3 992 458	3 423 099	3 167 610	-20.7	-7.5	-255 489
Viet Nam	2 081 551	2 607 214	2 678 406	28.7	2.7	71 192
Norway	2 348 154	2 293 462	2 033 560	-13.4	-11.3	-259 902
Philippines	2 155 951	1 948 101	1 865 213	-13.5	-4.3	-82 888
Malaysia	1 387 577	1 486 050	1 574 443	13.5	5.9	88 393
Chile Total	3 157 946	1 786 249	1 499 531	-52.5	-16.1	-286 718
Excluding anchoveta	2 109 785	1 246 154	1 162 095	-44.9	-6.7	-84 059
Morocco	1 074 063	1 349 937	1 431 518	33.3	6.0	81 581
Republic of Korea	1 746 579	1 640 669	1 377 343	-21.1	-16.0	-263 326
Thailand	1 830 315	1 317 217	1 343 283	-26.6	2.0	26 066
Mexico	1 401 294	1 315 851	1 311 089	-6.4	-0.4	-4 762
Myanmar[a]	1 159 708	1 107 020	1 185 610	2.2	7.1	78 590
Iceland	1 281 597	1 318 916	1 067 015	-16.7	-19.1	-251 901
Spain	939 384	967 240	905 638	-3.6	-6.4	-61 602
Canada	914 371	823 155	831 614	-9.1	1.0	8 459
Taiwan Province of China	960 193	989 311	750 021	-21.9	-24.2	-239 290
Argentina	879 839	795 415	736 337	-16.3	-7.4	-59 078
Ecuador	493 858	643 176	715 357	44.9	11.2	72 181
United Kingdom	631 398	704 502	701 749	11.1	-0.4	-2 753
Denmark	735 966	868 892	670 207	-8.9	-22.9	-198 685
Total: 25 major countries	65 451 506	66 391 560	63 939 966	-2.3	-3.7	-2 451 594
Total: other 170 countries	14 326 675	14 856 282	15 336 882	7.1	3.2	480 600
World total	79 778 181	81 247 842	79 276 848	-0.6	-2.4	-1 970 994
Share of 25 major countries	82.0%	81.7%	80.7%			

[a] Production figures for 2015 and 2016 are estimates.
Source: FAO (2018b).

Table 2
Fishing areas and capture production

Fishing area code	Fishing area name	Production (tons) Average 2005–2014	2015	2016	Percentage variation 2005–2014 (average) to 2016	2015 to 2016	Variation, 2015 to 2016 (tons)
21	Atlantic North-West	2 041 599	1 842 787	1 811 436	-11.3	-1.7	-31 351
27	Atlantic North-East	8 654 911	9 139 199	8 313 901	-3.9	-9.0	-825 298
31	Atlantic Western Central	1 344 651	1 414 318	1 563 262	16.3	10.5	148 944
34	Atlantic Eastern Central	4 086 427	4 362 180	4 795 171	17.3	9.9	432 991
37	Mediterranean Sea and Black Sea	1 421 025	1 314 386	1 236 999	-13.0	-5.9	-77 387
41	Atlantic South-West	2 082 248	2 427 872	1 563 957	-24.9	-35.6	-863 915
47	Atlantic South-East	1 425 775	1 677 969	1 688 050	18.4	0.6	10 081
51	Indian Ocean Western	4 379 053	4 688 848	4 931 124	13.9	5.2	242 276
57	Indian Ocean Eastern	5 958 972	6 359 691	6 387 659	7.2	0.4	27 968
61	Pacific North-West	20 698 014	22 057 759	22 411 224	7.7	1.6	353 465
67	Pacific North-East	2 871 126	3 164 604	3 092 529	7.7	-2.3	-72 075
71	Pacific Western Central	11 491 444	12 625 068	12 742 955	10.9	0.9	117 887
77	Pacific Eastern Central	1 881 996	1 675 065	1 656 434	-12.0	-1.1	-18 631
81	Pacific South-West	613 701	551 534	474 066	-22.8	-14.0	-77 468
87	Pacific South-East	10 638 882	7 702 885	6 329 328	-40.5	-17.8	-1 373 557
18, 48, 58, 88	Arctic and Antarctic	188 360	243 677	278 753	48.0	14.4	35 076
	World total	79 778 184	81 247 842	79 276 848	19.0	-43.9	-1 970 994

Source: FAO (2018b).

New estimates of world fish catches, reconstructed using data from 1950 to 2010 to include catches (and discards) missing from official statistics, suggested that world annual landings were underestimated by at least one third and that catches were declining faster than previously thought (Pauly and Zeller, 2016; Zeller and others, 2018). According to that method, a significant proportion of unreported fishing comprised discarded fishes and catches by vessels used for illegal, unreported or unregulated fishing, recreational fishers or small-scale fisheries. Reconstructed catches for FAO regions revealed especially large differences compared with recorded landings in the western Atlantic Ocean, the Mediterranean Sea and the Indian Ocean (Palomares and Pauly, 2019b).

With regional variations, the level of global fisheries employment in 2017 (40.4 million people) exhibited a small increase compared with 2012 (less than 3 per cent) (FAO, 2019a). With regard to target 2.3 of the Sustainable

Development Goals, which promotes, inter alia, access by small-scale fisheries to productive resources, services and markets (indicator 2.3.1), progress was noted in the development of targeted regulatory and institutional frameworks. However, more than 20 per cent of fishing States, in particular those in Oceania and South Asia, exhibited only low to medium levels of implementation of such frameworks (United Nations, 2019a).

It was estimated that small-scale fisheries employed more than 90 per cent of the world's 120 million people involved in capture fisheries (about 50 per cent of whom were women) (World Bank and others, 2012; FAO, 2015; 2019a). Despite their significant contribution to global catches, small-scale fisheries were marginalized, with increasing pressure from both industrialized (and often subsidized) fleets and other ocean uses (Schuhbauer and Sumaila, 2016; Bundy and others, 2017; Ding and others, 2017; Willmann and others, 2017; Cohen and others, 2019). Climate changes were expected to have an adverse impact on those involved with small-scale fisheries, and adaptive strategies were identified, including the need for alternative livelihoods (Shaffril and others, 2017).

Capture fisheries remained a key source of nutrition and employment for millions of people, but it was estimated that more than 820 million were still undernourished (FAO, 2019b). Between 90 and 95 per cent of small-scale fisheries landings were destined for local human consumption, thus making a sizable contribution to food security and nutrition (World Bank and others, 2012; Golden and others, 2016; Basurto and others, 2017; Johnson and others, 2018).

The application of information technologies to help to expand opportunities for small-scale fisheries in areas such as safety, the sharing of local knowledge, capacity-building and governance were outlined in the Voluntary Guidelines for Securing Sustainable Small-Scale Fisheries in the Context of Food Security and Poverty Eradication (FAO, 2015), which were considered central to achieving the fishery-relevant Sustainable Development Goals (Said and Chuenpagdee, 2019). Implementation was expected to take time, but a growing use of human rights approaches provided opportunities for the empowerment of such fisheries (Song and Soliman, 2019). Research efforts such as the Too Big to Ignore global partnership were organized to help to focus attention on small-scale fisheries (Too Big to Ignore, 2020), and 2022 was proclaimed the International Year of Artisanal Fisheries and Aquaculture by the General Assembly.[4]

Subsidies exacerbated problems of overcapacity and overfishing, especially where illegal, unreported or unregulated fishing was involved. Some subsidies in well-managed fisheries were beneficial, such as investments in stock assessments. In 2018, annual world fishery subsidies were estimated to be $35.4 billion, compared with $41.4 billion a decade earlier (2009 data expressed in 2018 dollars), but the decline was not considered significant (Sumaila and others, 2019a). Most subsidies were provided by developed States (Schuhbauer and others, 2017). Capacity-enhancing (detrimental) subsidies increased in proportion to the total, comprising 63 per cent of all subsidies (about $22 billion) versus 57 per cent a decade ago (Sumaila and others, 2019a).

Progress was made in proposing guidelines for assessing fisheries and accounting for their contributions in data-poor environments (Cai and others, 2019). FAO introduced a methodology for indicator 14.7.1 of the Sustainable Development Goals to measure sustainable fisheries as a percentage of gross domestic product (FAO, 2020c). A more comprehensive indicator to include illegal, unreported or

[4] See General Assembly resolution 72/72.

unregulated fishing, resource rents and trade in fisheries services also was under development.

2.2. High seas fisheries

Many of the world's most valuable capture fisheries were focused on highly migratory apex predators that straddled adjacent exclusive economic zones or migrated between such zones and the high seas (Sumaila and others, 2015). Fisheries for species groups such as tunas, billfish and sharks were targeted by national fleets within their own exclusive economic zones or by international fleets licensed to enter foreign zones or on the high seas. The use of ocean space by longline fisheries, for example, overlapped by more than 75 per cent with the known spatial distributions of commercially valuable sharks (Queiroz and others, 2019). High seas capture fisheries landings grew from about 0.5 million to 4.3 million tons between 1950 and 2014 (Cheung and others, 2019).

Since the 1950s, industrial fishing expanded significantly, with increases in landings from inshore waters, the high seas (especially the large pelagic zones) and the polar regions (United Nations, 2017; Watson and Tidd, 2018). High seas yields peaked in 1989 at 5.2 million tons but declined slightly in the past three decades. Although the high seas encompass 60 per cent of the global ocean, capture fishery yields comprised only about 5 per cent of world marine yields of both fishes and invertebrates. The contribution of high seas fisheries to global seafood supply was therefore of minor importance to food security during that time (Schiller and others, 2018).

Vessels flagged to high-income States comprised 97 per cent of industrial fishing vessels on the high seas (McCauley and others, 2018). Longline fishing was reported to account for 84–87 per cent of hours fished on the high seas (Crespo and others, 2018). More than 80 per cent of the effort was attributable to vessels from only five States. From 1950 to 2014, the distance fished from port by industrial fishing vessels more than doubled, but at the same time there was a decline from 25 to 7 tons (catch) per million metres (distance) travelled (Tickler and others, 2018).

Nearly 95 per cent of total ice-free ocean areas were exploited by industrial fishing, but since their peaks in 1996, the total industrial catch declined by 18 per cent and the industrial catch per unit of area fished declined by 22 per cent (Tickler and others, 2018). Fishing intensity (effort per month) by longline vessels was found to increase in boreal regions during the summer months, with intensity linked to environmental predictors (Crespo and others, 2018).

3. Invertebrate landings

Marine invertebrate harvests grew from about 12.4 million tons in 2012 to 12.5 million tons in 2017, representing a growth rate of only 0.1 per cent per year. Marine invertebrate landings involved a number of different types of organisms, including molluscs (squid, octopus and shellfish), crustaceans (shrimps, prawns, crabs, lobsters and krill), echinoderms (sea urchins and sea cucumbers) and tunicates, and yields of those groups represented about 15.5 per cent of world marine capture fisheries landings in 2017 (FAO, 2019a).

4. Levels of by-catch and side effects

There were few time series available to document trends in by-catch (ICES, 2018a). Owing to regulatory restrictions or poor quality, non-target fish caught or damaged were often discarded. In 2019, it was estimated that global discard levels accounted for 10.8 per cent of world catches (2010–2014 data), which translated to 9.1 million tons (ranging from 6.7 million to 16.1 million tons) (Pérez Roda and others, 2019). Progress is being made in policies and management measures in order not only to manage impacts on target species but also to include effects on other species (ICES, 2019).

5. Post-harvest fish losses

Post-harvest fish losses were harvested fish that lost part of their value owing to deterioration in quality, rendering them inedible or unmarketable (Diei-Ouadi and Mgawe, 2011). Such losses were an issue primarily for small-scale fisheries, where storage capacity, processing and transportation modes were limited. The most recent global estimate of non-discarded post-harvest fish losses was 10 million–12 million tons per year (Manning, 2010). More recent studies of such losses were limited to local fisheries, especially in Africa and Asia, where it was found that losses were reduced for older fishers and those exhibiting higher levels of education and from larger households (e.g., Adelaja and others, 2018).

6. Potential for fisheries enhancement

Fish stock propagation, more commonly known as fisheries enhancement, comprised a set of management approaches involving the use of aquaculture technologies, programmes in marine ranching, the construction of artificial reefs, and egg and larval releases to restore fish stocks that exhibited depressed populations. The science was still in its infancy but showed some potential to increase fishery yields beyond those achievable by the exploitation of wild stocks alone, although an understanding of the ecological consequences was inchoate (Taylor and others, 2017).

7. Marine protein and oils in agriculture and aquaculture

Fishmeal was used as feed, and fish oil was used as a feed additive for aquaculture and livestock. Whole fishes as a raw material are "reduced" by cooking, pressing and heating to yield ratios of whole fishes to fishmeal of approximately 22–23 per cent and of whole fishes to fish oil of approximately 4–5 per cent. Worldwide landings of whole fishes for such purposes stood at 14.3 million tons in 2016 (FAO, 2018b). In 2016, world production of fishmeal was 4.4 million tons and that of fish oil was 0.9 million tons. In 2016, 69 per cent of fishmeal and 75 per cent of fish oil production was for aquaculture. In agriculture, 23 per cent of fishmeal was used for swine production, 5 per cent for poultry and 3 per cent for other purposes.

Aquafeed for shrimp or fin fish culture (see chap. 16) comprised fishmeal and fish oil, oilseeds (especially soybeans) and by-products from the processing of other fish products (Silva and others, 2018). Estimates of total aquafeed production were 40.1 million tons in 2018 (Alltech, 2019), making the component attributed to fish reduction less than 15 per cent of the total. The proportion was expected to fall below 10 per cent by 2020 (Fry and others, 2016). In 2016, 19 per cent of world fishmeal production was derived from fishery by-products (Institute of Aquaculture, 2016). The proportion was projected to increase to 38 per cent by 2025 (FAO, 2018b).

Commercial harvests of krill were undertaken in the Southern Ocean and a few other regions to supply fishmeal and fish oil (European Market Observatory for Fishery and Aquaculture Products, 2018). Fisheries for mesopelagic fish were explored for the same end uses, but the costs of harvesting the fishes were considered steep and the ecological consequences of exploiting them had not yet been fully evaluated (Hidalgo and Browman, 2019).

8. Illegal, unreported or unregulated fishing

Illegal, unreported or unregulated fishing weakened efforts to manage fisheries sustainably, increasing risks to 4.3 billion people who depended on fisheries for protein (FAO, 2016d), exacerbating poverty, augmenting food insecurity and potentially hindering efforts to achieve some of the targets of the Sustainable Development Goals (FAO, 2016b). In 2016, illegal, unreported or unregulated fishing was thought to be responsible for annual catches of up to 26 million tons, with a gross landed value of up to $23 billion (FAO, 2016c). Several international legal instruments contained measures relevant to eliminating subsidies for illegal, unreported or unregulated fishing, and indicator 14.6.1 tracked the implementation of those instruments at the State, regional and global levels (see figure II). Across the world, the overall score for the degree of implementation of the applicable instruments was moderate (falling within band 3). Negotiations on the use of international trade measures to eliminate subsidies for illegal, unreported or unregulated fishing (and to prohibit certain other forms of subsidies) continued at WTO, with the expectation that agreement would be reached during 2020 (WTO, 2020).

Figure II
Indicator for Sustainable Development Goal target 14.6

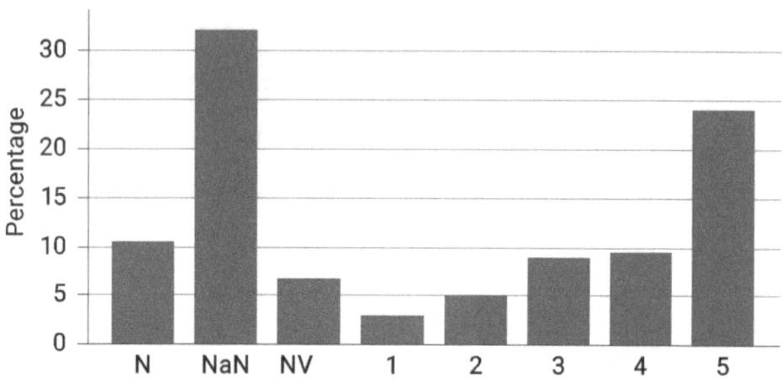

Note: Target 14.6 seeks, by 2020, to prohibit certain forms of fisheries subsidies that contribute to overcapacity and overfishing, eliminate subsidies that contribute to illegal, unreported or unregulated fishing and refrain from introducing new such subsidies, recognizing that appropriate and effective special and differential treatment for developing and least developed countries should be an integral part of the WTO fisheries subsidies negotiation. More specifically, the figure shows, as at 30 June 2020, the proportion of 199 FAO jurisdictions (mainly States) with varying degrees of implementation of international legal instruments aimed at eliminating subsidies that contribute to illegal, unreported or unregulated fishing. Based upon responses to a questionnaire, the *x* axis represents scores relating to the degree of implementation of applicable legal instruments, ranging from not applicable (N), e.g., in the case of a land-locked country, no response (NaN), or calculation method unknown, or not validated by a national statistical system for global reporting (NV) to very low (1), low (2), moderate (3), high (4) and very high (5). There were six applicable legal instruments; to determine the underlying scores attributed to each State, the instruments were assigned weights by FAO that depended on their relevance to target 14.6.
Abbreviations: N, not applicable; NaN, not a number; NV, not validated.
Source: FAO (2020d).

Where catches by vessels used for illegal, unreported or unregulated fishing contributed to illicit trade in seafood, significant economic and social consequences ensued. For example, the diversion of fish from legitimate trade led to estimated worldwide annual losses in economic contributions to States of $26 billion–$50 billion and tax revenue losses to States of $2 billion–$4 billion (Sumaila and others, 2020).

Illegal fishing also was linked to seafood fraud (Miller and Sumaila, 2016) and found to be associated with trafficking in both drugs and persons and forced labour (United Nations, 2017; Tickler and others, 2019). The International Labour Organization (ILO) estimated that a substantial proportion of the world's 21 million people trapped in forced labour were involved in the global fishing industry, including aquaculture, although the exact numbers were difficult to determine (FAO and ILO, 2013; ILO, 2016; Cavalli and others, 2019). Forced labour in fisheries in developed States was thought to be rare, but consumers in developed States were found to have purchased seafood from producers who utilized forced labour (Tickler and others, 2019).

In June 2016, the Agreement on Port State Measures to Prevent, Deter and Eliminate Illegal, Unreported and Unregulated Fishing,[5] which was the first binding international agreement to target illegal, unreported or unregulated fishing specifically, entered into force. Its effective implementation was expected to contribute to the long-term conservation and sustainable use of living marine resources and marine ecosystems (FAO, 2016b). Worldwide, as at 30 June 2020, there were 61 States parties to the Agreement. Its main objective was to prevent, deter and eliminate illegal, unreported or unregulated fishing by preventing vessels engaged in such fishing from landing catches in States parties' ports. The Agreement is therefore expected to reduce the incentives for such vessels to continue to operate and also block fishery products derived from such fishing from reaching national and international markets.

9. Outlook

Empirical evidence, along with modelling advances in fisheries science, demonstrated that effective management could improve fish stocks, increasing yields and resource rents and providing increased food security in developing States. The absence of effective and enduring governance in some of the world's fisheries, however, showed that they continued to be affected adversely by overexploitation, ongoing subsidization, illegal, unreported or unregulated fishing, illicit trading, by-catches and discarding, habitat damage due to bottom-trawling, post-harvest fish losses and gear abandonments. While the proportion of the world's marine capture fisheries characterized as "maximally sustainably fished" continued to grow, so did the proportion of those considered to be "overfished".

Significant efforts, including international negotiations under the auspices of WTO, continued to be made to prohibit certain fishing vessel subsidies and to eliminate subsidies for illegal, unreported or unregulated fishing. Furthermore, the Agreement on Port State Measures entered into force with the aim of mitigating landings of illegal, unreported or unregulated catches, but its adoption by States was incomplete.

Global climate change had already led to shifts in the distribution and abundance of fish populations, which were expected to continue or

[5] See www.fao.org/port-state-measures/resources/detail/en/c/1111616.

accelerate. Even with appropriate governance leading to stock rebuilding, scientists expected the adverse effects of climate change to impede progress towards sustainability in marine capture fisheries.

10. Key knowledge gaps

Alterations in the structures and functions of marine ecosystems as a consequence of anthropogenic forcings, including overfishing, nutrient pollution and climate change, were becoming more common. Especially with regard to the latter, there was limited understanding of the extent to which changing climate contributed to the redistribution of commercially important stocks or led to potentially irreversible shifts in marine ecosystem structures and processes. Developing States that depended on fisheries for food security, nutrition and exports were expected to be affected more severely than States with more diversified economies, but that hypothesis needed closer study.

Better understanding was needed of the potential for commercial stocks to migrate into the central Arctic Ocean (see chap. 7) and of the commercial values and ecological significance of other as yet unexploited stocks in deep-sea environments, such as in the mesopelagic zone.

Improvements in fisheries governance, including applications of effective management tools, were predicted by scientists to result in increases in biomass and biodiversity, thereby potentially allowing ocean ecosystems to adapt to global climate change, but there was little scientific consensus on whether recovered ecosystems could assume their former roles.

11. Key capacity-building gaps

The rebuilding of fish stocks remained a high priority for States and international organizations, but financial resources for undertaking scientific stock assessments and administering effective conservation and management measures needed further support and reinforcement in many fisheries, especially those of developing States. With appropriate governance, however, the most sanguine studies concluded that the median time required to rebuild overfished stocks could be less than a decade, and, if reforms were implemented, thereby enabling sustainable management, a large proportion of overfished stocks could be considered healthy by the middle of the twenty-first century.

References

Adelaja, Olusumbo Adeolu, and others (2018). Assessment of post-harvest fish losses Croaker *Pseudotolithus elongatus* (Bowdich, 1825), Catfish *Arius heudeloti* (Valenciennes, 1840) and Shrimp *Nematopalaemon hastatus* (Aurivillius, 1898) in Ondo State, Nigeria. *Aquaculture and Fisheries*, vol. 3, No. 5, pp. 209–216. https://doi.org/10.1016/j.aaf.2018.05.002.

Alltech (2019). 2019 Alltech Global Feed Survey estimates world feed production increased by 3 per cent to 1.103 billion metric tons. www.alltech.com/press-release/2019-alltech-global-feed-survey-estimates-world-feed-production-increased-3-percent.

Armitage, Derek, and others (2017). *Governing the Coastal Commons*. Taylor and Francis.

Barange, M., and others (2014). Impacts of climate change on marine ecosystem production in societies dependent on fisheries. *Nature Climate Change*, vol. 4, No. 3, pp. 211–216.

Basurto, Xavier, and others (2017). *Improving our knowledge on small-scale fisheries: data needs and methodologies*. FAO Fisheries and Aquaculture Proceedings, No. 56. Rome: FAO. www.fao.org/3/a-i8134e.pdf.

Berkes, Fikret, and Helen Ross (2013). Community resilience: toward an integrated approach. *Society and Natural Resources*, vol. 26, No. 1, pp. 5–20.

Bryndum-Buchholz, Andrea, and others (2018). Twenty-first-century climate change impacts on marine animal biomass and ecosystem structure across ocean basins. *Global Change Biology*, vol. 25, No. 2, pp. 459–472.

Bundy, Alida, and others (2017). Strong fisheries management and governance positively impact ecosystem status. *Fish and Fisheries*, vol. 18, No. 3, pp. 412–439.

Cai, Junning, and others (2019). *Understanding and measuring the contribution of aquaculture and fisheries to gross domestic product (GDP)*. FAO Fisheries and Aquaculture Technical Paper, No. 606. Rome: FAO. www.fao.org/3/CA3200EN/ca3200en.pdf.

Cavalli, Lissandra, and others (2019). Scoping global aquaculture occupational safety and health. *Journal of Agromedicine*. https://doi.org/10.1080/1059924X.2019.1655203.

Cheung, William W.L., and others (2019). Future scenarios and projections for fisheries on the high seas under a changing climate. Working Paper. London: International Institute for Environment and Development. http://pubs.iied.org/16653IIED.

Cohen, Philippa, and others (2019). Securing a just space for small-scale fisheries in the blue economy. *Frontiers in Marine Science*, vol. 6, art. 171.

Costello, Christopher, and others (2016). Global fishery prospects under contrasting management regimes. *Proceedings of the National Academy of Sciences of the United States of America*, vol. 113, No. 18, pp. 5125–5129. https://doi.org/10.1073/pnas.1520420113.

Crespo, Guillermo Ortuño, and Daniel C. Dunn (2017). A review of the impacts of fisheries on open-ocean ecosystems. *ICES Journal of Marine Science*, vol. 74, No. 9, pp. 2283–2297. https://doi.org/10.1093/icesjms/fsx084.

Crespo, Guillermo Ortuño, and others (2018). The environmental niche of the global high seas pelagic longline fleet. *Science Advances*, vol. 4, No. 8. https://doi.org/10.1126/sciadv.aat3681.

Crona, B.I., and others (2015). Using social-ecological syndromes to understand impacts of international seafood trade on small-scale fisheries. *Global Environmental Change*, vol. 35, pp. 162–175. https://doi.org/10.1016/j.gloenvcha.2015.07.006.

Cullis-Suzuki, Sarika, and Daniel Pauly (2010). Failing the high seas: a global evaluation of regional fisheries management organizations. *Marine Policy*, vol. 34, No. 5, pp. 1036–1042. https://doi.org/10.1016/j.marpol.2010.03.002.

Delpeuch, Claire, and Barbara Hutniczak (2019). Encouraging policy change for sustainable and resilient fisheries, No. 127. https://doi.org/10.1787/31f15060-en.

Diei-Ouadi, Yvette, and Yahya I. Mgawe (2011). *Post-harvest fish loss assessment in small-scale fisheries: a guide for the extension officer*. FAO Fisheries and Aquaculture Technical Paper, No. 559. Rome: FAO. www.fao.org/3/i2241e/i2241e.pdf.

Ding, Qi, and others (2017). Vulnerability to impacts of climate change on marine fisheries and food security. *Marine Policy*, vol. 83, pp. 55–61. https://doi.org/10.1016/j.marpol.2017.05.011.

Eigaard, Ole R., and others (2017). The footprint of bottom trawling in European waters: distribution, intensity, and seabed integrity. *ICES Journal of Marine Science*, vol. 74, No. 3, pp. 847–865. https://doi.org/10.1093/icesjms/fsw194.

European Market Observatory for Fishery and Aquaculture Products (2018). *Blue Bioeconomy: Situation Report and Perspectives*. Brussels: Directorate General for Maritime Affairs and Fisheries, European Commission. www.eumofa.eu/documents/20178/84590/Blue+bioeconomy_Final.pdf.

Food and Agriculture Organization (FAO) (2015). Voluntary Guidelines for Securing Sustainable Small-Scale Fisheries in the Context of Food Security and Poverty Eradication. Rome. www.fao.org/voluntary-guidelines-small-scale-fisheries/ihh/en.

________ (2016a). *Global Implications of Illegal, Unreported and Unregulated (IUU) Fishing. Rep. No. NIC WP 2016-02*. Rome.

________ (2016b). *Illegal, Unreported and Unregulated Fishing*. Rome. www.fao.org/3/a-i6069e.pdf.

________ (2016c). *The FAO Agreement on Port State Measures (PSMA) to Prevent, Deter and Eliminate Illegal, Unreported and Unregulated Fishing*. Rome. www.fao.org/port-state-measures/en.

________ (2016d). *The State of World Fisheries and Aquaculture 2016*. Rome. www.fao.org/3/a-i5555e.pdf.

________ (2018a). *Implementation of the 1995 FAO Code of Conduct for Responsible Fisheries*. Rome. www.fao.org/fishery.

________ (2018b). *The State of World Fisheries and Aquaculture 2018: Meeting the Sustainable Development Goals*. Rome. www.fao.org/state-of-fisheries-aquaculture.

________ (2019a). *Fishery and Aquaculture Statistics 2017*. Rome. www.fao.org/fishery/static/Year book/YB2017_USBcard/index.htm.

________ (2019b). *The State of Food Security and Nutrition in the World: Safeguarding against Economic Slow-Downs and Downturns*. Rome. www.fao.org/state-of-food-security-nutrition/en.

________ (2020a). *The State of World Fisheries and Aquaculture 2020*. Rome. www.fao.org/publica tions/sofia/2020/en.

________ (2020b). *Sustainable Development Goals: Indicator 14.4.1: Proportion of fish stocks within biologically sustainable levels*. Rome. www.fao.org/sustainable-development-goals/indicators /1441/en.

________ (2020c). *Sustainable Development Goals: Indicator 14.7.1: Sustainable fisheries as a percentage of GDP in small island developing States, least developed countries and all countries*. Rome. www.fao.org/sustainable-development-goals/indicators/1471/en.

________ (2020d). *Sustainable Development Goals: SDG Indicator 14.6.1: Progress by countries in the degree of implementation of international instruments aiming to combat illegal, unreported and unregulated fishing*. Rome. www.fao.org/sustainable-development-goals/indicators/14.6.1/en.

Food and Agriculture Organization of the United Nations (FAO) and International Labour Organization (ILO) (2013). *Guidance on Addressing Child Labour in Fisheries and Aquaculture*. Rome. ISBN 978-92-5-107709-2.

Fry, Jillian P., and others (2016). Environmental health impacts of feeding crops to farmed fish. *Environment International*, vol. 91, pp. 201–214. https://doi.org/10.1016/j.envint.2016.02.022.

Garcia, Serge M., and others, eds. (2018). *Rebuilding of marine fisheries – Part 1: Global review*. FAO Fisheries and Aquaculture Technical Paper, No. 630/1. Rome: FAO. www.fao.org/3/ca0161en/CA0161EN.pdf.

Golden, Christopher D., and others (2016). Nutrition: fall in fish catch threatens human health. *Nature News*, vol. 534, No. 7607, p. 317.

Harrison, J. (2019). Key challenges relating to the governance of regional fisheries. In *Strengthening International Fisheries Law in an Era of Changing Oceans*, Richard Caddell and Erik J. Molenaar, eds. New York: Hart Publishing.

Hidalgo, Manuel, and Howard I. Browman (2019). Developing the knowledge base needed to sustainably manage mesopelagic resources. *ICES Journal of Marine Science*, vol. 76, No. 3, pp. 609–615. https://doi.org/10.1093/icesjms/fsz067.

Hiddink, Jan Geert, and others (2017). Global analysis of depletion and recovery of seabed biota after bottom trawling disturbance. *Proceedings of the National Academy of Sciences of the United States of America*, vol. 114, No. 31, pp. 8301–8306. https://doi.org/10.1073/pnas.1618858114.

Hilborn, Ray, and Chris Costello (2018). The potential for blue growth in marine fish yield, profit and abundance of fish in the ocean. *Marine Policy*, vol. 87, pp. 350–355. https://doi.org/10.1016/j.marpol.2017.02.003.

Hilborn, Ray, and others (2020). Effective fisheries management instrumental in improving fish stock status. *Proceedings of the National Academy of Sciences of the United States of America*, vol. 117, No. 4, pp. 2218–2224. https://doi.org/10.1073/pnas.1909726116.

Ingeman, Kurt E., and others (2019). Ocean recoveries for tomorrow's Earth: hitting a moving target. *Science*, vol. 363, No. 6425. https://doi.org/10.1126/science.aav1004.

Institute of Aquaculture (2016). *Project to Model the Use of Fisheries By-Products in the Production of Marine Ingredients, with Special Reference to the Omega 3 Fatty Acids EPA and DHA*. Stirling, Scotland: University of Stirling and IFFO The Marine Ingredients Organisation.

Intergovernmental Panel on Climate Change (2019). Summary for policymakers. In *IPCC Special Report on the Ocean and Cryosphere in a Changing Climate,* Hans-Otto Pörtner and others, eds. Monaco: IPCC, 51st session, working groups I and II (24 September 2019).

International Council for the Exploration of the Sea (ICES) (2018a). *Report from the Working Group on Bycatch of Protected Species (WGBYC). Reykjavik, Iceland, 1–4 May 2018.* ICES CM 2018/ACOM:25.

________ (2018b). *Report of the Working Group on Ecosystem Effects of Fishing Activities (WGECO). San Pedro Del Pinatar, Spain, 12–19 April 2018.* ICES CM 2018/ACOM:27.

________ (2019). *Ecosystem Overviews.* www.ices.dk/community/advisory-process/Pages/Ecosystem-overviews.aspx.

International Labour Organization (ILO) (2016). *Fishers first-good practices to end labour exploitation at sea*. Geneva. ISBN: 978-92-2-131290-1.

Johnson, Derek S., and others (2018). *Social Wellbeing and the Values of Small-Scale Fisheries*. Springer.

Kramer, Daniel B., and others (2017). Coastal livelihood transitions under globalization with implications for trans-ecosystem interactions. *PLOS One*, vol. 12, No. 10, pp. 1–18. https://doi.org/10.1371/journal.pone.0186683.

Kroodsma, David A., and others (2018). Tracking the global footprint of fisheries. *Science*, vol. 359, No. 6378, p. 904. https://doi.org/10.1126/science.aao5646.

Lam, Vicky W.Y., and others (2016). Projected change in global fisheries revenues under climate change. *Scientific Reports*, vol. 6, No. 1, art. 32607. https://doi.org/10.1038/srep32607.

Long, Rachel D., and others (2015). Key principles of marine ecosystem-based management. *Marine Policy*, vol. 57, pp. 53–60. https://doi.org/10.1016/j.marpol.2015.01.013.

Lotze, Heike K., and others (2019). Global ensemble projections reveal trophic amplification of ocean biomass declines with climate change. *Proceedings of the National Academy of Sciences of the United States of America*, vol. 116, No. 26, pp. 12907–12912. https://doi.org/10.1073/pnas.1900194116.

Macfadyen, G., and others (2019). *IUU Fishing Index*. Hampshire, United Kingdom of Great Britain and Northern Ireland: Poseidon Aquatic Resource Management Limited and The Global Initiative Against Transnational Organized Crime. www.iuufishingindex.net.

Manning, P. (2010). *Fisheries and Aquaculture Topics: Food Security and Fisheries. Topics Fact Sheets.* Rome: FAO.

Marshall, Kristin N., and others (2018). Ecosystem-based fisheries management for social–ecological systems: renewing the focus in the United States with next generation fishery ecosystem plans. *Conservation Letters*, vol. 11, No. 1, p. e12367. https://doi.org/10.1111/conl.12367.

McCauley, Douglas J., and others (2018). Wealthy countries dominate industrial fishing. *Science Advances*, vol. 4, No. 8. https://doi.org/10.1126/sciadv.aau2161.

Miller, Dana D., and U. Rashid Sumaila (2016). Chapter 4: IUU Fishing and Impact on the Seafood Industry. In *Seafood Authenticity and Traceability*, Amanda M. Naaum and Robert H. Hanner, eds., pp. 83–95. San Diego: Academic Press. https://doi.org/10.1016/B978-0-12-801592-6.00004-8.

Neubauer, Philipp, and others (2013). Resilience and recovery of overexploited marine populations. *Science*, vol. 340, No. 6130, pp. 347–349.

Neuenhoff, Rachel D., and others (2019). Continued decline of a collapsed population of Atlantic cod (*Gadus morhua*) due to predation-driven Allee effects. *Canadian Journal of Fisheries and Aquatic Sciences* vol. 76, pp. 168–184. https://doi.org/10.1139/cjfas-2017-0190.

Palomares, Maria L.D., and Daniel Pauly (2019a). On the creeping increase of vessels' fishing power. *Ecology and Society* vol. 24, No. 3, art. 31. https://doi.org/10.5751/ES-11136-240331.

________ (2019b). Coastal fisheries: the past, present, and future. In *Coasts and Estuaries: The Future*, Eric Wolanski and others, eds., pp. 569–576. Amsterdam: Elsevier.

Patrick, Wesley S., and Jason S. Link (2015). Myths that continue to impede progress in ecosystem-based fisheries management. *Fisheries*, vol. 40, No. 4, pp. 155–160. https://doi.org/10.1080/03632415.2015.1024308.

Pauly, Daniel, and Dirk Zeller (2016). Catch reconstructions reveal that global marine fisheries catches are higher than reported and declining. *Nature Communications*, vol. 7, No. 1, art. 10244. https://doi.org/10.1038/ncomms10244.

Pentz, Brian, and others (2018). Can regional fisheries management organizations (RFMOs) manage resources effectively during climate change? *Marine Policy*, vol. 92, pp. 13–20. https://doi.org/10.1016/j.marpol.2018.01.011.

Pérez, Roda, and others (2019). *A third assessment of global marine fisheries discards*. FAO Fisheries and Aquaculture Technical Paper, No. 633. Rome: FAO.

Pinsky, Malin L., and others (2018). Preparing ocean governance for species on the move. *Science*, vol. 360, No. 6394, pp. 1189–1191. https://doi.org/10.1126/science.aat2360.

Priede, Imants G. (2017). *Deep-Sea Fishes: Biology, Diversity, Ecology and Fisheries*. Cambridge University Press. https://doi.org/10.1017/9781316018330.

Queiroz, Nuno, and others (2019). Global spatial risk assessment of sharks under the footprint of fisheries. *Nature*, vol. 572, pp. 461–466. https://doi.org/10.1038/s41586-019-1444-4.

Remesan, M.P., and others (2019). A review on techniques and challenges in the harvest of mesopelagics. *Fishery Technology*, vol. 56, pp. 243–253.

Rousseau, Yannick, and others (2019). Evolution of global marine fishing fleets and the response of fished resources. *Proceedings of the National Academy of Sciences of the United States of America*, vol. 116, No. 25, pp. 12238–12243. https://doi.org/10.1073/pnas.1820344116.

Said, Alicia, and Ratana Chuenpagdee (2019). Aligning the sustainable development goals to the small-scale fisheries guidelines: a case for EU fisheries governance. *Marine Policy*, vol. 107, art. 103599. https://doi.org/10.1016/j.marpol.2019.103599.

Sala, Enric, and others (2018). The economics of fishing the high seas. *Science Advances*, vol. 4, No. 6. https://doi.org/10.1126/sciadv.aat2504.

Samhouri, Jameal F., and others (2017). Defining ecosystem thresholds for human activities and environmental pressures in the California Current. *Ecosphere*, vol. 8, No. 6, p. e01860. https://doi.org/10.1002/ecs2.1860.

Schiller, Laurenne, and others (2018). High seas fisheries play a negligible role in addressing global food security. *Science Advances*, vol. 4, No. 8. https://doi.org/10.1126/sciadv.aat8351.

Schuhbauer, Anna, and U. Rashid Sumaila (2016). Economic viability and small-scale fisheries: a review. *Ecological Economics*, vol. 124, pp. 69–75. https://doi.org/10.1016/j.ecolecon. 2016.01.018.

Schuhbauer, Anna, and others (2017). How subsidies affect the economic viability of small-scale fisheries. *Marine Policy*, vol. 82, pp. 114–121. https://doi.org/10.1016/j.marpol.2017.05.013.

Selkoe, Kimberly A., and others (2015). Principles for managing marine ecosystems prone to tipping points. *Ecosystem Health and Sustainability*, vol. 1, No. 5, pp. 1–18. https://doi.org/10.1890/EHS14-0024.1.

Shaffril, Hayrol Azril Mohamed, and others (2017). Adapting towards climate change impacts: strategies for small-scale fishermen in Malaysia. *Marine Policy*, vol. 81, pp. 196–201. https://doi.org/10.1016/j.marpol.2017.03.032.

Silva, Catarina Basto, and others (2018). Life cycle assessment of aquafeed ingredients. *The International Journal of Life Cycle Assessment*, vol. 23, No. 5, pp. 995–1017. https://doi.org/10.1007/s11367-017-1414-8.

Song, Andrew M., and Adam Soliman (2019). Situating human rights in the context of fishing rights: contributions and contradictions. *Marine Policy*, vol. 103, pp. 19–26. https://doi.org/10.1016/j.marpol.2019.02.017.

Sumaila, U. Rashid, and others (2012). Benefits of rebuilding global marine fisheries outweigh costs. *PLOS One*, vol. 7, No. 7.

Sumaila, U. Rashid, and others (2015). Winners and losers in a world where the high seas is closed to fishing. *Scientific Reports*, vol. 5, No. 1, art. 8481. https://doi.org/10.1038/srep08481.

Sumaila, U. Rashid, and others (2019a). Updated estimates and analysis of global fisheries subsidies. *Marine Policy*, vol. 109, art. 103695. https://doi.org/10.1016/j.marpol.2019.103695.

Sumaila, U. Rashid, and others (2019b). Benefits of the Paris Agreement to ocean life, economies, and people. *Science Advances*, vol. 5, No. 2, p. eaau3855. https://doi.org/10.1126/sciadv.aau3855.

Sumaila, U. Rashid, and others (2020). Illicit trade in marine fish catch and its effects on ecosystems and people worldwide. *Science Advances*, vol. 6, p. eaaz3801. https://doi.org/10.1126/sciadv.aaz3801.

Sustainable Development Solutions Network (2019). *"Target 14.4". Indicators and a Monitoring Framework: Launching a Data Revolution for the Sustainable Development Goals*. New York. https://indicators.report.

Tai, Travis C., and others (2017). Ex-vessel fish price database: disaggregating prices for low-priced species from reduction fisheries. *Frontiers in Marine Science*, vol. 4, art. 363. https://doi.org/10.3389/fmars.2017.00363.

Taylor, Matthew D., and others (2017). Fisheries enhancement and restoration in a changing world. *Fisheries Research*, vol. 186, pp. 407–412. https://doi.org/10.1016/j.fishres.2016.10.004.

Tickler, David, and others (2018). Far from home: distance patterns of global fishing fleets. *Science Advances*, vol. 4, No. 8. https://doi.org/10.1126/sciadv.aar3279.

Tickler, David, and others (2019). Modern slavery and the race to fish. *Nature Communications*, vol. 9, No. 1, art. 4643. https://doi.org/10.1038/s41467-018-07118-9.

Too Big to Ignore (2020). *Global Partnership for Small-Scale Fisheries Research*. http://toobigtoignore.net.

United Nations (2017). *The First Global Integrated Marine Assessment: World Ocean Assessment I*. Cambridge: Cambridge University Press.

United Nations, Economic and Social Council (2019a). Special edition: progress towards the Sustainable Development Goals. Report of the Secretary-General. 8 May 2019 (E/2019/68). New York. https://unstats.un.org/sdgs/files/report/2019/ secretary-general-sdg-report-2019--EN.pdf.

United Nations, Statistics Division (2019b). *The Sustainable Development Goals Report 2019*. New York. https://unstats.un.org/sdgs/report/2019/goal-14.

Van Gemert, Rob, and Ken H. Andersen (2018). Challenges to fisheries advice and management due to stock recovery. *ICES Journal of Marine Science*, vol. 75, No. 6, pp. 1864–1870. https://doi.org/10.1093/icesjms/fsy084.

Watson, Reg A., and A. Tidd (2018). Mapping nearly a century and a half of global marine fishing, 1869–2015. *Marine Policy*, vol. 93, pp. 171–177. https://doi.org/10.1016/j.marpol.2018.04.023.

Willmann, Rolf, and others (2017). A human rights-based approach in small-scale fisheries: evolution and challenges in implementation. In *The Small-Scale Fisheries Guidelines*, S. Jentoft and others, eds., pp. 763–787. Springer.

World Bank (2017). *The Sunken Billions Revisited: Progress and Challenges in Global Marine Fisheries*. Washington, D.C. https://doi.org/10.1596/978-1-4648-0919-4.

World Bank, and others (2012). *Hidden Harvest: The Global Contribution of Capture Fisheries (English)*. Washington, D.C.: World Bank. https://doi.org/10.1596/978-1-4648-0919-4.

World Trade Organization (WTO) (2020). *Negotiations on fisheries subsidies 2020*. Geneva. www.wto.org/english/tratop_e/rulesneg_e/fish_e/fish_e.htm.

Zeller, Dirk, and others (2018). Global marine fisheries discards: a synthesis of reconstructed data. *Fish and Fisheries*, vol. 19, No. 1, pp. 30–39. https://doi.org/10.1111/faf.12233.

Chapter 16
Changes in aquaculture

Contributors: Rohana Subasinghe (convener), Pedro Barón, Malcolm Beveridge, Enrique Marschoff (lead member) and Doris Oliva.

Keynote points

- Global aquaculture production in 2017 (animals and plants) was recorded as 111.9 million tons, with an estimated first-sale value of $249.6 billion. Since 2000, world aquaculture has ceased to enjoy the high annual growth rates of the 1980s and 1990s (11.3 and 10.0 per cent, respectively). Nevertheless, it continues to grow at a faster rate than other major food production sectors. Annual growth declined to a moderate 5.8 per cent during the period from 2000 to 2016, although double-digit growth still occurred in a small number of countries, in particular in Africa, from 2006 to 2010. Fish produced by this rapidly growing sector is high in protein and contains essential micronutrients, sometimes essential fatty acids, which cannot easily be substituted by other food commodities.
- The United Nations predicts that the global population will reach 8.5 billion in 2030. This will inevitably increase the pressure on food sectors to increase production and reduce losses and waste. Production increases must be able to ensure sustainability, given a context in which key resources, such as land and water, are likely to be scarcer and the impact of climatic change will intensify. The aquaculture sector is no exception. Success in achieving the long-term goal of economic, social and environmental sustainability of the aquaculture sector, so as to ensure its continued contribution of nutritious food to keep the world healthy, will depend primarily on continued commitments by Governments to provide and support a good governance framework for the sector. As the sector further expands, intensifies and diversifies, it should recognize relevant environmental and social concerns and make conscious efforts to address them in a transparent manner, backed by scientific advice.

1. Current status and major improvements

The present section provides an assessment of major global changes and improvements in the aquaculture sector over the past decade, and outlines its current status.

1.1. Production and species

Aquaculture is expanding faster than other types of food production, although no longer at the growth rates of the 1980s and 1990s (11.3 and 10.0 per cent, respectively, excluding aquatic plants). Average annual growth declined to 5.8 per cent during the period from 2000 to 2016, although higher rates of growth occurred in several countries, in particular in Africa, from 2006 to 2010 (Food and Agriculture Organization of the United Nations (FAO), 2018a). Global production in 2016 included 80 million tons of food fishes, 30.1 million tons of aquatic plants and 37,900 tons of non-food products. Food production included 54.1 million tons of finfishes, 17.1 million tons of molluscs, 7.9 million tons of crustaceans and 938,500 tons of other animals. China, the major aquaculture producer in 2016, has produced more than the rest of the world combined since 1991. The other major producers in 2016 were India, Indonesia, Viet Nam, Bangladesh, Egypt and Norway. Aquatic plants (28 million tons) included seaweeds and a much smaller volume of microalgae. China and Indonesia were the major producers of aquatic plants in 2016 (FAO, 2018b). Ornamental fish and plant species are not included in the present review.

1.2. People and nutrition

Global official statistics indicate that 59.6 million people were engaged in the primary sector of capture fisheries and aquaculture in 2016, with 19.3 million people engaged in aquaculture and 40.3 million in fisheries (FAO, 2018b). In addition to the primary producers, many people are engaged in the aquaculture value chain. The sector supports the livelihoods, including family members, of 540 million people, or 8 per cent of the world population (FAO, 2017a). Women accounted for 19 percent of all people directly engaged in the primary sector in 2014 (FAO, 2016).

Aquaculture's contribution to human nutrition has been fully recognized (Chan and others, 2017; High-Level Panel of Experts on Food Security and Nutrition, 2014). Aquaculture improves the nutrition of the rural poor, especially mothers and young children (Thilsted and others, 2016), although there are concerns that the growth of the sector and the intensification of its production methods may result in decreased availability of certain fatty acids and micronutrients (Bogard and others, 2017). Considering the increasing global population and the importance of a healthy diet, Béné and others (2016) stressed that access to fish is a key issue in creating healthy populations, especially among the rural poor, worldwide.

1.3. Inputs and resources

Land and water are the most important resources for aquaculture development. Gentry and others (2017) estimated that 11,400,000 km^2 of coastline are suitable for fishes, and more than 1,500,000 km^2 could be developed for bivalves. The challenge is to secure suitable land and water resources for the development of aquaculture at the national level.

Good quality seeds and optimal feeds are essential. Most animal species are cultured with external feeds, and feeding the ever-expanding aquaculture sector has been a concern. In 2016, about 55.6 million tons of farmed fishes (including Indian carps) and crustaceans depended on external feeds (composed of fresh ingredients, farm-made or commercially manufactured) (FAO, 2018b).

In 2005, aquaculture consumed about 4.2 million tons of fishmeal (18.5 per cent of total aquafeeds by weight). By 2015, this had been reduced to 3.35 million tons (7 per cent of total aquafeeds by weight). Even with increasing production globally, the use of fishmeal for aquafeeds will decrease further to 3.33 million tons by 2020 (5 per cent of total aquafeeds by weight for that year). Efforts towards making sustainable feeds by replacing fishmeal and fish oils with plant-based feed can have an impact on levels of omega-3 fatty acids and the nutritional value of farmed fishes. The industry can make strategic use of fish oils in fish feed by feeding these essential compounds to farmed fishes at key life stages. Nevertheless, for aquaculture to grow, aquafeed production is expected to continue growing at a similar rate, to 69 million tons by 2020 (Hasan, 2017). Considering past trends and predictions, aquaculture sustainability is more likely to be closely linked with the sustained supply of terrestrial animal and plant proteins, oils and carbohydrate sources for aquafeeds (Troell and others, 2014). The aquaculture sector should therefore strive to ensure sustainable supplies of terrestrial and plant-based feed ingredients, including algae and processing waste, that do not compete directly with use for feeding people directly.

1.4. Biosecurity

Diseases continue to challenge global aquaculture and are one of the primary deterrents to the aquaculture development of many species. Thus, investment, along with a focus on biosecurity and health, have been on the increase worldwide (Subasinghe and others, 2019). Biosecurity in aquaculture consists of practices that minimize the risk of introducing an infectious disease and spreading it to the animals at a facility and the risk that diseased animals or infectious agents will leave a facility and spread disease to other sites and to other susceptible species. These practices also reduce stress on the animals, thus making them less susceptible to disease.

The long list of aquatic diseases and pathogens includes acute hepatopancreatic necrosis disease, which recently devastated shrimp aquaculture in Asian countries (e.g. China, Malaysia, the Philippines and Thailand). The causative agent is a virulent strain of *Vibrio parahaemolyticus*, a bacterium commonly found in coastal waters. Revenue loss due to the disease in South-East Asia has been estimated at over $4 billion. Countries must monitor other emerging diseases, such as *Enterocytozoon hepatopenaei* in shrimps and tilapia lake virus (*Tilapia tilapinevirus*), which could potentially have a severe impact on the sector if not addressed in a timely manner (FAO, 2017a). New molecular diagnostic tools are now being applied to the identification of disease agents and their distribution patterns in hatchery, farmed and wild fishes throughout the world. A recently developed microarray has also been used to look at the impacts of pathogen carrier status (sea lice and the infectious hematopoietic necrosis virus) on wild salmons.

While research aimed at finding vaccines is progressing, the emerging issue that countries face is the misuse and abuse of antimicrobials and other drugs, which result in residues and resistant pathogens. Prudent use of antimicrobials and a better understanding of the role of good husbandry management and microbiota in culture systems are important to reduce antimicrobial use and the resulting welfare implications in aquaculture production. Following the approval by the World Health Organization of the global action plan on antimicrobial resistance,[1] countries are encouraged to develop national action plans on aquatic antimicrobial resistance and to integrate them into the global action plan (FAO, 2017a).

1.5. Technology

Remarkable improvements have been made in genetics and breeding, in relation to both finfishes and shrimps. Specific pathogen-free (SPF) and specific pathogen-resistant shrimps (*Penaeus monodon* and *P. vannamei*), genetically improved farmed tilapia, some carp species with better growth performance and commercial-scale production of various species of grouper, pompano and cobia could be classed as success stories (FAO, 2017a). Technological improvements in feeds, nutrition, health management and disease control are contributing to intensification, expansion and sustainability (FAO, 2017a). The adoption of genetic improvement programmes is slow, even for some major aquaculture species. Such programmes are expensive to initiate, but there is evidence that public–private partnerships can be effective in building and sustaining long-term programmes (FAO, 2019). When deciding to introduce a species for culture, potential negative environmental and socioeconomic impacts should always be considered, along with the potential for developing native species culture (Wurmann, 2019).

Over the past few years, SPF *P. monodon* and *P. vannamei* have been more available in Asia and Latin America. However, the use and misuse of the term SPF has been and will continue to be a concern among aquaculture stakeholders (Alday-Sanz and others, 2018). The life cycles of important crab and lobster species have been experimentally closed but commercial production of their seed is still rudimentary.

Attempts have been made to use recirculating aquaculture systems for salmon, with some

1 World Health Organization, document WHA68/2015/REC/1, annex 3.

positive outcomes. Such systems are becoming the standard for smolt and post-smolt production in Chile and Norway. The approximate investment cost is $60 million for a complete system (FAO, 2017b). Other emerging technologies that help minimize disease and reduce waste are closed and semi-enclosed cage systems, currently being developed and deployed for salmon farming in Norway (Nilssen and others, 2017).

The transgenic AquAdvantage Atlantic salmon was under review by the Food and Drug Administration (FDA) of the United States of America for more than a decade. After an exhaustive and rigorous process, the FDA determined that AquAdvantage salmon is as safe to eat as any non-genetically engineered Atlantic salmon, and as nutritious. Approval for production and consumption was finally granted in November 2015 in the United States, and for sale in Canada by Health Canada in 2016.

2. Aquaculture and the environment

Many countries emphasize environmental sustainability and social responsibility. In addition to laws, regulations and voluntary codes aimed at ensuring environmental integrity, some of the means of achieving this goal include innovative, less polluting techniques proposed by the ecosystem approach to aquaculture, which emphasizes management for sustainability (FAO, 2010) and provides a planning and management framework to effectively integrate aquaculture into local planning (Brugère and others, 2018). Although efforts related to intensification have resulted in decreased use of land and fresh water per unit of fish produced (FAO, 2017a), they have also led to an increase in the use of energy and feed, and in pollution, per unit of farmed fish (Hall and others, 2011).

Although aquaculture has been accused of having negative environmental and social impacts (Bushmann and Fortt, 2005; Isla Molleda and others, 2016) and suffers from a biased perception on the part of the public, it has, from an ecological efficiency and environmental impact point of view, clear benefits over other forms of animal food production for human consumption. Life-cycle assessment is useful to determine environmental impacts and ensure environmentally sustainable development (Bohnes and Laurent, 2019). Farmed finfish is similar in feed conversion efficiency to poultry and much more efficient than beef. Recent estimates indicate that demand for feed crops and land for aquaculture will be less than for alternative food production systems, even if over one third of protein production comes from aquaculture, by 2050 (Froehlich and others, 2018). Filter-feeding carps and molluscs are even more efficient producers of animal protein, as they require no human-managed feeds and can improve water quality. Because aquaculture is relatively new, it offers great scope for innovation to increase resource efficiency (Waite and others, 2014). Where resources are stretched, the relative benefits of policies that promote aquaculture over other forms of livestock production should be considered.

In general, the environmental performance of aquaculture has improved significantly over the past decade. If aquaculture production doubles by 2030, the sector must improve its productivity and environmental performance for growth to be sustainable (Waite and others, 2014). In order to achieve "sustainable intensification", aquaculture must: (a) advance socioeconomic development; (b) provide safe, affordable and nutritious food; (c) increase production of fish relative to the amount of land, water, feed and energy used; and (d) minimize environmental impacts, fish diseases and escapes (FAO, 2017a).

3. Aquaculture and society

The importance of fishes and fishery-based activities to food security in less developed countries is particularly prominent. In 2016, Asia accounted for 85.7 per cent of the global population engaged in fisheries and aquaculture (FAO, 2018a), which represents an increase of more than 1 per cent since 2014. More than 19 million people (32 per cent of all people employed in the sector) were engaged in fish farming, and 95.9 per cent of all aquaculture activities were being conducted in Asia. The statistics clearly indicate the important and increasing contribution of aquaculture to that continent's regional food and nutrition security, as well as its socioeconomic development.

There are several major reviews on the subject (Allison, 2011; Béné and others, 2016). Fishes provide more than 4.5 billion people with at least 15 per cent of their animal protein intake. The nutritional properties of fishes make them important to the health of consumers in developed and developing countries. Fishes are efficient converters of feed into high quality food and their carbon footprint is lower than that of other animal production systems. Fisheries and aquaculture value chains contribute substantially to the income and employment, and therefore indirectly to the food security, of more than 10 per cent of the world's population, principally in developing countries and emerging economies (FAO, 2017a).

The 80 million tons of aquatic animals produced in 2016 contributed 46 per cent to total aquatic animal production and a little over 54 per cent to total fish consumption in the same year. Per capita food fish consumption was estimated at 20.3 kg in 2016, compared with 19.5 kg in 2013 (FAO, 2018b). An estimated 18.7 million people were employed in aquaculture in 2015 (FAO, 2017a).

The culture and use of small indigenous fish species with high nutritional value in human nutrition is recognized and is being practised (Castine and others, 2017). However, with the intensification of aquaculture production methods, and with the increasing use of plant-based feedstuffs, care must be taken to ensure that the nutrient contents of farmed aquatic animal products are as high as possible (Beveridge and others, 2013; Bogard and others, 2017).

4. Key remaining knowledge gaps

The rapid growth of intensive aquaculture, in some cases not well planned, has caused concern about environmental impact, human health and social issues. Although the lion's share of production originates in Asia, opposition to aquaculture development is strongest in some developed countries (Froehlich and others, 2017), where aquaculture is still a relatively new industry competing with well-established activities. The world's knowledge regarding the impact of climate change on aquaculture needs to be improved. Further research and investigation is necessary to improve seeds, feeds and health management. The increasing dependence of developed countries on farmed seafood imports from developing countries and insecurity regarding product environmental, social and safety credentials have sparked considerable public debate. Scientific uncertainties and conflicting information on the issues relating to seafood consumption have further confused the public. The establishment and application of third-party certification systems, covering the environmental, social and food safety concerns related to seafood, have begun to ease this situation. More research is needed to communicate the nutritional and health benefits of increased consumption of seafood. Determination of the nutritional profiles of cultured fishes and wild-caught products and quantification of the health benefits of socioeconomic improvements through aquaculture need further attention.

With a growing world population, annual supply from the aquaculture sector must surpass supply from capture fisheries and reach 62 per cent in 2030 in order to maintain current consumption levels. This presents tremendous challenges to the sector, to policymakers and to the aquaculture community at large. Improving perceptions will be instrumental in achieving this goal (Vannuccini and others, 2018). Better information and exchange thereof would help in allaying concerns, dispelling myths and resolving ambiguities. To improve public awareness of aquaculture, the industry needs a more open, broader dialogue that will increase transparency. To communicate the benefits of aquaculture more effectively, it must collaborate more with stakeholder groups viewed as credible by the public. While significant social and environmental issues are still to be addressed, it is important to put aquaculture in a wider perspective by comparing its costs and benefits with those of other animal production systems and with its potential contribution to sustainable food security, given forecasted demographic pressures. However, a holistic view, with a balanced evaluation of the risks and benefits of aquaculture, has been lacking, thus impeding the development of policies that reflect production realities (Bacher, 2015).

5. Key remaining capacity-building gaps

Capacity development is an integral part of aquaculture development. The Fisheries and Aquaculture Department of FAO has been conducting training on many aspects of capacity development in member countries for years. Sustainable development requires, inter alia, decent infrastructure, technology, policies and training. While technology to improve the efficiency of production systems is essential, the development of human resources, in terms of both quality and quantity, is pivotal to sustaining the industry, all the more so in view of the changing paradigms that are affecting the sector. Some of the key trends and challenges reflect an ever-increasing global call for sustainable development that is socially and environmentally acceptable, irrespective of the economic status of a particular nation.

To boost sustainable aquaculture development, countries must improve extension services. The training of extension workers must be modified to incorporate and reinforce information delivery methods and mechanisms, as well as practical farming techniques, which will enable them to better help farmers to improve their production systems and practices to increase production and profit. New models and players are needed in the extension field, as information technology and media, farmer associations, development agencies, private sector suppliers and others will probably enjoy greater prominence, thus broadening the training experience. The goal should be to improve extension services and ensure a more effective use of resources.

Numerous donor and development agencies have helped to expand aquaculture capacity in developing countries in the past five years. Many developing and developed countries have allocated resources to improve national aquaculture capacity. Numerous Governments have provided basic aquaculture extension support and some limited research and development services. However, the level of State support is inadequate in many countries. By contrast, private sector engagement in capacity development in aquaculture has been improving, with notable success in many countries.

6. Outlook

The major growth in aquatic production is expected to come from aquaculture and is projected to reach 109 million tons in 2030, an increase of 37 per cent over 2016 levels. However, it is estimated that the annual growth rate of aquaculture will slow from 5.7 per cent in the period from 2003 to 2016 to 2.1 per cent in the period from 2017 to 2030, mainly because of a reduced rate of growth in Chinese production, offset in part by an increase in production in other countries (FAO, 2018a). The share of farmed aquatic animal species in global fishery production (for food and non-food uses), which was 47 per cent in 2016, is projected to exceed that of wild species in 2020 and to grow to 54 per cent by 2030.

Over 87 per cent of the increase in aquaculture production in 2030 will come from Asian countries. Asia will continue to dominate world aquaculture production, contributing 89 per cent of total production in 2030. China will remain the world's leading producer, but its share of total production will decrease from 62 per cent in 2016 to 59 per cent in 2030. Production is projected to continue to expand on all continents, with variations in the range of species and products across countries and regions (World Bank, 2013).

Millions of people engaged in fisheries and aquaculture are struggling to maintain reasonable livelihoods. These are the people who are most vulnerable to certain climate change impacts, such as extreme weather conditions, storms, floods and rising sea levels, and particular attention needs to be paid to them when designing adaptation measures if the sector is to continue to contribute to meeting the global goals of poverty reduction and food security (FAO, 2018a).

The 2030 Agenda for Sustainable Development[2] emphasizes people, planet, prosperity, peace and partnership. The 2030 Agenda and its Sustainable Development Goals are highly relevant for policymaking, planning and management for the sustainable development of aquaculture. If developed appropriately, aquaculture will contribute to the achievement of many of the Goals, including Goal 14, and in particular target 14.7, on increasing, by 2030, the economic benefits to small island developing States and least developed countries from the sustainable use of marine resources, including through the sustainable management of fisheries, aquaculture and tourism.

A recent analysis shows that most of the available international guidance focusing on aquaculture development broadly meets the expectations set out in the Goals. Existing international commitments and calls for sustainable aquaculture development, such as in the FAO Code of Conduct for Responsible Fisheries and its associated Technical Guidelines, the 2000 Bangkok Declaration and Strategy and the 2010 Phuket Consensus, and the FAO Blue Growth Initiative for Small Island Developing States,[3] which includes the ecosystem approach to fisheries and aquaculture, are generally well aligned with the 2030 Agenda and will support delivery of the Goals (FAO, 2017a).

If no concerted effort is made to increase the rate of growth of aquaculture, FAO forecasts an apparent fish supply-demand gap in the early to mid-2020s. The study by Golden and others (2017) suggests that aquaculture is unlikely to contribute substantially to human nutrition in nutritionally vulnerable nations. The need for more integrated efforts to develop policies addressing both fisheries and aquaculture for human well-being has been discussed above. There is a need to rethink and redesign strategies aimed at future aquaculture development worldwide.

2 See General Assembly resolution 70/1.
3 See www.fao.org/3/a-i3958e.pdf.

References

Alday-Sanz, Victoria, and others (2018). Facts, truths and myths about SPF shrimp in Aquaculture. *Reviews in Aquaculture*, vol. 12, No. 1.

Allison, E.H. (2011). Aquaculture, fisheries, poverty and food security. Working Paper, No. 2011–65. Penang, Malaysia: WorldFish Center.

Bacher, Kathrin (2015). Perceptions and misconceptions of aquaculture: a global overview. GLOBEFISH Research Programme, vol. 120. Rome: FAO.

Béné, Christophe, and others (2016). Contribution of fisheries and aquaculture to food security and poverty reduction: assessing the current evidence. *World Development*, vol. 79, pp. 177–196.

Beveridge, Malcolm C.M., and others (2013). Meeting the food and nutrition needs of the poor: the role of fish and the opportunities and challenges emerging from the rise of aquaculture. *Journal of Fish Biology*, vol. 83, No. 4, pp. 1067–1084.

Bogard, Jessica R., and others (2017). Higher fish but lower micronutrient intakes: temporal changes in fish consumption from capture fisheries and aquaculture in Bangladesh. *PloS One*, vol. 12, No. 4.

Bohnes, Florence Alexia, and Alexis Laurent (2019). LCA of aquaculture systems: methodological issues and potential improvements. *The International Journal of Life Cycle Assessment*, vol. 24, No. 2, pp. 324–337.

Brugère, Cecile, and others (2018). The ecosystem approach to aquaculture 10 years on: a critical review and consideration of its future role in blue growth. *Reviews in Aquaculture*, vol. 11, No. 3, pp. 493–514.

Castine, Sarah and others (2017). Homestead pond polyculture can improve access to nutritious small fish. *Food Security: The Science, Sociology and Economics of Food Production and Access to Food*, vol. 9, No. 4, pp. 785-801.

Chan, Chin Yee, and others (2017). Fish to 2050 in the ASEAN region. Working Paper, No. 2017–01. Penang, Malaysia: WorldFish Center and Washington D.C.: International Food Policy Research Institute.

Food and Agriculture Organization of the United Nations (FAO) (2010). Aquaculture Development: 4. Ecosystem Approach to Aquaculture. FAO Technical Guidelines for Responsible Fisheries, No. 5, Suppl. 4. Rome.

________ (2016). *The State of World Fisheries and Aquaculture 2016: Contributing to Food Security and Nutrition for All*. Rome.

________ (2017a). Aquaculture, the Sustainable Development Goals (SDGs)/Agenda 2030 and FAO's common vision for sustainable food and agriculture. Working document, COFI:AQ/IX/2017/5. Ninth Session of the Committee on Fisheries, Subcommittee on Aquaculture. Rome.

________ (2017b). *World Aquaculture 2015: A Brief Overview*. FAO Fisheries and Aquaculture Circular, No. 1140. Rome.

________ (2018a). *Impacts of Climate Change on Fisheries and Aquaculture: Synthesis of Current Knowledge, Adaptation and Mitigation Options*. Manuel Barange and others, eds. FAO Fisheries and Aquaculture Technical Paper, No. 627. Rome.

________ (2018b). *The State of World Fisheries and Aquaculture 2018: Meeting the Sustainable Development Goals*. Rome.

________ (2019). *The State of the World's Aquatic Genetic Resources for Food and Agriculture*. Rome.

Froehlich, Halley E., and others (2017). Public perceptions of aquaculture: evaluating spatiotemporal patterns of sentiment around the world. *PloS One*, vol. 12, No. 1.

Froehlich, Halley E., and others (2018). Comparative terrestrial feed and land use of an aquaculture-dominant world. *Proceedings of the National Academy of Sciences*, vol. 115, No. 20, pp. 5295–5300.

Gentry, Rebecca R., and others (2017). Mapping the global potential for marine aquaculture. *Nature Ecology and Evolution*, vol. 1, No. 9, pp. 1317–1324.

Golden, Christopher D., and others (2017). Does aquaculture support the needs of nutritionally vulnerable nations? *Frontiers in Marine Science*, vol. 4, art. 159.

Hall, Stephen J., and others (2011). Blue Frontiers: Managing the Environmental Costs of Aquaculture. Penang, Malaysia: WorldFish Center.

Hasan, Mohammad R. (2017). Feeding global aquaculture growth. FAO Aquaculture Newsletter, No. 56.

High-Level Panel of Experts on Food Security and Nutrition (2014). Sustainable fisheries and aquaculture for food security and nutrition: report by the High-Level Panel of Experts on Food Security and Nutrition of the Committee on World Food Security. Rome.

Isla Molleda, Mercedes, and others (2016). Development of mariculture in Cuba: impacts and challenges to achieve sustainable management preserving coastal ecosystems (in Spanish). *Áreas Naturales Protegidas Scripta*, vol. 2, No. 1, pp. 7-26.

Nilssen, Arve, and others (2017). Effective protection against sea lice during the production of Atlantic salmon in floating enclosures. *Aquaculture*, vol. 466, pp. 41–50.

Subasinghe, Rohana, and others (2019). Vulnerabilities in aquatic animal production. *Scientific and Technical Review of the Office International des Epizooties*, vol. 38, No. 2 (in press).

Thilsted, Shakuntala Haraksingh, and others (2016). Sustaining healthy diets: the role of capture fisheries and aquaculture for improving nutrition in the post-2015 era. *Food Policy*, vol. 61, pp. 126–131.

Troell, Max, and others (2014). Does aquaculture add resilience to the global food system? *Proceedings of the National Academy of Sciences*, vol. 111, No. 37, pp. 13257–13263.

Vannuccini, Stefania, and others (2018). Understanding the impacts of climate change for fisheries and aquaculture: global and regional supply and demand trends and prospects. In *Impacts of Climate Change on Fisheries and Aquaculture: Synthesis of Current Knowledge, Adaptation and Mitigation Options*. M. Barange and others, eds. FAO Fisheries and Aquaculture Technical Paper, No. 627. Rome: FAO.

Waite, Richard, and others (2014). Improving productivity and environmental performance of aquaculture. Working paper, installment 5 of *Creating a Sustainable Food Future*. Washington, D.C.: World Resources Institute.

World Bank (2013). Fish to 2030: Prospects for Fisheries and Aquaculture. World Bank Report, No. 83117-GLB.

Wurmann, Carlos (2019). Aquaculture in Latin America and the Caribbean: progresses, opportunities and challenges. *AquaTechnica*. vol. 1, No. 1.

Chapter 17 Changes in seaweed harvesting and use

Contributors: Hilconida Calumpong (convener and lead member), Franciane Pellizzari, Renison Ruwa (co-lead member) and Noemí Solar-Bacho

Keynote points

- As of 2012, about 80 per cent of seaweeds were either consumed directly, such as kelps, or processed for phycocolloids, such as carrageenan, for use in the food industry. The rest were used widely in pet food and in industrial, cosmetic and medical applications. World production of seaweeds steadily rose from 2012 to 2017 at a rate of about 2.6 per cent annually, or about 1.8 million tons (wet weight) per year, owing mostly to demand from farming and aquaculture, with an estimated value of about $12 billion.
- China remains the top producer of seaweeds, followed by Indonesia. The Philippines is still the world's third largest producer, despite being struck by typhoons every year; Filipino seaweed farmers have become resilient and can revive their farming operations immediately. The Republic of Korea ranks fourth and has made a concerted effort to increase exports to North America through marketing campaigns.
- The top species farmed are still the carrageenophytes, *Kappaphycus alvarezii* and *Eucheuma* spp. (accounting for 85 per cent of world's carrageenan production), which are grown in the Indo-Pacific region; alginate-producing kelps (*Saccharina* and *Undaria*), which are cold-water species, are the major species harvested.
- Emerging applications of seaweeds in agriculture include their use for the reduction of methane production in farmed animals, but such applications are still incipient because of issues relating to bromoforms, which can have environmental consequences.
- Production has been affected negatively in typhoon-vulnerable areas.

1. Introduction

The present chapter deals only with seaweed harvesting, uses by human society and ecosystem services. The taxonomy and ecological role of seaweeds and how they are affected by other components of the marine environment are covered in the present Assessment in chapter 6G on marine plants and macroalgae.

Seaweeds are macroalgae and belong to three main groups: *Rhodophyta* (red), *Phaeophyta*[1] (brown) and *Chlorophyta* (green). They are of economic importance for many countries as food for direct human consumption or as food in the aquaculture of commercial species, for the production of phycocolloids (e.g. agar, carrageenan, alginates) and for use in the manufacture of different products of commercial interest, mainly in the processed food and pharmaceutical industries (Buschmann and others, 2017; Kim and others, 2017; also Park and others, 2018, for a historical review).

According to the baseline review of status, as provided in chapter 14 of the first *World Ocean Assessment* (United Nations, 2017), red, brown and green seaweeds were harvested in commercial quantities from the wild in about 37 countries and farmed in more than 27 countries. About 96 per cent of total worldwide production, amounting to around 26 million tons (wet weight) in 2012 and valued at about $6 billion, came from mariculture. China was the top producer by volume, accounting for at least 50 per cent of total world production from 2003 to 2012. In 2007, the Philippines was overtaken by Indonesia as the second largest

1 Recently placed in the division Ochrophyta, kingdom Chromista (see chap. 6G, sect. 5.1).

producer, a spot it has held since then due to its vast farming areas and improved farming technology. Chile was the top supplier from wild stock harvests, followed by China, Norway and Japan. As of 2012, about 80 per cent of seaweeds were either consumed directly, such as kelps, or processed for phycocolloids, such as carrageenan, for use in the food industry. The rest were used widely in pet food and in industrial, cosmetic and medical applications. Seaweeds were also used as animal feed additives, fertilizer, water purifiers and prebiotics in aquaculture. The top species farmed were the red seaweeds, *Kappaphycus alvarezii* and *Eucheuma* spp., as sources of carrageenan, accounting for 33 per cent of production, while 20 per cent was contributed by alginate-producing brown seaweeds called kelps (such as *Laminaria* from wild harvesting). Seaweed harvests from wild stocks were reported to be considerably affected by overharvesting and climatic changes. The kelps were reported to be most affected by surface seawater heating and abrupt changes in temperature, since reproduction will not occur above 20°C. Kelp die-backs were reported in Norway and France and along the coastlines of other European countries. Seaweed farming has been seriously affected by the bacterial "ice-ice" disease – so named because it causes the seaweed body to become translucent – which specifically targets *Kappaphycus alvarezii*. The increase in disease has been attributed to low genetic diversity and monocrops of cultured stocks. Reported environmental and ecological impacts of commercial-scale seaweed harvesting include habitat destruction, damage to substrata and changes in particle size distribution in sediments, disturbance of birds and wildlife, disruption of food webs and localized faunal and floral changes in biodiversity, often affecting fishers' harvests. Direct effects on seaweed populations include increased growth rates and the covering of available substrata by algae other than kelps.

In terms of the socioeconomic impacts of seaweed farming, small-scale farmers appear to benefit most, as such farming offers substantial employment opportunities relative to other forms of aquaculture. However, small-scale farmers were found to be disadvantaged compared with large-scale growers due to their lack of farm and financial management skills and their dependence on processors for their materials.

2. Documented changes in the state of seaweed production and uses (2012–2017)

World production of seaweed has steadily risen since the first Assessment baseline, mostly as a result of farming and aquaculture (see figure I). From more than 24.6 million tons (wet weight) in 2012, farmed seaweed production rose to close to 32 million tons (wet weight) in 2017 (Food and Agriculture Organization of the United Nations (FAO), 2019), accounting for 96.6 per cent of total world production and representing an annual increase of about 1.8 million tons (wet weight). This production is now valued at $11.85 billion (FAO, 2019).

China remains the top supplier, with production of about 1 million tons per year and growing, and now accounts for more than 54 per cent of world production. This represents a steady 1 per cent increase every year since 2012. Indonesia ranked second and, although that country's production jumped 66 per cent in 2013, it remained more or less steady through 2017. The Philippines is the world's third largest producer of seaweed. Despite the country being struck by typhoons every year, Filipino seaweed farmers have become resilient and can revive their farming operations immediately.

Figure I
World seaweed production from aquaculture by country or region, 2003–2017

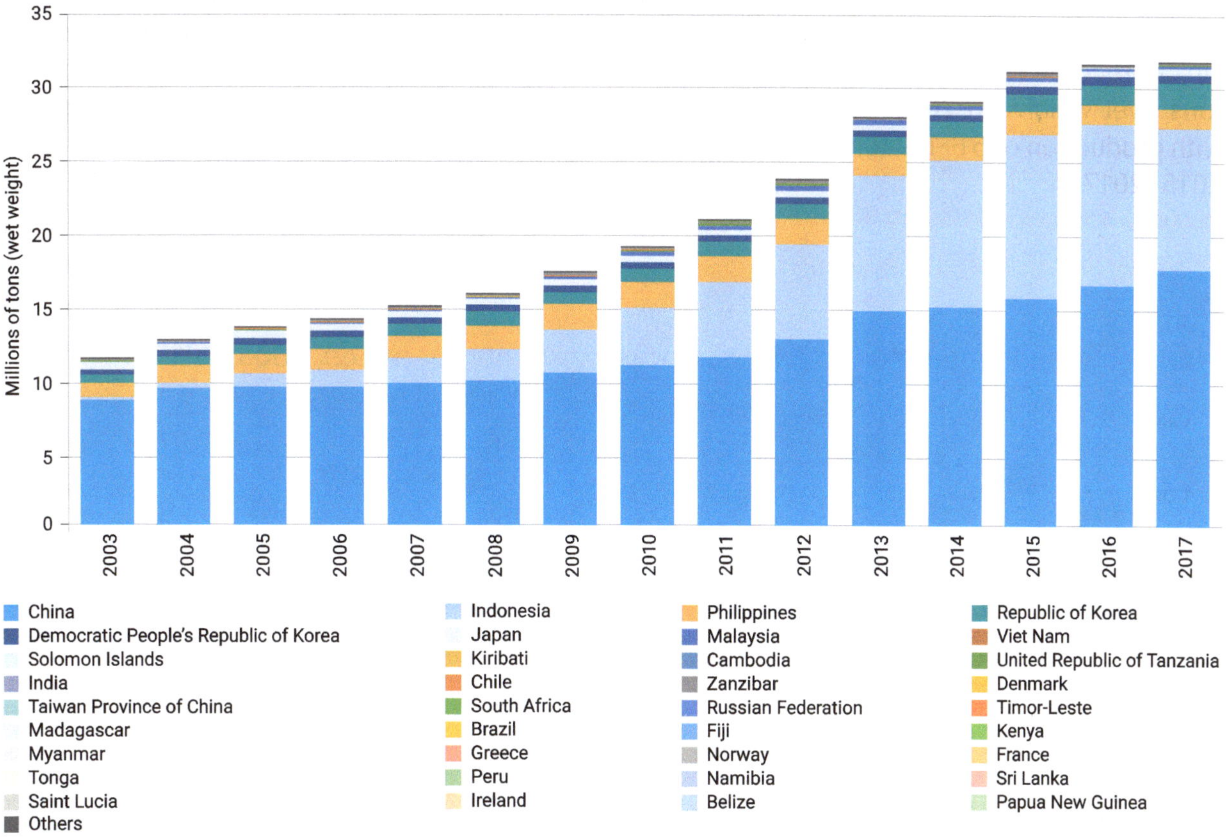

Source: Data for 2003–2012 are from FAO (2014) and for 2013–2017 from FAO (2019), tables 5 and 6.

Trono and Largo (2019) reported that, apart from typhoons, the steady decline in seaweed production in that country was caused by epiphytism, loss of genetic diversity due to culture methods used, and political unrest in the main farming areas in the southern Philippines. Recent data on edible seaweed from the United States of America were generated by Piconi and Veidenheimer (2020). The Government of the Republic of Korea has made a very concerted effort to grow new markets in North America.

The carrageenan-producing *Kappaphycus alvarezii* and *Eucheuma* spp. continued to be the major species farmed, with an increase in production from 8.3 million tons (wet weight) in 2012 to 12.3 million tons (wet weight) in 2016. Kelp production also increased from 5.7 million tons (wet weight) in 2012 to 8.4 million tons (wet weight) in 2016 (see figure II).

Figure II
World seaweed production, by group, 2003–2016

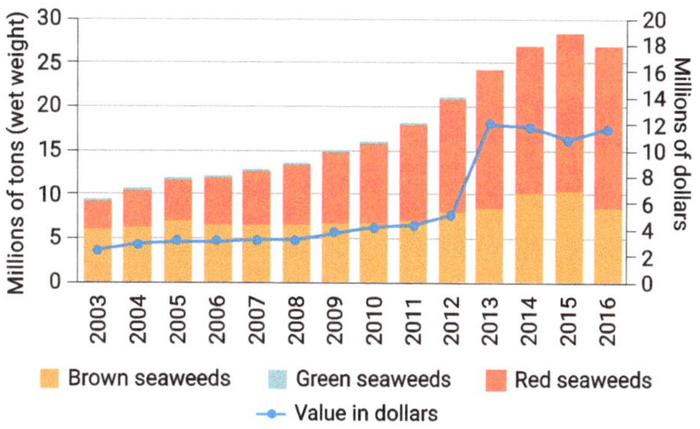

Source: Data for 2003–2012 are from FAO (2014) and for 2013–2016 from FAO (2018). Monetary values for 2013–2016 are from FAO (2019), tables 5 and 6.

Papua New Guinea has increased production over the past seven years, from 100 tons (wet weight) in 2010 to 4,300 tons (wet weight) in 2017. New producers include Cambodia, which recorded production of 2,000–2,200 tons (wet weight) in 2015–2017, and Norway, with production of 51–149 tons (wet weight) in 2015–2017.

The trend in production from the harvesting of wild stocks has been more or less the same over the five-year period since the 2012 baseline (see figure III), with Chile, China, Norway and Japan still the top four producers, in that order. Indonesia replaced France in fifth place with a harvest in 2017 (about 50,000 tons) that was six times the size of the 2012 harvest (7,600 tons).

Figure III
World seaweed production from wild stocks by country or region, 2003–2017

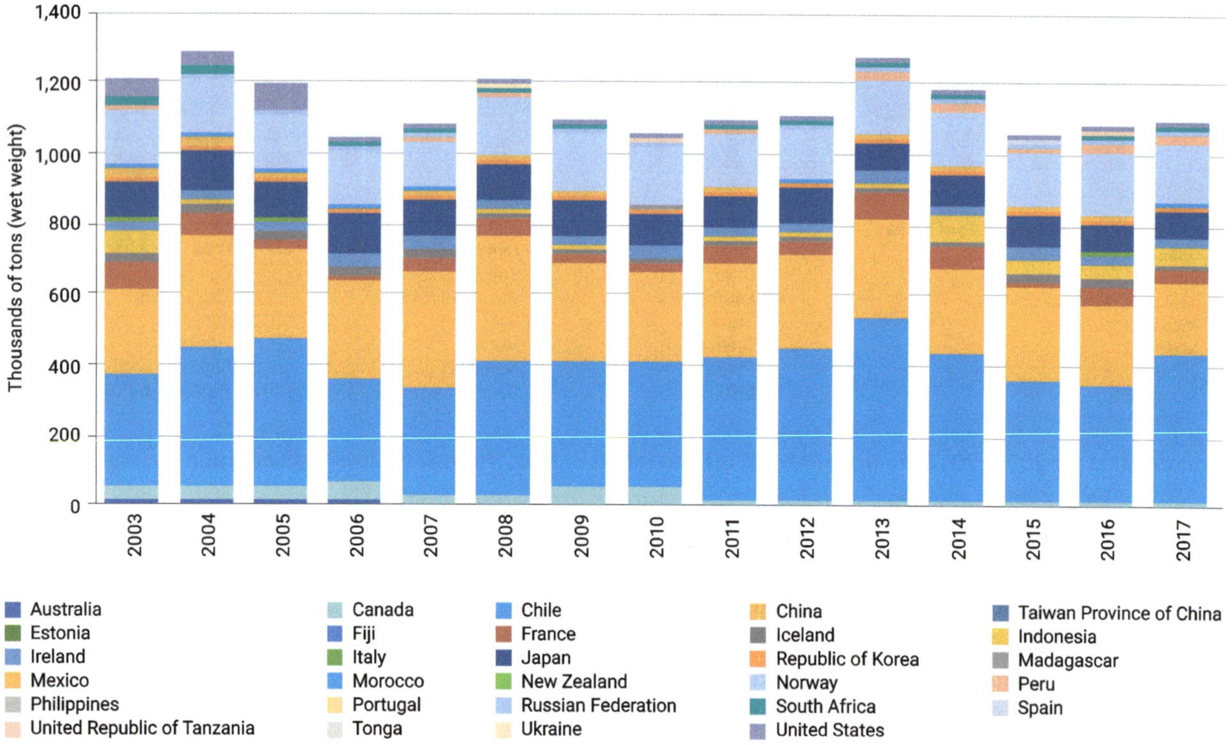

Source: Data for 2003–2012 are from FAO (2014) and for 2013–2017 from FAO (2019), table A-6.

3. Consequences of changes in seaweed harvesting and use for communities, economies and well-being

Buschmann and others (2017) have predicted that seaweed (as well as microalgae) production could account for 18 per cent of the global alternative protein market, or 56 million tons of protein, by 2054.

There is increasing national consumption of seaweed and seaweed-based products worldwide, thus augmenting local incomes. This is due to new innovations in dining, such as gourmet restaurants and bakeries featuring seaweed-enhanced dishes and new health trends for people with different food needs, such as vegans, diabetics and athletes looking for foods that are high in vegetable protein and soluble fibre, as well as in minerals, essential amino acids and vitamins (Bradford, 2014; Ibáñez and Herrero, 2017; Kim and others, 2017).

In countries where capitalization is a main factor in large-scale production, such as Brazil, the increase in seaweed production depends partly on associations and cooperatives. Nearshore seaweed farms can often be beset with problems, such as coliform contamination, siltation and other anthropogenic activities that affect coastal regions.

4. Key region-specific changes and consequences

Although seaweed production is concentrated in three major regions, namely, the Indian Ocean, the North Pacific and the South Pacific, production is increasing in other regions. In the South Atlantic, for example, specifically Brazil, *Kappaphycus alvarezii* and *Gracilaria* spp. are cultivated on a family scale, as promoted by Government agencies and international organizations, for agar extraction for the commercial market (Simioni and others, 2019). Wild harvesting of *Sargassum* for agricultural purposes occurs in some regions.

Unlike Argentina, Brazil and Mexico, which have only small-scale processing plants for the production of algae, Chile is the only country in its region (the South Pacific) where the harvesting, cultivation and processing of seaweeds occurs on a commercial scale. Most of the *Gracilaria* algae that it produces (50 per cent of global production) is acquired by Chinese processors (Ramírez and others, 2018). As a result of the opening of new markets and the facilitation of international trade, seaweeds are no longer marine products of relatively low economic value (i.e. commodities) but rather export goods with a high trading value. New legislation for the conservation of marine resources and the restriction of wild harvesting, and management policies allowing unions and granting cooperative rights to sea plots in order to promote cultivation, have led to a transformation at the national level to governance that promotes sustainability. This has been particularly important for the artisanal sector in social and economic matters (Gelcich and others, 2015; Gallardo and others, 2018).

5. Outlook

With regard to the Sustainable Development Goals[2] in general, and Goal 14 in particular, seaweed farming and harvesting are relevant to the following targets: 14.1 on reducing marine pollution, since they require no fertilizer inputs and recycle nutrients; 14.2 on sustainably managed and protected marine and coastal ecosystems; 14.3 on reducing ocean acidification, by absorbing atmospheric carbon dioxide; 14.4 on reducing overexploitation of fisheries, by reducing fishing in capture fisheries; 14.5 on conservation of marine and coastal areas; and 14.b by supporting small-scale artisanal fisheries. Seaweed farming and harvesting also contribute to achieving the other Goals, including but not limited to Goal 2 on achieving food security and Goal 8 on sustained and inclusive economic growth, especially since women and children are involved in the process.

Bjerregaard and others (2016) discussed seaweed aquaculture for food security, income generation and environmental health in tropical developing countries. Buschmann and others (2017) state that seaweeds could be the "ultimate sustainable crop", which would lead to the growth of the aquaculture industry required to sustain the world's food supply. Without the constraints of arable land (since the sea covers 71 per cent of the planet's surface), fertilizer and freshwater inputs, mariculture of seaweeds or "phyconomy" (see Hurtado and others, 2019), coupled with "new aquaculture" technologies, could provide the 14 per cent per year growth rate required of the industry to secure global food security by 2050. Seaweeds supply not only food for human consumption but also raw materials for feeds, nutraceuticals and pharmaceuticals. They also provide a carbon sink to help to combat climate change.

Apart from the projected continuing increase in seaweed production for traditional and current uses, emerging applications in agriculture could help cattle-producing countries to reduce global warming. For example, the red alga *Asparagopsis*, added to cattle feed, has been observed to considerably reduce methane "belching" in cattle by about 26 per cent (Roque and others, 2019).

Seaweed farming is pursuing eco-friendly certification for sustainable production. The Seaweed Standard will contribute to the health of the world's aquatic ecosystems by promoting environmentally sustainable and socially responsible use of seaweed resources.

6. Key remaining knowledge and capacity-building gaps

Cottier-Cook and others (2016) discussed the safeguarding of the future of the global seaweed aquaculture industry. Duarte and others (2017) discussed how seaweed farming might play a role in climate change mitigation and adaptation. Further scientific approaches will be necessary to answer these questions.

Buschmann and others (2017) identified many knowledge gaps with regard to large-scale production, economics and climate change. The biology of many seaweed species is still unknown and, even for those which are already harvested or farmed, there are aspects of their biology which are still not well understood. Advanced production models need the above phyconomic information and, for

2 See General Assembly resolution 70/1.

offshore farms, information on the effects of climate change is especially important. The establishment of offshore farms will need long-term data on typhoons and sea surface temperatures, as well as oceanographic data. Large-scale phyconomic production would also need information to generate appropriate economic and financial models, such as new applications, markets and "externalities". Artisanal farmers and harvesters still face the age-long problems of capitalization, lack of healthy and vigorous planting materials, and pricing variabilities.

Currently, institutions in five countries are working together to address some of these knowledge and capacity-building gaps, with a focus on safeguarding the seaweed industry, especially in developing countries (GlobalSeaweedSTAR).

References

Bjerregaard, R., and others (2016). Seaweed aquaculture for food security, income generation and environmental health in tropical developing countries. Washington, D.C.: World Bank Group.

Bradford, M. (2014). *Algas: las verduras del mar – los nutritivos tesoros marinos para la salud y el paladar,* 8th ed. Barcelona: Océano Ambar.

Buschmann, A.H., and others (2017). Seaweed production: overview of the global state of exploitation, farming and emerging research activity. *European Journal of Phycology*, vol. 52, No. 4, pp. 391–406.

Cottier-Cook, E.J., and others (2016). Safeguarding the future of the global seaweed aquaculture industry. Policy Brief. United Nations University, Institute for Water, Environment and Health and Scottish Association for Marine Science.

Duarte, C. M., and others (2017). Can seaweed farming play a role in climate change mitigation and adaptation? *Frontiers of Marine Science*. vol. 4, art. 100.

Food and Agriculture Organization of the United Nations (FAO) (2014). *FAO Yearbook: Fishery and Aquaculture Statistics 2012*. Rome.

________(2018). *The State of World Fisheries and Aquaculture 2018: Meeting the Sustainable Development Goals*. Rome.

________(2019). *FAO Yearbook: Fishery and Aquaculture Statistics 2017*. Rome.

Gallardo Fernández, G.L., and others (2018). *Granjeras del mar: luchas y sueños en Coliumo*. Historia del área de manejo del sindicato, No. 2. Santiago: Andros Impresores.

Gelcich, Stefan, and others (2015). Exploring opportunities to include local and traditional knowledge in the recently created "Marine Management Plans" policy of Chile. In *Fishers' Knowledge and the Ecosystem Approach to Fisheries: Applications, Experiences and Lessons in Latin America*. Johanne Fischer and others, eds. FAO Fisheries and Aquaculture Technical Paper, No. 591. Rome: FAO.

GlobalSeaweedSTAR. www.globalseaweed.org.

Hurtado, A.Q., and others (2019). Phyconomy: the extensive cultivation of seaweeds, their sustainability and economic value, with particular reference to important lessons to be learned and transferred from the practice of eucheumatoid farming, *Phycologia*, vol. 58, No. 5, pp. 472–483.

Ibáñez, E., and M. Herrero (2017). *Las algas que comemos*. ¿Qué sabemos de? Series. Madrid: CSIC.

Kim J.K., and others (2017). Seaweed aquaculture: cultivation technologies, challenges and its ecosystem services. *Algae*, vol. 32, No. 1, pp. 1–13. Available at https://doi.org/10.4490/algae.2017.32.3.3.

Park, M., and others. (2018). Application of open water integrated multi-trophic aquaculture to intensive monoculture: a review of the current status and challenges in Korea. *Aquaculture*, vol. 497, pp. 174–183. Available at https://doi.org/10.1016/j.aquaculture.2018.07.051

Piconi, P., and R. Veidenheimer (2020). *Edible Seaweed Analysis Report*. Rockland, Maine, United States: Island Institute.

Ramírez, M., and others (2018). *Flora marina bentónica de Quintay*. Santiago: RIL Editores and Universidad Andres Bello.

Roque, B.M., and others (2019). Inclusion of *Asparagopsisarmata* in lactating dairy cows' diet reduces enteric methane emission by over 50 per cent. *Journal of Cleaner Production*, vol. 234, pp. 132–138.

Simioni, C., and others (2019). Seaweed resources of Brazil: what has changed in 20 years? *Botanica Marina*, vol. 62, No. 5, pp. 433–441.

Trono, G.C., and D.B. Largo (2019). The seaweed resources of the Philippines. *Botanica Marina*, vol. 62, No. 5,pp. 483–498.

United Nations (2017). Chapter 14: Seaweeds. In *The First Global Integrated Marine Assessment: World Ocean Assessment I*. Cambridge: Cambridge University Press.

Chapter 18 **Changes in seabed mining**

Contributors: James R. Hein and Pedro Madureira (joint conveners), Maria João Bebianno (co-lead member), Ana Colaço, Luis M. Pinheiro, Richard Roth, Pradeep Singh, Anastasia Strati (co-lead member) and Joshua T. Tuhumwire (lead member).

Keynote points

- The present chapter provides an update to chapter 23 of the first *World Ocean Assessment* (United Nations, 2017a) in terms of shallow-water aggregate, placer deposits, ironsand deposits and phosphorite deposits. It focuses on exploration licences for deepwater seabed mineral resources, the number of which has increased significantly since the first Assessment.
- New technologies to reduce impacts on the marine environment are now envisaged for the exploitation of placer deposits, traditionally mined by dredging. Prospects for mining phosphorite deposits have faced opposition from stakeholders and have yet to become a reality.
- Seabed mineral deposits covered in the present chapter (polymetallic nodules, polymetallic sulphides and cobalt-rich ferromanganese crusts) are being considered for mining and the object of 30 contracts for exploration awarded by the International Seabed Authority (ISA).
- One driver for those activities is that deepwater seabed mineral resources contain diverse rare and critical metals that would support the implementation of Sustainable Development Goals adopted by the United Nations in 2015.[1]
- The environmental impacts of the exploitation of those seabed mineral resources are a scientific community focus, and regulations are now being developed by ISA.
- A lack of information on biodiversity, connectivity and ecosystem services exists, and a robust collection of baseline ecological data is necessary for predictions related to the future deepwater seabed mining activities, given the risk of irreversible damage to deep-sea ecosystems.
- ISA has considered various financial models for the commercial mining of polymetallic nodules. Metal prices are difficult to predict, which can create significant risk that may delay commercial mining.
- Deepwater seabed mineral resources are typically located far from human communities and the social impacts of their exploitation may be less than those of terrestrial mining. However, significant concerns exist about loss of biodiversity and ecosystem services, including the role of the deep ocean in climate regulation. Those legitimate concerns constitute the basis for a "social licence to operate".

1. Introduction

1.1. Links to the first *World Ocean Assessment*

Chapter 23 of the first *World Ocean Assessment* focused on marine mining, in particular established extractive industries, which are predominantly confined to nearshore areas, where shallow-water aggregate and placer deposits and somewhat deeper-water phosphate deposits are found (United Nations, 2017b). At the time of its publication, there were no commercially developed deepwater seabed mining (DSM) deposits, but an assessment of mining leases and exploration activity was included. Since the first Assessment, the number of deepwater (i.e., depths greater than 200 m below the ocean surface) seabed exploration licences has increased both within national jurisdictions

[1] See General Assembly resolution 70/1.

of coastal, island and archipelagic States and beyond, in the Area (the seabed, ocean floor and subsoil thereof beyond the limits of national jurisdiction) under ISA administration. In the twenty-first century, deepwater seabed test mining was carried out for the first time in 2017. That was done by Japan within its exclusive economic zone (EEZ) at a water depth of 1,600 m (Ministry of Economy, Trade and Industry (METI), 2017). The present chapter focuses on the nascent deepwater seabed mining industry and mineral deposits, and the term "seabed" is used hereinafter for "deepwater seabed".

The first Assessment focused on the environmental impacts of dredging activities and contained a list of references for some mining operations. However, an environmental baseline for DSM could not be provided, and it was considered that available data did not allow for an adequate understanding of environmental, social and economic aspects. Data on potential environmental impacts are still scarce and can differ greatly between mineral extraction from nearshore and that from deep ocean seabed mining sites. Information on the economic benefits and, to some extent, the social impacts of mining is becoming progressively more accessible, thanks to several initiatives promoting an increase in transparency in the extractive industries.

In 2015, the General Assembly adopted the 2030 Agenda for Sustainable Development, which includes 17 Sustainable Development Goals to be addressed on the basis of a global partnership. DSM activities may have implications for the achievement of Goals 1, 5, 7 to 10, 12 to 14 and 17.

1.2. Drivers, challenges and opportunities for seabed mining

Many drivers, challenges and opportunities exist for DSM and have been discussed in many scientific papers and popular media (Hein and others, 2013; Banerji, 2019; Koschinsky and others, 2018). One key issue related to drivers is how to ensure the supply of critical materials to support infrastructure development and the provision of goods for an expanding middle class in developing societies, as well as the transition to urbanization in those societies. Another question is how rare and critical materials, which are abundant in seabed mineral deposits, will be sourced to support green technologies (e.g., wind turbines, electric vehicles and solar cells), which are viewed by some stakeholders as solutions to achieve a low-carbon future and combat global climate change (Graedel and others, 2015; Kim and others, 2015; McLellan and others, 2016; Zweibel, 2010; World Bank, 2017a). DSM has been suggested as offering a potential partial solution to those important issues (World Bank, 2017a).

Many unique characteristics of future seabed mines have been pointed out as additional drivers for this new industry (Hein and others, 2013; Petersen and others, 2016). They include the high grades (concentrations) and tonnages of rare and critical metals in seabed mineral deposits; the fact that marine-based mine sites will not entail the construction of roads, sea floor ore transport systems, water and electrical transport systems, buildings, waste dumps or other infrastructure on the sea floor; and, most importantly, the fact that no overburden will need to be removed before mining can take place because the deposits of interest are exposed on the sea floor. All of those characteristics could lessen environmental impacts.

There are, however, many challenges to DSM. The most significant challenge is obtaining a sufficient understanding of the various ecosystems that characterize seabed mineral deposit environments, as well as the knowledge needed to avoid, reduce and mitigate the environmental impacts of resource extraction. Other challenges include social licence, which can be addressed through transparency and communications. Challenges for the industry include how to improve mining engineering and environmental safeguards, and the development of green metallurgical processing technologies.

The ever-present volatility of metal prices and markets, as well as competition with land-based mines, will also create significant challenges.

DSM is being heavily regulated even before extraction has begun. This offers an opportunity right from the start to apply a precautionary approach and adaptive management, supported by real-time monitoring.

1.3. Overview

Marine aggregate exploitation is still the major offshore mining activity, as an alternative to mitigate the huge negative impacts of both legal and illegal beach and onshore sand mining (Torres and others, 2017). Section 2 contains an update on this activity, before considering other nearshore shallow-water deposits (placer diamonds, placer tin, ironsand deposits and phosphorite deposits) and seabed mining. Seabed deposits of current economic interest are polymetallic nodules (PNs), sea floor massive sulphides (SMSs) (or polymetallic sulphides) and cobalt-rich ferromanganese crusts (CFCs) (see figure I). The marine environment and the need to gather sufficient data and information on the environmental impacts that may arise from their exploitation are addressed in section 3. Envisaged economic and social impacts related to DSM are discussed in section 4. Lastly, section 5 briefly identifies the major capacity-building needs.

Figure I
Global permissive areas for deepwater seabed mineral deposits

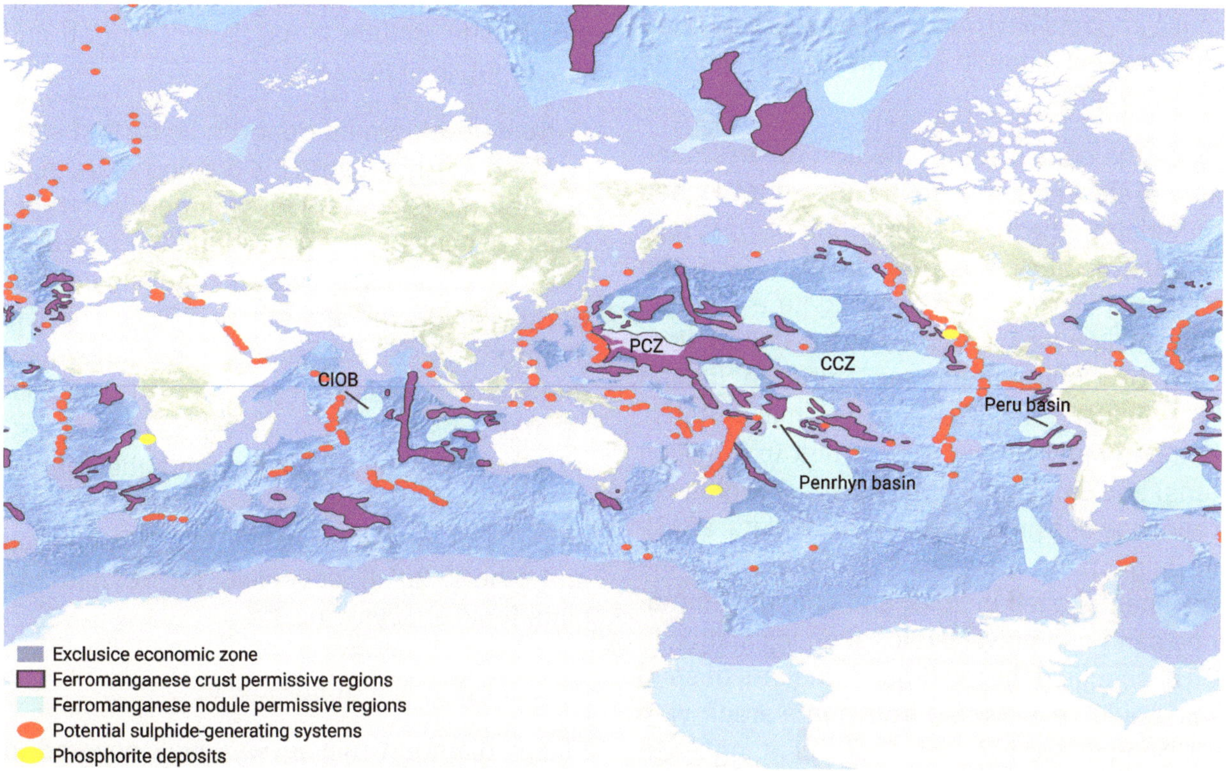

Note: Red indicates the location of hydrothermal vent sites (after Beaulieu, 2015), which are potential sea floor massive sulphide-generating systems; such sulphides have not been found at all of those sites. In the area of the Prime Crust Zone (Hein and others, 2009), the permissive areas for cobalt-rich ferromanganese crusts and polymetallic nodules overlap; nodule fields occur between seamounts and ridges in much of the western Prime Crust Zone. Location of the three phosphorite deposits discussed in the present chapter are indicated with yellow-filled circles. The four well-known polymetallic nodule fields are also indicated: the Clarion-Clipperton Fracture Zone, the Peru basin, the Penrhyn basin and the central Indian Ocean basin (modified from Hein and others, 2013). The dark grey area around Antarctica is not an exclusive economic zone but simply represents the 200 nautical mile extent.
Abbreviations: CCZ, Clarion-Clipperton Zone; CIOB, central Indian Ocean basin; PCZ, Prime Crust Zone.

2. Changes in scale and significance of sea floor mining

The first Assessment emphasized the state of active marine mining, which was and still is limited to nearshore shallow-water deposits. Some updates are provided below, but little has changed since the first Assessment in this field of activity.

2.1. Current state of changes

2.1.1. Aggregate, sand and gravel update

The first Assessment contained a thorough overview of aggregate mining and identified the large negative impacts of beach sand exploitation, especially with regard to coastal vulnerability and resilience to flooding, storm surges, tsunamis and rising sea levels. All those impacts have led to growing global interest in the exploitation of offshore aggregate as an alternative.

Since the conclusion of the first Assessment, aggregates continue to be the most mined materials in the marine environment, generally at water depths of less than 50 m. In 2016, the Netherlands led marine aggregate extraction (12.5 million tons), followed by the United Kingdom of Great Britain and Northern Ireland (11.9 million tons), Germany (10 million tons), France (7 million tons), Denmark (6.6 million tons) and Belgium (6.6 million tons) (Union européenne des producteurs de granulats, 2018). In Belgium, no gravel was exploited from the continental shelf in 2017, and changes to marine sand and gravel exploitation legislation in 2014 specified the maximum amounts of sand that could be extracted from some areas, with an annual decrease of 1 per cent between 2014 and 2019 (International Council for the Exploration of the Sea (ICES), 2018). In Finland, there was no marine extraction in 2017, but permits have been issued to extract 8 million m^3 of sand up to 2027 off the shore of Helsinki and from the mouth of the Iijoki River (ICES, 2018). Since the first Assessment, there has been an increase in the extraction of sand and gravel in the United States of America, where they are used as a source material for storm damage coastal restoration projects, in particular along the Atlantic and Gulf of Mexico coasts. For the United States Atlantic region alone, total aggregate extraction in 2018 was 17.45 million m^3, 97 per cent of which was used for beach replenishment (coastal restoration) public sector projects (ICES, 2019). China identified abundant marine aggregates, mainly in the East China Sea, the Taiwan Strait and the northern continental shelf of the South China Sea, which have been roughly estimated at 1.6 x 10^{12} tons (Qin and others, 2014). Other countries already identified in the first Assessment with significant offshore aggregate mining activity include India, Japan, Kiribati and the Republic of Korea.

Current development trends indicate that sand demand will increase in the coming years at an accelerated rate, associated largely with rapid urban expansion, putting additional strain on the limited sand deposits and causing conflict worldwide (Torres and others, 2017). There is therefore a need for technological innovation that would minimize impacts on the environment (Gavriletea, 2017), as well as for integrated studies to better understand the marine environment and the time of recovery from aggregate mining impacts (e.g., Gonçalves and others, 2014), especially for benthic and planktonic ecosystems.

2.1.2. Placer diamonds update

Placer diamond mining was well covered in the first Assessment, but a few updates are warranted. About 75 per cent of the diamond production in Namibia is from its offshore placer deposits. Debmarine, a 50/50 joint venture between De Beers and Namibia, is in the process of building a new mining vessel (SS *Nujoma*) that will increase offshore production by about 500,000 carats per year.[2] This custom-made ship will be ready for operation in 2022 and equipped with new technologies that will provide higher efficiency and productivity. Placer diamond mining off Namibia has now reached water depths of up to 200 m.

2.1.3. Placer tin update

Placer deposits in riverbeds and valleys and on the sea floor make up nearly 80 per cent of the world's tin resources (Kamilli and others, 2017). The most extensive area for onshore and offshore placers is in the enormous tin belt of South-East Asia. In 2017, Indonesia became the second largest producer of tin in the world, mined both onshore and offshore in nearly equal amounts, and it is the largest producer of offshore tin. Based on the 2018 annual report of the PT Timah Tbk mining enterprise,[3] total tin production in Indonesia increased from 24,121 tons in 2016 to 33,444 tons in 2018, the highest level of production since 2012. Indonesia's reserve of 800,000 tons is second only to that of China, which stands at 1,100,000 tons (United States Geological Survey (USGS), 2019), and Timah estimates that the tin resources of Indonesia total 1,043,633 tons. Timah is exploring in its offshore mining the use of borehole mining technology, which the company thinks can increase tin ore production with a much lower environmental impact. This would be a significant development, as offshore placer tin deposits are mined by dredging methods, which have an environmental impact on benthic, mid-water and pelagic ecosystems.

For comparison, in 2018, mine production in Malaysia was only 4,000 tons, but the reserves are estimated at 250,000 tons (USGS, 2019). Historically, Malaysia has produced 55 per cent of the tin used worldwide (Kamilli and others, 2017).

2.1.4. Ironsand deposit update

Ironsand is a sand containing grains of iron oxides (usually magnetite), typically found along coastal areas. The sand is mined for the iron to be used in the steel industry. The first Assessment contained a case study of ironsand deposits that occur off the coast of New Zealand at water depths of 20–42 m. A mining permit was granted to Trans-Tasman Resources Limited (TTR) in May 2014 for up to 50 million tons of ore per year for 20 years, which would be mined over an area of 66 km^2, as a first step in a regulatory process to allow extraction. As reported in the first Assessment, the decision-making committee of the New Zealand Environmental Protection Agency refused to grant an environmental permit to mine in June 2014, on the basis of inadequate environmental data. However, in August 2018, the committee granted an environmental permit to mine up to 50 million tons a year of ironsand for 35 years, on the basis of a revised application. That decision was subsequently appealed by environmental and fishing groups, and the New Zealand High Court ruled in August 2018 that no mining would take place and sent it back to the committee for further consideration based on the Court's criteria concerning correct legal tests for adaptive management. TTR appealed the High Court's ruling to the Court of Appeals, and has now appealed it to the Supreme Court, where the case currently resides.

Three additional companies have been granted exploration permits for ironsand in the New Zealand maritime zones. Cass Offshore Minerals Limited has an ironsand exploration permit off the coast of New Plymouth, the same general region as the mining permit granted to TTR. In May 2018, Ironsands Offshore Mining Limited was granted permission for exploration inside a marine sanctuary off the coast of New Plymouth, the same region as the TTR

2 See www.mining-technology.com/features/giant-mining-vessels-how-high-quality-gems-are-exploited-from-the-sea.

3 See www.timah.com.

and Cass permits. Pacific Offshore Mining has an exploration permit for iron-titanium sands (ilmenite) off the Bay of Plenty, east of the North Island.

2.1.5. Phosphorite deposits: Chatham Rise (New Zealand), Don Diego (Mexico) and Namibian Marine Phosphate Sandpiper and other projects (Namibia)

Phosphorite is a sedimentary rock or sediment with enough content of phosphate minerals to be of economic interest. Phosphate is used as a fertilizer in agriculture and in the chemical industry, for example as phosphoric acid used in most soft drinks. Mining has not yet started at any of the three licence areas listed in the present section (see figure I for locations).

Chatham Rock Phosphate (CRP) has held a mining permit since December 2013 and applied for an environmental consent in June 2014, but the decision-making committee appointed by the New Zealand Environmental Protection Agency turned down the application in February 2015. CRP expects completing the Environmental Protection Agency reapplication and hearing by the end of 2021. It plans to mine 820 km^2 for up to 1.5 million tons of phosphate annually at water depths of up to 450 m. CRP is currently considering the possibility of extracting rare earth elements as an important potential by-product.

The proposed Mexican Don Diego phosphate project was covered in the first Assessment, at which time Odyssey Marine Exploration had submitted an environment impact assessment for approval to the Secretariat of Environment and Natural Resources of Mexico (SEMARNAT). The application to develop the phosphate project through its subsidiary, Exploraciones Oceánicas, was denied in April 2016. In 2018, the decision was appealed to the Mexican administrative tribunal, which ruled that the decision had failed to consider the extensive environmental mitigation procedures proposed, but SEMARNAT reinstated its previous decision. The project is currently in various stages of negotiation.

Namibian Marine Phosphate Limited (NMP) was awarded a mining licence (ML170) in July 2011, and it provided an environmental impact assessment and environmental management programme in 2012. It received an environmental clearance certificate for mining in September 2016; however, the certificate was retracted two months later, following protests by various stakeholders. NMP appealed to the High Court of Namibia. In May 2018, the company was successful in its appeal and the High Court set aside the retraction.[4] NMP operations will be at water depths of 190–345 m over an area of about 2,200 km^2, 60 km off the coast of Namibia. Other companies also hold permits for areas off the coast of Namibia, including CRP.

2.1.6. Seabed mining

It was stated in the first Assessment that SMS mining might begin in the Manus Basin, Bismarck Sea, Papua New Guinea EEZ, in 2017. However, owing to an inability to raise the necessary funds, this component of the company has been curtailed.[5]

The Pacific Island States are working on the development and adoption of seabed mining legislation for areas within national jurisdiction. The publication of national regulatory legislation has been supported by a number of initiatives, including the ongoing Pacific Maritime Boundaries Consortium and the European Union-funded Deep Sea Minerals Project (2011–2016) of the Pacific Community.

ISA is currently administering 30 exploration contracts.[6] At the time of reporting, Africa was

4 See https://namiblii.org/na/judgment/high-court-main-division/2018/122.
5 See https://dsmf.im.
6 See https://isa.org.jm/index.php/exploration-contracts.

the only continent without countries sponsoring exploration activities in the Area. Draft regulations on exploitation of marine mineral resources in the Area are currently under discussion within ISA, and the ISA Council considers their adoption to be a matter of urgency.[7]

2.1.6.1. Polymetallic nodules

PNs form predominantly on the sediment-covered abyssal sea floor of the global ocean at water depths of about 3,500–6,500 m (Kuhn and others, 2017) (see figures I, II.C and II.D). The economic interest in those deposits is focused on nickel, copper, cobalt and manganese, although molybdenum, titanium, lithium, zirconium and the rare earth elements and yttrium also occur in high concentrations (Hein and others, 2013; Kuhn and others, 2017).

At the time of writing, 18 contracts for PNs had entered into force, with 16 located in the North-East Pacific Clarion-Clipperton Fracture Zone (see figure III), one in the North-West Pacific Ocean and one in the central Indian Ocean basin. The exploration area allocated to a contractor can reach a maximum size of 150,000 km^2 but may not exceed 75,000 km^2 after eight years from the date of the contract.[8]

Apart from the Clarion-Clipperton Zone, high prospective areas occur in the Peru basin and Penrhyn-Samoa basins. Although most nodule fields are in the Area, important PN deposits can also be found within the EEZs of, inter alia, the Cook Islands, Kiribati, Niue and American Samoa (United States) (Hein and others, 2005; 2015).

2.1.6.2. Sea floor massive sulphides or polymetallic sulphides

High-temperature hydrothermal circulation systems occur in all ocean basins along mid-ocean ridge spreading centres and along volcanic arcs and back-arc spreading centres (see figure I). The highest-temperature products are SMSs and sulphate deposits in focused-flow systems, such as chimneys, and lower-temperature hydrothermal manganese and iron oxide deposits in diffuse-flow systems (see figures II.E and II.F). Deposits may form at water depths of 200–5,000 m, deeper-water deposits being generally located along spreading centres and shallower-water deposits along volcanic arcs. High concentrations of copper, zinc, gold and silver occur in some SMS deposits at all those settings. The tonnages of actively forming deposits are generally poorly known but usually small. Tonnages and grades of inactive, off-axis SMS deposits are even more poorly known but are likely to be of higher tonnage and comparable with some land-based counterparts (German and others, 2016; Jamieson and others, 2017).

Hydrothermal SMS occurrences are common within the EEZs of many Pacific States, such as Fiji, Japan, New Zealand, Solomon Islands, Tonga and Vanuatu, as well as those of Norway and Portugal in the Atlantic and Saudi Arabia and the Sudan in the Red Sea. The last of those corresponds to the metalliferous mud deposit of Atlantis II Deep, which may be the only SMS deposit similar in scale to the large terrestrial deposits (up to 90 million tons) (Hoagland and others, 2010).

In the Area, seven contracts for exploration for SMSs have entered into force since 2011: three in the Atlantic Ocean and four in the Indian Ocean. The area covered by each contract is comprised of not more than 100 blocks, arranged in five or more clusters; each block being approximately 10 km by 10 km and no greater than 100 km^2. The exploration area may not exceed 2,500 km^2 by the end of the tenth year from the date of the contract.[9]

7 See document ISBA/24/C/8/Add.1, para. 7.
8 See document ISBA/19/C/17, regulation 25.
9 See document ISBA/16/A/12/Rev.1, regulations 12 and 27.

Figure II
Sea floor and mineral deposit photographs of crusts, nodules and polymetallic sulphides

(A) Sea floor covered with cobalt-rich crust in the Marshall Islands area, West Pacific (cruise USGS F10-89-CP), water depth of approximately 1,650 m, field of view of about 3x3 m.

(B) 12 cm-thick crust (CD29-2, USGS cruise F7-86-HW) from the Johnston Island area, central Pacific, recovered from a water depth of 2,225 m, scale bar represents 10 cm.

(C) Seabed covered with polymetallic nodules in the Clarion-Clipperton Zone nodule field, water depth of approximately 5,000 m, field of view of about 3x3 m.

(D) Polymetallic nodules from the Clarion-Clipperton Zone, North-East Pacific Ocean; each nodule is 3 cm in diameter, approximate water depth of 5,000 m.

(E) Seabed with an active black smoker from the North-East Pacific Ocean, approximate water depth of 2,200 m, field of view of about 4x4 m (National Oceanic and Atmospheric Administration, Ring of Fire).

(F) Cross-section through a large zinc sulphide chimney showing a yellow silica conduit through which the hot fluids flowed; collected at 377 m water depth, Mariana volcanic arc, West Pacific (JAMSTEC, cruise NT10-12).

Figure III
International Seabed Authority Clarion-Clipperton Fracture Zone contract map overlaying Europe to show scale

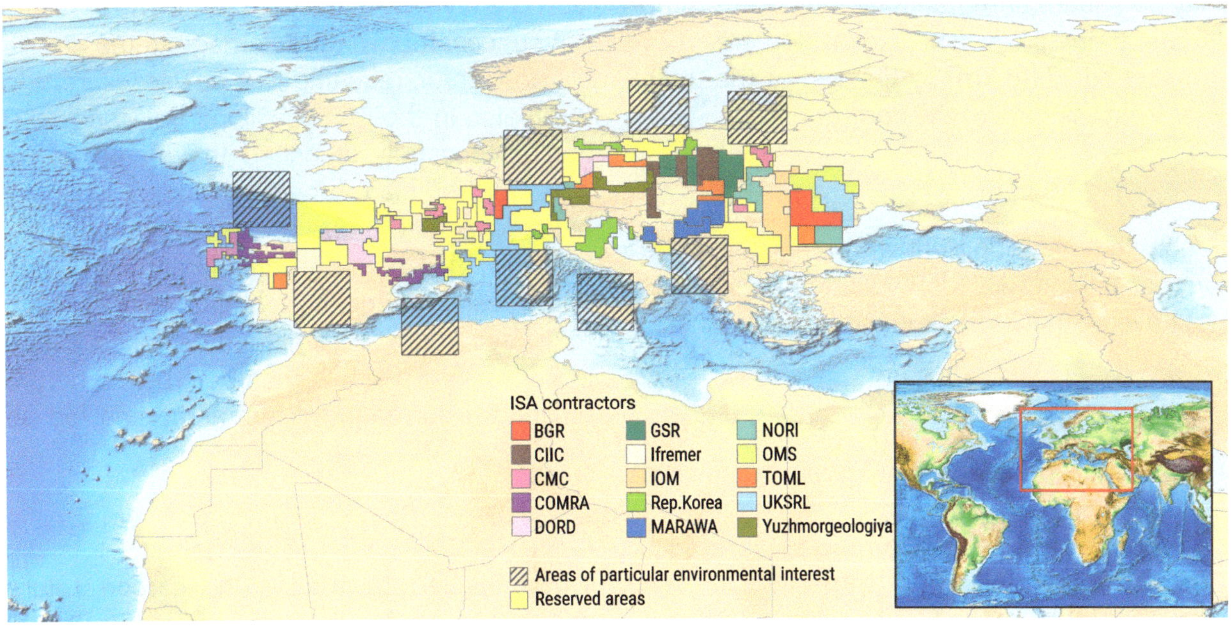

Source: International Seabed Authority, 2019: Clarion-Clipperton Zone.

Abbreviations: BGR, Federal Institute for Geosciences and Natural Resources of Germany; CIIC, Cook Islands Investment Corporation; CMC, China Minmetals Corporation; COMRA, China Ocean Mineral Resources Research and Development Association; DORD, Deep Ocean Resources Development Co. Ltd.; GSR, Global Sea Mineral Resources NV; Ifremer, Institut français de recherche pour l'exploitation de la mer; IOM, Interoceanmetal Joint Organization; Rep. of Korea, Government of the Republic of Korea; MARAWA, Marawa Research and Exploration Ltd.; NORI, Nauru Ocean Resources Inc.; OMS, Ocean Mineral Singapore Pte. Ltd.; TOML, Tonga Offshore Mining Limited; UKSRL, UK Seabed Resources Ltd.

2.1.6.3. Cobalt-rich ferromanganese crusts

CFCs form on the flanks and summit of seamounts, ridges and plateaux where rock is exposed at the sea floor (see figure II.A and II.B). Many thousands of such edifices occur in the ocean basins, and they are especially abundant in the Pacific Ocean (see figure I). CFCs are found at water depths ranging from about 400–7,000 m. In addition to cobalt, nickel and manganese, CFCs contain a wide array of rare and critical metals of economic interest and that have applications in emerging and next-generation technologies, especially tellurium, niobium, the rare earth elements and yttrium, scandium and platinum group metals (Hein and others, 2013; 2017). Based on grade, tonnage, topography, age of oceanic crust and oceanographic conditions, the best areas within the Area and national jurisdictions for CFC exploration and future mining are located within the central Pacific Prime Crust Zone, as defined by Hein and others (2009; 2013), including the EEZs of the Bonin Islands (Japan), the Commonwealth of the Northern Mariana Islands (United States), the Izu Islands (Japan), the Johnston Atoll (United States) and the Marshall Islands. Seamounts and ridges within the huge Prime Crust Zone are about half in EEZs and half in the Area. A smaller resource potential is located in the Pacific EEZs of French Polynesia (France), Kiribati, Niue and Tuvalu. Seamounts in the North-East Atlantic Ocean (EEZs of Portugal and Spain) also show metal grades and tonnages that warrant further study.

Exploration for CFCs in the Area is currently active under five contracts established with ISA: four in the western part of the Prime Crust Zone and one in the South-West Atlantic

Ocean. The area covered by each contract for exploration is comprised of not more than 150 blocks arranged in clusters; each block may be square or rectangular in shape and no greater than 20 km^2. The exploration area may not exceed 1,000 km^2 by the end of the tenth year from the date of the contract.[10]

2.2. Technological developments

Technology development with regard to CFCs lags far behind that for SMSs and PNs and is not covered in the present section.

2.2.1. Sea floor massive sulphides

Since the first Assessment, several seabed test-mining operations have been conducted in situ, the most complete of which was a two-month operation performed by Japan Oil, Gas and Metals National Corporation in the summer of 2017 in the EEZ of Japan at a depth of 1,600 m, near Okinawa Prefecture (METI, 2017). The operation involved a pilot test of the full system envisaged for the recovery of SMSs from the sea floor (see figure IV.D). Three mining production tools designed by Nautilus Minerals to conduct extraction of the Solwara 1 SMS deposit off the coast of Papua New Guinea were used for the first time. The machines, built by the company Soil Machine Dynamics, underwent submerged trials in an enclosed onshore excavation on Motukea Island.[11] Other SMS-mining tools have been developed or are under development, for example the Bauer BC40 SMS trench cutter mining tool (see figure IV.C).

2.2.2. Polymetallic nodules

An in situ mining test for PNs was scheduled in 2019 in the Clarion-Clipperton Zone and inside the contract areas of the Federal Institute for Geosciences and Natural Resources and of Global Sea Mineral Resources NV (GSR), sponsored by Germany and Belgium, respectively, at a water depth of around 4,500 m. The test focused on the nodules prototype collector (Patania II) developed by the DEME corporation, of which GSR is a division (see figure IV.A). The test was not successful owing to damage to the umbilical connector, resulting in a power failure.[12] In 2017, also in the Clarion-Clipperton Zone, GSR had successfully launched the pre-prototype collector Patania I. Other PN-mining tools have been developed or are under development, for example, the Korea Research Institute of Ships and Ocean Engineering tandem nodule mining tools (see figure IV.B) that are designed to collect PNs, which would then be crushed before entering a buffering system and being fed up the riser pipe.

2.3. Future directions

The transition towards a low-carbon future committed to by most Governments could invigorate interest in seabed mining and the search for new metal sources. The majority of DSM activities may occur in the Area, which includes most of the abyssal plains, as well as most sections of the mid-ocean ridges and seamounts forming the seabed. This should promote a paradigm shift in the mining industry. Mining of the seabed will be mostly monitored by the international community within the ISA framework, which currently comprises 168 members. However, many questions still remain open that need to be addressed at the global level. For example, how will this potential economic activity affect the production of land-based mining, often an important source of income in many developing countries? And how and at what level will DSM have an impact on the environment in the short, medium and long terms? To answer at least the latter question, substantial technological developments are still needed to foster in situ monitoring of the marine environment and acquire representative spatial and time series data.

10 See document ISBA/18/A/11, regulations 12 and 27.
11 See https://dsmobserver.com/2017/07/nautilus-png-submerged-trials.
12 See www.deme-gsr.com/news/article/patania-ii-technical-update.

Figure IV
Examples of new prototype deepwater seabed mining tools

(A) Global Seabed Mineral Resources (GSR; Belgium) Patania II mining machine for polymetallic nodules (photograph courtesy of GSR).
(B) Korea Research Institute of Ships and Ocean Engineering mining machine for polymetallic nodules (photograph by Hein).
(C) Bauer Maschinen GmbH, BC40 polymetallic sulphide trench cutter mining tool.
(D) Japan Oil, Gas and Metals National Corporation mining machine for polymetallic sulphides, tested in an Okinawa trough sulphide field in the summer of 2017.

3. Environmental aspects

3.1. Advances in knowledge and environmental impacts

The deepwater environment covers more than 90 per cent of the ocean area and includes a range of ecosystems and habitats at the sea floor level and in the water column (Ramirez-Llodra and others, 2011; Gollner and others, 2017). The different types of mineral resources at the sea floor level are located in various geological and oceanographic settings that, as a result, host various types of habitats and communities.

Regulations designed to avoid, reduce and mitigate impacts on resource- and non-resource-associated fauna would typically address physical impacts, noise, light and particle plumes. Over the past decade, several projects and initiatives have identified potential impacts of DSM, such as the extent and impact of sediment or water plumes away from the areas directly mined and their potential toxicity (Managing Impacts of Deep-Sea Resource Exploitation, 2016). Some of the expected impacts on ecosystems (see table below) include limiting connectivity between populations, interference with the life cycle of species, behavioural changes, loss of species and habitat, impacts on ecosystem structure and functioning and impacts on the water column chemistry. Several deep-sea species and ecosystems show vulnerable characteristics, such as maturation at a relatively old age, slow growth rates, long life expectancies and low or unpredictable recruitment.

Recent work has highlighted the role of all sea floor resources as critical habitats for communities. A wide range of fauna habitats are associated with PNs, which are the dominant hard substrate on the Clarion-Clipperton Zone abyssal plain (Vanreusel and others, 2016; Simon-Lledó and others, 2019a). Active hydrothermal vent ecosystems are unusual, fragmented habitats colonized by endemic chemosynthetic organisms and mostly rare species (Van Dover and others, 2018). SMSs associated with inactive vent fields are not well studied, but the existing literature identifies the presence of cold-water corals and sponges, whose dependence on microbial communities are as yet to be determined (Boschen and others, 2016; Van Dover, 2019). CFCs on seamounts host various ecosystems depending on location and water depth, which include cold water corals and sponges and other habitat-forming species (Rowden and others, 2010).

At a recent workshop, held with the aim of evaluating the nature of mid-water mining plumes and their potential effects on mid-water ecosystems, it was reported that seabed mining activities might affect mid-water organisms in a number of ways but that the scale of potential perturbations was still unclear, and further research on pelagic fauna was called for, especially in the mid-water bathyal and abyssal pelagic realms (Drazen and others, 2019).

Although there has been no commercial-scale mining, deep-sea experiments simulating mining activities have been carried out. The first commercial test mining for PNs was carried out in 1970. Since then, there have been a number of small-scale commercial test mining or scientific disturbance events designed to simulate mining. The results of the simulated mining impacts set a lower bound on the likely intensity of mining disturbance effects and the timescales required for benthic community recovery (Jones and others, 2017, and references therein). Those designed to mimic PN mining provided insights into recovery processes following small-scale disturbance events, which occurred up to 26 years ago, on abyssal plains (Gollner and others, 2017; Jones and others, 2017). Results of those studies showed that large sessile fauna experience very slow recovery after disturbance (see also Vanreusel and others, 2016) and consistently showed major impacts and lack of faunal recovery even over decadal time periods (Jones and others, 2017). While the impacts on the nodule-dwelling fauna were entirely expected, as nodules take millions of years to form, there are also important impacts on the organisms inhabiting the sediments in and near disturbed tracks (Simon-Lledó and others, 2019b).

Seabed mining pressures, potential impacts on different habitats and ecosystem services that might be affected[a]

<table>
<tr><th>Pressure</th><th>Potential impact</th><th>Affected ecosystem services</th><th>Habitat</th></tr>
<tr><td>Extraction of sea floor substrate</td><td>– Loss of benthic fauna by direct removal
– Changes in sediment composition
– Habitat loss or degradation
– Stress induced on fauna</td><td rowspan="6">Supporting
– Nutrient cycling
– Circulation
– Chemosynthetic production
– Secondary production
– Biodiversity

Regulating
– Carbon sequestration
– Biological regulation
– Nutrient regeneration
– Biological habitat formation
– Bioremediation and detoxification

Provisioning
– CO_2 storage
– Fisheries
– Natural products</td><td>– Benthopelagic
– Benthic</td></tr>
<tr><td>Extraction plume</td><td>– Loss of or damage to benthic species by smothering of organisms (from macrofauna to microorganisms)
– Behavioural changes in animals
– Changes in sediment composition
– Changes in seabed morphology</td><td>– Benthopelagic
– Benthic</td></tr>
<tr><td>Dewatering plume</td><td>– Clogging of feeding, sensorial or breathing structure
– Mechanical damage to tissues
– Stress</td><td>– Pelagic
– Benthopelagic
– Benthic</td></tr>
<tr><td>Release of substances from sediments (extraction and dewatering plume)</td><td>– Toxicity
– Nutrient release
– Turbidity</td><td>– Pelagic
– Benthopelagic
– Benthic</td></tr>
<tr><td>Underwater noise</td><td>Disturbance of animals</td><td>– Pelagic
– Benthopelagic
– Benthic</td></tr>
<tr><td>Underwater light</td><td>Disturbance of animals</td><td>– Pelagic
– Benthopelagic
– Benthic</td></tr>
</table>

[a] See also, for example, Thurber and others, 2014.

3.2. Policies and legislation: new regulations and policies and international, regional and national developments

For nearshore deposits, an increase in sand and gravel extraction is foreseen for the next decades, possibly extending to water depths greater than 50 m. Tighter environmental regulation, accompanied by the development of more environmentally friendly extraction technologies, is also expected for marine aggregate exploitation (e.g., Ellis and others, 2017; Kaikkonen and others, 2018).

Environmental management standards and guidelines for DSM are in their infancy (Jones and others, 2019). ISA has adopted a "Mining Code" to regulate prospecting and exploration activities and will establish regulations for exploitation of minerals in the Area. The development of those regulations will move in parallel with that of standards and guidelines aimed at defining environmental objectives and establishing environmental thresholds. A crucial tool for the definition of environmental objectives is the adoption of regional environmental management plans in areas where exploration contracts exist. The first environmental management plan was developed by ISA in 2011 for PNs in the Clarion-Clipperton Zone and adopted in 2012.[13] Several workshops have been held or will soon be convened to develop criteria that will support the establishment of new regional environmental management plans.[14]

As can be ascertained from the comparative study of existing national legislation on DSM, as at 5 June 2018, a total of 31 States had provided ISA with information on or the texts of national legislation relevant to DSM activities.[15]

3.3. Data, information and knowledge gaps

The definition and accurate quantification of mining impacts on the water column and at the seabed level can be approached using specific environmental indicators that determine what represents good environmental conditions and appropriate thresholds for impacts. Currently, there is a significant lack of information on deep-sea ecosystems, deep-sea species basic life history and biological traits, the characteristics of future mining technologies and the response of deep-sea organisms to mining impacts. Thus, there may be unforeseen consequences to mining. The knowledge gaps can be grouped into three categories: biodiversity, connectivity, and functions and services (Miller and others, 2018; Thornborough and others, 2019). Information is still lacking on the basic components of each ecological system, the interactions among those components and the relationships of ecosystems to environmental gradients. This baseline ecological information is necessary to allow predictions of how biodiversity, species connectivity and ecosystem functions and services will respond to change.

---

13 See documents ISBA/17/LTC/7 and ISBA/18/C/22.

14 See document ISBA/24/C/3.

15 Available at www.isa.org.jm/national-legislation-database.

4. Economic and social impacts

4.1. Economic impacts

The economics of DSM is intimately linked to the state of mining technology and the increased demand for metals in cutting-edge technology applications. Of the three seabed mineral deposit types considered in the present chapter, PNs are the closest to being mined. This is due to a combination of relative ease of retrieval, as a result of the discrete nature of the nodules, and of expected growth in demand, in particular for cobalt and nickel for new green energy technologies. Consequently, the economic focus in the present section is placed specifically on nodules.

4.1.1. Economics of seabed mining for polymetallic nodules

Commercial mining activities for PNs will depend not only on the overall economics of the system, but also on the economics of individual stakeholders. While the potential revenue from the sales of metals would be sufficient financially to justify the substantial investments and operating costs associated with DSM, the first call on that revenue in relation to the Area has to be the administrative expenses of ISA. The remaining funds may be used for meeting other obligations under part XI of the United Nations Convention on the Law of the Sea[16] and the Agreement relating to the implementation of Part XI of the United Nations Convention on the Law of the Sea of 10 December 1982,[17] including the equitable sharing of benefits, in accordance with article 140 and article 160, paragraph 2 (g), of the Convention, and compensation to developing land-based producer countries, should the impact of DSM on metal prices affect them. Funding for environmental and regulatory monitoring and remediation should also be provided. Mining operations will occur only if the revenue remaining after payments to ISA (or their equivalents for DSM in areas under national jurisdiction) can cover operating costs and provide sufficient returns to entice investment. Initial investigations of the economics of DSM suggest that revenues may be capable of reaching such levels, but issues remain, including the level of funds required for meeting obligations under part XI of the Convention, liabilities for environmental damage and returns required by investors.

4.1.2. Metal revenues

While PNs contain many metals, currently only four occur at sufficiently high concentrations to justify the cost of extraction for metal processors. Manganese is by far the largest metal by mass, and is therefore an important part of the revenue stream, in spite of relatively low market prices. While cobalt, copper and nickel concentrations are lower, they command higher prices and thus provide significant sources of revenue.

Future metal prices are difficult to predict and may differ from those forecasts, an uncertainty that can create significant risk for investors. Cobalt and nickel are both expected to play significant roles in future energy storage solutions and, consequently, may experience high demand growth and upward trends in price. With regard to manganese, PN mining will add a large quantity of material onto a market of limited size and may itself put a significant downward pressure on prices.

4.1.3. Nodule collection, metal processing investments and operating costs

ISA has authority over activities at the marine mine site only, but costs beyond that jurisdiction must be considered when evaluating the

16 United Nations, *Treaty Series*, vol. 1833, No. 31363.
17 Ibid., vol. 1836, No. 31364.

economic factors that will have an impact on investments. For that reason, studies of the financial systems need to look at investments and costs at sea as well as downstream.

On the cost side, there is a need for large upfront investments and then ongoing operational expenditure. Economies of scale dictate a minimum size of operation that many experts think will involve the extraction and processing of between 1.5 and 3 million dry tons of nodules per year. A system generating 3 million dry tons per year would require approximately $4 billion in upfront investments, including about $300 million for exploration and feasibility studies, over $1.5 billion for the nodule collection equipment and a dedicated transport system, and over $2 billion for a metal processing plant. Annual operational expenses are estimated to be in the order of $1 billion per year, with roughly one third going towards nodule collection and the remaining two thirds towards metal processing. Metal revenues for an operation of 3 million dry tons per year, with consideration of metallurgical losses and industry forecasts for long-term metal prices, would yield approximately $2.5 billion in annual revenues.

4.1.4. Distribution of funds across stakeholders

It remains to be seen whether the level of revenue remaining after meeting the requirements of part XI of the Convention on the Law of the Sea will be sufficient to motivate all players to participate. It is expected that a market for nodules, where they are transferred from collectors to metal processors through a global trading clearance centre, will develop if and when they become available. However, until that market emerges, the system economics can be evaluated only by estimating the flow of revenues among all stakeholders. Nodule collectors will pay a royalty to ISA when they retrieve nodules from the seabed. Metal processors will pay nodule collectors for the resource (nodules). Metal processors will sell the final metal products on the global market to a variety of end consumers. These revenues will have to cover all their operational expenses, plus payment to collectors (and transport providers) to obtain nodules. Any excess will be subject to local corporate taxes. While ISA will not see those funds, local taxes could have a significant impact on the economics of other stakeholders and, therefore, must be considered when evaluating whether the system generates sufficient returns to justify investment.

Collectors will receive payments for the nodules from the metal processors but will have to cover their operational costs and provide royalty payments to ISA (or local authority, if the operation is within an EEZ) for the rights to exploit the nodules. They may also have to contribute to environmental sustainability funds and provide bonds as guarantees for unanticipated environmental damages. Any profits may be subject to taxes from their sponsor State and may be subject to additional royalty payments to ISA. ISA will receive funds in the form of royalty payments for the rights to the nodules. ISA may also be the guardian of any sustainability funds or environmental liability bonds. Royalty payments must be sufficient to compensate for relinquishing the rights to the nodules plus any other changes to the deep-sea environment.

A variety of royalty systems are under discussion at ISA, including fixed single or two-stage ad valorem systems, variable ad valorem systems where the rate changes with metal prices or other financial conditions of the market, and combination systems with a fixed ad valorem rate and an additional rate tied to profits. Ad valorem systems are those in which the royalty is tied to the value of the metal retrieved. Each system has different pros and cons, in particular with regard to which stakeholders bear the risks and benefit from the rewards of changes in metal prices and project costs, and the timing of revenues to each stakeholder.

4.1.5. Returns to investors and cash flows to the International Seabed Authority

The very high upfront investments will require the nodule collectors and metal processors to raise funds on global capital markets. It is estimated that financiers will invest only if their rate of return on investment is around 18 per cent for most reasonable future metal price and cost scenarios. By contrast, investments in traditional land mining often require rates of return above 15 per cent but involve considerably lower levels of technological risk. A variety of royalty systems and rates would leave sufficient revenues for the contractors to achieve those rates of return. It is, however, still unclear whether any of those would provide sufficient revenue to ISA to compensate for the removal of the nodules and changes to the deep-sea environment.

4.2. Social impacts

The potential social impacts of DSM activities are understood to be both complex and cumulative (Koschinsky and others, 2018). While DSM may result in a myriad of social impacts, there is a general view that the direct impact on society will be less than that of terrestrial mining (Roche and Bice, 2013). For example, terrestrial mining projects often result in community displacements, changes in land use and the need to construct infrastructure, such as roads and railways (World Bank, 2017b). Unsafe working conditions (e.g., occupational hazards) and risks to the safety and general health of communities living close to mine sites (e.g., as a result of on-site disasters or exposure to air and water pollution caused by mining) are also noteworthy impacts of terrestrial mining (International Resource Panel, 2020). Those factors will not play a part in DSM. Moreover, mineral deposits on the seabed tend to contain higher metal contents than those on land, and a shift of focus to the seabed as a supplemental source for metals would reduce the need to mine terrestrial sites more extensively (Sharma and Smith, 2019).

One pertinent consideration to ascertain the possible social impacts of DSM is the planned location of related activities. It is becoming increasingly clear that the areas typically associated with seabed mineral deposits are located far from human communities. Thus, issues of relocation or conflict pertaining to land use do not arise with respect to DSM, as opposed to terrestrial mining (Sharma and Smith, 2019). Furthermore, social impacts related to DSM in the Area will differ from those within national jurisdictions. However, it is acknowledged that DSM activities may come into conflict with other uses of the marine space, such as fisheries, shipping, deep-sea cables, nursing grounds and migratory routes.

Owing to the nature of DSM and the possibility of transboundary harm, the sections of society that could be directly affected by related activities include the communities in the territory where those activities are carried out and those in adjacent coastal States (Dunn and others, 2017). Because the Area and its mineral resources have been declared the common heritage of mankind, social impacts need to be considered as a whole (Hunter and others, 2018). Notwithstanding how far any activities in the Area would be from population centres, significant concerns remain as to how the loss of biodiversity and ecosystem services, including the role of the deep sea in climate regulation, could have a negative impact on society as a whole (Kaikkonen and others, 2018).

When considering social impacts, an approach that encompasses the benefits that society might be able to reap from DSM as well as potentially negative consequences would provide important information on which to base decisions. This approach might include the distribution of financial benefits through a benefit-sharing mechanism, as well as the introduction of an additional source of metal supply to meet current and future demands. It should be recognized that, while a new source of metal supply might be beneficial, there could be negative consequences, such as for countries the economy of which relies heavily on the export of metals obtained from terrestrial mining, and, in accordance with article 151, paragraph 10, of the Convention on the Law of the Sea and section 7, paragraph 1, of the annex to the Agreement relating to the implementation of part XI thereof, those consequences need to be studied and addressed.

In addition to the above, the concept of a "social licence to operate" deserves particular attention. This concept includes acceptance by society, in addition to the permission required from the regulator, that a commercial activity such as resource extraction may be undertaken (Owen and Kemp, 2013; Parsons and Moffat, 2014). Issues of transparency and the inclusion of broad stakeholder participation in decision-making are also of particular interest (Ardron and others, 2018; Madureira and others, 2016).

Lastly, in order to ensure that all external costs to society arising from DSM are internalized, the incorporation of the polluter pays principle into ISA regulatory framework might be considered as an approach (Lodge and others, 2019).

5. Capacity-building needs

There exists a crucial need for capacity-building in deep-sea biodiversity research and conservation, as well as offshore mineral deposit identification and assessment, especially in developing States. Exploration techniques and exploration technology relating to CFCs are significantly lagging behind those relating to SMSs and PNs.

Another key component now needed for offshore mining is an expanded collection of baseline data, especially in respect of the characterization of ecosystems and their components, as well as natural variations of environmental baselines, including for the shallow-water continental shelf and the deep sea. Lastly, there is also a clear need for transparent and inclusive regulatory capacity development to avoid, reduce and mitigate impacts to ecosystems and for long-term online monitoring of the impacts of mining.

In 2019, the African Group submitted to the ISA Assembly a document on training programmes for developing countries in which it underlined capacity-building and developmental needs.[18]

In a recent assessment report on the review of the capacity-building programmes and initiatives implemented by ISA,[19] the ISA secretariat detailed the work of the Authority with respect to capacity-building. The secretariat examined in that report the core capacity-building themes implemented by ISA thus far, namely, the contractor training programme, the Endowment Fund for Marine Scientific Research in the Area and the internship programme. That report, along with others, was the subject of the International Workshop on Capacity Development, Resources and Needs Assessment held in Kingston from 10 to 12 February 2020. A summary of the workshop is available on the ISA website.[20]

Strategic partnerships between the United Nations and regional institutions focused on the creation of platforms to strengthen international cooperation for capacity-building programmes address some particular issues faced by developing countries and help to create common ground for improving actions. Within the framework of ISA, the need

18 See document ISBA/25/A/8.
19 Available at www.isa.org.jm/files/2020-02/assessment.pdf.
20 Available at www.isa.org.jm/files/2020-02/outcomessummary_0.pdf.

to expand opportunities for developing States to participate in activities in the Area is recognized.[21] While programmes for the training of personnel of ISA and developing States remain a contractual obligation for entities that have entered into exploration contracts with the Authority, monitoring the positive impacts and new opportunities that those programmes may have created for those countries represents a challenge.

References

Ardron, Jeff A., and others (2018). Incorporating transparency into the governance of deep-seabed mining in the Area beyond national jurisdiction. *Marine Policy*, vol. 89, pp. 58–66.

Banerji, A. (2019). India plans deep dive for seabed minerals. *Marine Technology Magazine*, 2019.

Beaulieu, S.E. (2015). *InterRidge Global database of Active Submarine Hydrothermal Vent Fields*. Prepared for InterRidge, Version 3.3, kml file produced 16 September 2015. Available at http://vents-data.interridge.org.

Boschen, Rachel E., and others (2016). Seafloor massive sulfide deposits support unique megafaunal assemblages: implications for seabed mining and conservation. *Marine Environmental Research*, vol. 115, pp. 78–88.

Dunn, D.C., and others (2017). Adjacency: How legal precedent, ecological connectivity, and Traditional Knowledge inform our understanding of proximity. https://nereusprogram.org/wp-content/uploads/2018/09/BBNJ-Policy-brief-adjacency_v5.pdf.

Drazen J.C., and others (2019). Report of the workshop Evaluating the nature of midwater mining plumes and their potential effects on midwater ecosystems. Research Ideas and Outcomes 5, e33527. https://doi.org/10.3897/rio.5.e33527

Ellis, J., and others (2017). Environmental management frameworks for offshore mining: the New Zealand approach. *Marine Policy*, vol. 85, pp. 178–192.

Gavriletea, Marius Dan (2017). Environmental impacts of sand exploitation. Analysis of sand market. *Sustainability*, vol. 9, No. 7, art. 1118.

German, Christopher R., and others (2016). Hydrothermal exploration of mid-ocean ridges: where might the largest sulfide deposits be forming? *Chemical Geology*, vol. 420, pp. 114–126.

Gollner, Sabine, and others (2017). Resilience of benthic deep-sea fauna to mining activities. *Marine Environmental Research*, vol. 129, pp. 76–101.

Gonçalves, D.S., and others (2014). Morphodynamic evolution of a sand extraction excavation offshore Vale do Lobo, Algarve, Portugal. *Coastal Engineering*, vol. 88, pp. 75–87.

Graedel, Thomas E., and others (2015). On the materials basis of modern society. *Proceedings of the National Academy of Sciences*, vol. 112, No. 20, pp. 6295–6300.

Hein, James R., and others (2013). Deep-ocean mineral deposits as a source of critical metals for high- and green-technology applications: Comparison with land-based resources. *Ore Geology Reviews*, vol. 51, pp. 1–14.

Hein, James R., and others (2015). Critical metals in manganese nodules from the Cook Islands EEZ, abundances and distributions. *Ore Geology Reviews*, vol. 68, pp. 97–116.

Hein, James R., and others (2017). Arctic deep water ferromanganese-oxide deposits reflect the unique characteristics of the Arctic Ocean. *Geochemistry, Geophysics, Geosystems*, vol. 18, No. 11, pp. 3771–3800.

[21] See document ISBA/24/A/10.

Hein, James R., and others (2009). Seamount characteristics and mine-site model applied to exploration-and mining-lease-block selection for cobalt-rich ferromanganese crusts. *Marine Georesources and Geotechnology*, vol. 27, No. 2, pp. 160–176.

Hein, James R., and others (2005). Marine mineral resources of Pacific Islands-a review of the Exclusive Economic Zones of islands of US affiliation, excluding the State of Hawaii.

Hoagland, Porter, and others (2010). Deep-sea mining of seafloor massive sulfides. *Marine Policy*, vol. 34, No. 3, pp. 728–732.

Hunter, Julie, and others (2018). Broadening common heritage: Addressing gaps in the deep sea mining regulatory regime. *Harvard Environmental Law Review*, vol. 16. https://harvardelr.com/2018/04/16/broadening-common-heritage.

International Council for the Exploration of the Sea (ICES) (2018). *Interim Report of the Working Group on the Effects of Extraction of Marine Sediments on the Marine Ecosystem (WGEXT), 16–19 April 2018, Copenhagen, Denmark.* ICES CM 2018/HAPISG:05.

________(2019). *Working Group on the Effects of Extraction of Marine Sediments on the Marine Ecosystem (WGEXT).* ICES Scientific Reports, vol. 1, No. 87. http://doi.org/10.17895/ices.pub.5733.

International Resource Panel (2020). Mineral Resource Governance in the 21st Century: Gearing extractive industries towards sustainable development. Ayuk, E.T., and others, A Report by the International Resource Panel. United Nations Environment Programme, Nairobi.

Jamieson, John W., and others (2017). Seafloor Massive Sulfide Resources. In *Encyclopedia of Maritime and Offshore Engineering*, pp. 1–10. American Cancer Society. https://doi.org/10.1002/9781118476406.emoe579.

Jones, Daniel O.B., and others (2017). Biological responses to disturbance from simulated deep-sea polymetallic nodule mining. *PLoS One*, vol. 12, No. 2, e0171750.

Jones, Daniel O.B., and others (2019). Existing environmental management approaches relevant to deep-sea mining. *Marine Policy*, vol. 103, pp. 172–181.

Kaikkonen, Laura, and others (2018). Assessing the impacts of seabed mineral extraction in the deep sea and coastal marine environments: current methods and recommendations for environmental risk assessment. *Marine Pollution Bulletin*, vol. 135, pp. 1183–1197.

Kamilli, Robert J., and others (2017). Tin. Report 1802S. Professional Paper. Reston, United States. USGS Publications Warehouse. https://doi.org/10.3133/pp1802S.

Kim, Junbeum, and others (2015). Critical and precious materials consumption and requirement in wind energy system in the EU 27. *Applied Energy*, vol. 139, pp. 327–34. https://doi.org/10.1016/j.apenergy.2014.11.003.

Koschinsky, Andrea, and others (2018). Deep-sea mining: Interdisciplinary research on potential environmental, legal, economic, and societal implications. *Integrated Environmental Assessment and Management*, vol. 14, No. 6, pp. 672–691.

Kuhn, Thomas, and others (2017). Composition, formation, and occurrence of polymetallic nodules. In *Deep-Sea Mining*, pp. 23–63. Springer.

Lodge, Michael W., and others (2019). Environmental Policy for Deep Seabed Mining. In *Environmental Issues of Deep-Sea Mining*, pp. 347–379. Springer.

Madureira, Pedro, and others (2016). Exploration of polymetallic nodules in the Area: Reporting practices, data management and transparency. *Marine Policy*, vol. 70, pp. 101–107.

Managing Impacts of Deep-Sea Resource Exploitation (2016). www.eu-midas.net.

McLellan, Benjamin C., and others (2016). Critical minerals and energy–impacts and limitations of moving to unconventional resources. *Resources*, vol. 5, No. 2. https://doi.org/10.3390/resources5020019.

Miller, K.A., and others (2018). An overview of seabed mining including the current state of development, environmental impacts, and knowledge gaps. *Frontiers in Marine Science*, vol. 4, art. 418. http://doi.org/10.3389/fmars.2017.00418.

Ministry of Economy, Trade and Industry (METI) of Japan (2017). www.meti.go.jp/english/index.html.

Owen, John R., and Deanna Kemp (2013). Social licence and mining: A critical perspective. *Resources Policy*, vol. 38, No. 1, pp. 29–35.

Parsons, Richard, and Kieren Moffat (2014). Constructing the meaning of social licence. *Social Epistemology*, vol. 28, Nos. 3–4, pp. 340–363.

Petersen, Sven, and others (2016). News from the seabed–Geological characteristics and resource potential of deep-sea mineral resources. *Marine Policy*, vol. 70, pp. 175–187.

Qin, Ya-Chao, and others (2014). Offshore aggregates resources on the northern continental shelf of the East China Sea. *Resource Geology*, vol. 65, No. 1, pp. 39–46. https://doi.org/10.1111/rge.12052

Ramirez-Llodra, Eva, and others (2011). Man and the Last Great Wilderness: Human Impact on the Deep Sea. *PLOS ONE*, vol. 6, No. 8, pp. 1–25. https://doi.org/10.1371/journal.pone.0022588.

Roche, Charles, and Sara Bice (2013). Anticipating social and community impacts of deep sea mining. *Deep Sea Minerals and the Green Economy, Secretariat of the Pacific Community, Suva*, pp. 59–80.

Rowden, Ashley A., and others (2010). A test of the seamount oasis hypothesis: seamounts support higher epibenthic megafaunal biomass than adjacent slopes. *Marine Ecology*, vol. 31, pp. 95–106.

Sharma, Rahul, and Samantha Smith (2019). Deep-sea mining and the environment: an introduction. In *Environmental Issues of Deep-Sea Mining*, pp. 3–22. Springer.

Simon-Lledó, Erik, and others (2019a). Biological effects 26 years after simulated deep-sea mining. *Scientific Reports*, vol. 9, art. 8040. https://doi.org/10.1038/s41598-019-44492-w.

Simon-Lledó, Erik, and others (2019b). Ecology of a polymetallic nodule occurrence gradient: Implications for deep-sea mining. *Limnology and Oceanography*, vol. 64, No. 5, pp. 1883–94. https://doi.org/10.1002/lno.11157.

Thornborough, K.J., and others (2019). Towards an ecosystem approach to environmental impact assessment for deep-sea mining. In *Environmental Issues of Deep-Sea Mining*, pp. 63–94. Springer, Cham.

Thurber, Andrew R., and others (2014). Ecosystem function and services provided by the deep sea. *Biogeosciences*, vol. 11, No. 14, pp. 3941–3963.

Torres, Aurora, and others (2017). A looming tragedy of the sand commons. *Science*, vol. 357, No. 6355, pp. 970–971.

Union européenne des producteurs de granulats (2018). *A Sustainable Industry for a Sustainable Europe Annual Review 2017–2018*. Brussels: European Aggregates Association.

United Nations (2017a). *The First Global Integrated Marine Assessment: World Ocean Assessment I*. Cambridge: Cambridge University Press.

________ (2017b). *The First Global Integrated Marine Assessment: World Ocean Assessment I*. Cambridge: Cambridge University Press.

United States Geological Survey (USGS) (2019). *Mineral Commodity Summaries 2019*. United States Geological Survey.

Van Dover, Cindy Lee, and others (2018). Scientific rationale and international obligations for protection of active hydrothermal vent ecosystems from deep-sea mining. *Marine Policy*, vol. 90, pp. 20–28.

Van Dover, Cindy Lee, and others (2019). Inactive Sulfide Ecosystems in the Deep Sea: A Review. *Frontiers in Marine Science*, vol. 6, art. 461. https://doi.org/10.3389/fmars.2019.00461.

Vanreusel, Ann, and others (2016). Threatened by mining, polymetallic nodules are required to preserve abyssal epifauna. *Scientific Reports*, vol. 6, art. 26808.

World Bank (2017a). The growing role of minerals and metals for a low carbon future. World Bank Publications, Washington, D.C. http://documents.worldbank.org/curated/en/207371500386458722/pdf/117581-WP-P159838-PUBLIC-ClimateSmartMiningJuly.pdf.

________ (2017b). Precautionary management of deep sea minerals. World Bank Publications, Washington, D.C. http://documents1.worldbank.org/curated/en/349631503675168052/pdf/119106-WP-PUBLIC-114p-PPDSMbackgroundfinal.pdf.

Zweibel, Ken (2010). The impact of tellurium supply on cadmium telluride photovoltaics. *Science*, vol. 328, No. 5979, pp. 699–701. https://doi.org/10.1126/science.1189690.

Chapter 19
Changes in hydrocarbon exploration and extraction

Contributors: Amardeep Dhanju (convener), Arsonina Bera, Kacou Yebowe Seraphim, Alan Simcock (co-lead member) and Joshua T. Tuhumwire (lead member).

Keynote points

- Since the first *World Ocean Assessment* (United Nations, 2017a), the offshore oil and gas sector has continued to expand globally, in particular in deep and ultradeep waters. The use of tension leg platforms, spars and floating production, storage and offloading (FPSO) systems are key to such expansion.
- In the next decade, frontier regions such as the eastern Mediterranean, the east coast of South America (Brazil and Guyana), and the west coast of Africa could be the major growth drivers for offshore oil and gas exploration and production.
- There is an upward trend in decommissioning activity, in particular in mature regions, such as the North Sea and the Gulf of Mexico.
- Exploration and production practices continue to evolve to minimize potential impacts on the surrounding environment.
- The creation of regulatory capacity to manage offshore resources effectively, especially in frontier regions, requires significant commitment and long-term institutional investment.
- Technological innovation and sophisticated industrial capability built over decades by the offshore oil and gas sector are benefiting the emergence of the marine renewable energy (MRE) industry.
- A major thrust to the offshore hydrocarbon sector since the first Assessment is technological advancement in analysing offshore exploration and production data to enhance operational and financial efficiencies.

1. Introduction

1.1. Scope

Chapter 21 of the first Assessment (United Nations, 2017b) provided a baseline state for the offshore hydrocarbon industry with regard to exploration and production trends, social and economic aspects, emerging technologies and potential future trends. It also covered environmental impacts associated with resource development and production activities and highlighted capacity gaps to assess impacts.

The present chapter contains an assessment of the current state of the global offshore hydrocarbon sector and presents some of the advances made in the field since the first Assessment. It describes exploration, production and decommissioning trends, includes an in-depth assessment of economic, social and environmental aspects, including potential impacts, and covers capacity-building gaps, in particular in emerging economies, and the crucial role of the offshore hydrocarbon industry in facilitating the MRE industry at the global level. Its content also relates to chapters 6D, 8, 9, 20, 21 and 26 of the present Assessment.

The present chapter is linked to five Sustainable Development Goals:[1] Goal 8 (Promote sustained, inclusive and sustainable economic growth, full and productive employment and decent work for all), Goal 9 (Build resilient infrastructure, promote inclusive and sustainable industrialization and foster innovation), Goal 12 (Ensure sustainable consumption and production patterns), Goal 13 (Take urgent action to combat climate change and its impacts) and Goal 14 (Conserve and sustainably use the oceans, seas and marine resources for sustainable development).

1 See General Assembly resolution 70/1.

1.2. Overview of global offshore hydrocarbon resources and production trends

Global crude oil production has grown steadily and exceeded 100 million barrels per day in 2018, while natural gas production has increased more rapidly, to 113.7 billion MBtu (million British thermal units) in 2016 (International Energy Agency, 2019).[2] Onshore oil and gas production continues to dominate, yet offshore oil production, which had been steady at around 27 million barrels per day for a decade, is showing an upward trend (Clemente, 2018). Meanwhile, offshore natural gas production has grown steadily in the past decade by 35 billion MBtu, with gains off the coast of Brazil and Australia, in the eastern Mediterranean and, most significantly, in the Persian Gulf, with the development of the massive North Field off the coast of Qatar (Davis, 2018). Natural gas production is projected to increase mainly from activities in shallow waters, while oil production increases will rely largely on drilling in deep and ultradeep water areas.

Offshore oil is produced in more than 50 countries, the most important producers being Saudi Arabia, the United States of America, Brazil, Mexico and Norway. More recently, large untapped resources have been discovered off the east coast of South America. According to the Organization of the Petroleum Exporting Countries (OPEC),[3] offshore oil production from Brazil and Guyana will compensate for falling production in other regions, although production in the exclusive economic zone of the United States in the Gulf of Mexico, the oldest offshore oil and gas producing region, may hold steady with discoveries in deep and ultradeep waters (OPEC, 2019).[4]

Figure I

Past and projected crude oil output in selected offshore producing areas

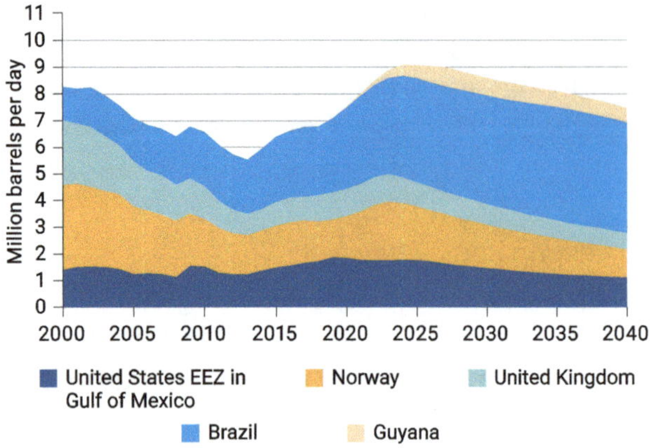

Source: OPEC (2019).

1.3. Advances in knowledge and capacity

New exploration and development in offshore areas remain a major source of increasing global oil and gas production. Technological advances in the past decade have encouraged exploration in deep and ultradeep waters further away from shore and enabled the discovery of significant new reserves. The water depth capabilities for offshore exploration increased from about 3,050 m to more than 3,350 m between 2010 and 2018, while production capability using floating platforms reached almost 2,900 m in 2018, up from 2,438 m in 2010 (Barton and others, 2019). Such technological advances have in part enabled the expansion of the offshore oil and gas sector to new regions, including the eastern Mediterranean and areas off the coast of Guyana.

There have also been advances in understanding the potential environmental and social impacts of exploration and production activities

2 Converted to 5.3 million barrels of oil equivalent per day.

3 OPEC member countries in 2020: Algeria, Angola, Congo, Equatorial Guinea, Gabon, Iran (Islamic Republic of), Iraq, Kuwait, Libya, Nigeria, Saudi Arabia, United Arab Emirates and Venezuela (Bolivarian Republic of).

4 Shallow water is generally understood as extending to no more than 300 m deep, and deep water as comprised between 300 and 1,500 m, while depths greater than 1,500 m are considered ultradeep water.

on the surrounding environment and in the development of new approaches to mitigate impacts. For example, the United Kingdom of Great Britain and Northern Ireland has created a Marine Noise Registry to record human activities that produce loud impulsive noise (10 Hz–10 kHz) in the seas around its territory.[5] This initiative intends to create baseline data and to quantify the pressure on the environment from anthropogenic activities associated with hydrocarbon exploration and development, including seismic surveys, sub-bottom profiling and pile driving. Similarly, the SERPENT project, which stands for "Scientific and Environmental ROV (remotely operated vehicle) Partnership using Existing iNdustrial Technology", is an example of international collaboration among the scientific community, environmental regulators and the oil and gas industry to gather and provide baseline information on ecosystems around offshore oil and gas installations using cutting-edge remotely operated vehicles that can operate in the deep ocean (SERPENT Project, 2020).

More recently, the offshore oil and gas industry has contributed to the MRE sector by providing expertise for the construction, maintenance and decommissioning of utility-scale offshore wind projects. The design and structural engineering concepts for the floating wind turbines, which can significantly expand the development of wind power in deeper waters associated with higher wind resources, are largely influenced by deepwater oil and gas installations (International Renewable Energy Agency, 2016).

2. Offshore hydrocarbon exploration, production and decommissioning

2.1. Offshore hydrocarbon technologies for survey and exploration

Oil and gas survey and exploration techniques locate hydrocarbon resources accumulated under impermeable rock formations. An initial assessment using seismic surveys evaluates the location of hydrocarbon-rich geologic plays (a group of oil- and gas-bearing rocks) that share a common history of hydrocarbon generation, migration and entrapment (Maloney, 2018; Bureau of OceanEnergy Management, 2017). This sets the stage for geological and geophysical surveys to obtain refined data on resource-bearing geological formations. Such surveys also provide an assessment of marine mineral, archaeological and benthic resources and any artificial structures buried and abandoned on the ocean floor.

Offshore seismic surveys use specialized vessels equipped with a combination of air guns and other acoustic sources. The equipment also includes hydrophones attached to a set of cables (streamers) towed behind the vessel. The acoustic sources produce a seismic pulse projected toward the ocean floor that reflects off the boundaries between various layers of rock. The reflected pulse is then recorded by the hydrophones and collected for analysis.

Recent advances in supercomputing and full waveform inversion technology are transforming resource estimation. Full waveform inversion, a new kind of processing technique applied to existing seismic data using supercomputers, creates a model of the subsurface rock layers in rich detail (Stratas Advisors,

5 See Joint Nature Conservation Committee, Marine Noise Registry Service. Available at https://mnr.jncc.gov.uk.

2019). Similarly, advances in four-dimensional seismic technology, coupled with superior computing power, now provide new insights into hydrocarbon reservoir characteristics, thus offering greater certainty to prospective resource developers.

2.2. Technological changes in drilling and production, including emerging technologies

Offshore drilling and production continue to benefit from significant technological advances. Sophisticated techniques now make it possible to drill multiple wells from a single drilling platform, while advances in real-time fibre-optic monitoring of the well bore is optimizing the reservoir performance and mitigating equipment failure risks (Beaubouef, 2019). Similarly, the use of predictive analytics and artificial intelligence tools is enhancing data analysis for detecting equipment breakdown and improving operational efficiency (Husseini, 2018).

The use of FPSO vessels enables drilling in areas further offshore and without ready access to a pipeline network to transport oil and gas onshore. It has also opened previously inaccessible hostile environments, in particular in the higher latitudes and in the Arctic, to exploration and development. FPSO vessels are equipped to store hydrocarbons onboard and periodically transfer their load to tankers for transportation onshore. They can also disconnect from their moorings in case of adverse weather conditions, such as cyclones and hurricanes. Once the reservoirs are depleted, an FPSO vessel can be redeployed to a new prospective site. The global market for FPSO vessels is currently boosted by large investments in deepwater exploration and development in such areas as the coast of Brazil (Rystad Energy, 2019). Meanwhile, FPSO vessel design is evolving to enhance safety, minimize complexity and reduce fabrication and operation costs (Barton, 2018).

Such technological advances have enabled exploration and production at uncharted depths and distance from shore. As of March 2019, the record for an ultradeep water exploration well was in depths of 3,400 m, off the coast of Uruguay, while the record for an operational production platform stood at 2,896 m, in the Gulf of Mexico (Barton and others, 2019).

2.3. Decommissioning techniques and trends

Although decommissioning regulations vary among jurisdictions, regulators increasingly require the complete removal of all drilling and production structures from the offshore environment. The Convention for the Protection of the Marine Environment of the North-East Atlantic[6] of 1992 (OSPAR) requires the removal of disused offshore installations, unless an exemption is provided to leave the entire installations or parts thereof in place (OSPAR Commission, 1992). Similarly, the Convention for the Protection of the Marine Environment and the Coastal Region of the Mediterranean[7] of 1995 provides the framework for decommissioning in the Mediterranean region and mandates the removal of all abandoned or disused installations. Other regions have adopted similar regulatory frameworks based on either regional conventions, such as the Protocol concerning Marine Pollution Resulting from Exploration and Exploitation of the Continental Shelf, under the Regional Organization for the Protection of the Marine Environment (ROPME) in the Middle East (ROPME, 1989) or, in the absence of a regional convention,

---

6 United Nations, Treaty Series, vol. 2354, No. 42279. The contracting parties to the Convention are Belgium, Denmark, the European Union, Finland, France, Germany, Iceland, Ireland, Luxembourg, the Netherlands, Norway, Portugal, Spain, Sweden, Switzerland and the United Kingdom of Great Britain and Northern Ireland.

7 United Nations, *Treaty Series*, vol. 1102, No. 16908.

the Guidelines and Standards for the Removal of Offshore Installations and Structures on the Continental Shelf and in the Exclusive Economic Zone of the International Maritime Organization (IMO) (IMO, 1989) which are based on article 60, paragraph 3, of the United Nations Convention on the Law of the Sea.[8] Regulations pertaining to pipelines vary. While some jurisdictions require complete removal, others deal with them on a case-by-case basis, depending on the hazards to fishing and navigation (International Association of Oil & Gas Producers, 2017). The Convention on the Prevention of Marine Pollution by Dumping of Wastes and Other Matter[9] of 1972 is the major international treaty protecting the marine environment from all sources of pollution, including the dumping of structures and wastes. A protocol to the Convention was adopted in 1996 to prohibit all abandonment (for the purpose of deliberate disposal) of human-made structures at sea, including the toppling of oil and gas platforms at site (IMO, 2020).

Decommissioning usually involves plugging the abandoned well, preparing the platform for removal by flushing and cleaning any residual hydrocarbons, cutting the pipes and cables between deck modules and mobilizing equipment, such as derrick barges and cranes, to dismantle and move the topside platform onshore for disposal. The process also involves removing the jacket or the foundation structure using heavy lifting equipment, a time-consuming and expensive process. Once on land, the structure is dismantled further for disposal or sold off as scrap.

The offshore decommissioning activity is largely concentrated in the North Sea, United States Gulf of Mexico and parts of the Asia-Pacific region. The steady depletion of legacy oil fields in the North Sea has created significant demand for decommissioning, which is expected to cost $32 billion between 2018 and 2022 (Wood Mackenzie, 2017). In the exclusive economic zone of the United States in the Gulf of Mexico, decommissioning is focused on platforms in shallow waters, while drilling and production move to deep and ultradeep waters.

Offshore platforms contribute hard structure to the marine environment and, in the process, provide food sources and complex physical habitat for a variety of organisms. Studies indicate higher levels of biological and fish productivity around platforms compared with natural reefs at similar depths (Shinn, 1974; Claisse and others, 2015). Because they recognize the ecological value of those structures, nations such as Brunei Darussalam and Malaysia are looking at converting obsolete platforms into artificial reefs in lieu of complete removal and disposal onshore, a process known as "rigs-to-reefs" programmes (Bull and Love, 2019). Rigs-to-reefs conversions are already being conducted in the United States, where obsolete platforms are "reefed" on a case-by-case basis in consultation with the coastal states. As of April 2018, 532 platforms previously installed on the United States outer continental shelf had been reefed in the Gulf of Mexico (Bureau of Safety and Environmental Enforcement, 2020).

In order to evaluate the decommissioning options, the State of California, United States, and other jurisdictions are proposing to use the net environmental benefit analysis (NEBA) as a tool to make decisions on reefing and removal options. NEBA is an analytical approach for comparing alternatives to a proposed action by including non-monetary environmental metrics, such as ecosystem services and values (Efroymson and others, 2004). It is conceivable that other jurisdictions may adopt NEBA or similar approaches to consider the environmental and ecosystem impacts of decommissioning options in a holistic manner.

---

8 United Nations, *Treaty Series*, vol. 1833, No. 31363.
9 Ibid., vol. 1046, No. 15749.

3. Economic, social, and environmental aspects of offshore hydrocarbon exploration, production and decommissioning

3.1. Economic and social impacts

Offshore oil and gas exploration and production are highly capital-intensive, with an estimated annual global investment expenditure of $155 billion in 2018 and projected investment of over $200 billion in 2021 (Sandøy, 2018). Engineering, procurement, construction and installation of drilling and production structures are the major areas of capital expenditure.

The specialized workforce in the offshore oil and gas sector draws significantly from a highly skilled global talent pool. Cities like Houston, United States, and Aberdeen, United Kingdom, have emerged as global hubs, not only serving the regional offshore industry but also providing expertise and services for projects around the world. The industry has also created a strong linkage with local communities, offering much valued business and employment opportunities, often in synergy with traditional activities. For example, shrimpers in the State of Louisiana, United States, rent out boats for offshore oil and gas activities during the lean fishing season (Priest, 2016), while some fishers supplement their income by working on the production platforms. According to the Office for Coastal Management of the United States National Oceanic and Atmospheric Administration (NOAA), offshore oil and gas activities in the United States contributed around $80 billion to the economy in 2016 and directly employed some 130,000 workers at an average salary of $153,000 per year, which is almost three times the national average wage (NOAA, 2018). Considering direct and indirect employment, more than 268,000 jobs are supported by oil and gas activities in the United States outer continental shelf (United States Department of the Interior, 2018). Meanwhile, in the United Kingdom, offshore oil and gas activities remain a significant source of skilled employment and supported some 259,900 jobs in 2018, which includes a significant number of indirect and induced jobs (Oil & Gas UK (OGUK), 2019). Offshore oil and gas activities in other regions also produce high levels of economic output and employ workers at above-average wage levels.

Offshore oil and gas production is maturing in many regions, especially in the North Sea and the shallow waters of the Gulf of Mexico. As the production declines and major oil reservoirs deplete beyond recovery, the industry expects to spend around $100 billion at the global level over the next decade on decommissioning activities (OGUK, 2018). This trend has the potential to create significant employment opportunities, some of which can offset the contraction of exploration- and production-related jobs.

3.2. Environmental impacts

Offshore oil and gas exploration and development practices have evolved significantly in terms of minimizing impacts on the surrounding environment, but operational and accidental discharges and other environmental impacts still occur. Operational discharges include chemicals that arise from drilling activities, produced water, drilling muds and cuttings, as well as small amounts of treated domestic and sanitary wastes. Noise, seabed disturbance and loss of biodiversity are frequent further significant impacts. In addition, the installation of pipelines and related infrastructure also contributes to certain discharges into the marine environment. The decommissioning of installations can also be carried out with more or less severe environmental impacts, depending on removal methodologies and subsequent environmental follow-up measures.

Produced water is a mix of oil and water from underground formations brought to the surface during production. The percentage of water, which is initially small, increases over time, while that of hydrocarbons decreases (Clark and Veil, 2009). The global average is estimated at three barrels of produced water for each barrel of oil (Khatib and Verbeek, 2002). Older wells, meanwhile, can display a ratio in excess of 50 barrels of produced water for each barrel of oil. According to a study by IFP Énergies Nouvelles, produced water is set to exceed 300 million barrels per day in 2020 at the global level, an increase of 20 per cent over 2008 levels. Most of the increase is expected from offshore oil and gas production (IFP Énergies Nouvelles, 2011).

Disposal options include injection into the same formation from where the oil is produced, treating the produced water to meet a certain quality standard and then either discharging it into the environment or using treated water in oil and gas field operations. While most of the treated produced water onshore is injected underground, in the offshore environment, it is discharged in the marine environment. Such discharges are often regulated by local or national water quality regulations, such as the Clean Water Act in the United States. The United States Department of Energy is currently investing $4.6 million to fund projects that would advance produced water treatment technologies (Department of Energy, 2019). Although the funded projects focus on land-based drilling, many advances will be relevant to offshore oil and gas production.

The emission of criteria pollutants related to platform or non-platform sources can have an impact on air quality in the vicinity of the drilling and production platforms. Platform sources comprise emissions from on-board equipment, such as boilers, natural gas engines and pneumatic pumps, while non-platform sources comprise emissions from pipe-laying operations, support and survey vessels and helicopters. In addition, open flaring of unwanted or excess gas from production platforms affects air quality. According to the World Bank, around 145 billion cubic metres of gas related to oil production was flared globally in 2018, equivalent to the total annual gas consumption of Central and South America (World Bank, 2019a). Multilateral initiatives such as the Global Gas Flaring Reduction Public-Private Partnership, led by the World Bank, aim at significantly reducing flaring at production sites. The Partnership promotes related research, disseminates best practices and works with national oil companies, regional and national Governments and international institutions to remove technical barriers to flaring reduction (World Bank, 2019b).

There have been significant improvements in oil spill forecasting and response and in the understanding of its impacts. Improvements in oil spill forecasting has been achieved through better visualization of the trajectory and fate of oil using expanded modelling suites, such as the General NOAA Operational Modelling Environment in the United States (NOAA, 2019). Similarly, project GRACE on integrated oil spill response actions and environmental effects, in the European Union, is investigating the hazardous impacts of oil spills and the environmental impacts of oil spill response technologies in cold climate conditions, such as the North Atlantic (Jørgensen and others, 2019). There are also advances in the use of satellites and other techniques for oil spill surveillance and monitoring; methods to evaluate the toxic effects of the spilled oil; and understanding impacts on corals, marine mammals and sea turtles to find the best ways to protect, rescue and restore marine wildlife and ecosystems affected by oil spills (NOAA, 2020).

4. Key knowledge and capacity-building gaps

4.1. Importance of long-term environmental monitoring and mitigation

The short-term impacts of oil and gas exploration and development on the marine environment have been studied extensively. However, the understanding of long-term effects is less complete. Long-term monitoring provides valuable insights into ecology, environmental changes and natural resource management (Lohner and Dixon, 2013). It also provides a systematic measurement of key environmental, social and economic indicators over time to design and implement effective policies and mitigation measures, while establishing a natural baseline for measuring trends over time. This baseline can then be used to assess changes due to ongoing drilling and production activities. Although establishing long-term monitoring programmes in the offshore environment is particularly challenging, the oil and gas industry and regulators are encouraging such programmes to assess changes and design effective mitigation strategies. For example, two observatory systems installed off the coast of Angola record long-term changes in the physical, chemical and biological environment caused by oil and gas development (Vardaro and others, 2013). Similarly, the long-term monitoring of the Flower Garden Banks National Marine Sanctuary in the Gulf of Mexico is one of the longest such programmes to monitor the health of coral reefs in the vicinity of operational oil and gas production facilities (NOAA, 2018). On a global scale, the SERPENT project mentioned above uses cutting-edge technology for the long-term monitoring of deep-sea corals habitats and other ecosystems. More such programmes are required to monitor long-term environmental impacts and ensure that resources are developed in an environmentally responsible manner.

4.2. Capacity-building gaps, especially in emerging economies

Offshore oil and gas exploration and production are expanding, sometimes to regions with minimal experience in managing those resources. Resource management in the offshore environment presents unique challenges for oil and gas resource managers to control access and encourage development. At its core, effective management requires the definition of property rights for offshore oil and gas resources within a country's exclusive economic zone.

National resource management systems are generally aimed at clarifying offshore jurisdiction, resolving multi-use conflict and implementing a regulatory framework for development in combination with laws on environmental protection, pollution prevention, health and safety standards, oil spill response and others. Regulatory frameworks tend to follow one of two approaches (Dagg and others, 2011), namely, prescriptive, whereby operators are told what to do; or performance- or goal-based, whereby goals to be achieved by operators are identified, but operators are allowed to choose how to achieve them.

These two approaches have benefits and drawbacks. Prescriptive regulations have the advantage of being relatively simple to implement and track, but they may stifle both innovation and creative solutions, owing to their emphasis on narrowly defined rules and regulations. Performance-based regulations, on the other hand, can create additional administrative burdens in terms of tracking regulations and verifying that goals have been met. The two approaches are often blended to create a hybrid regulatory system.

In creating a new offshore oil and gas regulatory framework, a jurisdiction can recalibrate its existing regulatory framework for land-based mineral development, while also adopting elements from jurisdictions with more established regulatory practices and significant experience in managing offshore oil and gas resources. This can be aided by capacity-building facilitated by multilateral institutions, such as the World Bank, and through exchanges of information among jurisdictions.

A regulatory framework could be subject to periodic reviews to assess economic impacts and other unintended consequences. This can be achieved by using the Regulatory Impact Analysis framework, which is used in many jurisdictions for routine assessment and is supported by such international entities as the Organization for Economic Cooperation and Development (OECD) (OECD, 2019). In addition, it is important to build good regulatory practices into the administration itself if the public or regulatory agencies are expected to carry out policies effectively and efficiently. Such practices require capacities to judge when, what and how much to regulate in order to allow for responsiveness to changing conditions and to ensure transparency, flexibility and policy coordination.

Creating capacity to manage offshore energy resources properly and effectively requires significant commitment and long-term institutional investment. The rewards, though, are commensurate, ensuring that resources are developed in a responsible manner and that economic benefits are distributed fairly.

5. Role of the offshore hydrocarbon industry in facilitating the marine renewable energy industry

The offshore oil and gas sector has built sophisticated industrial capability through technological innovation and decades of experience operating in some of the most challenging environments in the world. The emerging MRE industry, which includes wave, tidal, ocean current and offshore wind power, is now benefiting from the knowledge thus acquired. Offshore wind power, the most developed form of MRE, in particular, has made use of technology and skills perfected by the oil and gas sector. Wind turbine foundations and towers are engineered to withstand wave, wind, scouring and other forces that were first analysed when designing oil and gas platforms. Similarly, the experience gained from addressing the corrosive impact of salt water and sea spray on oil platforms has been applied to marinize and suitably modify terrestrial wind turbines for offshore installation (Breeze, 2016). Biofouling solutions for submerged oil and gas structures have been extensively researched and more recently applied to MRE structures. The installation of MRE transmission cables on the ocean floor has also drawn on technology and expertise first developed for laying submerged pipelines to serve offshore oil and gas platforms.

The large manufacturing infrastructure that serves the offshore oil and gas industry is now supporting the offshore wind industry. The jacket foundations of the first offshore wind project in the United States, off Block Island, were fabricated and supplied by a company in Louisiana with expertise in building structures for the offshore oil and gas industry in the Gulf of Mexico. Similarly, in the North Sea, extensive expertise in the oil and gas sector was harnessed to design and engineer floating offshore wind turbines for the Hywind Scotland project, where the installation of conventional bottom-founded turbines is not viable.

The oil and gas sector experience in maritime logistics is now shaping the MRE industry. In the United States, engineers have designed a marine vessel sufficiently versatile for both the installation of wind turbines and the decommissioning of oil and gas platforms (McGowan, 2018). Such initiatives enable significant cost saving for the development of MREs. The use of port infrastructure and service vessels are other examples of the leveraging of existing assets to facilitate the exploitation of new marine energy resources.

The MRE industry is considering the use of abandoned offshore oil and gas platforms to install wind turbines, although structural integrity concerns could impede such conversion plans. A potentially more viable option is the reconfiguration of abandoned platforms to convert MRE-generated electricity into hydrogen or synthetic gas, which could then be used to ride over low-wind or -wave periods and enhance the market potential of MRE projects. A simulated pilot project to test this concept was conducted by the Energy Delta Institute in the Netherlands in 2015 (Jepma and van Schot, 2016). Repurposing a platform has the additional benefits of delaying the expensive decommissioning costs while providing a new lease of life with positive economic returns. Another proposal envisions the supply of electricity produced by offshore wind turbines to oil and gas platforms for on-board operations, now commonly supplied by gas turbines located on board. A case study employing that approach in the North Sea concluded that this would lead to significant cost savings and a reduction in the emissions of criteria pollutants and greenhouse gases (Korpås and others, 2012). Subsequently, in 2019, Hywind Tampen, an 88 MW floating offshore wind project, was approved to provide electricity to oil and gas platforms in the North Sea (Oil & Gas Journal, 2020). Creating such synergies and leveraging oil and gas sector experience, expertise and infrastructure enable the burgeoning MRE sector to reduce costs and save time and resources.

6. Conclusion

Offshore oil and gas are an important contributor to global hydrocarbon production. Increasing global hydrocarbon demand, coupled with technological advances in offshore exploration and production, has pushed the industry to discover new reserves in ever deeper waters and challenging environments, often in areas with no prior resource development or in semi-closed seas, which are particularly vulnerable to environmental accidents. Global offshore hydrocarbon production, therefore, continues to increase, creating economic opportunities for coastal communities and much-needed leasing and royalty revenues for national Governments. It is important that new and existing offshore projects be managed in an environmentally responsible manner and that the decommissioning of obsolete facilities be undertaken in compliance with national regulations and regional marine environment conventions. A number of major trends have been observed since the first Assessment, including technological advances in collecting and analysing exploration and production data to enhance operational efficiencies, greater use of flexible platforms, such as FPSO systems, to expand production in unexplored areas, and a renewed push by the industry and regulators to minimize environmental impacts by deploying enhanced safety measures and using science to inform resource development.

References

Barton, Christopher M. (2018). FPSO market inches forward. *Offshore*, 1 August 2018. www.offshore-mag.com/field-development/article/16762275/fpso-market-inches-forward.

Barton, C., and others (2019). Worldwide progression of water depth capabilities for offshore drilling & production. *Offshore*, May 2019. https://digital.offshore-mag.com.

Beaubouef, Bruce (2019). Drilling technologies advance to meet challenging reservoir environments. *Offshore*, 25 September 2019. www.offshore-mag.com/drilling-completion/article/14040687/drilling-technologies-advance-to-meet-challenging-reservoir-environments.

Breeze, Paul (2016). *Wind Power Generation*. Academic Press.

Bull, Ann Scarborough, and Milton S. Love (2019). Worldwide oil and gas platform decommissioning: a review of practices and reefing options. *Ocean & Coastal Management*, vol. 168, pp. 274–306. https://doi.org/10.1016/j.ocecoaman.2018.10.024.

Bureau of Ocean Energy Management (2017). National Assessment of Undiscovered Oil and Gas Resources of the U.S. Outer Continental Shelf. OCS Report, BOEM 2017-085. www.boem.gov/sites/default/files/oil-and-gas-energy-program/Resource-Evaluation/Resource-Assessment/2016a-National-Assessment-of-Undiscovered-Oil-and-Gas-Resources.pdf

Bureau of Safety and Environmental Enforcement (2020). Rigs to Reefs. www.bsee.gov/what-we-do/environmental-focuses/rigs-to-reefs

Claisse, Jeremy T., and others (2015). Impacts from partial removal of decommissioned oil and gas platforms on fish biomass and production on the remaining platform structure and surrounding shell mounds. *PloS One*, vol. 10, No. 9, pp. e0135812.

Clark, C.E., and J.A. Veil (2009). Produced water volumes and management practices in the United States.

Clemente, Jude (2018). The quiet rise in U.S. offshore oil production. *Forbes*. 2018. www.forbes.com/sites/judeclemente/2018/04/10/the-quiet-rise-in-u-s-offshore-oil-production.

Dagg, Jennifer, and others (2011). Comparing the offshore drilling regulatory regimes of the Canadian Arctic, the US, the UK, Greenland and Norway. *The Pembina Institute*.

Davis, Carolyn (2018). Offshore Natural Gas Discoveries, Production Overtaking Oil. NGI's Daily Gas Price Index. 2018. www.naturalgasintel.com/articles/114290-offshore-natural-gas-discoveries-production-overtaking-oil?v=preview.

Department of Energy (2019). Department of Energy Invests $4.6M in Produced Water Treatment. Energy.Gov. 2019. www.energy.gov/fe/articles/department-energy-invests-46m-produced-water-treatment.

Efroymson, Rebecca A., and others (2004). A Framework for Net Environmental Benefit Analysis for Remediation or Restoration of Contaminated Sites. *Environmental Management*, vol. 34, No. 3, pp. 315–31. https://doi.org/10.1007/s00267-004-0089-7.

Husseini, Talal (2018). Big Data in oil and gas operations and other awesome tech advancements. *Offshore Technology: Oil and Gas News and Market Analysis*, blog, 22 October 2018. www.offshore-technology .com/features/big-data-in-oil-and-gas-tech.

IFP Énergies Nouvelles (2011). Water in fuel production: oil production and refining. Panorama. https://inis.iaea.org/collection/NCLCollectionStore/_Public/42/050/42050183.pdf?r=1

International Association of Oil & Gas Producers (2017). *Overview of International Offshore Decommissioning Regulations – Volume 1: Facilities*. Report 584. www.iogp.org/bookstore/product/overview-of-international-offshore-decommissioning-regulations-volume-1-facilities.

International Energy Agency (2019). Gas 2019: Analysis and forecasts to 2024. 2019. www.iea.org/reports/market-report-series-gas-2019.

International Maritime Organization (IMO) (1989). *1989 Guidelines and Standards for the Removal of Offshore Installations and Structures on the Continental Shelf and in the Exclusive Economic Zone.* IMO resolution A.672(16). https://cil.nus.edu.sg/wp-content/uploads/formidable/18/1989-Guidelines-and-Standards-for-the-Removal-of-Offshore-Installations-and-Structures-on-the-Continental-Shelf-and-in-the-Exclusive-Economic-Zone.pdf.

________ www.imo.org/en/OurWork/Environment/Pages/London-Convention-Protocol.aspx.

International Renewable Energy Agency (2016). Floating foundations: A game changer for offshore wind power. www.irena.org/-/media/Files/IRENA/Agency/Publication/2016/IRENA_Offshore_Wind_Floating_Foundations_2016.pdf.

Jepma, Catrinus, and Miralda van Schot (2016). Connect North Sea oil and gas platforms to offshore wind farms to produce green gas. Energypost.Eu, 22 January 2016. https://energypost.eu/connect-north-sea-oil-gas-platforms-offshore-wind-farms-produce-green-gas.

Jørgensen, Kirsten. S., and others (2019). The EU Horizon 2020 project GRACE: integrated oil spill response actions and environmental effects. *Environmental Sciences Europe*, vol. 31, No. 44. https://doi.org/10.1186/s12302-019-0227-8.

Khatib, Zara, and Paul Verbeek (2002). Water to Value – Produced Water Management for Sustainable Field Development of Mature and Green Fields. In *SPE-73853-MS*, p. 4. SPE: Society of Petroleum Engineers. https://doi.org/10.2118/73853-MS.

Korpås, Magnus, and others (2012). A case-study on offshore wind power supply to oil and gas rigs. *Energy Procedia*, vol. 24, pp. 18–26.

Lohner, Timothy W., and Douglas A. Dixon (2013). The value of long-term environmental monitoring programs: an Ohio River case study. *Environmental Monitoring and Assessment*, vol. 185, No. 11, pp. 9385–9396.

Maloney, Joseph (2018). What's on the Shelf? Assessing oil and gas resources on the OCS. *BOEM Ocean Science*, vol. 15, No. 2, www.boem.gov/Ocean-Science-Dec-Jan-Feb-Mar-2018.

McGowan, Elizabeth (2018). Oil industry expertise is helping to get offshore wind turbines in the water. Energy News Network, 21 June 2018. https://energynews.us/2018/06/21/northeast/oil-industry-expertise-is-helping-to-get-offshore-wind-turbines-in-the-water.

National Oceanic and Atmospheric Administration (NOAA) (2020). 8 Advances in Oil Spill Science in the Decade Since Deepwater Horizon. Office of Response and Restoration. https://blog.response.restoration.noaa.gov/8-advances-oil-spill-science-decade-deepwater-horizon

________ (2019). GNOME Suite for Oil Spill Modeling. Office of Response and Restoration. https://response.restoration.noaa.gov/gnome

________ (2018). *NOAA Report on the U.S. Ocean and Great Lakes Economy. Office of Coastal Management.* https://coast.noaa.gov/data/digitalcoast/pdf/econ-report.pdf.

Oil & Gas Journal (2020). Equinor, partners get green light for Hywind Tampen development. 8 April 2020. www.ogj.com/general-interest/article/14173631/equinor-partners-get-green-light-for-hywind-tampen-development

Oil & Gas UK (OGUK) (2018). *Decommissioning Insight 2018.* https://oilandgasuk.co.uk/wp-content/uploads/2019/03/OGUK-Decommissioning-Insight-Report-2018.pdf.

________ (2019). *Economic Report 2019.* https://oilandgasuk.co.uk/wp-content/uploads/2019/09/Economic-Report-2019-OGUK.pdf.

Organization for Economic Cooperation and Development (OECD) (2019). *Regulatory Impact Analysis.* www.oecd.org/regreform/regulatory-policy/ria.htm.

Organization of the Petroleum Exporting Countries (OPEC) (2019). *World Oil Outlook 2040.*

OSPAR Commission (1992). On the prevention and elimination of pollution from offshore sources. Annex III. www.ospar.org/site/assets/files/1169/pages_from_ospar_convention_a3.pdf.

Priest, Tyler (2016). Shrimp and Petroleum: The Social Ecology of Louisiana's Offshore Industries. *Environmental History*, vol. 21, No. 3, pp. 488–515. https://doi.org/10.1093/envhis/emw031.

Regional Organization for the Protection of the Marine Environment (ROPME) (1989). *Protocol Concerning Marine Pollution Resulting from Exploration and Exploitation of the Continental Shelf (1989)*. http://ropme.org/42_ROPME_PROTOCOLS_EN.clx.

Rystad Energy (2019). FPSO market is booming with Brazil fueling demand. www.rystadenergy.com/news events/news/press-releases/FPSO-market-is-booming-with-Brazil-fueling-demand.

Sandøy, Emil Varre (2018). Offshore oil and gas investments expected to grow starting in 2019. *Offshore*, 2 February 2018. www.offshore-mag.com/field-development/article/16762252/offshore-oil-and-gas-investments-expected-to-grow-starting-in-2019.

SERPENT Project (2020). Scientific and Environmental ROV Partnership using Existing Industrial Technology (SERPENT) project. www.serpentproject.com/

Shinn, Eugene A. (1974). Oil structures as artificial reefs. In *Proceedings of an International Conference on Artificial Reefs*, pp. 91–96. Texas A&M University.

Stratas Advisors (2019). Advances in Seismic Imaging Technology, Hart Energy. 2019. www.hartenergy.com/exclusives/advances-seismic-imaging-technology-177370.

United Nations (2017a). *The First Global Integrated Marine Assessment: World Ocean Assessment I*. Cambridge: Cambridge University Press.

________ (2017b). *The First Global Integrated Marine Assessment: World Ocean Assessment I*. Cambridge: Cambridge University Press.

United States Department of the Interior (2018). FY 2018 Economic Contributions, DOI Contributions by Bureau, Bureau of Ocean Energy Management. https://doi.sciencebase.gov/doidv/doi-bureau.html?bureau=Bureau%20of%20Ocean%20Energy%20Management

Vardaro, Michael F., and others (2013). A Southeast Atlantic deep-ocean observatory: first experiences and results. *Limnology and Oceanography: Methods*, vol. 11, No. 6, pp. 304–15. https://doi.org/10.4319/lom.2013.11.304.

Wood Mackenzie (2017). US $32 billion of decommissioning worldwide over the next five years: is the industry ready? www.woodmac.com/reports/upstream-oil-and-gas-us32-billion-of-decommission ing-worldwide-over-the-next-five-years-is-the-industry-ready-9599.

World Bank (2019a). Global Gas Flaring Reduction Partnership (GGFR). 2019. www.worldbank.org/en/topic/gas-flaring-reduction.

________ (2019b). Increased shale oil production and political conflict contribute to increase in global gas flaring. 2019. www.worldbank.org/en/news/press-release/2019/06/12/increased-shale-oil-production-and-political-conflict-contribute-to-increase-in-global-gas-flaring.

Chapter 20
Trends in inputs of anthropogenic noise into the marine environment

Contributors: Ana Širović (convener), Karen Evans (lead member), Carlos Garcia-Soto (co-lead member), John A. Hildebrand, Sergio M. Jesus and James H. Miller.

Keynote points

- The main anthropogenic noise sources in the ocean include vessels, industrial activity, including seismic exploration and renewable energy development, and sonar.
- Anthropogenic noise levels vary across space and time, the primary drivers being levels of human activity and propagation characteristics in the region. Noise does not persist once the sound source has been removed from the environment, although impacts can potentially persist.
- Areas with the highest levels of anthropogenic noise are those characterized by heavy industrial use, such as the Gulf of Mexico, the North Sea and the North Atlantic Ocean.
- Areas where anthropogenic noise is expected to increase include the Arctic, as the area opens up to shipping, and Africa, as investment in the region expands.
- Understanding of the impacts of anthropogenic noise on marine biodiversity is increasing, in parallel with a growing recognition of the need to monitor and possibly reduce the noise entering the marine environment.

1. Introduction

The last few decades have been characterized by an increased awareness of the importance of sound to marine life and a greater understanding of the potential impact of anthropogenic noise on such life. In the past 10 years, there has been an increased effort in some regions to develop guidelines and standards for monitoring and regulating the contribution of anthropogenic noise to the marine environment. While anthropogenic noise was not addressed as a stand-alone chapter in the first *World Ocean Assessment* (United Nations, 2017), it was the focus of a meeting of the United Nations Open-ended Informal Consultative Process on Oceans and the Law of the Sea.[1] Increasing awareness of its impacts warrants specific consideration in the present Assessment. The current chapter therefore presents a broad overview, including a description of the main sources of anthropogenic noise in the marine environment and the current state of knowledge on the status of such anthropogenic noise. In addition, as the main contributors of anthropogenic noise include shipping, energy generation, and oil and gas exploration and extraction, the chapters of the first Assessment addressing those activities are relevant here.

The United States Navy was an early source of ocean ambient noise data, making recordings that offer insight into ambient sound at frequencies below several hundred hertz (Hz) from the 1950s onwards (Ross, 2005). In addition to individual or small group research efforts, over the last decade, acoustic data has begun to be collected by ocean observing systems on a regional scale, starting with Neptune Canada, now part of Ocean Networks Canada, and the Australian Integrated Marine Observing System. Those observing systems began deploying hydrophones and collecting acoustic recordings in 2008 and 2009, respectively. More recently, the development of metrics and guidelines has also led to advances in impact assessments and modelling of ambient sound using alternative data sources that serve as proxies for major sources of anthropogenic noise, such as Automatic Information System (AIS) and impulsive noise registry data (e.g.,

1 See A/73/68.

Sertlek and others, 2019; United States National Oceanographic and Atmospheric Administration (NOAA) (2020) CetSound: Cetacean and sound mapping project).

At the same time, challenges remain in the measurement of ambient noise and modelling of acoustic propagation, as well as in the understanding of the impact that noise has on animal populations. Measurement challenges include the collection of calibrated data and the lack of standardization for both measurement and reporting. The American National Standards Institute/Acoustical Society of America and the International Organization for Standardization (ISO) have issued standards for measurement of underwater noise from ships, but the need for arrays of sensors to implement the standards has limited their application. The relatively high cost of deployment and recovery of underwater devices and even costlier installation of cabled systems are an additional impediment to data collection. From the modelling perspective, challenges include the lack of the fine-scale reliable data on environmental conditions needed for accurate models and the low spatial and temporal resolution of measured data for the validation of models. Finally, with regard to impact, work is under way to improve understanding of the hearing sensitivities of many species, in particular baleen whales, the cumulative effects of multiple noise sources and the impacts at the level of populations; however, practical difficulties remain.

2. Description of the environmental status

Sound is an efficient means of communication in the marine environment as sound waves travel very well through water, at speeds approximately five times higher than in the air. Nevertheless, the acoustic power is diminished as sound travels away from the source. Differences in absorption and spreading losses at different frequencies mean that lower sound frequencies travel further than higher frequencies. In addition, the properties of the environment affect sound propagation, ocean bottom and water properties affect the sound speed and bottom topography affects the direction of sound travel. In deep waters, special environmental conditions can result in the efficient propagation of sound in a deep channel or the convergence of sound at regular distances (Jensen and others, 2011). Unique propagation conditions, such as the waveguide effect or the Lloyd mirror effect, can contribute to the intensification of sound near the surface (Jensen and others, 2011), and bathymetric shielding can create large variability in sound intensity among nearby locations (McDonald and others, 2008).

Sound levels in the ocean, reported in units of decibels (dB), are calculated by referencing the measured sound pressure levels (in units of pascals) to one micropascal (dB re 1 μPa). Sound pressure levels are typically measured as instantaneous peak or peak-to-peak values or by calculating the root-mean-square of sound pressure for longer duration signals. Those differences in measurements result in sound pressure level differences of up to 4.5 dB. It should be noted that, since sound levels in air are calculated relative to 20 micropascals, ocean and air sound levels are not directly comparable. Higher acoustic impedance in water relative to air further contributes to a difference in measurements between those environments. As a result, a correction of 61.5 dB is required to compare airborne sound levels with those made underwater. When reporting noise levels, the calculation of power spectral density requires further normalization by the bandwidth of the signal and is thus typically reported in units of dB re 1 $\mu Pa^2/Hz$. Background ocean ambient sound levels in the absence of noise are not uniform across different frequencies, but range from 60 to 70 dB re 1 μPa²/Hz at frequencies below 100 Hz and decrease to below 40 dB re 1 μPa²/Hz at frequencies higher than 10 kilohertz (kHz) (Wenz, 1962). Particle motion, another component of sound waves, is more challenging to measure, but is an important consideration when evaluating the impact of sound on fish (Popper and Hawkins, 2019).

Major contributors to the ocean soundscape include geophysical sources, such as wind, waves, ice, volcanoes and earthquakes; biological sources, such as marine mammals, fishes and invertebrates; and anthropogenic sources. There are multiple sources of anthropogenic noise in the marine environment; the main ones include vessels (e.g., merchant ships, fishing vessels and recreational and cruise ships), industrial activity (e.g., offshore energy generation, including seismic exploration activity, coastal development and mining operations) and sonar (e.g., sonars used for fishing and for military and scientific purposes). In some cases, the production of sound is intentional and critical for the activity in question, as with seismic exploration and sonar, while in others it is incidental, as with shipping and coastal development. Anthropogenic noise levels are variable across space and time, two primary drivers being levels of human activity present and acoustic propagation characteristics in the region.

An overview of the main anthropogenic contributors to ocean ambient sound, the level for each source and the main frequency range is provided in the table below. Following the approach taken in other reviews of ocean noise, seismic survey activity is considered separately from other industrial activities, as it is a major contributor at low frequencies over large scales, with impacts that are substantially different from those of other industrial noise sources. A review of the impacts of noise on marine life is also provided. Among possible impacts considered here are physiological and behavioural effects, as well as impacts on mortality, when that was reported in the past. An important extension of those studies on the impact of noise on individuals, however, is an understanding of the consequences of acoustic disturbance at the level of populations, including cumulative effects (National Academies, 2017).

2.1. Marine traffic as a contributor to ocean noise

The dominant sources of sound emanating from marine vessels are cavitation and turbulence generated by propellers, but machinery is also a substantial component of the acoustic energy contribution, transmitted and radiated through the ship's hull (Ross, 1976). The flow noise generated as a ship advances through the water adds, at a lower level, to the vessel's contribution to ambient noise. The levels of contribution from the various components depend on a series of physical variables, including the ship's dimensions, tonnage, draft, load and speed, as well as wind and sea conditions, in as far as they interfere with the ship's movement in the water.

Marine traffic covers merchant shipping, cruise liners, military vessels, ferries, fishing boats and coastal boating for recreational purposes. Merchant shipping includes container ships, oil tankers, dry bulk carriers, general cargo ships and passenger liners. Different ship classes have distinct noise signatures that also depend on ship speed and length (Ross, 1976; McKenna and others, 2013). For example, a modern commercial container ship at a typical operating speed of 12 metres per second (m/s) has sound levels of 195 dB re 1μPa at 1 m with most acoustic energy below 100 Hz (Gassmann and others, 2017). In the case of smaller vessels (e.g., those below 20 m long, such as passenger and fishing boats, recreational high-speed boats, jet skis, etc.), radiated sound levels are lower (128–142 dB re

1 μPa at 1 m; Erbe, 2013) with a power spectrum including acoustic energy above 1 kHz (Erbe, 2013), resulting in propagation ranges shorter than those of commercial shipping.

Merchant shipping noise is often the main anthropogenic contributor to ocean noise at frequencies below 200 Hz (Wenz, 1962; Frisk, 2012; Roul and others, 2019). Globalization of the economy has resulted in a steep increase in merchant shipping throughout the world in the past 30 years. The global volume of seaborne trade has steadily increased (except in 1985 and 2009), reaching 10.7 billion tons in 2017 (United Nations Conference on Trade and Development (UNCTAD), 2018). Mean annual growth of 3.8 per cent was projected for the period 2018–2023; however, that could be affected by the COVID-19 pandemic. In addition to a steady increase in the volume of trade, vessels are also spending more time at sea, with an increase of 5 per cent recorded in 2017 (UNCTAD, 2018). The total gross tonnage has also increased in line with the volume of trade. Overall increases in merchant shipping are highly correlated with increases in ocean sound pressure levels, which rose by approximately 3 dB re 1 $\mu Pa^2/Hz$ per decade over the 10–50 Hz band throughout the last decades of the twentieth century (McDonald and others, 2006). That increase appears to have plateaued since the start of the twenty-first century (Frisk, 2012, and references therein).

The "distant shipping" component of ambient noise, which arises when signatures from individual vessels are indistinguishable in the data, but appear as increased acoustic energy at frequencies below 100 Hz (Wenz, 1962) at a given location and time, strongly depends on ship distribution at that moment. Shipping is unevenly distributed by latitude, with higher densities in the northern hemisphere along heavily used shipping lanes. As a result, high levels of ambient sound (80–90 dB re 1 $\mu Pa^2/Hz$ or more) at frequencies dominated by shipping (10–100 Hz) are typically found in the North Atlantic Ocean and the North Pacific Ocean (Ross, 2005; McDonald and others, 2006; Širović and others, 2013; 2016). In the Arctic, where shipping traffic is substantially lower, ambient noise at low frequencies is largely driven by environmental factors, such as sea ice cover and wind conditions (Roth and others, 2012). In coastal waters, near busy harbours and beaches, small and medium-sized fishing vessels, recreational boats and small ferries can also be important contributors to anthropogenic noise (Samuel and others, 2005; Merchant and others, 2012).

Ambient noise levels from distant shipping have not been linked to lethal, tissue-damaging or other direct physical injury in marine mammals (although see chap. 6D for other threats to marine mammals caused by shipping). Shipping and small craft noise has been associated with wide-ranging impacts on the survival, physiology and behaviour of individuals, with potential consequences for the survival of populations and communities across a number of marine taxa. In marine mammals, those include increased stress levels in North Atlantic right whales (*Eubalaena glacialis*) (Rolland and others, 2012); changes in the foraging behaviour of humpback whales (*Megaptera novaeangliae*) and their vocalizations during the breeding season (Blair and others, 2016; Tsujii and others, 2018); changes in harbour porpoise (*Phocoena phocoena*) behaviour (Dyndo and others, 2015); and changes in calling behaviour and the masking of or reduction in communication space (Parks and others, 2010; Putland and others, 2018). In other taxa, the impacts include increases in stress levels for a number of fish species (see, for example, Nichols and others, 2015; Simpson and others, 2016a), potentially resulting in an increased predation risk in some species (Simpson and others, 2016a), a reduced ability of fish and coral larvae to select suitable habitats (Simpson and others, 2008; 2016b) and the masking of and reduction in communication space (Putland and others, 2018; Weilgart, 2018 and references therein).

2.2. Seismic exploration as a contributor to ocean noise

The use of sound to image sub-sea floor geological structures is the predominant marine geophysical technique employed by the offshore oil and gas industry. Seismic reflection profiling provides information about potential oil and gas deposits several kilometres below the sea floor. To generate the high levels of sound needed to penetrate the solid earth, large arrays of airguns are towed behind survey vessels. Each airgun releases a volume of air under high pressure, creating a high intensity sound pressure wave. Typically, an array of airguns used in the seismic industry will include from 25 to 50 individual guns (Dragoset, 2000). The acoustic pressure signal of airgun arrays is focused vertically, producing a signal 12–15 dB stronger in the vertical direction for most arrays. The peak source level for those arrays is impossible to calculate at a standard 1 m reference but, according to a simplified estimate, if it is considered as a single source, it can reach 260 dB_{peak} re 1 µPa at 1 m (Turner and others, 2006). Seismic operations can be limited in duration (weeks to months) but, depending on bathymetry, can affect entire ocean basins as low frequency signals propagate over significant ranges.

Seismic surveys can also be conducted for research purposes, including outside of areas that are subject to commercial surveys, such as in the Southern Ocean. High resolution geophysical surveys are also conducted in coastal areas for the construction of critical infrastructure, such as bridges, ports and, more recently, offshore wind farms. Those surveys employ sound sources such as sparkers and Uniboom that are less powerful (210–230 dB re 1 µPa at 1 m) than airguns and operate in a higher frequency band (0.5–2.5 kHz; Gontz and others, 2006). While those surveys tend to be localized in both time and space, their impact may be relevant for sensitive inshore species and ecosystems.

Marine areas of all continents except Antarctica are undergoing active seismic exploration. The Gulf of Mexico has among the highest levels of activity in the world, with deepwater exploration the dominant source of low frequency ambient noise in that region (Wiggins and others, 2016). High activity has also occurred in the North Atlantic (Nieukirk and others, 2004), the South Atlantic (Miksis-Olds and Nichols, 2016; Haver and others, 2017) and the North Sea (Hildebrand, 2009). Seismic survey activity was increasing in the late 2000s and early 2010s owing to increasing prices of crude oil, in particular in such areas as the South Atlantic and the Mediterranean Sea (Maglio and others, 2016). The global average number of active seismic vessels increased from 40 in 2004 (Hildebrand, 2009) to 75 by 2014 (based on seismic crew records), with the highest levels of activity recorded in the Gulf of Mexico, Europe, Asia Pacific and Africa. However, following a decline in crude oil prices in 2015 and 2016, the number of active vessels had decreased to 58 by mid-2018 (GeoTomo, 2018).

The impacts on marine life of sound produced during seismic exploration surveys have been documented across a number of taxa, ranging from zooplankton to marine mammals. McCauley and others (2017) reported zooplankton depletion immediately following seismic operations, concurrent with an increase in dead zooplankton comprising a variety of species. Controlled experiments on scallop larvae showed that they exhibit significant developmental delays and developmental malformations if exposed to seismic airgun pulses (Aguilar de Soto and others, 2013), while adult scallops were observed to have disrupted reflexes (Day and others, 2016). Seismic operations may also be implicated in the stranding of giant squids (Guerra and others, 2004). Fish have been observed to exhibit behavioural and physiological changes as a result of seismic operations (Weilgart, 2018 and references therein), with changes in fish catch rates also reported (Løkkeborg,

1991; Løkkeborg and others, 2012). Seismic operations have been observed to have a negative impact on baleen whale communication (Di Iorio and Clark, 2009; Cerchio and others, 2014). While a number of impacts of seismic exploration on marine life have been observed, controlled exposure experiments have reported no observable impacts on the development and survival of southern rock lobster (*Jasus edwardsii*) embryos and Dungeness crab larvae (*Metacarcinus magister*) (Pearson and others, 1994; Day and others, 2016) and a limited effect on the copepod *Calanus finmarchicus* (Fields and others, 2019).

2.3. Industrial activity as a contributor to ocean noise

A comprehensive review of underwater noise from industrial activity was completed in 2003 by the National Research Council (NRC) of the United States of America. Below is a summary of the main findings of that report and of the research in the area of ocean industrial noise published since 2003. For the purposes of the present chapter, non-seismic oil and gas industry contributions have been separated from other industrial activity that contributes to marine noise.

2.3.1. Industrial noise from the oil and gas industry

As well as through seismic surveys, the purpose of which is to explore for oil and gas, the oil and gas industry also contributes noise during the drilling and production phases. Oil and gas industrial activities occur worldwide from latitudes 72° north to 45° south. Activities associated with seismic surveys and oil and gas production are present along the coastlines of all the continents of the world except Antarctica (NRC, 2003). The noise levels associated with oil and gas production and associated activities, such as the installation of pipelines, the generation of energy on platforms, pipeline flow and support vessel activity, are typically much lower than those associated with seismic surveying (Richardson and others, 1995). The impacts of that production noise can be restricted to areas near facilities, but persist during the active life of the facility, which can last for years (ibid.). Based on data collected along the North Slope of Alaska and the adjoining coast of Canada, ships actively engaged in drilling activity have high radiated sound levels with a maximum broadband source pressure level calculated from the root-mean-square of pressure across the 10 Hz–10 kHz band of about 190 dB_{rms} re 1 µPa at 1 m (Richardson and others, 1995).

2.3.2. Other industrial and construction contributions to ocean noise

The range of activities in this category is extremely broad. Pile-driving and power-generating wind turbines are often found in deeper waters, while dredging, coastal development and associated construction, shipyards and daily harbour functions located near the shore contribute noise in shallow waters. Deep seabed mining is still largely limited in scope because of prohibitive costs (Miller and others, 2018; Thompson and others, 2018), but may expand in future. The compound impact of various industrial activities, for example, a combination of terrestrially based, shoreline or nearshore sound sources, on the marine environment is poorly understood. Nevertheless, that broad range of industrial activities produces a range of source levels and acoustic patterns described in detail below.

Pile-driving typically consists of thousands of impacts by large hammers occurring about once a second to drive stabilizing structures for above-water structures into the seabed. Pile-driving noise source levels are substantial, with peak source levels ranging from 226 to 248 dB_{peak} re 1 µPa at 1 m (Bailey and others, 2014; Miller and others, 2017). There are a number of techniques for reducing propagated noise levels from pile-driving, including the use of freely rising bubble screens (Würsig and others, 2000), fixed air bubble screens (Rustemeier and others, 2011) and Helmholtz resonator screens (Lee and others, 2012). Deployment of those techniques has the potential to reduce received sound levels away from the activity by up to 20 dB, although average reductions are in the order of 5 dB (Buehler and others, 2015).

Operating offshore wind farms produce noise levels of about 150 dB re 1 µPa at 1 m (Nedwell and Howell, 2004; Hildebrand, 2009). That can represent a 5–25 dB increase in overall ambient sound levels at nearby locations (within approximately 1 km) (Norro and others, 2011). As with oil and gas facilities, the noise associated with wind farm construction, largely stemming from pile-driving activities, is limited in duration, but can affect large areas of the ocean. Once the wind farms are operational, noise generated by the operation will affect a smaller area, but will last throughout its exploitation.

In recent years, there has been renewed interest in commercial operations for extracting economically valuable metals from the deep sea, including in hydrothermal vent locations worldwide, with exploration undertaken in the Mid-Atlantic Ridge area around the Azores (see also chap. 18). The levels of sound those activities contribute to the deep sea are unknown.

Anthropogenic noise from dredging consists of sound from ship-borne machinery and mechanical motion, for example from suction and earth-moving devices, as well as the possible use of explosives. Noise levels recorded during dredging range from approximately 163 dB to 190 dB re 1 µPa at 1 m, depending on the type of dredging operation (Greene, 1985; Nedwell and others, 2008; Robinson and others, 2011; Reine and others, 2012; McQueen and others, 2020).

Those various industrial activities can have differing impacts on marine life. Impulsive noise such as that created by pile-driving has been observed to disrupt harbour porpoise habitat use (Carstensen and others, 2006) and has the potential to cause hearing impairment in marine mammals and fish close to the noise source (Madsen and others, 2006; Casper and others, 2013). The noise generated by pile-driving has been observed to increase metabolic rate in some fish and mussel species (Spiga and others, 2016; Bruintjes and others, 2017), as well as to alter fish swimming and schooling behaviour (Mueller-Blenkle and others, 2010; Herbert-Read and others, 2017) and elicit responses in squid (Jones and others, 2020). Vibrations of the seabed resulting from experiments designed to simulate pile-driving have also been observed to have a negative impact on the growth and body condition of bottom-dwelling mussels (Roberts and others, 2015). While fish and marine mammals can detect sounds from operating wind farms at distances of a few kilometres, it is not known if those sounds cause any disruptions to their biological functioning, although they were shown to disrupt crab settlement (Pine and others, 2012).

2.3.3. Ocean noise from sonar

Different types of sonars are used for mapping the ocean bottom and detecting and localizing various objects in the water column (e.g., plankton, fish or submarines). Sonar is used by the military, the commercial, charter and recreational fishing communities, and the scientific research community, among others. The type of use is different within each of those groups.

Sonar use in the military is primarily focused on anti-submarine warfare and involves two types of sonar: low-frequency active (LFA) sonar and mid-frequency active (MFA) sonar. LFA sonar operates in the 100–500 Hz band, with an overall source level of 230–240 dB re 1 µPa at 1 m, allowing detection over long ranges (hundreds of kilometres). MFA sonar operates at frequencies of 2–8 kHz, has a source level of 235 dB re 1 µPa at 1 m (Hildebrand, 2009) and operates over ranges of tens of kilometres. The

United States Navy has four ships dedicated to LFA sonar use, and there are approximately 300 MFA sonars in active service in the world's navies (Hildebrand, 2009).

In non-military uses, the sonars most frequently encountered on vessels include "fish finders" and other echo sounders, called multibeam sonars and side-scan sonars, operating at single or multiple frequencies. Sonars not used for military purposes generally operate at lower source levels than military sonars and, in most cases, their beams are directed downwards under the vessel track, or across the track in the case of multibeam sonars. The typical operating frequency of a fish finder is between 15 and 200 kHz. The multibeam mapping sonars typically used by the research community operate at frequencies ranging from 12 kHz for deepwater systems to 400 kHz for shallow-water systems, with narrow directional beams (approximately 1 degree) and source levels between 232 and 245 dB re 1 μPa at 1 m (Hildebrand, 2009).

The use of LFA sonar has been restricted by some countries owing to concerns about its impact on divers and marine mammals (Miller and others, 2000), although it has been reported that LFA sonar does not affect the behaviour of herring (Doksæter and others, 2012). The use of MFA sonar has been implicated in the stranding of multiple species of cetaceans (Balcomb and Claridge, 2001). Beaked whales appear to be particularly sensitive to that type of sonar, which has been associated with both physiological damage (Fernández and others, 2005) and behavioural changes in several beaked whale species (Tyack and others, 2011; DeRuiter and others, 2013; Moretti and others, 2014). Overall, however, responses vary by population, and there is some indication that beaked whales regularly exposed to MFA sonar may acclimate to the sound (Bernaldo de Quirós and others, 2019). Presence of MFA sonar has been observed to alter the behaviour of baleen whales (Goldbogen and others, 2013) and multiple odontocete species (Sivle and others, 2012). Beaked whales also appear to be sensitive to other forms of sonar, with observed changes in their behaviour documented in the presence of an echo sounder deployed for scientific purposes (Cholewiak and others, 2017).

Main sources of anthropogenic noise

Industry/sector	Sound source	Sound type	Source level (dB re 1 μPa at 1 m)	Frequency of main energy (kHz)
Commercial shipping				
Medium-sized ships (50–100 m)	Propeller/cavitation	Continuous	165–180[a]	< 1
Large vessels (e.g., supertankers and container ships)	Propeller/cavitation	Continuous	180–219[a]	< 0.2
Resource exploration and exploitation				
Oil and gas	Seismic airgun	Impulsive	220–262[c]	0.05–0.1
	Drilling	Continuous	124–190[a]	0.1–1
Renewable energy	Impact pile-driving	Impulsive	220–257[c]	0.1–2
	Operational wind farm	Continuous	144[a]	< 0.5
Navy	Low-frequency sonar	Impulsive	240[b]	0.1–0.5
	Mid-frequency sonar	Impulsive	223–235[b]	2.8–8.2
	Explosions (e.g., ship shock trials and exercises)	Impulsive	272–287[a]	0.006–0.02
Fishing	Propeller/cavitation	Continuous	160–198[a]	< 1–10
	Deterrent/harassment device	Impulsive	132–200[b]	5–30
	Sonar (echo sounder)	Impulsive	185–210[b]	20–260
Dredging	Propeller/cavitation, cutting, pumping, grabbing and digging	Mainly continuous	163–188[a]	0.1–0.5
Marine scientific research (e.g., research vessel)	Propeller/cavitation	Continuous	165–180[a]	< 1
Recreational activities (e.g., recreational craft and speedboat)	Propeller/cavitation	Continuous	160–175[a]	1–10
Tourism (e.g., whale and dolphin watching and cruise ships)				
Vessels (<50m– >100m)	Propeller/cavitation	Continuous	160–190[a]	< 0.2–10
Harbour construction	Impact pile-driving (e.g., sheet piling)	Impulsive	200[b]	0.1–0.5

Source: United Nations document A/73/68, annex.

[a] Root-mean-square sound pressure level.

[b] Peak sound pressure level.

[c] Peak-to-peak sound pressure level.

Main sources of anthropogenic noise

Industry/sector	Sound source	Sound type	Source level (dB re 1 µPa at 1 m)	Frequency of main energy (kHz)
Commercial shipping				
Medium-sized ships (50–100 m)	Propeller/cavitation	Continuous	165–180[a]	< 1
Large vessels (e.g., supertankers and container ships)	Propeller/cavitation	Continuous	180–219[a]	< 0.2
Resource exploration and exploitation				
Oil and gas	Seismic airgun	Impulsive	220–262[c]	0.05–0.1
	Drilling	Continuous	124–190[a]	0.1–1
Renewable energy	Impact pile-driving	Impulsive	220–257[c]	0.1–2
	Operational wind farm	Continuous	144[a]	< 0.5
Navy	Low-frequency sonar	Impulsive	240[b]	0.1–0.5
	Mid-frequency sonar	Impulsive	223–235[b]	2.8–8.2
	Explosions (e.g., ship shock trials and exercises)	Impulsive	272–287[a]	0.006–0.02
Fishing	Propeller/cavitation	Continuous	160–198[a]	< 1–10
	Deterrent/harassment device	Impulsive	132–200[b]	5–30
	Sonar (echo sounder)	Impulsive	185–210[b]	20–260
Dredging	Propeller/cavitation, cutting, pumping, grabbing and digging	Mainly continuous	163–188[a]	0.1–0.5
Marine scientific research (e.g., research vessel)	Propeller/cavitation	Continuous	165–180[a]	< 1
Recreational activities (e.g., recreational craft and speedboat)	Propeller/cavitation	Continuous	160–175[a]	1–10
Tourism (e.g., whale and dolphin watching and cruise ships)				
Vessels (<50m– >100m)	Propeller/cavitation	Continuous	160–190[a]	< 0.2–10
Harbour construction	Impact pile-driving (e.g., sheet piling)	Impulsive	200[b]	0.1–0.5

Source: United Nations document A/73/68, annex.

[a] Root-mean-square sound pressure level.

[b] Peak sound pressure level.

[c] Peak-to-peak sound pressure level.

3. Description of economic and social consequences and other economic or social changes

During the discussions of the United Nations Open-ended Informal Consultative Process on Oceans and the Law of the Sea on anthropogenic underwater noise in 2018, the importance of addressing the socioeconomic impacts of such noise was stressed. It has been shown, for example, that the presence of seismic airgun surveys reduces catches of gadid and sebastid fishes (Hirst and Rodhouse, 2000). That may result in short-term economic loss for concerned fisheries during seismic surveys. The impacts of noise on species that are of particular social, economic and cultural relevance may have socioeconomic effects on coastal communities, in particular if they alter the availability of commercially or recreationally important marine species. A similar decline of social and economic benefits may be expected in association with the displacement of marine mammals that are the focus of tourism activities. In addition, the displacement of marine animals may affect traditional and cultural practices of indigenous communities that rely on artisanal fishing and subsistence hunting. The area of interactions between anthropogenic noise and its impact on social and economic factors has not been well studied in the past, but an increased interest in anthropogenic noise in the ocean may lead to a greater focus on the human consequences of the increase in noise.

While anthropogenic underwater noise may be most obviously connected to the achievement of Sustainable Development Goal 14 (Conserve and sustainably use the oceans, seas and marine resources for sustainable development), it is also linked to a number of other Goals.[2] Ensuring access to affordable, reliable, sustainable and modern energy for all (Goal 7) is likely to lead to localized, short-term increases in anthropogenic noise levels in the ocean during the construction of offshore wind farms, but could result in an overall reduction in anthropogenic noise associated with a decrease in the need to exploit fossil fuels. The successful implementation of Goal 11, on sustainable cities and communities, and Goal 12, on responsible consumption and production, could ultimately affect overall anthropogenic noise in the ocean if the achievement of those goals results in changes in global shipping.

4. Key region-specific changes and consequences

4.1. Arctic Ocean

The opening up of shipping channels in the Arctic as a result of decreases in sea ice caused by climate change has started to result in increased ship traffic through the Arctic Basin (Eguíluz and others, 2016). While it is still a rather uncommon path, the Arctic is likely to become a more common shipping and tourism route in the future, as sea ice continues to recede (Smith and Stephenson, 2013). The consequences for local Arctic communities and marine animals of changes in shipping and, in particular, associated changes in soundscapes to more anthropogenically driven ones are largely unknown (Ho, 2010). Oil exploration in the Chukchi Sea began in the mid-2000s, but further exploration and development were abandoned when the region's reserves were found to be insufficient to warrant additional investment (Shell, 2015). Offshore oil and gas development in the Canadian Arctic is

2 See General Assembly resolution 70/1.

currently not allowed, with a review of the ban due in 2021 (Nunatsiaq, 2016).

4.2. North Atlantic Ocean, Baltic Sea, Black Sea, Mediterranean and North Sea

The North Atlantic is a busy shipping route all year round (Vettor and Soares, 2015). Seismic exploration noise is seasonally present in the polar areas of the North Atlantic (Klinck and others, 2012; Haver and others, 2017). A rapid expansion of offshore wind farm development in the North and Baltic Seas has resulted in the presence of nearly 90 operational wind farms, as of 2018, and continued development in the future is predicted (Xu and others, 2020; Rusu, 2020), which will result in substantial increases in noise during the building phase (Miller and others, 2017). The main noise hotspots in the Mediterranean are the areas around major harbours. In addition, the Ionian Sea and the Adriatic Sea, as well as coasts along north-western Africa and in the eastern Mediterranean, have seen a recent increase in oil and gas exploratory surveys (Maglio and others, 2016). An increase in seismic activity in the Black Sea is also a possibility (Broad, 2014).

4.3. Gulf of Mexico, South Atlantic Ocean and Wider Caribbean

The number of vessels conducting seismic surveys has decreased in the Gulf of Mexico, but expanded off the Atlantic coast of South America (GeoTomo, 2018; United States Energy Information Administration (USEIA), 2020), potentially increasing noise levels at low frequency over the past decade. Large discoveries of offshore oil by Guyana (Cummings, 2018) may lead to higher levels of seismic exploration and industrial activity in the area. Noise associated with vessel traffic is ubiquitous throughout the Caribbean (Heenehan and others, 2019).

4.4. Indian Ocean, Arabian Sea, Bay of Bengal, Red Sea, Gulf of Aden and Persian Gulf

Development in Africa, including an increased number of new ports, is contributing to a rapid expansion in shipping in the region (Tournadre, 2014), which is in turn increasing anthropogenic noise in areas that were previously relatively noise-free. Seismic exploration continues offshore from Australia (Paumard and others, 2019).

4.5. North Pacific Ocean

New offshore wind projects are being developed off Japan, the Republic of Korea, Taiwan Province of China, and China (Yang and others, 2018; Li and Yuan, 2019). As part of that process, Japan is also starting to define acoustic monitoring parameters. Similarly, offshore wind projects have been proposed, but not yet permitted or constructed, off the west coast of the United States (Bureau of Ocean Energy Management, 2020). Some areas along the west coast of the United States, as well along the Hawaiian island chain, are designated as marine sanctuaries and could be protected from direct development.

4.6. South Pacific Ocean

Seismic exploration continues offshore from Australia and New Zealand (e.g., Cheong and Evans, 2018; Urosevic and others, 2019). Otherwise, the South Pacific remains relatively free from anthropogenic noise sources, with little shipping and industrial development.

4.7. Southern Ocean

The Southern Ocean has seen an increase in cruise ship traffic in recent years, both in the Antarctic Peninsula region, which has had some cruise ship traffic in the past, and in Eastern Antarctica and the Ross Sea, both previously unexplored (Sánchez and Roura, 2016). Overall, however, the region has had few anthropogenic noise sources, with little shipping and industrial development (Dziak and others, 2015).

5. Outlook

Anthropogenic noise in the ocean is largely driven by shipping, oil and gas exploration, and, at the more local or regional level, coastal development. Population growth, migration to coastal areas, increased industrialization and tourism and other developments will result in an increase in activities that contribute to anthropogenic noise, unless accompanied by mitigation efforts. A number of such efforts have been initiated. The Scientific Committee of the International Whaling Commission (IWC) has endorsed the goal of reducing ocean ambient sound by 3 dB in the next decade and 10 dB over the next 30 years. IWC is actively engaged with the International Maritime Organization (IMO) on discussions regarding strategies to achieve those reductions. One step may be to reduce noise from shipping, a major anthropogenic noise contributor at low frequencies in the open ocean (Wenz, 1962; Frisk, 2012; Roul and others, 2019). Shipping noise can be reduced by modifying propeller blades to make them quieter and by isolating engines and other noise contributors on the vessel so that the noise generated by them does not propagate through the ship into the ocean. Those technologies already exist but need wider implementation. Alternative measures being considered that can be implemented without technological advancements include decreasing ship speed or diverting ship traffic away from sensitive areas for marine life, such as marine sanctuaries, parks or reserves. In the oil and gas industry, new alternatives to the use of airguns in exploration surveys, such as marine vibrator technology, are being investigated. Even with new technological advances, adequate protection of the marine environment cannot be reached without a consensus on a global approach that fills the knowledge gaps related to anthropogenic noise impacts. Taking those considerations into account, for example, in 2014, IMO adopted the Guidelines for the reduction of underwater noise from commercial shipping to address adverse impacts on marine life.

The importance of anthropogenic noise has been acknowledged by various United Nations entities. In June 2018, anthropogenic noise was the main topic of the nineteenth meeting of the United Nations Open-ended Informal Consultative Process on Oceans and the Law of the Sea. The presentations and discussions during the meeting covered, inter alia, a review of the sources of anthropogenic noise, the effects and socioeconomic impacts of noise, and cooperation and coordination among States to address anthropogenic noise. Among other things, it was noted that the application of a precautionary approach to the management of noise impacts had been proposed at both the regional and global levels and that cross-sectoral cooperation was needed for identifying and mitigating impacts.[3]

Given that sound is a form of energy, its introduction into the marine environment is regarded by many as a form of contamination, owing to its potentially deleterious effects. In its resolution 12.14, the Conference of the Parties to the Convention on the Conservation of Migratory Species of Wild Animals recognized the impact of anthropogenic underwater noise on marine species and encouraged further study and mitigation of such noise. It also endorsed guidelines on environmental impact assessment for marine noise-generating activities

[3] See A/73/124.

that had been developed in collaboration with the secretariats of the Agreement on the Conservation of Cetaceans in the Black Sea, Mediterranean Sea and contiguous Atlantic area and the Agreement on the Conservation of Small Cetaceans of the Baltic, North East Atlantic, Irish and North Seas. Furthermore, it welcomed related technical support information (Prideaux, 2017).[4]

A number of States have been developing their own guidelines for managing ocean noise. The European Union has a mandate from its member States to measure and report anthropogenic noise under descriptor 11 of the Marine Strategy Framework Directive adopted in June 2008. The aim of the Directive is to achieve good environmental status by 2020, with each member State determining how that might be achieved. Under the Directive, there has been a proliferation across the region of ocean noise-targeted projects, including noise registers or databases with specifications on impulsive noise activity. Examples of those registers include the Baltic Marine Environment Protection Commission impulsive noise register and the Agreement on the Conservation of Cetaceans of the Black Sea, Mediterranean Sea and neighbouring Atlantic Area noise register for the Mediterranean and Black Seas. Canada is building the Marine Environmental Research Infrastructure for Data Integration and Application Network,[5] a database on underwater acoustics and vessel tracking, including visualization and analytical tools to provide information to managers, the public and researchers. In the United States, measures for comprehensively managing the impact of noise on marine species are set out in the National Oceanographic and Atmospheric Administration Ocean Noise Strategy (Gedamke and others, 2016), which also includes the use of mapping tools to assist in evaluating the impacts of anthropogenic noise on cetaceans (NOAA, 2020). Those national efforts to document noise sources should result in an increased ability to map variability in sound levels across the region. At the same time, such initiatives are leading to increased efforts to standardize data collection and measurements. For example, the International Quiet Ocean Experiment, a collaborative international science programme aimed at promoting research, observation and modelling to improve the understanding of ocean soundscapes and the effects of sound on marine organisms, has established working groups on data collection and data management standardization.

Sound has also recently been identified as an Essential Ocean Variable by the Biology and Ecosystems Panel of the Global Ocean Observing System (GOOS) (GOOS, 2020). Ocean sound is recognized as a cross-disciplinary variable as it includes such geophysical sources as wind, bubbles, ice, earthquakes and volcanoes. That global recognition and incorporation of observing systems into new initiatives should contribute to an increase in monitoring of anthropogenic noise, as well as to a better understanding of its contributions to ambient sound and of possible changes in soundscapes over time, in particular in relation to changing ocean use and climate change.

High levels of noise in the ocean can have a variety of consequences for marine life. A theoretical framework to evaluate the consequences of acoustic disturbances at the level of populations is available for marine mammals, but should be applicable to other taxa as well (Pirotta and others, 2018). Such an approach can be used for management purposes, but also offers a framework to investigate the proximate mechanisms of phenomena that induce changes at the individual level and guide future data collection and

---

4 Detailed information on the *CMS Family Guidelines on Environmental Impact Assessments for Marine Noise-generating Activities* is available at www.cms.int/guidelines/cms-family-guidelines-EIAs-marine-noise.

5 See https://meridian.cs.dal.ca.

model development. Considering that those consequences occur among commercially and recreationally important species, as well as those that are relied on for subsistence, there is potential for negative social and economic impacts. For example, a reduction in the recruitment of commercially important fishes (Simpson and others, 2008) may lead over time to a reduction in catches, and higher mortality may decrease fishery yields. For species that are the focus of tourism activities, those activities themselves, for example whale watching, may result in increased noise and can cause impacts (Erbe, 2002; Holt and others, 2009).

6. Key remaining knowledge gaps

Several challenges remain in evaluating the relative increases and possible impacts of anthropogenic noise in the ocean. A fundamental problem is the lack of knowledge regarding baseline ocean ambient noise. Given that no recordings are available from time periods prior to human activities, there is limited understanding of the marine soundscapes that marine life evolved with or the extent to which they might have adapted to anthropogenic noise inputs. The best proxy are regions outside the influence of human development and activity, which may exist in isolated basins, such as areas of the Southern Ocean, or were present until recently in parts of the Arctic. However, on the basis of best estimates, many regions of the ocean have ambient noise levels at low frequency (10–200 Hz) at least 20–30 dB higher than primordial levels.

Another major gap is in the understanding of the impact of noise on marine ecosystems. To date, most work has been focused on the impact of a single stressor on a particular species, the result of which may not be directly applicable to populations (Gill and others, 2001). It is unclear, and very difficult to study, how the combination of noise and other stressors (e.g., shifting food webs, changing water temperatures and habitat destruction) affect marine populations. A framework has been developed to assess the consequences of disturbance on populations, but often too many key parameter values are missing to enable an evaluation at the population level (King and others, 2015). For example, very little is known about the hearing response of large baleen whales. In addition, environments can be subject to multiple sources of noise over large scales, with the potential to affect multiple species at the same time, which can compound any effects (Shannon and others, 2016). At the current stage, the precautionary approach has been followed in many regulations that are based on insufficient data. However, it will be essential to expand the ability to integrate effects and impacts across different scales and sources in order to allow for a realistic assessment of the impact of anthropogenic noise on marine animals.

Finally, substantial effort is needed to standardize monitoring approaches, measurements and archival frameworks or systems for acoustic recording approaches and associated collected data. The American National Standards Institute/Acoustical Society of America standard (2009) and the ISO standard (2016) for measurement of underwater noise from ships in deep water require multiple sound measurements by arrays of sensors and, in practice, have been rarely applied. Among other work currently under way, ISO is developing standards on soundscape measures and monitoring, which will include underwater data, and standards are being developed through the Acoustical Society of America standards procedures regarding towed array systems and data archiving. In future, standards for other parts of the acoustic monitoring effort, such as fixed recordings, calibrations and ambient sound data, should also be developed.

7. Key remaining capacity-building gaps

Thus far, the monitoring and modelling of anthropogenic noise have been concentrated in areas of North America and Europe, with some concentrated monitoring also taking place off the coast of Australia. However, across-the-board capacity-building in the area of the Indian Ocean and its adjacent seas, including monitoring, impact assessment and development of management frameworks, would help to increase understanding of the changes taking place in the environment. Since sound travels broadly across ocean basins, and anthropogenic noise sources are found worldwide, there is a need for increased collaboration and cooperation across all States and regions, as well as greater sharing of information and technology. One example of differences in technological availability relates to AIS for ship tracking. Knowledge of ship positions is essential for accurate mapping of underwater noise. AIS is a localization and identification system developed for ship collision avoidance that, over time, has been adopted and mandated across vessels of a broad range of sizes. Ships are most comprehensively monitored in the developed world, owing to relatively good spatial coverage by AIS receivers. The move to satellite-based AIS that is under way will enable broader data coverage, and timely international collaboration to use those data might be an opportunity to bridge some capacity gaps in modelling across States. Enhanced cooperation and collaboration activities with developing States would facilitate the sharing of best practices and best available technologies necessary to build national and regional programmes, not only to monitor the effects of anthropogenic underwater noise, but also to provide the information needed for well-informed policy decisions.

References

Aguilar de Soto, Natacha, and others (2013). Anthropogenic noise causes body malformations and delays development in marine larvae. *Scientific Reports*, vol. 3, No. 1, p. 2831. https://doi.org/10.1038/srep02831.

American National Standards Institute/Acoustical Society of America (ANSI/ASA) (2009). *Quantities and Procedures for Description and Measurement of Underwater Sound from Ships–Part 1: General Requirements*. American National Standards Institute/Acoustical Society of America New York.

Bailey, Helen, and others (2014). Assessing environmental impacts of offshore wind farms: lessons learned and recommendations for the future. *Aquatic Biosystems*, vol. 10, No. 1, p. 8.

Balcomb, Kenneth C. III, and Diane E. Claridge (2001). A mass stranding of cetaceans caused by naval sonar in the Bahamas. *Bahamas Journal of Science*, vol. 8, No. 2, pp. 2–12.

Bernaldo de Quirós, Y., and others (2019). Advances in research on the impacts of anti-submarine sonar on beaked whales. *Proceedings of the Royal Society B*, vol. 286, No. 1895, art. 20182533.

Blair, Hannah B., and others (2016). Evidence for ship noise impacts on humpback whale foraging behaviour. *Biology Letters*, vol. 12, No.8, p. 20160005.

Broad, William J. (2014). In taking Crimea, Putin gains a sea of fuel reserves. *The New York Times*, 17 May 2014.

Bruintjes, Rick, and others (2017). The impact of experimental impact pile driving on oxygen uptake in black seabream and plaice. *Proceedings of Meetings on Acoustics*, vol. 27, No. 1, art. 010042. https://doi.org/10.1121/2.0000422.

Buehler, D., and others (2015). Technical guidance for assessment and mitigation of the hydroacoustic effects of pile driving on fish. *Technical Report No. CTHWANP-RT-15-306.01.01.*

Bureau of Ocean Energy Management (2020). California Activities. www.boem.gov/renewable-energy/state-activities/california-activities.

Carstensen, J., and others (2006). Impacts of offshore wind farm construction on harbour porpoises: acoustic monitoring of echolocation activity using porpoise detectors (T-PODs). *Marine Ecology Progress Series*, vol. 321, pp. 295–308.

Casper, Brandon M., and others (2013). Effects of exposure to pile driving sounds on fish inner ear tissues. *Comparative Biochemistry and Physiology, Part A*, vol. 166, No. 2, pp. 352–360. http://doi.org/10.1016/j.cbpa.2013.07.008

Cerchio, Salvatore, and others (2014). Seismic surveys negatively affect humpback whale singing activity off northern Angola. *PloS One*, vol. 9, No. 3. e86464.

Cheong, Sei-Him, and Breanna Evans (2018). Acoustic ground truthing of seismic noise in Chatham Rise, New Zealand. *Journal of the Acoustical Society of America*, vol. 143, No. 3, p. 1974. https://doi.org/10.1121/1.5036504.

Cholewiak, Danielle, and others (2017). Beaked whales demonstrate a marked acoustic response to the use of shipboard echo sounders. *Royal Society Open Science*, vol. 4, No. 12, art. 170940.

Cummings, Anthony R. (2018). How Guyana's Oil Discovery Rekindled a Border Controversy. *Journal of Latin American Geography*, vol. 17, No. 3, pp. 183–211.

Day, Ryan D., and others (2016). Seismic air gun exposure during early-stage embryonic development does not negatively affect spiny lobster *Jasus edwardsii* larvae (Decapoda: Palinuridae). *Scientific Reports*, vol. 6, p. 22723.

Day, Ryan D., and others (2017). Exposure to seismic air gun signals causes physiological harm and alters behavior in the scallop *Pecten fumatus*. *Proceedings of the National Academy of Sciences*, vol. 114, No. 40, pp. E8537–E8546.

DeRuiter, Stacy L., and others (2013). First direct measurements of behavioural responses by Cuvier's beaked whales to mid-frequency active sonar. *Biology Letters*, vol. 9, No. 4, p. 20130223.

Di Iorio, Lucia, and Christopher W. Clark (2009). Exposure to seismic survey alters blue whale acoustic communication. *Biology Letters*, vol. 6, No. 1, pp. 51–54.

Doksæter, Lise, and others (2012). Behavior of captive herring exposed to naval sonar transmissions (1.0–1.6 kHz) throughout a yearly cycle. *Journal of the Acoustical Society of America*, vol. 131, No. 2, pp. 1632–1642.

Dragoset, Bill (2000). Introduction to air guns and air-gun arrays. *The Leading Edge*, vol. 19, No. 8, pp. 892–897.

Dyndo, Monika, and others (2015). Harbour porpoises react to low levels of high frequency vessel noise. *Scientific Reports*, vol. 5, p. 11083.

Dziak, Robert P., and others (2015). Sources and Levels of Ambient Ocean Sound near the Antarctic Peninsula. *PLOS ONE*, vol. 10, No. 4, pp. 1–23. https://doi.org/10.1371/journal.pone.0123425.

Eguíluz, Victor M., and others (2016). A quantitative assessment of Arctic shipping in 2010–2014. *Scientific Reports*, vol. 6, No. 1, p. 30682. https://doi.org/10.1038/srep30682.

Ehizuelen, Michael Mitchell Omoruyi (2017). More African countries on the route: the positive and negative impacts of the Belt and Road Initiative. *Transnational Corporations Review*, vol. 9, No. 4, pp. 341–359.

Erbe, Christine (2002). Underwater noise of whale-watching boats and potential effects on killer whales (*Orcinus orca*), based on an acoustic impact model. *Marine Mammal Science*, vol. 18, No. 2, pp. 394–418.

________ (2013). Underwater noise of small personal watercraft (jet skis). *The Journal of the Acoustical Society of America*, vol. 133, No. 4, pp. EL326–EL330.

Fernández, Antonio, and others (2005). "Gas and fat embolic syndrome" involving a mass stranding of beaked whales (Family Ziphiidae) exposed to anthropogenic sonar signals. *Veterinary Pathology*, vol. 42, No. 4, pp. 446–457.

Fields, David M., and others (2019). Airgun blasts used in marine seismic surveys have limited effects on mortality, and no sublethal effects on behaviour or gene expression, in the copepod *Calanus finmarchicus*. *ICES Journal of Marine Science*, vol. 76, No. 7, pp. 2033–44. https://doi.org/10.1093/icesjms/fsz126.

Frisk, George V. (2012). Noiseonomics: The relationship between ambient noise levels in the sea and global economic trends. *Scientific Reports*, vol. 2, p. 437.

Gassmann, Martin, and others (2017). Deep-water measurements of container ship radiated noise signatures and directionality. *The Journal of the Acoustical Society of America*, vol. 142, No. 3, pp. 1563–1574.

Gedamke, Jason, and others (2016). *Ocean Noise Strategy Roadmap*. Washington, D.C.: National Oceanographic and Atmospheric Administration.

GeoTomo (2018). Seismic Crew Count – World seismic crew summary: May 2018. https://geotomo.com/seismicCrewCount.dmx.

Gill, Jennifer A., and others (2001). Why behavioural responses may not reflect the population consequences of human disturbance. *Biological Conservation*, vol. 97, No. 2, pp. 265–268.

Global Ocean Observing System (GOOS) (2020). Essential Ocean Variables. www.goosocean.org/index.php?option=com_content&view=article&id=170&Itemid=114

Goldbogen, Jeremy A., and others (2013). Blue whales respond to simulated mid-frequency military sonar. *Proceedings of the Royal Society B: Biological Sciences*, vol. 280, No. 1765, p. 20130657.

Gontz, A.M., and others (2006). Shallow-water seismic surveys-how much noise are we introducing into the ocean? In OCEANS 2006, pp. 1–5. IEEE.

Greene, C.R. (1985). Characteristics of waterborne industrial noise, 1980-1984. In Behavior, Disturbance Responses, and Distribution of Bowhead Whales *Balaena Mysticetus* in the Eastern Beaufort Sea, 1980–84, W.J. Richardson, ed., pp. 197–253. OCS Study MMS 85-0034, LGL Ecological Research Associates, Bryan, Texas, United States, for U.S. Minerals Management Service, Reston, Virginia, United States, NTIS PB87-124376.

Guerra, A., and others (2004). A review of the records of giant squid in the north-eastern Atlantic and severe injuries in *Architeuthis dux* stranded after acoustic explorations. ICES CM, vol. 200, p. 29.

Haver, Samara M., and others (2017). The not-so-silent world: Measuring Arctic, Equatorial, and Antarctic soundscapes in the Atlantic Ocean. *Deep Sea Research Part I: Oceanographic Research Papers*, vol. 122, pp. 95–104. https://doi.org/10.1016/j.dsr.2017.03.002.

Heenehan, Heather, and others (2019). Caribbean Sea soundscapes: monitoring humpback whales, biological sounds, geological events and anthropogenic impacts of vessel noise. *Frontiers in Marine Science*, vol. 6, art. 347.

Herbert-Read, James E., and others (2017). Anthropogenic noise pollution from pile-driving disrupts the structure and dynamics of fish shoals. *Proceedings of the Royal Society B: Biological Sciences*, vol. 284, No.1863, p. 20171627.

Hildebrand, John A. (2009). Anthropogenic and natural sources of ambient noise in the ocean. *Marine Ecology Progress Series*, vol. 395, pp. 5–20.

Hirst, Andrew G., and Paul G. Rodhouse (2000). Impacts of geophysical seismic surveying on fishing success. *Reviews in Fish Biology and Fisheries*, vol. 10, No. 1, pp. 113–118.

Ho, Joshua (2010). The implications of Arctic sea ice decline on shipping. *Marine Policy*, vol. 34, No. 3, pp. 713–715.

Holt, Marla M., and others (2009). Speaking up: Killer whales (*Orcinus orca*) increase their call amplitude in response to vessel noise. *The Journal of the Acoustical Society of America*, vol. 125, No. 1, pp. EL27–EL32.

International Organization for Standardization (ISO) (2016). *ISO 17208-1:2016, I. Underwater Acoustics – Quantities and Procedures for Description and Measurement of Underwater Sound from Ships – Part 1: Requirements for Precision Measurements in Deep Water Used for Comparison Purposes*. Geneva.

Jensen, Finn B., and others (2011). Computational Ocean Acoustics. New York: Springer.

Jones, Ian T., and others (2020). Impulsive pile driving noise elicits alarm responses in squid (*Doryteuthis pealeii*). *Marine Pollution Bulletin*, vol. 150, 110792. http://doi.org/10.1016/j.marpolbul.2019.110792

King, Stephanie L., and others (2015). An interim framework for assessing the population consequences of disturbance. Methods in Ecology and Evolution, vol. 6, No. 10, pp. 1150–1158.

Klinck, Holger, and others (2012). Seasonal presence of cetaceans and ambient noise levels in polar waters of the North Atlantic. *The Journal of the Acoustical Society of America*, vol. 132, No. 3, EL176–EL181.

Lee, Kevin M., and others (2012). Mitigation of low-frequency underwater anthropogenic noise using stationary encapsulated gas bubbles. In *Proceedings of Meetings on Acoustics ECUA2012*, 17: p.070011. ASA.

Li, Aitong, and Yuan Xu (2019). The governance for offshore wind in Japan. *Energy Procedia*, vol. 158, pp. 297–301. https://doi.org/10.1016/j.egypro.2019.01.092.

Løkkeborg, Svein (1991). Effects of a geophysical survey on catching success in longline fishing.

Løkkeborg, Svein, and others (2012). Sounds from seismic air guns: gear-and species-specific effects on catch rates and fish distribution. *Canadian Journal of Fisheries and Aquatic Sciences*, vol. 69, No. 8, pp. 1278–1291.

Madsen, Peter T., and others (2006). Wind turbine underwater noise and marine mammals: implications of current knowledge and data needs. *Marine Ecology Progress Series*, vol. 309, pp. 279–295.

Maglio, Alessio, and others (2016). Overview of the noise hotspots in the ACCOBAMS area. *Final Report to the ACCOBAMS Secretariat*.

McCauley, Robert D., and others (2017). Widely used marine seismic survey air gun operations negatively impact zooplankton. *Nature Ecology & Evolution*, vol. 1, No. 7, art. 0195.

McDonald, Mark A., and others (2008). A 50 year comparison of ambient ocean noise near San Clemente Island: A bathymetrically complex coastal region off Southern California. *The Journal of the Acoustical Society of America*, vol. 124, No. 4, pp. 1985–1992.

McDonald, Mark A., and others (2006). Increases in deep ocean ambient noise in the Northeast Pacific west of San Nicolas Island, California. *The Journal of the Acoustical Society of America*, vol. 120, No. 2, pp. 711–718.

McKenna, Megan F., and others (2013). Relationship between container ship underwater noise levels and ship design, operational and oceanographic conditions. *Scientific Reports*, vol. 3, p. 1760.

McQueen, Andrew D., and others (2020). Ecological risk assessment of underwater sounds from dredging operations. *Integrated Environmental Assessment and Management*, vol. 16, No. 4, pp. 481–493.

Merchant, Nathan D., and others (2012). Assessing sound exposure from shipping in coastal waters using a single hydrophone and Automatic Identification System (AIS) data. *Marine Pollution Bulletin*, vol. 64, No. 7, pp. 1320–1329.

Miksis-Olds, Jennifer L., and Stephen M. Nichols (2016). Is low frequency ocean sound increasing globally? *The Journal of the Acoustical Society of America*, vol. 139, No. 1, pp. 501–11. https://doi.org/10.1121/1.4938237.

Miller, James H., and others (2017). Overview of underwater acoustic and seismic measurements of the construction and operation of the Block Island Wind Farm. *The Journal of the Acoustical Society of America*, vol. 141, No. 5, p. 3993.

Miller, Kathryn A., and others (2018). An overview of seabed mining including the current state of development, environmental impacts, and knowledge gaps. *Frontiers in Marine Science*, vol. 4, art. 418.

Miller, Patrick J.O., and others (2000). Whale songs lengthen in response to sonar. *Nature*, vol. 405, No. 6789, p. 903.

Moretti, David, and others (2014). A risk function for behavioral disruption of Blainville's beaked whales (*Mesoplodon densirostris*) from mid-frequency active sonar. *PloS One*, vol. 9, No. 1, e85064.

Mueller-Blenkle, Christina, and others (2010). Effects of pile-driving noise on the behaviour of marine fish.

National Academies (2017). *Approaches to Understanding the Cumulative Effects of Stressors on Marine Mammals*. The National Academies Press.

National Oceanographic and Atmospheric Administration (NOAA) (2020). CetSound: Cetacean and sound mapping. https://cetsound.noaa.gov/cetsound.

National Research Council (NRC) (2003). *Ocean Noise and Marine Mammals*. Washington, D.C.: The National Academies Press. https://doi.org/10.17226/10564.

Nedwell, J., and D. Howell (2004). A review of offshore windfarm related underwater noise sources. *Cowrie Report*, vol. 544, pp. 1–57.

Nedwell, J.R., and others (2008). Modelling and measurement of underwater noise associated with the proposed Port of Southampton capital dredge and redevelopment of berths 201/202 and assessment of the disturbance to salmon. *Subacoustech Report, 805R0444*.

Nichols, Tye A., and others (2015). Intermittent noise induces physiological stress in a coastal marine fish. *PLoS One*, vol. 10, No. 9, e0139157.

Nieukirk, Sharon L., and others (2004). Low-frequency whale and seismic airgun sounds recorded in the mid-Atlantic Ocean. *The Journal of the Acoustical Society of America*, vol. 115, No. 4, pp. 1832–1843.

Norro, A., and others (2011). Characterisation of the operational noise generated by offshore wind farms in the Belgian part of the North Sea. In Offshore Wind Farms in the Belgian Part of the North Sea. Selected Findings from the Baseline and Targeted Monitoring, S. Degraer, Robin Brabant, and B. Rumes, eds., pp. 17–26. Royal Belgian Institute of Natural Sciences, Management Unit of the North Sea Mathematical Models. Marine ecosystem management unit.

Nunatsiaq News (2016). Trudeau bans future oil, gas activity in Canadian Arctic. https://nunatsiaq.com/stories/article/65674trudeau_bans_future_oil_gas_activity_in_canadian_arctic.

Parks, Susan E., and others (2010). Individual right whales call louder in increased environmental noise. Biology Letters, vol. 7, No. 1, pp. 33–35.

Paumard, Victorien, and others (2019). Imaging past depositional environments of the North West Shelf of Australia: lessons from 3D seismic data. In Sedimentary Basins of Western Australia V: Proceedings of the Petroleum Exploration Society of Australia Symposium, Perth, Western Australia, 2019, Myra Keep and Steven J. Moss, eds. Petroleum Exploration Society of Australia.

Pearson, Walter H., and others (1994). Effects of seismic energy releases on the survival and development of zoeal larvae of Dungeness crab (*Cancer magister*). *Marine Environmental Research*, vol. 38, No. 2, pp. 93–113.

Pine, Matthew K., and others (2012). Turbine sound may influence the metamorphosis behaviour of estuarine crab megalopae. *PLoS One*, vol. 7, No. 12. e51790.

Pirotta, Enrico, and others (2018). Understanding the population consequences of disturbance. *Ecology and Evolution*, vol. 8, No. 19, pp. 9934–46. https://doi.org/10.1002/ece3.4458.

Popper, Arthur N., and Anthony D. Hawkins (2019). An overview of fish bioacoustics and the impacts of anthropogenic sounds on fishes. *Journal of Fish Biology*, vol. 94, No. 5, pp. 692–713.

Prideaux, G., (2017). Technical Support Information to the CMS Family Guidelines on Environmental Impact Assessments for Marine Noise-generating Activities, Convention on Migratory Species of Wild Animals, Bonn.

Putland, Rosalyn L., and others (2018). Vessel noise cuts down communication space for vocalizing fish and marine mammals. *Global Change Biology*, vol. 24, No. 4, pp. 1708–1721.

Reine, Kevin J., and others (2012). Characterization of underwater sounds produced by a hydraulic cutterhead dredge fracturing limestone rock. DOER Technical Notes Collection—erdctn-doer-e34. Vicksburg, Mississippi, United States: U.S. Army Engineer Research and Development Center.

Richardson, W. John, and others (1995). *Marine Mammals and Noise*. San Diego: Academic Press. https://doi.org/10.1016/B978-0-08-057303-8.50003-3.

Roberts, Louise, and others (2015). Sensitivity of the mussel *Mytilus edulis* to substrate-borne vibration in relation to anthropogenically generated noise. *Marine Ecology Progress Series*, vol. 538, pp. 185–195.

Robinson, Stephen P., and others (2011). Measurement of underwater noise arising from marine aggregate dredging operations.

Rolland, Rosalind M., and others (2012). Evidence that ship noise increases stress in right whales. *Proceedings of the Royal Society B: Biological Sciences*, vol. 279, No. 1737, pp. 2363–2368.

Ross, Donald (1976). *Mechanics of Underwater Noise / Donald Ross*. New York: Pergamon Press.

________ (2005). Ship sources of ambient noise. *IEEE Journal of Oceanic Engineering*, vol. 30, No. 2, pp. 257–261.

Roth, Ethan H., and others (2012). Underwater ambient noise on the Chukchi Sea continental slope from 2006–2009. *Journal of the Acoustical Society of America*, vol. 131, No. 1, pp. 104–110.

Roul, Soubhagya, and others (2019). Ambient noise estimation in territorial waters using AIS data. *Applied Acoustics*, vol. 148, pp. 375–380. http://doi.org/10.1016/j.apacoust.2018.07.036

Rustemeier, J., and others (2011). Testing of bubble curtains to mitigate hydro sound levels at offshore construction sites (2007 to 2011). www.rave-offshore.de/files/downloads/konferenz/konferenz-2012/Session4/4.4_Grieszmann.pdf.

Rusu, E. (2020). An evaluation of the wind energy dynamics in the Baltic Sea, past and future projections. *Renewable Energy*, vol. 160, pp. 350–362.

Samuel, Y., and others (2005). Underwater, low-frequency noise in a coastal sea turtle habitat. *Journal of the Acoustical Society of America*, vol. 117, No. 3, pp. 1465–1472.

Sánchez, Rodolfo A., and Ricardo Roura (2016). Supervision of Antarctic shipborne tourism: a pending issue? In *Tourism in Antarctica: A Multidisciplinary View of New Activities Carried Out on the White Continent*, Monika Schillat and others, eds., pp. 41–63. Cham, Switzerland: Springer International Publishing. https://doi.org/10.1007/978-3-319-39914-0_3.

Sertlek, Hüseyin Özkan, and others (2019). Source specific sound mapping: spatial, temporal and spectral distribution of sound in the Dutch North Sea. *Environmental Pollution*, vol. 247, pp. 1143–1157.

Shannon, Graeme, and others (2016). A synthesis of two decades of research documenting the effects of noise on wildlife. *Biological Reviews*, vol. 91, No. 4, pp. 982–1005.

Shell (2015). Shell updates on Alaska exploration. www.shell.com/media/news-and-media-releases/2015/shell-updates-on-alaska-exploration.html.

Simpson, Stephen D., and others (2008). Settlement-stage coral reef fishes prefer the higher frequency audible component of reef noise. *Animal Behaviour*, vol. 75, pp. 1861–1868. 10.1016/j.anbehav.2007.11.004.

Simpson, Stephen D., and others (2016a). Anthropogenic noise increases fish mortality by predation. *Nature Communications*, vol. 7, art. 10544.

Simpson, Stephen D., and others (2016b). Small-boat noise impacts natural settlement behavior of coral reef fish larvae. In *The Effects of Noise on Aquatic Life II*, pp. 1041–1048. Springer.

Širović, Ana, and others (2013). Ocean noise in the tropical and subtropical Pacific Ocean. *Journal of the Acoustical Society of America*, vol. 134, No. 4, pp. 2681–89. https://doi.org/10.1121/1.4820884.

Širović, Ana, and others (2016). Ocean ambient sound south of Bermuda and Panama Canal traffic. *Journal of the Acoustical Society of America*, vol. 139, No. 5, pp. 2417–2423.

Sivle, Lise Doksæter, and others (2012). Changes in dive behavior during naval sonar exposure in killer whales, long-finned pilot whales, and sperm whales. *Frontiers in Physiology*, vol. 3, art. 400.

Smith, Laurence C., and Scott R. Stephenson (2013). New Trans-Arctic shipping routes navigable by mid-century. Proceedings of the National Academy of Sciences, vol. 110, No. 13, pp. E1191–E1195. https://doi.org/10.1073/pnas.1214212110.

Spiga, Ilaria, and others (2016). Influence of pile driving on the clearance rate of the blue mussel, *Mytilus edulis* (L.). In *Proceedings of Meetings on Acoustics 4ENAL*, vol. 27: p.040005. ASA.

Thompson, Kirsten F., and others (2018). Seabed mining and approaches to governance of the deep seabed. *Frontiers in Marine Science*, vol. 5, art. 480.

Tournadre, J. (2014). Anthropogenic pressure on the open ocean: The growth of ship traffic revealed by altimeter data analysis. *Geophysical Research Letters*, vol. 41, No.22, pp. 7924–32. https://doi.org/10.1002/2014GL061786.

Tsujii, Koki, and others (2018). Change in singing behavior of humpback whales caused by shipping noise. *PloS One*, vol. 13, No. 10. e0204112.

Turner, Stephen, and others (2006). Preliminary acoustic level measurements of airgun sources from Conoco Phillips' 2006 seismic survey in Alaskan Chukchi Sea. *JASCO Research, Victoria, British Columbia, Canada.*

Tyack, Peter L., and others (2011). Beaked whales respond to simulated and actual navy sonar. *PloS One*, vol. 6, No. 3. e17009.

United Nations (2017). *The First Global Integrated Marine Assessment: World Ocean Assessment I*. Cambridge: Cambridge University Press.

United Nations Conference on Trade and Development (UNCTAD) (2018). *Review of Maritime Transport 2018*. UNCTAD/RMT/2018.

United States Energy Information Administration (USEIA) (2020). Maximum U.S. Active Seismic Crew Counts. www.eia.gov/dnav/pet/pet_crd_seis_s1_m.htm.

Urosevic, M., and others (2019). Seismic Exploration of Mineral Resources in Western Australia with Distribute Acoustic Sensing, vol. 2019, No. 1, pp. 1–5. https://doi.org/10.3997/2214-4609.201902377.

Vettor, Roberto, and C. Guedes Soares (2015). Detection and Analysis of the Main Routes of Voluntary Observing Ships in the North Atlantic. *Journal of Navigation*, vol. 68, No. 2, pp. 397–410. https://doi.org/10.1017/S0373463314000757.

Weilgart, Lindy S. (2018). The impact of ocean noise pollution on fish and invertebrates. Report for OceanCare, Switzerland. www.oceancare.org/wp-content/uploads/2017/10/OceanNoise_FishInvertebrates_May2018.pdf.

Wenz, Gordon M. (1962). Acoustic Ambient Noise in the Ocean: Spectra and Sources. *Journal of the Acoustical Society of America*, vol. 34, No. 12, pp. 1936–56. https://doi.org/10.1121/1.1909155.

Wiggins, Sean M., and others (2016). Gulf of Mexico low-frequency ocean soundscape impacted by airguns. *Journal of the Acoustical Society of America*, vol. 140, No. 1, pp. 176–183.

Würsig, B., and others (2000). Development of an air bubble curtain to reduce underwater noise of percussive piling. *Marine Environmental Research*, vol. 49, No. 1, pp. 79–93.

Xu, W., and others (2020). Proliferation of offshore wind farms in the North Sea and surrounding waters revealed by satellite image time series. *Renewable and Sustainable Energy Reviews*, vol. 133, p. 110167.

Yang, Chun-Mei, and others (2018). Observation and comparison of tower vibration and underwater noise from offshore operational wind turbines in the East China Sea Bridge of Shanghai. *Journal of the Acoustical Society of America*, vol. 144, No. 6, EL522.

Chapter 21
Developments in renewable energy sources

Contributors: Takvor Soukissian (convener), Joan Bondareff, Valerie Cummins, Amardeep Dhanju, Carlos Garcia-Soto (co-lead member), Lars Golmen, Osman Keh Kamara (co-lead member), Jimmy Murphy, Eric Mwangi Njoroge, Anastasia Strati (lead member) and Georges Vougioukalakis.

Keynote points

- The offshore wind sector is expanding globally to regions with no utility-scale (grid) installations at present. The use of floating platforms is a step change enabling the industry to open up large areas with deeper waters.
- In 2019, 28.3 gigawatts (GW) of installed capacity from the offshore wind sector was deployed globally, with 22 GW off Europe, primarily in the North Sea, 5.9 GW off China and 0.4 GW in other markets.
- In the next decade, Asia and the United States of America could be major growth drivers for the development and installation of offshore wind power.
- Wave and ocean current energy projects have not yet achieved full commercialization at utility scale, and tidal energy projects are still rare.
- Progress in energy storage could make a significant contribution to the development of offshore wind power and other marine renewable energy (MRE) technologies.
- Proper siting of MRE projects could minimize conflicts with other ocean uses and potential impacts on the marine environment.

1. Introduction

The present chapter covers advances in knowledge and capacity made in recent years for the various types of marine renewable energy (MRE) at the global level. For the purpose of the present chapter, MRE as a category includes offshore wind energy, tidal and ocean current energy, wave energy, ocean thermal energy, osmotic power, marine biomass energy and offshore solar and geothermal energy. The chapter has linkages with chapters 6F, 8A, 9, 19, 20, 26, 27 and 28 of the present Assessment.

1.1. Climate change and the clean energy challenge

Fossil fuel energy use accounts for a large proportion of global anthropogenic greenhouse gas emissions. In 2019, global energy consumption increased by 0.6 per cent,[1] while total emissions of energy-related carbon dioxide (CO_2) fell by 3.2 per cent (International Energy Agency (IEA), 2020). However, global average atmospheric CO_2 was 409.8 parts per million, the highest level for 800,000 years (Dlugokencky and Tans, 2020), while the global average temperature was about 1.1°C, with a standard error of 0.1°C, above pre-industrial levels (World Meteorological Organization (WMO), 2020).

In view of the current status of greenhouse gas emissions, it is very likely that the agreed temperature thresholds of 1.5°C or 2°C above pre-industrial levels will be exceeded. As clearly highlighted in the special report entitled *Global Warming of 1.5°C* of the Intergovernmental Panel on Climate Change (2018), for global warming not to exceed 1.5°C, global net anthropogenic emissions of CO_2 would need to fall by about 45 per cent from 2010 levels by 2030, reaching "net zero" around 2050. That means that any remaining emissions would need to be balanced by removing CO_2 from the air. Reducing greenhouse gas emissions is therefore an important step towards climate change mitigation. In order to move in that

1 Enerdata, "Consumption", Global Energy Statistical Yearbook 2020. Available at https://yearbook.enerdata.net/total-energy/world-consumption-statistics.html.

direction, many States are taking measures to increase the development of renewable energy sources such as MRE in order to meet national clean energy and climate change goals. MRE is also linked to Sustainable Development Goal 7, in which affordable and clean energy is recognized as a key driver for development.[2]

2. State of marine renewable energy at the global level

2.1. Advances in knowledge and capacity between 2010 and 2020

The ocean has the potential to be a major source of renewable energy. In addition to climate change mitigation, MRE can contribute to socioeconomic development, energy security and energy access in remote coastal regions (Edenhofer and others, 2011). In 2019, global installed offshore wind capacity increased by 4.7 GW, a 19.8 per cent increase on 2018, to a total of 28.3 GW. The global capacity of other types of MRE reached 531 megawatts (MW), 90 per cent of which came from two tidal barrages in France and the Republic of Korea (International Renewable Energy Agency (IRENA), 2020a).

The various types of MRE technologies are evolving and developing at different speeds: bottom-fixed offshore wind technology is mature and technically advanced, floating offshore wind technology is on the cusp of being commercialized and tidal energy converters have reached the commercial stage, while other MRE technologies are currently at the development stage.[3] The emerging offshore wind markets include India, Japan, the Republic of Korea and the United States.[4] There has been a significant increase in wind turbine rated capacities, with turbines of up to 12 MW capacity expected to be on the market in 2021.[5]

2.2. Regional advances

2.2.1. Offshore wind energy

The global technical potential for offshore wind energy is estimated by IEA (in collaboration with Imperial College London) at more than 120,000 GW (IEA, 2019). Europe, with cumulative capacity of 21.98 GW in 2019, dominates the sector. The main countries developing offshore wind energy are the United Kingdom of Great Britain and Northern Ireland (1.7 GW installed in 2019 and 9.9 GW in total), Germany (1.1 GW installed in 2019 and 7.5 GW in total), and China (1.3 GW installed in 2019 and 5.9 GW in total) (IRENA, 2020a).

There have been significant developments in the sector. The largest offshore wind farm in the world, the Hornsea One Project in the United Kingdom, was completed in 2020 with an installed capacity of 1.2 GW. In 2019, the Haliade-X 12 MW prototype developed by the United States wind turbine manufacturer GE became the largest wind turbine ever built. As the size of both offshore wind turbines and offshore wind farms continues to increase, concerns about potential environmental impacts, impacts on fisheries and issues related to human use of areas near or within wind farms are becoming more important.

Offshore wind energy is demonstrating the viability, both technical and economic, of

2 See General Assembly resolution 70/1.

3 See European Commission, "New technologies in the ocean energy sector", 29 October 2018.

4 See Global Wind Energy Council, "The growth of the global offshore wind market will be driven by Asia", 23 September 2019.

5 See GE, "GE Renewable Energy unveils the first Haliade-X 12 MW, the world's most powerful offshore wind turbine", 22 July 2019.

utility-scale projects in different marine environments. The globally weighted levelized cost of energy (LCOE)[6] for utility-scale projects has fallen by 28.6 per cent since 2010, thus driving installation around the world (IRENA, 2020b). Moreover, the sector estimates that LCOE values in the order of €50/MWh are achievable by 2030.[7] The main drivers for cost reduction include the use of larger, more efficient turbines in bigger offshore wind farm developments, the reduced capital cost of financing projects, and the certainty of a long pipeline of projects, which has allowed the supply chain to invest and innovate. Fixed platforms are viable for water depths up to 60 m, but the industry also plans to operate in deeper waters using floating platforms within the next decade. Many coastal countries worldwide see floating wind power as a future major contributor to the achievement of renewable energy production targets. The world's first utility-scale floating offshore wind farm began production in 2017 (off the coast of Peterhead, Scotland, United Kingdom) using the Hywind concept developed by the Norwegian company Equinor (Musial and others, 2019; see figure below). This marks an important milestone for the offshore wind industry in the development of projects in deeper water and further from the coast.

The success of offshore wind power in reducing installation and production costs, combined with existing expertise in onshore wind energy, has made offshore wind the leading MRE technology.

World's first commercial wind farm comprising floating wind turbines

Photographer: Øyvind Gravås/Woldcam; image provided by Equinor.

6 The levelized cost of energy is the present value of the average minimum price of produced electrical energy required to offset the total cost of its production (construction, operation and maintenance, and fuel costs). It is levelized over the lifetime of a generating plant.

7 See also Kerry Chamberlain, "Offshore wind opex set to fall 40% by 2030 as suppliers dig deep", Reuters Events, 25 October 2017.

2.2.2. Tidal and ocean current energy

Global tidal energy capacity (combined theoretical tidal range and stream resource) is estimated at 3 terawatt (TW) (Lewis and others, 2011; Scottish Enterprise, 2018), while the worldwide potential of ocean currents is estimated at 450 GW.[8] Tidal stream energy requires flow speeds greater than 2.0 m/s to be exploitable (Encarnacion and others, 2019). The funnelling effects of bays, estuaries and inlets can provide a viable tidal or current energy resource. Such locations as the Bay of Fundy in Canada, Cook Strait in New Zealand and the Pentland Firth in Scotland are known for their significant potential and have been targeted for development. Early commercial ventures, such as the 240-MW La Rance tidal energy station in France and the 254-MW Sihwa Lake tidal power station in the Republic of Korea, harnessed tidal energy by impoundment through the construction of barrages.

Although various tidal projects have been proposed, in particular on the west coast of the United Kingdom, progress with regard to construction has been slow, mainly because tidal barrages can affect ecosystems and water quality (Kadiri and others, 2012). Very high capital costs are another deterrent. The tidal energy industry has therefore focused primarily on extracting energy from fast-flowing tidal streams using horizontal axis tidal turbines, which have progressed from single prototype deployments to small-scale arrays (Encarnacion and others, 2019). The environmental monitoring programme implemented for the deployment of the first large-scale commercial tidal stream generator (SeaGen) became the road map for future tidal projects (Savidge and others, 2014). The first grid-connected tidal array with three 100 kW turbines has been operating successfully in the Shetland Islands since 2016,[9] while the MeyGen project, also in Scotland, is the largest tidal energy array currently deployed, with 6 MW.[10] However, since 2016, the industry has largely stalled, in particular in the United Kingdom. In addition, the high-profile collapse of OpenHydro created significant negative publicity for the industry.[11] As of 2020, tidal energy has yet to make a significant leap towards the installation of utility-scale projects.

2.2.3. Wave energy

The world's theoretical wave power resource is estimated at 2.11 TW, and sites with values around 30 kW/m (or even lower) are usually considered to be commercially viable for wave energy extraction, depending on the technology (Sandberg and others, 2016). The locations with the largest wave power resource are between latitudes 40° and 60° (Gunn and Stock-Williams, 2012). For instance, wave energy sites off the coast of Ireland present annual average power density levels of more than 80 kW/m.

As of 2019, the sector was still not close to commercialization, but progress had been made in assessing the difficulties involved in extracting wave power at a reasonable cost. Significant challenges are the hostile environment in which wave energy converters produce power and the need to design technologies that can reliably operate over the lifetime of a commercial project. A large number of different wave energy conversion concepts and devices are under development, but such variety has resulted in a lack of convergence and overall focus within the sector. However, since 2015, multiple full-scale wave energy converters have been deployed by such developers as

8 See Ocean Energy Council, "Ocean Current Energy". Available at www.oceanenergycouncil.com/ocean-energy/ocean-current-energy.
9 See Yasmin Ali, "World's first grid connected baseload tidal power station", Microgrid Knowledge, 27 November 2018.
10 See Simec Atlantis Energy, "MeyGen".
11 See Offshore Energy, "OpenHydro another casualty of innovation 'valley of death', EMEC says", 27 July 2018.

Wello Oy[12] and SeaBased,[13] and a wave energy conversion device developed by Ocean Energy will be deployed in Hawaii.[14]

2.2.4. Salinity and thermal gradient energy

Salinity gradient energy depends on the differences in the salinity of seawater masses and is obtained when fresh water and salt water are mixed. Estimates of the theoretical resources available globally range between 647 GW and 1,183 GW (IRENA, 2014; Alvarez-Silva and others, 2016). Pressure-retarded osmosis and reverse electrodialysis are the most promising technologies to date (Schaetzle and Buisman, 2015). Pressure-retarded osmosis technology was first utilized in 2009 in Norway (Chae and Kim, 2018) while reverse electrodialysis technology was first used in 2014 in a pilot plant in southern Italy (Tedesco and others, 2017).

Thermal gradient energy can be harnessed from the temperature differences between different seawater masses at various depths (Rau and Baird, 2018). Estimates of theoretical ocean thermal energy conversion (OTEC) potential range from 1 to 3 TW, or up to 7 TW when desalination is also included (Scottish Enterprise, 2018). The minimum required temperature differential between seawater masses is on the order of 20°C, which occurs in areas extending between latitudes 30° north and 30° south (Breeze, 2019). The countries most active in the ocean thermal energy conversion sector are China, France, Japan, Malaysia, the Netherlands, Oman, the Philippines, the Republic of Korea and the United States (Edenhofer and others, 2011; Lewis and others, 2011). Several ocean thermal energy conversion projects are under development or already operational, including a 100 kW onshore installation in Kailua-Kona, Hawaii, United States, which was connected to the grid in 2015 (Patel, 2015), an onshore prototype installed in 2012 in Réunion, France,[15] and a 250 kW plant operating on the island of Kumejima, Japan, since 2013.[16] The nutrient-rich deep water can also be used to enhance mariculture and onshore farming, creating significant additional revenues. Ocean thermal energy conversion and wave and ocean current energies are energy sources with important potential for the African continent.

2.2.5. Marine biomass energy

Marine biomass energy involves the use of marine algae and other viable organic matter for the production of biofuels. The use of marine biomass could circumvent many of the constraints associated with terrestrial biomass energy production, including competition with food crops for agricultural land and the use of energy-intensive fertilizers and pesticides in farming. Interest in biomass energy is also driven by the high productivity of marine ecosystems compared with terrestrial ecosystems (Sheehan and others, 1998; Perlack and others, 2005) and the versatility of marine biomass, which can adapt to a wide range of salinity and light intensity conditions.

The marine biofuel production cycle has two components: the continuous cultivation of marine biomass on a sufficiently large scale to feed the biofuel production cycle, and the conversion of marine biomass into biofuels. Giant kelp is considered to be one of the most prolific organisms on Earth, with growth rates up to 60 cm per day.[17] Efforts are currently under way off the Pacific coast of the United States

12 See https://wello.eu.
13 See https://seabased.com/projects.
14 See Association of Energy Engineers, Hawaii Chapter, Blog Archives, "Navy's wave energy test site: Ocean energy deployment", 27 February 2020. Available at https://aeehawaii.org/blog///wave_article.
15 See Ocean Energy Europe, "OTEC".
16 See OTEC Okinawa, Renewable Energy for the Future, "Related projects". Available at http://otecokinawa.com/en/Project/index.html.
17 See Oceana, "Giant kelp".

to develop an open ocean cultivation system for giant kelp, which can then be converted to biocrude (Buck, 2019). Even though marine biomass remains a promising source of energy, the production of biofuel derived from it has not yet been scaled up to the industrial level. Moreover, further research is needed on the calculation of the carbon intensity of marine biofuels, taking into account, inter alia, the absorption of CO_2 through photosynthesis in the cultivation system and the corresponding emissions during biofuel combustion.

2.2.6. Emerging sources of marine renewable energy

Emerging MRE sources include offshore solar energy and ocean floor geothermal energy. Offshore solar energy is at a nascent stage of development, but it has significant commercial potential (Wang and others, 2019). Ocean floor geothermal energy, in contrast to inland geothermal power generation technology, is still at the conceptual stage (Shnell, 2009; Shnell and others, 2015; Pedamallu and others, 2018).

Offshore solar energy is based on floating solar systems that are designed to withstand the harsh environmental conditions at sea.[18] Given that the offshore environment allows full advantage to be taken of solar irradiation during the day, it seems to be an ideal alternative for the solar industry. Although offshore solar systems are more expensive to install than land-based ones, they are usually more efficient, since panels are in direct contact with seawater, reducing thermal losses and lowering the panel temperature (Trapani and Redón Santafé, 2015; Sahu and others, 2016; Ranjbaran and others, 2019; Spencer and others, 2019). The first floating solar power farm for the marine environment was installed in 2014 in Maldives.[19] Japan, the Netherlands, Singapore and the United Arab Emirates are interested in developing offshore solar farms. Floating solar power farms in inland water bodies are already operating, or are under development or consideration, in many other countries, including Australia, Brazil, China, India, Japan and the Republic of Korea (World Bank Group and others, 2019).

The use of geothermal energy is currently limited to areas on land that host geothermal resources (Tester and others, 2006; Saibi and others, 2013). However, vast amounts of geothermal resources in a supercritical state (fluids at a very high temperature and pressure) can be found in the ocean floor, for example, in mid-ocean volcanic ridges (Hiriart and Hernandez, 2010). The benefits of offshore geothermal energy include the use of seawater as an unlimited geothermal fluid and, because of its cold temperature, as a limitless condenser for the heat exchanger system (Banerjee and others, 2018). Offshore geothermal power plants require no land space or extension of the energy field and, when compared with land-based plants, have potential for further development, although they are not profitable under the current financial framework (Karason and others, 2013).

Current initiatives, such as the Marsili project in Italy and the hydrothermal vents project in the Gulf of California, produce power using steam from an underwater volcano and from hydrothermal vents, respectively. Additional locations for potential offshore geothermal exploration have been found in Iceland and Indonesia (Karason and others, 2013; Prabowo and others, 2017). In the Netherlands, the Exploration Working Programme for Ultra-Deep Geothermal Heat is exploring the viability of offshore geothermal projects in order to assess further investment (Heijnen and others, 2019).

---

18 See Kosatka.Media, "High-wave offshore panels soon a reality", 22 July 2019.

19 See Swimsol, "Recent Swimsol solar energy projects". Available at https://swimsol.com/solar-projects-offshore-solarsea-and-rooftop.

3. Potential environmental impacts of marine renewable energy development

Electricity generation based on MRE can contribute towards the reduction of greenhouse gas emissions, water pollution, particulate matter and waste products, as well as help with climate change mitigation. However, given that any human intervention in the marine environment has inevitable impacts on the surrounding biotic and abiotic systems, it is vital to mitigate or avoid potential negative impacts and increase potential positive impacts, and environmental impact assessments are integral to the evaluation of such impacts (Mendoza and others, 2019). The magnitude and temporal extent of the environmental impacts depends on the size and scale of the project, its location and the type of MRE technology used; for example, simulation studies have shown that small arrays of wave energy converters have minimal effects on the physical environment. A practical way to assess the environmental impact of an MRE installation is to consider the interactions between the environmental stressors introduced by the installation (e.g., collision risk or underwater noise) and the receptors (i.e., the ecosystem elements, such as seabirds or marine mammals). The receptors discussed below are benthic and pelagic habitats, fish and fisheries, marine birds and bats, marine mammals and the oceanographic system and coastal morphology.

The fact that offshore wind projects have been operating since 1991 has enabled the accumulation of experience exists regarding their environmental effects. For instance, an extensive environmental monitoring programme was started in the Belgian part of the North Sea in 2008, when the first offshore wind turbines became operational, and reports describing their environmental impacts have been published annually, up to 2019.[20] However, the impacts of other MRE devices have not been studied in detail, owing to the scarcity of operating wave energy converters and tidal and ocean current turbines; as a result, there are limited baseline and post-installation data with regard to those devices (Copping and Hemery, 2020). Reviews of the environmental impacts of MRE installations can be found in Bray and others, 2016; Willsteed and others, 2017; International Council for the Exploration of the Sea (ICES), 2019; and Copping and Hemery, 2020.

3.1. Benthic and pelagic habitats

The underwater infrastructure of MRE installations, including foundations and anchors, mooring systems and cables, may affect benthic habitats (e.g., reefs, coralligenous formations and seagrass meadows) and pelagic habitats by causing changes in their function and characteristics. Such changes arise as a result of damage (e.g., during cable installation or scouring around the device and mooring foundations) and the creation of habitats (through artificial reef and reserve effects and biofouling) (Copping and Hemery, 2020). Owing to the introduction of a hard substrate, the installed infrastructure plays an important role in the creation of new habitats (replacing previous habitats or restoring damaged ones), which may also attract new species to a site; that issue should be considered within the framework of the specific management objectives for a given installation. Additional indirect effects are described in Copping and Hemery (2020).

Although further research is certainly required, properly designed artificial reefs can have positive impacts for the marine environment.

20 See Kelle Morau, "Offshore wind farms and the marine ecosystem: 10 years of monitoring", Royal Belgian Institute of Natural Sciences, 15 June 2020.

Wind turbine foundations can be utilized as artificial reefs, enhance connectivity among marine protected areas and allow sustainable aquaculture (Bishop and others, 2017; Boero and others, 2017; Roa-Ureta and others, 2019; Glarou and others, 2020). Moreover, the anticipated environmental benefits of not completely removing an offshore wind farm when it is decommissioned are significant, as the remaining substructures can lead to biodiversity enhancement, provide reef habitats and offer protection from bottom trawling (Topham and others, 2019).

Much more research is needed regarding the interactions of floating solar energy and ocean floor geothermal energy with aquatic habitats. Floating solar panels, which are subject to biofouling, may have environmental effects on species dependent on solar radiation (including corals, seagrass and kelp forests) and cause changes in biodiversity (Sahu and others, 2016; Pimentel Da Silva and Branco, 2018). Changes in fluid concentrations as a result of ocean floor geothermal energy use may cause large-scale environmental impacts, such as habitat loss and degradation (Pedamallu and others, 2018).

3.2. Fish and fisheries

The underwater infrastructure of MRE installations may pose a collision risk for fish. Such risk is variable and depends, inter alia, on fish abundance, water velocity and turbine rotational frequency. However, it is still unknown whether there have been actual collisions of fish with underwater turbines, and such occurrences would be difficult to observe. Consequently, the collision outcomes, such as injuries or death, are unknown, and further research on sublethal and non-contact effects is necessary (Copping and Hemery, 2020). In addition, there is a lack of relevant information on fish behaviour with respect to MRE underwater structures. Large marine animals are likely to face entanglement issues (Taormina and others 2018).

Underwater transmission cables connecting MRE projects to land-based electrical substations induce electromagnetic fields. The organisms that may be affected by such fields are those with specific electroreceptors for orientation, mating, navigation and hunting purposes, such as elasmobranchs, marine mammals and invertebrates. Factors that determine the potential vulnerability of marine organisms to electromagnetic fields are: (a) the volume or size of the electrical current being carried by the cable; (b) the cable design; and (c) the distance of marine organisms from the power cable (Snyder and others, 2019). Evidently, more research is needed to understand whether electromagnetic fields are harmful to the few species that can detect them. In Copping and Hemery (2020), it is noted that preliminary evidence indicates that the risk of electromagnetic fields from small numbers of MRE devices could be retired.

Finally, further research is needed regarding the potential environmental interactions of MRE with fisheries, keeping in mind that some major offshore wind markets, such as Denmark and the United Kingdom, allow commercial fishing within offshore wind farms. In the context of marine biomass energy exploitation, potential impacts on fisheries and hazards to protected species should be considered in relation to any large-scale production of macroalgae (Langton and others, 2019).

3.3. Marine birds and bats

Birds are considered to be at risk from MRE development. The physical presence of offshore wind farms may pose threats to seabirds at the individual and population levels, while the magnitude of impact is determined by numerous factors, including bird species, site characteristics and conditions, and seasonal variations. The most important effects are bird collisions, both lethal and sublethal, barrier effects with respect to movement (mainly displacement from foraging sites), avoidance, attraction and habitat loss. Dierschke and others (2016) note that the extent to which seabirds are displaced from, or attracted to, offshore wind farms is uncertain. Specifically, an analytic study of 20 offshore wind farms in European seas revealed that the behavioural responses of seabirds were varied, ranging from strong avoidance to strong attraction. On the other hand, many species showed little behavioural response, while some species used the offshore wind farm structures for dry roosting. The increase in food availability due to the artificial reef effect seems to have an important influence on several species. There is also evidence that large-bodied birds avoid offshore wind turbines (Fox and Petersen, 2019). Nevertheless, long-term monitoring campaigns are needed to fill the gaps in the understanding of how birds, including seabirds, behave around wind turbines and to provide robust estimates of the numbers of bird collisions at such turbines. Proper siting, and the shutting down of turbines on demand, may decrease bird fatalities during the operation of offshore wind farms (Marques and others, 2014; Best and Halpin, 2019).

Owing to the limited number of studies relating to the direct interactions of diving seabirds with tidal turbines, there is a lack of evidence showing that such interactions will occur or that tidal turbines will harm individual seabirds or populations. The most up-to-date published information on the effects of MRE development on seabirds is presented in Copping and Hemery (2020), while recommendations are provided in Isaksson and others (2020).

Finally, the potential impacts of offshore wind farms on bats are poorly understood. As bats have been observed offshore, impacts similar to those of onshore wind farms may therefore be expected (Arnett and others, 2016).

3.4. Marine mammals

Although collisions of marine mammals with moving parts of MRE devices (e.g., blades of a tidal turbine) have not been observed, the possibility of such collisions and the consequences thereof, which are still unknown, remain an active area of research. Entanglement of marine mammals in mooring lines, cables and anchors is another emerging topic of investigation. The risk of injury and mortality of marine mammals caused by entanglement is considered low for single devices; however, a combination of modelling results and field observations will enhance the assessment of that risk. Knowledge gaps and uncertainties include the scaling of collision risk from a single turbine to arrays and the translation of individual collision risk to population-level risk (Copping and Hemery, 2020).

Underwater noise emitted by operational MRE devices is unlikely to cause acoustic injury to marine animals, and there is a low probability of its causing changes in behaviour. On the other hand, underwater noise generated during the construction phase of an MRE installation can have significant impacts. For instance, underwater noise generated during pile-driving operations (for piled bottom-fixed offshore wind farms) can mask the echolocation sounds used by some marine mammals for navigation, hunting and communication, and may potentially impair fish and mammal hearing. Those problems can be addressed through restrictions on pile-driving operations during, for example, the migration of marine mammals or through noise mitigation measures (Koschinski and Lüdemann, 2013). In that respect, floating wind technology and bottom-fixed offshore wind farm foundations that do not require piling – such as gravity-based foundations and suction buckets – compare favourably with piled bottom-fixed foundations. Additional sources of underwater noise include increased vessel traffic during construction and decommissioning activities, the rotation of the turbine itself, the displacement of fluid by turbine blades and such operations as underwater explosions, rock-dumping and dredging.

3.5. Oceanographic system and coastal morphology

The development of MRE in large-scale arrays has the potential to alter the physical processes driven by waves, currents and tides. On the basis of results from numerical model simulations, changes in water circulation, wave height, current speed, salinity, sediment transport and water quality are encountered within and around the area of MRE installations. Up to 2020, there have been few field and laboratory studies to quantify the impacts of MRE devices. The alteration of hydrographic characteristics and the physical presence of large-scale MRE installations, especially when sited nearshore, may also affect neighbouring coastal areas, including by increasing flooding risk (Cazenave and others, 2016; Soukissian and others, 2017).

In conclusion, minimizing environmental impacts while ensuring energy generation at a competitive cost is necessary for the successful deployment of MRE projects. In that context, more real data and coordinated studies are needed to gain a full picture of the environmental impacts of various types of MRE devices.

4. Socioeconomic benefits and impacts from marine renewable energy deployment

4.1. Socioeconomic benefits

MRE has the potential to drive regional and local economic development by providing access to reliable energy in coastal areas and in non-interconnected islands and island States (Kuang and others, 2016). The presence of MRE in the energy mix can reduce vulnerability to volatile energy prices and fluctuating availability.

4.1.1. Creation of new jobs

MRE development can provide economic opportunities and employment in coastal areas (Hoegh-Guldberg and others, 2019). Ocean Energy Systems[21] has set a global target of 300 GW for MRE, excluding offshore wind, by 2050, which could save up to 5.2 billion tons of CO_2 by that year and create 680,000 direct jobs (Huckerby and others, 2016).

In 2018, the onshore and offshore wind energy sectors employed 1.16 million people (REN21, 2019). In 2019, the global offshore wind sector received investment of $29.9 billion, of which China received the highest proportion ($14 billion) (Frankfurt School and UNEP Centre/BloombergNEF, 2020). Offshore wind farms are more labour-intensive than onshore wind farms, which can result in the economic revitalization of coastal communities (IRENA, 2019).

4.1.2. Synergies with other marine sectors

Aquaculture and MRE could be synergetic sectors. Aquaculture sites are mainly located in areas of low energetic conditions; thus, an MRE installation could provide an ideal environment for aquaculture development in its lee. Moreover, the multifunctional co-location of those two sectors (with the sharing of the

21 Ocean Energy Systems is an intergovernmental collaboration between countries, founded in 2001, which operates under a framework established by IEA. It promotes the development of ocean energy around the world. See www.ocean-energy-systems.org.

same infrastructure, for example) can be facilitated through marine spatial planning (see chap. 26), as well as technical advances in the design of more robust fish cages, technological developments in automation, advances in mooring systems and benefit-sharing (where MRE arrays provide shelter to fish farms).

Moreover, abandoned oil and gas platforms can be converted into production and storage units that convert electricity from offshore wind farms into hydrogen and synthetic gas (Jepma and van Schot, 2016; see also chap. 19). Synergies may also arise between the MRE sector and other marine industries, such as transport and operations, supply and manufacture, new materials and mining (Huckerby and others, 2016), and shoreline protection and marine conservation efforts (LiVecchi and others, 2019).

4.2. Potential adverse socioeconomic impacts

Considerable challenges will have to be faced in order to achieve the deployment of MRE, as a new energy source, on a significant scale. Apart from the higher energy cost of MRE installations compared with land-based installations, social acceptance also needs to be addressed. MRE installations may meet strong opposition from other maritime sectors and local coastal communities that are reluctant to share marine space (Dalton and others, 2015; Lange and others, 2018). Important issues arising from the interactions of fisheries and offshore wind farms include the loss of fishing grounds and displacement, gear damage, inadequate compensation schemes and the need for a more dynamic engagement of fishers in planning processes (Gray and others, 2016). MRE installations may also be a cause for concern for the coastal tourism sector because of the potential visual disturbance. Studies conducted on the French Mediterranean coast, as well as in North Wales, United Kingdom, and New Zealand, revealed the opposition of coastal communities to offshore wind farms and wave energy installations, especially in places of high scenic beauty (Devine-Wright and Howes, 2010; Westerberg and others, 2013; Brownlee and others, 2015). Potential conflicts regarding safe navigation and operation of marine vessels may also arise when MRE installations are close to existing maritime transport routes.

In conclusion, the potential environmental and socioeconomic risks underline the importance of extensive stakeholder engagement, robust environmental impact assessments and risk analysis before planning and siting MRE projects.

5. Key remaining knowledge and capacity-building gaps

5.1. Cost reduction

Cost reduction is the most important issue that the MRE industry has to address. Bottom-fixed offshore wind farms may be approaching cost parity with conventional electricity generation sources in some markets; however, no other MRE technology is close to becoming commercially viable without further research and development, targeted innovation and significant financial incentives. The reduction of MRE costs is necessary to attract investors and advance the sector's development. Cost reduction can be achieved on the basis of the following pillars (SI Ocean, 2013; Smart and Noonan, 2018):

- **Scale and volume.** Larger MRE devices and array installations decrease the manufacturing and installation costs, while the larger-scale production of MRE devices reduces the overall individual component cost.

- **Experience and generation of knowledge.** Knowledge generation is important for MRE capacity-building and cost reduction. New knowledge acquired through experience and learning-by-doing will foster the integration of MRE into relevant State policies. The sharing of data and information, exchanges of experiences, research and development and lessons learned are important drivers of cost reduction.
- **Innovation.** Targeted innovation (in the research and development phase of an MRE concept or in the context of actual industrial MRE projects) will reduce costs and increase the yield and reliability of MRE devices.
- **Energy storage.** Accurate short-term forecasting and energy storage are relevant to the issues of intermittent electricity generation and stochastic fluctuation, respectively. Current technologies for energy storage consist of electrochemical systems (e.g., batteries and fuel cells, and hydrogen energy storage), electrical storage (e.g., supercapacitor energy storage and magnetic systems), mechanical systems (e.g., flywheels and water pumps) and thermal systems (Ould Amrouche and others, 2016; Olabi, 2017). Pumped hydroelectric energy storage is the most mature of those technologies and the largest in scale (see also Wang and others, 2019).

5.2. Environmental monitoring and mitigation measures

The environmental monitoring of marine organisms and metocean (oceanographic and meteorological) characteristics is essential for identifying and quantifying variability in the marine environment from the design to the decommissioning of an MRE installation, while the mapping of the ocean floor may contribute significantly to the proper siting of MRE installations (Mulcan and others, 2015).

The establishment of environmental baselines (e.g., mapping and characterization of the seabed, including sediment composition and shallow and deep geology) and the monitoring of biotic elements are necessary to address any adverse impact on biodiversity of the activities in question. In that context, there is a need to define standards for the analysis of environmental monitoring data for MRE development sites and to identify the area over which biological effects may occur to inform baseline data collection.[22] It is also necessary to set thresholds, determine changes in species abundance, diversity, distribution and behaviour and readjust management actions (Foley and others, 2015). The MRE technologies used and the stressors introduced in the marine environment should be considered when designing monitoring procedures. Predictive models can be a supplementary tool, ideally when combined with in situ observations.

Metocean data can be obtained from in situ measurements, outputs from numerical models and remote sensing instruments. Long-term data are required for the preliminary estimation of the available MRE resource and the metocean climate characteristics in the area of the installation. Short (up to 3 days) and medium-term (3–7 days) forecasting of metocean conditions is also important for operational planning activities. During the operation phase, reliable short-term forecasts of the expected power production are required for large-scale power integration.

22 See, for example, United States Department of Commerce, National Oceanic and Atmospheric Administration, "Takes of marine mammals incidental to specified activities; taking marine mammals incidental to construction of the Vineyard Wind Offshore Wind Project", Federal Register, vol. 84, No. 83, 30 April 2019. Available at www.govinfo.gov/content/pkg/FR-2019-04-30/pdf/2019-08666.pdf.

5.3. Strategic considerations for development of marine renewable energy, including funding

The development of national energy strategies may involve a number of objectives. In that context, some critical factors to be considered include reducing the cost of MRE and enhancing its large-scale integration into electric power systems; leveraging a diversity of MRE sources and determining their geographical distribution; reducing barriers to deployment, including siting conflicts and permitting processes; and attracting significant investment in the sector.

Furthermore, the World Conservation Congress of the International Union for Conservation of Nature (IUCN), at its sixth session, asked States and competent authorities to implement a strategy for the development of offshore renewable energy that takes environmental issues into account and to subject that strategy to rigorous strategic environmental assessment (IUCN, 2016). That commitment is completely in line with Sustainable Development Goal 7.[3]

The full development of MRE can enhance the diversity of low-carbon energy options and provide viable alternatives to fossil fuel. Traditional commercial funding sources are often insufficient to achieve that goal, so innovative strategies are required. Private-public partnerships are considered critical for the development of MRE. For example, the European Commission has set up the Ocean Energy Forum, bringing together industry, finance, academia and public authorities to identify solutions and make investment more attractive. In the United States, the Business Network for Offshore Wind[23] is promoting the offshore wind industry.

The importance of the public sector's support is not confined to the funding of the early stages of development of new technologies. Equally, if not more, important is its role in creating a favourable private investment environment through financial and fiscal incentives, renewable portfolio standards, offsets or feed-in tariffs. Investment in new technologies is generally limited to States with the financial means to accept the risks associated with technologies that are not commercially viable. However, developing countries could invest in those MRE technologies that are more mature.

6. Anticipated future trends

Although considerable progress has been made towards the exploitation of MRE, the industry is still in the early stages of development, except for the offshore wind sector. As in general wave energy and tidal energy are not yet commercially viable, the immediate target is to encourage more offshore deployments of single prototypes or small-scale arrays. Such deployments, if successful, will build confidence in the sector and encourage the investments required to develop large-scale farms.

Technological advances are also required to improve power take-off performance and reliability, along with control systems to maximize power absorption. The survivability, reliability and cost-reduction potential of wave and tidal technologies offset the significant investment risk.

In Europe, ambitious LCOE reduction targets for offshore wind, wave and tidal energy have been established under the Strategic Energy Technology Plan (European Commission

23 See www.offshorewindus.org/about-us.

Directorate-General for Energy and others, 2018). The goal for offshore wind energy is to reduce LCOE to a no-subsidies point for fixed offshore wind and to less than €120/MWh for floating offshore wind by 2025. The corresponding targets for wave and tidal energy are €200/MWh and €150/MWh, respectively. Worldwide support by national Governments would allow the industry to develop the critical mass that would in turn generate large cost reductions. The corresponding LCOE projections for salinity gradient energy and ocean thermal energy conversion are €80/MWh and €150–200/MWh, respectively (Ocean Energy Europe, 2016).

A recent trend for increased open sea deployments of wave, tidal and ocean current devices has been to focus on niche markets. Local MRE options may offer a solution for energy needs in off-grid areas and remote coastal and island communities (e.g. small island developing States), including for desalination and aquaculture (LiVecchi and others 2019; Rusu and Onea, 2019).[24] In such applications, wave and tidal energy have the potential to be competitive with diesel generators. In most cases, wave and tidal energy devices would be smaller in size than utility-scale devices, so a high capital outlay would not be required. Working towards utility scale by incrementally scaling up devices and array size may provide the pathway to the commercialization of wave and tidal energy.

The offshore wind sector is expected to expand globally, including in areas where no offshore wind farms are currently operational. In the next decade, Asia and the United States are expected to make significant progress, with growth in offshore wind energy also accelerating in nascent markets. The use of floating platforms is a step change for the industry. Floating wind energy is on the cusp of commercial deployment, and there are new technologies at earlier stages of development with the potential for offshore deployment. For example, multi-turbine platforms may offer an alternative to continuous increases in wind turbine size. High-altitude wind concepts, such as autonomous kites or unpiloted aircraft, and hybrid platforms combining various types of MRE technology on a single platform are also moving along the development process.

Acknowledgements: We would like to acknowledge the substantial contribution of Nikolaos Koukouzas with respect to offshore geothermal energy.

References

Alvarez-Silva, O.A., and others (2016). Practical global salinity gradient energy potential. *Renewable and Sustainable Energy Reviews*, vol. 60, pp. 1387–1395.

Arnett, E.B., and others (2016). Impacts of wind energy development on bats: a global perspective. In *Bats in the Anthropocene: Conservation of Bats in a Changing World*, Voigt, C.C., and T. Kingston, eds. Springer International Publishing.

Banerjee, A., and others (2018). Evaluation of possibilities in geothermal energy extraction from oceanic crust using offshore wind turbine monopiles. *Renewable and Sustainable Energy Reviews*, vol. 92, pp. 685–700.

24 See United States Department of Energy, Office of Energy Efficiency and Renewable Energy, "Powering the Blue Economy".

Best, B.D., and P.N. Halpin (2019). Minimizing wildlife impacts for offshore wind energy development: Winning tradeoffs for seabirds in space and cetaceans in time. *PLOS ONE*. vol. 14, No. 5, e0215722.

Bishop, M.J., and others (2017). Effects of ocean sprawl on ecological connectivity: impacts and solutions. *Journal of Experimental Marine Biology and Ecology*, vol. 492, pp. 7–30.

Boero, F., and others (2017). CoCoNet: towards coast to coast networks of marine protected areas (from the shore to the high and deep sea), coupled with sea-based wind energy potential. *SCIRES-IT – SCIentific RESearch and Information Technology*, vol. 6 (Supplement), pp. 1–95.

Bray, L., and others (2016). Expected effects of offshore wind farms on Mediterranean marine life. *Journal of Marine Science and Engineering*, vol. 4, No. 1, 18.

Breeze, P. (2019). Chapter 14 - Marine Power Generation Technologies. In *Power Generation Technologies*, P. Breeze, ed. Third Edition, pp. 323–349. Newnes.

Brownlee, M.T.J., and others (2015). Place attachment and marine recreationists' attitudes toward offshore wind energy development. *Journal of Leisure Research*, vol. 47, No. 2, pp. 263–284.

Buck, H.J. (2019). Marine cultivation technology opening the door to the rich sources of clean energy in our oceans. *Science Focus*. www.sciencefocus.com/planet-earth/marine-cultivation-technology-opening-the-door-to-the-rich-sources-of-clean-energy-in-our-oceans.

Cazenave, P.W., and others (2016). Unstructured grid modelling of offshore wind farm impacts on seasonally stratified shelf seas. *Progress in Oceanography*, vol. 145, pp. 25–41.

Chae, S.H., and J.H. Kim (2018). Recent issues relative to a low salinity pressure-retarded osmosis process and suggested technical solutions. In *Membrane-Based Salinity Gradient Processes for Water Treatment and Power Generation*, S. Sarp and N. Hilal, eds., pp. 273–295. Elsevier.

Copping, A., and L.G. Hemery, eds., (2020). OES-Environmental 2020 State of the Science Report: Environmental Effects of Marine Renewable Energy Development Around the World. Report for Ocean Energy Systems (OES). https://tethys.pnnl.gov/sites/default/files/publications/2020-State-of-the-Science-Report-LR-Tabs.pdf.

Dalton, G., and others (2015). Economic and socio-economic assessment methods for ocean renewable energy: Public and private perspectives. *Renewable and Sustainable Energy Reviews*, vol. 45, pp. 850–878.

Devine-Wright, P., and Y. Howes (2010). Disruption to place attachment and the protection of restorative environments: a wind energy case study. *Journal of Environmental Psychology*, vol. 30, No. 3, pp. 271–280.

Dierschke, V., and others (2016). Seabirds and offshore wind farms in European waters: Avoidance and attraction. *Biological Conservation*, vol. 202, pp. 59–68.

Dlugokencky Ed, and Pieter Tans (2020). Trends in Atmospheric Carbon Dioxide, NOAA/GML. www.esrl.noaa.gov/gmd/ccgg/trends.

Edenhofer, O., and others (2011). Renewable Energy Sources and Climate Change Mitigation: Special Report of the Intergovernmental Panel on Climate Change. New York: Cambridge University Press.

Encarnacion, J.I., and others (2019). Design of a horizontal axis tidal turbine for less energetic current velocity profiles. *Journal of Marine Science and Engineering*, vol. 7, No. 7, 197.

European Commission, Directorate-General for Energy and others (2018). SET Plan Delivering Results: The Implementation Plans. Research & Innovation Enabling the EU's Energy Transition. European Union. https://setis.ec.europa.eu/sites/default/files/setis%20reports/setplan_delivering_results_2018.pdf.

Foley, M.M., and others (2015). Using ecological thresholds to inform resource management: current options and future possibilities. *Frontiers in Marine Science*, vol. 2, art. 95.

Fox, A.D., and I.K. Petersen (2019). Offshore wind farms and their effects on birds. *Dansk Ornitologisk Forenings Tidsskrift*, vol. 113, No. 3, pp. 86–101.

Frankfurt School-UNEP Centre/BNEF (2020). Global Trends in Renewable Energy Investment 2020. www.fs-unep-centre.org/wp-content/uploads/2020/06/GTR_2020.pdf.

Glarou, and others (2020). Using artificial-reef knowledge to enhance the ecological function of offshore wind turbine foundations: implications for fish abundance and diversity. *Journal of Marine Science and Engineering*, vol. 8, 332.

Gray, M., and others (2016). Changes to Fishing Practices Around the UK as a Result of the Development of Offshore Windfarms–Phase 1. The Crown Estate. www.thecrownestate.co.uk/media/2600/final-published-ow-fishing-revised-aug-2016-clean.pdf.

Gunn, K., and C. Stock-Williams (2012). Quantifying the global wave power resource. *Renewable Energy*, vol. 44, pp. 296–304.

Heijnen, L., and others (2019). Ultra-Deep Geothermal Program in the Netherlands. In *European Geothermal Congress*. The Hague, Netherlands: European Geothermal Energy Council, p. 6.

Hiriart, G., and I. Hernandez (2010). Electricity Generation from Hydrothermal Vents. *Geothermal Resources Council Transactions*, vol. 34, pp. 1033–1038.

Hoegh-Guldberg, O., and others (2019). The ocean as a solution to climate change: five opportunities for action. https://oceanpanel.org/sites/default/files/2019-10/HLP_Report_Ocean_Solution_Climate_Change_final.pdf.

Huckerby, J and others (2016). An International Vision for Ocean Energy. Version III. Ocean Energy Systems Technology Collaboration Programme. https://testahemsidaz2.files.wordpress.com/2017/03/oes-international-vision.pdf.

ICES (2019). Working Group on Marine Benthal Renewable Developments (WGMBRED). ICES Scientific Reports. Denmark: International Council for the Exploration of the Sea. www.ices.dk/sites/pub/Publication%20Reports/Expert%20Group%20Report/HAPISG/2019/Working%20Group%20on%20Marine%20Benthal%20and%20Renewable%20Energy%20Developments.pdf.

International Energy Agency (IEA) (2019). Offshore Wind Outlook 2019. World Energy Outlook Special Report. International Energy Agency.

________ (2020). Global CO_2 emissions in 2019, IEA, Paris. www.iea.org/articles/global-co2-emissions-in-2019.

Intergovernmental Panel on Climate Change (IPCC) (2018). Global Warming of 1.5°C. An IPCC Special Report on the impacts of global warming of 1.5°C above pre-industrial levels and related global greenhouse gas emission pathways, in the context of strengthening the global response to the threat of climate change, sustainable development, and efforts to eradicate poverty, V. Masson-Delmotte and others, eds. Intergovernmental Panel on Climate Change. In press.

International Renewable Energy Agency (IRENA) (2014). Salinity Gradient Energy: technology brief. Abu Dhabi: International Renewable Energy Agency. www.irena.org/publications/2014/Jun/Salinity-Gradient.

________ (2019). Renewable Energy and Jobs. Annual Review 2019. Abu Dhabi: International Renewable Energy Agency. www.irena.org/-/media/Files/IRENA/Agency/Publication/2019/Jun/IRENA_RE_Jobs_2019-report.pdf.

________ (2020a). Renewable Capacity Statistics 2020. Abu Dhabi: International Renewable Energy Agency. www.irena.org/-/media/Files/IRENA/Agency/Publication/2020/Mar/IRENA_RE_Capacity_Statistics_2020.pdf.

________ (2020b). Renewable Power Generation Costs in 2019. Abu Dhabi: International Renewable Energy Agency. www.irena.org/-/media/Files/IRENA/Agency/Publication/2020/Jun/IRENA_Power_Generation_Costs_2019.pdf.

Isaksson, Natalie, and others (2020). Assessing the effects of tidal stream marine renewable energy on seabirds: A conceptual framework. *Marine Pollution Bulletin*, vol. 157, 111314.

International Union for Conservation of Nature (IUCN) (2016). Development of Offshore Renewable Energy and Biodiversity Conservation. IUCN Resolutions, Recommendations and Other Decisions. World Conservation Congress. International Union for Conservation of Nature and Natural Resources, Honolulu, Hawaii, United States.

Jepma, C., and M. van Schot (2016). Connect North Sea oil and gas platforms to offshore wind farms to produce green gas. Energypost.Eu. https://energypost.eu/connect-north-sea-oil-gas-platforms-offshore-wind-farms-produce-green-gas.

Kadiri, M., and others (2012). A review of the potential water quality impacts of tidal renewable energy systems. *Renewable and Sustainable Energy Reviews*, vol. 16, No. 1, pp. 329–341.

Karason, B., and others (2013). Utilization of Offshore Geothermal Resources for Power Production. In Proceedings of Thirty-Eighth Workshop on Geothermal Reservoir Engineering, Stanford University, Stanford, California, (SGP-TR-198), p. 10.

Koschinski, S., and K. Lüdemann (2013). Development of Noise Mitigation Measures in Offshore Wind Farm Construction. Federal Agency for Nature Conservation. www.cbd.int/doc/meetings/mar/mcbem-2014-01/other/mcbem-2014-01-submission-noise-mitigation-en.pdf.

Kuang, Y., and others (2016). A review of renewable energy utilization in islands. *Renewable and Sustainable Energy Reviews*, vol. 59, pp. 504–513.

Lange, M., and others (2018). Governance challenges of marine renewable energy developments in the US–Creating the enabling conditions for successful project development. *Marine Policy*, vol. 90, pp. 37–46.

Langton, R., and others (2019). An Ecosystem Approach to the Culture of Seaweed. NOAA Tech. Memo. NMFS-F/SPO-195. https://spo.nmfs.noaa.gov/sites/default/files/TMSPO195.pdf.

Lewis, A., and others (2011). Ocean Energy. In *IPCC Special Report on Renewable Energy Sources and Climate Change Mitigation*, Edenhofer, O., and others, eds. Cambridge University Press.

LiVecchi, A., and others (2019). Powering the Blue Economy; Exploring Opportunities for Marine Renewable Energy in Maritime Markets. U.S. Department of Energy, Office of Energy Efficiency and Renewable Energy. Washington, D.C. www.energy.gov/sites/prod/files/2019/09/f66/73355-v2.pdf.

Marques, A.T., and others (2014). Understanding bird collisions at wind farms: An updated review on the causes and possible mitigation strategies. *Biological Conservation*, vol. 179, pp. 40–52.

Mendoza, E., and others (2019). A framework to evaluate the environmental impact of OCEAN energy devices. *Renewable and Sustainable Energy Reviews*, vol. 112, pp. 440–449.

Mulcan, A., and others (2015). Marine Benthic Habitats and Seabed Suitability Mapping for Potential Ocean Current Energy Siting Offshore Southeast Florida. *Journal of Marine Science and Engineering*, vol. 3, pp. 276–298.

Musial, W.D., and others (2019). 2018 Offshore Wind Technologies Market Report. National Renewable Energy Lab.(NREL), Golden, CO (United States). www.energy.gov/sites/prod/files/2019/09/f66/2018%20Offshore%20Wind%20Technologies%20Market%20Report.pdf.

Ocean Energy Europe (2016). European Commission Issue Paper on Ocean Energy Industry Response. Technical Report. https://setis.ec.europa.eu/system/files/tpoandoee_input_act1and2_ocean.pdf.

Olabi, A.G. (2017). Renewable energy and energy storage systems. *Energy*, vol. 136, pp. 1–6.

Ould Amrouche, S., and others (2016). Overview of energy storage in renewable energy systems. *International Journal of Hydrogen Energy*, vol. 41, No. 45, pp. 20914–20927.

Patel, S. (2015). Largest OTEC Facility Inaugurated in Hawaii. *Power Magazine*. www.powermag.com/largest-otec-facility-inaugurated-in-hawaii.

Pedamallu, L.R.T., and others (2018). Environmental Impacts of Offshore Geothermal Energy. *Geothermal Resources Council Transactions*, vol. 42, p. 10.

Perlack, R.D., and others (2005). Biomass as feedstock for a bioenergy and bioproducts industry: The technical feasibility of a billion-ton annual supply. US Department of Energy and US Department of Agriculture. https://www1.eere.energy.gov/bioenergy/pdfs/final_billionton_vision_report2.pdf.

Pimentel Da Silva, G.D., and D.A.C. Branco (2018). Is floating photovoltaic better than conventional photovoltaic? Assessing environmental impacts. *Impact Assessment and Project Appraisal*, vol. 36, pp. 390–400.

Prabowo, T.R., and others (2017). A new idea: The possibilities of offshore geothermal system in Indonesia marine volcanoes. In *IOP Conference Series: Earth and Environmental Science*, vol. 103, 012012. IOP Publishing.

Ranjbaran, P., and others (2019). A review on floating photovoltaic (FPV) power generation units. *Renewable and Sustainable Energy Reviews*, vol. 110, pp. 332–347.

Rau, G.H., and J.R. Baird (2018). Negative-CO_2-emissions ocean thermal energy conversion. *Renewable and Sustainable Energy Reviews*, vol. 95, pp. 265–272.

REN21 (2019). Renewables 2019 Global Status Report. Paris: REN21 Secretariat. www.ren21.net/wp-content/uploads/2019/05/gsr_2019_full_report_en.pdf.

Roa-Ureta, R.H., and others (2019). Modelling long-term fisheries data to resolve the attraction versus production dilemma of artificial reefs. *Ecological Modelling*, vol. 407, 108727.

Rusu, E., and F. Onea (2019). An assessment of the wind and wave power potential in the island environment. *Energy*, vol. 175, pp. 830–846.

Sahu, A., and others (2016). Floating photovoltaic power plant: A review. *Renewable and Sustainable Energy Reviews*, vol. 66, pp. 815–824.

Saibi, H., and others (2013). Geothermal energy. In *Handbook of Sustainable Engineering*, Kauffman J., and K.M. Lee, eds. Springer.

Sandberg, A., and others (2016). Critical factors influencing viability of wave energy converters in off-grid luxury resorts and small utilities. *Sustainability*, vol. 8, No. 12, 1274.

Savidge, G., and others (2014). Strangford Lough and the SeaGen Tidal Turbine. In *Marine Renewable Energy Technology and Environmental Interactions. Humanity and the Sea*, Shields M., and A. Payne, eds. Springer.

Schaetzle, O., and C.J.N. Buisman, (2015). Salinity Gradient Energy: Current State and New Trends. *Engineering,* vol. 1, No. 2, pp. 164–166.

Scottish Enterprise (2018). Marine Renewable Energy, Subsea Engineering Opportunity, International Market Insights Report Series. p. 10.

Sheehan, J., and others (1998). Look back at the US department of energy's aquatic species program: biodiesel from algae; close-out report. National Renewable Energy Lab., Golden, CO. (US). www.nrel.gov/docs/legosti/fy98/24190.pdf.

Shnell, J. (2009). Global Supply of Clean Energy from Deep Sea Geothermal Resources. *Geothermal Resources Transactions*, pp. 137–142.

Shnell, J., and others (2015). Energy from Ocean Floor Geothermal Resources. In *Proceedings World Geothermal Congress 2015*, Melbourne, Australia, p. 6.

SI Ocean (2013). Ocean Energy: Cost of Energy and Cost Reduction Opportunities. https://oceanenergy-sweden.se/wp-content/uploads/2018/03/130501-si-ocean-cost-of-energy-report.pdf.

Smart, G., and M. Noonan (2018). Tidal Stream and Wave Energy Cost Reduction and Industrial Benefit: Summary Analysis. Report by ORE Catapult. www.marineenergywales.co.uk/wp-content/uploads/2018/05/ORE-Catapult-Tidal-Stream-and-Wave-Energy-Cost-Reduction-and-Ind-Benefit-FINAL-v03.02.pdf.

Snyder, D.B., and others (2019). Evaluation of Potential EMF Effects on Fish Species of Commercial or Recreational Fishing Importance in Southern New England. OCS Study BOEM 2019-049. https://espis.boem.gov/final%20reports/BOEM_2019-049.pdf.

Soukissian, T.H., and others (2017). Marine renewable energy in the Mediterranean Sea: status and perspectives. *Energies*, vol. 10, 1512.

Spencer, R.S., and others (2019). Floating photovoltaic systems: Assessing the technical potential of photovoltaic systems on man-made water bodies in the continental United States. *Environmental Science & Technology*, vol. 53, No. 3, pp. 1680–1689.

Taormina, B., and others (2018). A review of potential impacts of submarine power cables on the marine environment: Knowledge gaps, recommendations and future directions. *Renewable and Sustainable Energy Reviews*, vol. 96, pp. 380–391.

Tedesco, M., and others (2017). Towards 1 kW power production in a reverse electrodialysis pilot plant with saline waters and concentrated brines. *Journal of Membrane Science*, vol. 522, pp. 226–236.

Tester, J.W., and others (2006). The future of geothermal energy. Impact of Enhanced Geothermal Systems (EGS) on the United States in the 21st Century. Massachusetts Institute of Technology, Cambridge, MA. https://www1.eere.energy.gov/geothermal/pdfs/future_geo_energy.pdf.

Topham, Eva. and others (2019). Challenges of decommissioning offshore wind farms: Overview of the European experience. In *Journal of Physics: Conference Series*, vol. 1222, 012035. IOP Publishing.

Trapani, K., and M. Redón Santafé (2015). A review of floating photovoltaic installations: 2007–2013. *Progress in Photovoltaics: Research and Applications*, vol. 23, No. 4, pp. 524–532.

Wang, Z., and others (2019). A review of marine renewable energy storage. *International Journal of Energy Research,* vol. 43, No. 12, pp. 6108–6150.

Westerberg, V., and others (2013). The case for offshore wind farms, artificial reefs and sustainable tourism in the French Mediterranean. *Tourism Management*, vol. 34, pp. 172–183.

Willsteed, E., and others (2017). Assessing the cumulative environmental effects of marine renewable energy developments: Establishing common ground. *Science of the Total Environment*, vol. 577, pp. 19–32.

World Meteorological Organization (WMO) (2020). WMO Statement on the State of the Global Climate in 2019. WMO-No. 1248. Switzerland: World Meteorological Organization. https://library.wmo.int/doc_num.php?explnum_id=10211.

World Bank Group and others (2019). Where Sun Meets Water: Floating Solar Market Report. Washington, D.C.: World Bank. http://documents1.worldbank.org/curated/en/579941540407455831/pdf/Floating-Solar-Market-Report-Executive-Summary.pdf.

Chapter 22
Invasive species

Contributors: Thomas W. Therriault (convener), Marnie L. Campbell, Alan Deidun, Bella S. Galil, Chad L. Hewitt, Graeme Inglis, Henn Ojaveer (lead member), Chul Park (co-lead member), Bing Qiao, Renison Ruwa (co-lead member) and Evangelina Schwindt.

Keynote points

- Globally, about 2,000 marine non-indigenous species (NIS) have been introduced to new locations through human-mediated movements. A few of those have economic value, but most have had negative ecological, socioeconomic or human health impacts. With increased trade and climate change, biological invasions are likely to increase.
- NIS can pose significant biosecurity and biodiversity hazards. Large-scale NIS surveys with broad taxonomic coverage are lacking, as are studies documenting the range of potential impacts in recipient environments.
- Major invasion vectors (i.e., ballast water, biofouling, aquaculture, trade in live specimens, canals and plastic or other debris) lack characterization and understanding at the global, and often regional, levels and, other than for the management of ballast water and sediments, there is an absence of regulation. Given the multi-vector nature of both the introduction and the spread of NIS, there is a need for comprehensive and integrated legal instruments with robust enforcement to mitigate the movement of species and holistic monitoring programmes that can detect them.
- Better tools are urgently required to assess the potential risks of NIS under changing environmental conditions, to identify the native species and ecosystems most at risk and to determine the best way to respond (i.e., through early detection and rapid response). That is especially true for species with no previously documented invasion history.

1. Introduction

Invasion by non-indigenous species (NIS) is a major driver of biodiversity change that can reduce biodiversity, alter community structure and function, diminish fisheries and aquaculture production and impact human health and well-being. It is exacerbated by climate change, including extreme events, and other human-induced disturbances (Bax and others, 2003; MEA, 2005; Ojaveer and others, 2018). NIS are those species, including microbes, that have overcome a natural dispersal barrier to become established in a new biogeographical area outside their native range as an intentional or unintentional result of human-mediated activities (Carlton, 1999). Those species can then spread in the newly invaded area, either naturally or by means of additional human-mediated activities, through a wide range of invasion vectors (i.e., the physical means by which individuals are moved, including biofouling, aquaculture, trade in live specimens and canals) (Carlton and Ruiz, 2005; Richardson and others, 2011). Invasion pathways represent a combination of processes and opportunities that allow individuals to be moved from a source location to a recipient (non-native) one and include some elements of invasion vectors (the term "invasion pathway" has sometimes been used interchangeably with "invasion vector") (Carlton and Ruiz, 2005; Richardson and others, 2011). Species that undergo distributional changes owing to ecosystem regime shifts or in response to climate change in their native range are not considered to be NIS, and neither are cryptogenic species (those whose native range is unknown) (Carlton, 1996). A subset of all NIS, often identified as "invasive alien species", have significant biological, economic or human health impacts (Williamson, 1996; UNEP, 2002). Given that it is often impossible to predict which NIS will become invasive in which area and under which circumstances, the precautionary approach has been followed in the present chapter, which therefore covers all NIS from marine and estuarine systems.

NIS are drivers of change in invaded ecosystems. They are influenced by the ecosystems that they are invading and the activities and events that have allowed them to be moved from their native range. Moreover, there is increased recognition that NIS are a critical component of multiple stressors, especially in coastal marine habitats, and that developments in the global economy and improved transportation are contributing to the spread of NIS (MEA, 2005). Marine ecosystems that are already stressed or degraded as a result of other human-caused impacts, such as overfishing, eutrophication, ocean acidification and habitat alteration, have been shown to be favourable to the establishment of NIS (Crooks and others, 2011). Thus, changes in native biodiversity (including in relation to species included in the appendices to the Convention on International Trade in Endangered Species of Wild Fauna and Flora),[1] productivity (including fisheries), harmful algal blooms and ecosystem structure and function (chaps. 6, 7, 10 and 15) can all directly affect marine invasion success, including where NIS are pathogens. In addition, expected increases in artificial habitats (chap. 14) that allow fouling species to become established in otherwise unsuitable environments may facilitate the introduction and the spread of NIS, the range of which is also extended by human-mediated activities such as marine transport and shipping, aquaculture- and fishing-related movements and stocking, habitat restoration, canals and diversions, marine debris and litter (especially plastics, which do not degrade rapidly and can thus persist as a transport vector) and research activities (chap. 16) (Ruiz and others, 1997; Carlton and others, 2017; Galil and others, 2018; Therriault and others, 2018).

NIS have the potential to affect, directly or indirectly, the biota and ecosystems that support healthy and productive human communities. Although NIS unintentionally introduced or escaped to the wild after an intentional introduction have been occasionally exploited (e.g. the Pacific oyster (*Crassostrea gigas*), the Red Sea prawn (*Penaeus pulchricaudatus*), the Asian tiger shrimp (*P. monodon*), the blue swimming crab (*Portunus segnis*) and the Manila clam (*Ruditapes philippinarum*)), the longer-term impacts tend to be negative, with reduced native diversity. Impacts also extend to coastal communities, directly or indirectly, by reducing the overall productivity and resilience of marine systems that traditionally support sustainable fisheries or aquaculture (Molnar and others, 2008; Schröder and de Leaniz, 2011).

For an improved understanding of invasions at the global scale, there is a need for validated, detailed georeferenced inventories of NIS accessible in searchable databases that can be used to better understand the distribution of such species and the potential mechanisms by which their range is extended. Currently, there is limited, incomplete or no understanding of NIS in many locations around the world, including in relation to the date of their first arrival (or detection) and the likely introduction vectors. Although progress has been made in terms of biodiversity assessments (Costello and others, 2010; Narayanaswamy and others, 2013), especially with advances in molecular techniques (Darling and others, 2017), critical gaps remain with respect to NIS. Specifically, not only does the taxonomy need to be fully resolved for each species, especially where NIS and sibling native species overlap, but an understanding of the native range of such species is also required. Similarly, there is a need for an improved geospatial and temporal understanding of invasion vectors and pathways. Although some regional studies have been conducted in relation to ballast water, there is

1 United Nations, *Treaty Series*, vol. 993, No. 14537.

in general limited information on the NIS transported by many invasion vectors. In addition, there is an incomplete understanding of, inter alia, the characteristics, routes, frequency and intensity of important invasion pathways. Collectively, such information is essential to inform NIS policy and management.

2. Documented baseline and changes in non-indigenous species

Since the first *World Ocean Assessment* (United Nations, 2017) did not contain a formal assessment of the status of NIS and related trends, it is not possible to evaluate changes since its publication. However, there are multiple lines of evidence confirming that NIS continue to spread globally, with new introductions reported in new locations, as a result of a general lack of management and control. Although the International Convention for the Control and Management of Ships' Ballast Water and Sediments, 2004,[2] came into force in September 2017 (International Maritime Organization (IMO), 2019), the degree to which it has been implemented globally and its effectiveness in reducing marine invasions at the regional level are not clear. However, the current experience-building phase may provide important information for future assessments. Similarly, some States have implemented the International Council for the Exploration of the Sea (ICES) Code of Practice on the Introductions and Transfers of Marine Organisms (ICES, 2005) to reduce the threat posed by NIS when intentionally introduced to new areas for cultivation, but invasions have still occurred. Recognizing the growing importance of hull fouling as a vector, ICES has recommended four actions to evaluate and mitigate biofouling introductions (ICES, 2019). However, there are still many invasion vectors that are not globally regulated at present (see below).

Globally, the information available on NIS is quite variable spatially, temporally and taxonomically. NIS are not routinely surveyed or monitored in many locations. There are also strong biases in the breadth and depth of taxonomic coverage and expertise, with significantly better information available on larger, more conspicuous species (i.e., fishes and large crustaceans) than on smaller, less conspicuous ones (i.e., worms and other small invertebrates).

It is important to note that the consequences of marine invasions can take a considerable time to manifest and are notoriously difficult to quantify. There are often time lags between when an NIS is introduced to a new location and when the species is detected or impacts are noted. Furthermore, important pre-invasion baseline data are often not available. Thus, it is difficult to attribute observed ecosystem changes to NIS specifically, especially when so many other external stressors are affecting marine ecosystems. However, if global or regional baseline inventories are established, as suggested by Tsiamis and others (2019) for European Union countries, it will be possible to gain a better understanding of both the changes in NIS over space and time and their impacts on ecosystems and human well-being, recognizing that critical validation of those inventories will be required to ensure that they are fit for purpose. The first comprehensive region-specific analysis of baseline status and trends for multiple taxonomic groups is provided below (see sect. 4).

2 International Maritime Organization, document BWM/CONF/36, annex.

3. Consequences for human communities, economies and well-being

Not only do NIS affect the realization of Sustainable Development Goal 14 (Conserve and sustainably use the oceans, seas and marine resources for sustainable development) by contributing to the degradation of coastal habitats and the ecosystem goods and services associated with them, but they may also directly or indirectly affect that of many other Goals[3] (see International Council for Science (ICSU) and others, 2017). The achievement of Goal 1 (End poverty in all its forms everywhere) may be hindered by the continued spread of NIS that negatively affect fisheries and aquaculture directly or indirectly by altering the structure and function of ecosystems, especially in the case of small island developing States and least developed countries, which lack NIS regulations, policies, and monitoring and early detection and rapid response plans. Similarly, NIS could jeopardize the achievement of Goal 2 (End hunger, achieve food security and improved nutrition and promote sustainable agriculture) by compromising seafood safety and security by means of the same mechanisms. In many cases, NIS, especially those with the potential to affect human health, can be considered as a biological contaminant. Thus, the continued global spread of NIS, especially human pathogens such as *Vibrio cholerae*, also affects the achievement of Goal 3 (Ensure healthy lives and promote well-being for all at all ages). Some NIS have the potential to dramatically alter marine coastal environments and communities and, as such, could negatively influence the achievement of Goal 6 (Ensure availability and sustainable management of water and sanitation for all). There is growing evidence that many biofouling marine NIS are able to exploit anthropogenic structures, including docks, oil platforms and wind farms. As growing energy demands result in the development of coastal and offshore infrastructure, NIS could also hinder the achievement of Goal 7 (Ensure access to affordable, reliable, sustainable and modern energy for all). Sustainable growth in fisheries and aquaculture could be compromised in areas where NIS continue to spread unchecked. Thus, NIS also have the potential to compromise the achievement of Goal 8 (Promote sustained, inclusive and sustainable economic growth, full and productive employment and decent work for all) and Goal 9 (Build resilient infrastructure, promote inclusive and sustainable industrialization and foster innovation).

Good ocean governance, associated with Goal 16 (Promote peaceful and inclusive societies for sustainable development, provide access to justice for all and build effective, accountable and inclusive institutions at all levels) could play an important role in improving the understanding of marine NIS and their impacts globally. Such governance could include the development of a reporting framework or database that would allow the ever-changing distributions of NIS to be documented, so as to allow informed management or policy development in areas beyond national jurisdictions. Furthermore, there are many marine ecosystems in respect of which even basic information on NIS is lacking (see sects. 2 and 4). In that regard, global partnerships and capacity-building may be possible under Goal 17 (Strengthen the means of implementation and revitalize the Global Partnership for Sustainable Development). If progress on achieving the Sustainable Development Goals is slow, then the spread and impacts of NIS could be exacerbated. For example, without progress on Goal 13 (Take urgent action to combat climate change and its impacts), the few marine

---

3 See General Assembly resolution 70/1.

ecosystems that currently have only a limited number of NIS, such as the Arctic Ocean and the Southern Ocean (see sect. 4), are likely to see invasions proceed at a much faster rate as those environments become more suitable for a wide variety of taxa, and abiotic and biotic barriers to invasion are degraded or removed.

NIS are also addressed by other global policy documents, especially those pertaining to biodiversity, given the negative relationship between the two. For example, the Convention on Biological Diversity[4] recognizes the threat of NIS and article 8 (h) thereof provides that each contracting party shall, as far as possible and as appropriate, prevent the introduction of, control or eradicate those alien species that threaten ecosystems, habitats or species. The Intergovernmental Science-Policy Platform on Biodiversity and Ecosystem Services has also recognized the negative impacts of NIS around the world and has started a process for the assessment of those species.

Some NIS have the potential to impair human health and well-being. For example, introduced *Vibrio* bacteria and harmful algal species (dinoflagellates, diatoms and cyanobacteria) that create toxins can have a negative impact on marine biota and human consumers. Their effects are expected to worsen as they benefit from climate change (Ruiz and others, 2000; Paerl and Huisman, 2009). In the highly invaded Mediterranean, nine venomous and poisonous NIS from the Indian Ocean or the Western Indo-Pacific pose human health risks (Galil, 2018). In addition, the Indo-Pacific lionfish *Pterois volitans* produces a toxin that is dangerous to humans, although it rarely results in death. However, only fragmentary information is available concerning the spatial and temporal trends in those impacts on human health, as underdiagnosis and underreporting hamper the quantitative assessment of the global incidence of medically treated cases, and ignorance of the extent and severity of, and trends in, those emerging public health risks may hinder risk analyses.

Some NIS, whether introduced intentionally or not, have provided economic benefits, but there is often a trade-off between such benefits and the ecological consequences. For example, the Pacific oyster has been introduced in coastal environments around the world, including in North America, South America, Africa, Australia and Europe, resulting in economic opportunities with global production in excess of 4 million tons (Shatkin, 1997; Food and Agriculture Organization of the United Nations (FAO), 2019). However, in many places, that species has spread beyond culture locations and has had a negative impact in some areas on native biodiversity and ecosystem functioning, and human well-being (Molnar and others, 2008; Herbert and others, 2016). The Atlantic salmon (*Salmo salar*) has also been used to create economic opportunities in countries around the world, but large-scale escape events can have negative ecological and socioeconomic impacts (Schröder and de Leaniz, 2011). In the Barents Sea, the red king crab (*Paralithodes camtschaticus*) was introduced intentionally for fisheries but has rapidly spread to adjacent waters and increased in abundance, thus creating conflicts among various user groups and having a negative impact on biodiversity and ecosystem functioning, especially in coastal fjords (Falk-Petersen and others, 2011). The establishment of fisheries of NIS has longer-term implications, especially given the push to ensure that fisheries are sustainable. Furthermore, some NIS, such as the salt marsh grass (*Spartina alterniflora*), which was intentionally introduced to China as an ecosystem engineer, have significantly changed the ecosystems that they have invaded (Wan and others, 2009). Schlaepfer and others (2011) suggest that some NIS may provide ecological or conservation benefits, but predicting those is often complex and dependent on context.

[4] United Nations, *Treaty Series*, vol. 1760, No. 30619.

4. Key region-specific baselines, changes and consequences

4.1. Arctic Ocean

Although basin-wide assessments of NIS in the Arctic Ocean are lacking, there appear to be relatively few invaders at present (Molnar and others, 2008; Chan and others, 2013). However, with rapid environmental changes, including increased temperatures and reduced sea ice, those waters could become suitable for a number of potential invaders in the future (Ware and others, 2016; Goldsmit and others, 2018). Furthermore, those environmental changes could lead to changes in the presence of human-mediated invasion vectors in the Arctic Ocean, especially marine transport, which could result in increased propagule pressure in the future (Miller and Ruiz, 2014).

4.2. North Atlantic Ocean, Baltic Sea, Black Sea, Mediterranean and North Sea

The Mediterranean has a long history of invasions, with 22 NIS recorded before 1900 (Galil, 2012). By the early 2000s, country-level NIS inventories had been initiated and, as of 2011, a total of 787 NIS were listed as being present in European Union marine waters (Macaronesia included), with the highest number (242) reported in the western Mediterranean (Tsiamis and others, 2019; see also Gómez, 2019, regarding 52 microalgal species). However, the omission of data from the eastern and southern Mediterranean induced a major bias, since the number of NIS is substantially greater in the eastern than in the western Mediterranean (over 400 NIS recorded along the coast of Israel alone). There are 727 metazoan NIS in the entire Mediterranean, and the number is rapidly increasing (Galil and others, 2018) (see figure below), while, as of 2018, 173 NIS and cryptogenic species had been reported in the Black Sea. Despite the growing awareness of the role played by the Suez Canal in Mediterranean invasions, measures to mitigate probable NIS propagule increases have yet to be considered for the "New Suez Canal" project, which was launched in 2014 to substantially increase the depth and width of the original canal (Galil and others, 2017). Thus, the main invasion vectors for the Mediterranean include the introduction of Red Sea biota through the Suez Canal; shipping, both commercial and recreational; mariculture; and the aquarium trade. Although the latter vectors contribute fewer NIS, some have had disproportionate impacts, including the green alga (*Caulerpa taxifolia*) introduced with aquarium spillover (Meinesz and Hesse, 1991) and the brown alga (*Fucus spiralis*) introduced in the packaging of fishing bait (Sancholle, 1988).

Changes in non-indigenous species reports over time for the Mediterranean

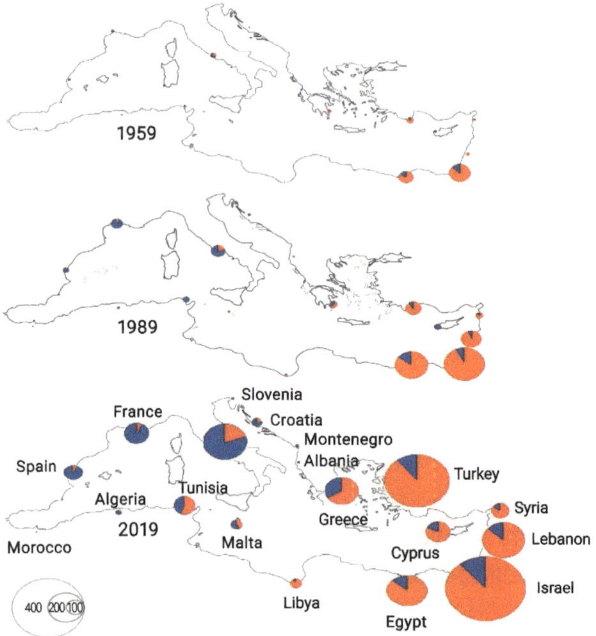

Source: Agnese Marchini and Bella Galil.
Note: Red indicates species introduced through the Suez Canal and blue represents species introduced by other vectors.

Since the beginning of the twenty-first century, the apparent rate of introductions into the Baltic Sea has been 3.2 species per year, almost twice as high as the 1.4 species per year recorded between 1950 and 1999 (ICES, 2018). Ballast water and hull fouling are the main vectors for primary introductions, followed by the natural spread of NIS introduced by rivers and the North Sea. Most NIS in the Baltic Sea originate from North America, the Ponto-Caspian region and East Asia but introductions of subtropical NIS have recently been increasing, such that a total of 174 NIS and cryptogenic species have been recorded in the Baltic Sea (AquaNIS, 2019; Ojaveer and others, 2017; ICES, 2018). However, there remains considerable uncertainty about the direction and magnitude of the impacts of even the most widespread NIS on the structure and dynamics of Baltic Sea ecosystems (Ojaveer and Kotta, 2015).

Although there is some overlap in the studies, NIS reported in the eastern Atlantic include at least 80 species in the North Sea (Reise and others, 2002); 90 in waters around the United Kingdom of Great Britain and Northern Ireland (Minchin and others, 2013); 104 in French Atlantic waters (Goulletquer and others, 2002); and more than 100 in the English Channel (Dauvin and others, 2019). There are at least 189 NIS reported in the western Atlantic (Ruiz and others, 2015) but their number is likely to be higher. For policy and management, validated regional lists are required.

4.3. South Atlantic Ocean and Wider Caribbean

Records of NIS in the South Atlantic Ocean and Wider Caribbean are incomplete both spatially and temporally. The earliest historical compilations are from South Africa, where 12 NIS were reported in the early 1990s, including two global invaders, the European green crab (*Carcinus maenas*) and the blue (Gallo) mussel (*Mytilus galloprovincialis*) (Griffiths and others, 1992). Mead and others (2011) reassessed NIS occurrences in the region and identified 86 NIS, singling out ballast water and ship fouling as the main vectors. Apart from South Africa, the South-East Atlantic coast remains largely unstudied with regard to NIS, although a recent study from Angola reported 29 NIS (Barros Pestana and others, 2017). In the South-West Atlantic, the earliest compilations, which were for Argentina and Uruguay, identified 31 NIS, including one intentionally introduced species (the Pacific oyster) (Orensanz and others, 2002). A recent reassessment for that region identified more than 120 NIS from diverse taxonomic groups (from viruses to plants and fishes), including 33 new detections since 2002 (Schwindt and others, 2020) and, as in the case of South Africa, ships were the main vector for species introductions. The most recent surveys from Brazil identified 73 NIS (Lopes and others, 2009; Teixeira and Creed, 2020), along an extensive coastline with a long history of shipping, which suggests that that number could be underestimating the true richness of NIS. A data gap exists for the North Atlantic coast of South America (from French Guiana to Guyana), where there has been little attention to NIS (Schwindt and Bortolus, 2017), and no extensive compilations are available for the wider Caribbean region, although smaller-scale information is available for the Bolivarian Republic of Venezuela, where 22 NIS have been identified (Pérez and others, 2007), and Colombia, with 16 NIS recorded (Gracia and others, 2011). The lionfish *Pterois volitans* is one of the most problematic and studied NIS in the Caribbean region. Similarly, two invasive sun corals, *Tubastraea coccinea* and *T. tagusensis*, have spread rapidly in the tropical Western Atlantic and the Gulf of Mexico, outcompeting, overgrowing and replacing native corals (Creed and others, 2017).

4.4. Indian Ocean, Arabian Sea, Bay of Bengal, Red Sea, Gulf of Aden and Persian Gulf

Regional records of NIS are incomplete, both spatially and temporally. Despite the size and diversity of the Indian Ocean, studies on marine NIS in that area are scarce, mostly qualitative and geographically scattered, resulting in significant knowledge gaps (Indian Ocean Commission, 2016). For example, two red algae (*Eucheuma denticulatum* and *Kappaphycus alvarezii*) native to the Philippines were introduced for mariculture along the East African coastline (Kenya, Mozambique and the United Republic of Tanzania), resulting in deleterious impacts (Bergman and others, 2001; Halling and others, 2013). *K. alvarezii* was also introduced along the western coast of India and has spread into the Gulf of Mannar Biosphere Reserve, where it has had an impact on native corals (Chandrasekaran and others, 2008). As elsewhere, intentional introductions have been attributed to mariculture activities developed to address food insecurity and to the aquarium trade, for economic benefit, while unintentional introductions are mostly due to maritime shipping activities or transport on floating objects (Indian Ocean Commission, 2016; Anil and others 2003).

4.5. North Pacific Ocean

The North Pacific Ocean is large and biogeographically diverse and, as in other regions, NIS reporting is incomplete. However, as of 2012, at least 747 NIS had been reported in the 23 ecoregions studied (which include Hawaii, United States of America, and the northern Central Indo-Pacific), a similar number to that reported in the Mediterranean. More than 70 per cent of those NIS belong to four phyla, namely, Arthropoda (224), Chordata (tunicates and fishes) (114), Mollusca (110) and Annelida (89) (Lee and Reusser, 2012; Kestrup and others, 2015). While 32 per cent of them were native elsewhere in the North Pacific Ocean, 48 per cent were native to regions outside the North Pacific Ocean and 20 per cent were cryptogenic (Lee and Reusser, 2012; Kestrup and others, 2015). The North-East Pacific (368 NIS) and Hawaii (347 NIS) had similar numbers of invaders, while lower numbers were observed in the North-West Pacific (208) and the northern Central Indo-Pacific (75), possibly owing to different levels of sampling effort. Furthermore, it is important to note that, as there is no systematic survey effort in at least 27 other ecoregions in the North Pacific Ocean, predominately in South-East Asia (Spalding and others, 2007), the number of NIS is expected to be higher for the North Pacific Ocean as a whole. Some more comprehensive studies have been conducted at smaller spatial scales or focused on specific taxonomic groups. For example, there are at least 6 planktonic and 10 algal NIS in the Bohai Sea and port locations in China (Qiao, 2019) not previously reported in baseline surveys (Liu, 2008; Wang and Li, 2006), and San Francisco Bay has more than 234 NIS (Cohen and Carlton, 1998).

As in the case of other regions, ballast water discharges, hull fouling, intentional stocking, aquaculture escapes, aquaculture-associated species and the aquarium and plant trade were all important vectors for the North Pacific. Intentional stocking and aquaculture escapes were more prominent vectors in the North-West Pacific than in the North-East Pacific or Hawaii, which probably reflects the larger scale of aquaculture efforts in Asia. Another difference between the North-East and North-West Pacific was the greater importance of aquaculture-associated NIS in the North-East Pacific (about 42 per cent of NIS), probably reflecting the large number introduced through the import of the Atlantic oyster (*Crassostrea virginica*) from the Atlantic coast of North America and the Pacific oyster from Asia, which resulted in many "hitchhikers" becoming established outside their native range. Increased regulation in recent decades has been effective in reducing the number of inadvertent aquaculture-related movements of NIS. In 2011 the great east Japan earthquake and the resulting tsunami provided a unique vector for species indigenous to Japan to be transported across the North Pacific to Hawaii and North America (Carlton and others, 2017; Therriault and others, 2018).

4.6. South Pacific Ocean

There have been no synthetic assessments of the status of marine bioinvasions across the geographically, culturally and ecologically diverse area of the South Pacific. Most existing information comes from literature and field studies undertaken since the late 1990s in Australia, New Zealand and Chile. A literature review combined with NIS surveys in 41 Australian shipping ports between 1995 and 2004 identified 132 NIS throughout Australia (Sliwa and others, 2009), with 100 NIS detected in Port Phillip Bay alone (Hewitt and others, 2004). There were more NIS in southern temperate Australia than in tropical northern Australia (Hewitt, 2002) but such patterns are confounded by poorer taxonomic resolution in the tropical environments and by the larger urban centres and longer history of shipping in southern Australia (Hewitt and Campbell, 2010). Forty-three similar baseline surveys conducted in New Zealand between 2001 and 2007 (Seaward and others, 2015), combined with published records, museum holdings and submissions to the Marine Invasives Taxonomic Service (Cranfield and others, 1998; Kospartov and others, 2010), show that, as of March 2018, 377 NIS had been recorded in that country's marine waters (214 species are considered established in recipient systems, while the remaining 163 have been recorded only from vessels or transient structures or were failed introductions). Forty-six new NIS were recorded between 2010 and 2018, only 15 of which appear to have become established (Seaward and Inglis, 2018).

At least 53 marine NIS have been reported in Chile (1 seagrass, 15 algae, 26 invertebrates and 11 fishes) (Castilla and Neill, 2009; Turon and others, 2016). However, that is likely to be an underestimate, as there appear to have been few studies of biofouling assemblages in ports and harbours, where introduced species tend to be more abundant. For example, 53 NIS marine invertebrates were recently reported in the Galapagos Islands, Ecuador (Carlton and others, 2019), of which 30 species (57 per cent) were first recorded in fouling plate and shoreline surveys undertaken around shipping docks and infrastructure. Cárdenas-Calle and others (2019) have identified 6 NIS in mainland Ecuador.

There is limited information about the distribution and impact of NIS in the Pacific Island Countries and Territories, as relatively few systematic studies have been done in the region. Surveys undertaken in American Samoa, United States, in 2002 identified 17 NIS, most of which were restricted to Pago Pago harbour and were species known to occur across a broad geographical range (Coles and others, 2003). Forty NIS have been identified in Guam, United States (Paulay and others, 2002) and a preliminary survey of fouling assemblages in Malakal Harbour, Palau, identified 11 NIS (Campbell and others, 2016), in each case comprising mostly ascidians, bryozoans, hydroids and bivalve molluscs. Six NIS, comprising five invertebrates and one alga, have been recorded from the remote Palmyra Atoll, United States (Knapp and others, 2011). Nuisance blooms of fucoid algae, possibly spread by shipping, have been reported in Tahiti, France, (Stiger and Payri, 1999) and Tuvalu (De Ramon N'Yeurt and Iese, 2013).

More than 80 per cent of known NIS in Australia and New Zealand have been associated with incidental transport in ballast water or biofouling (Hewitt and Campbell, 2010; Kospartov and others, 2010) while deliberate introductions of aquaculture species have accounted for less than 2 per cent of records. Introductions of aquaculture species have been more numerous in Chile and Peru (Castilla and Neill, 2009), as well as in the Pacific Island Countries and Territories, throughout which at least 38 NIS have already been transported deliberately over the past 50 years in attempts to establish fisheries or small-scale aquaculture ventures (Eldredge, 1994). In the 1970s and 1980s, the green mussel (*Perna viridis*), sourced from the Philippines, was successively introduced to New Caledonia (France), Fiji, Tonga, the Society Islands (France), Samoa and the Cook Islands (Baker and others, 2007).

4.7. Southern Ocean

The Antarctic Circumpolar Current acts as a strong barrier to natural dispersal that has probably contributed to the uniqueness of Southern Ocean communities. Furthermore, the Southern Ocean has limited shallow-water continental shelves and a poorly described fauna (Brandt and others, 2007). It appears that the most likely vectors for NIS to those waters would either be direct human-mediated transport, such as shipping, or indirect transport by means of longer-distance rafting on artificial marine debris (Lewis and others, 2003; Barnes and others, 2006; Hughes and Ashton, 2017). In addition, any NIS that reached those environments would face challenging environmental conditions. However, with increased rates of climate change, they may become more prone to invasions. To date, only the North Atlantic spider crab (*Hyas araneus*) appears to have been introduced to the Southern Ocean by human activities (Tavares and de Melo, 2004), but it is likely that that will change in the future. Potential future invaders include the blue mussel (Lee and Chown, 2007), the predatory sea star (*Asterias amurensis*) (Byrne and others, 2016) and the kelp (*Undaria pinnatifida*) (James and others, 2015). Owing to its relatively low biodiversity, simple ecosystem structure and unique assemblages dominated by soft-bodied organisms, the Southern Ocean system may be especially vulnerable to introductions of NIS, in particular predatory species that could have a significant impact.

5. Outlook

While introductions of NIS continue as a result of human activities, there are many regions where temporal analyses have not been possible because information on NIS is either very poorly documented or completely lacking. Furthermore, climate change will add to other drivers of ocean change, including water pollution, severe storm events and overfishing, to potentially increase the abundance, ranges and impacts of NIS by altering recipient ecosystems in which native species will be increasingly stressed and by changing human-mediated connectivity through shifts in vectors and pathways. About 40 per cent of the world's population lives in coastal communities, increasing pressure on coastal marine ecosystems through multiple activities and their consequences that contribute to the introduction and spread of NIS, including shipping, boating, marine farming, land-based pollution and marine litter, coastal installations and development, energy production and multiple extraction activities (oil and gas, sediments and fish). It has been predicted that, in regions such as the Arctic, changing environmental conditions will increase the likelihood of new invaders from a variety of taxa (e.g., Goldsmit and others, 2018). They may also lead to changes in shipping patterns, with traffic expected to increase along the Northern Sea Route and become possible along the Northwest passage, which could in turn increase the supply of propagules (Miller and Ruiz, 2014).

Despite the risks posed by NIS, they are substantially underrepresented in existing databases and registries, such that many of the challenges inherent in dealing with such species stem from the limited or incomplete nature of the knowledge base. The magnitude and breadth of that knowledge gap is difficult to assess. It varies among taxa, habitats and regions, and owes much to the inaccessibility of marine ecosystems, caused by such factors as the higher costs of research relative to other ecosystems, the lack of expertise and the lack of interest in NIS that do not benefit or interfere with human needs. Generally, impacts are not well documented unless the NIS is profitable or highly destructive. Thus, the impacts of the vast majority of marine NIS have not been quantitatively or experimentally studied across sufficiently large time periods and spatial scales and remain unknown, as do their cumulative and synergetic connections with other drivers of change affecting the marine environment (Ojaveer and others, 2015).

Vector management is the most effective strategy for preventing the translocation of plants and animals, thereby reducing the introduction and spread of marine NIS. Given the lack of effective control of propagule transfer by the major vectors, management is limited to eradication, removal and control efforts that are frequently futile. NIS that are known or suspected to cause harm, and are identified while they are spatially confined, should be removed in order to mitigate long-term, ongoing management costs. Once NIS have spread widely, eradication or removal is virtually impossible, and attempts to reduce the population to an economically or ecologically acceptable level over the long term are rarely successful (Forrest and Hopkins, 2013). Legislation, regulations and policies to date have been reactive and fragmentary, often following disastrous and costly NIS outbreaks. The United Nations Convention on the Law of the Sea[5] was the first global legally binding instrument that addressed the intentional or accidental introduction of marine species. While guidelines for preventing the introduction of unwanted aquatic organisms and pathogens from ships' ballast water and sediment discharges were established in 1991, and the International Convention for the Control and Management of Ships' Ballast Water and Sediments[6] entered into force in 2017, the management of ships' biofouling is not yet required, despite the IMO guidelines adopted in 2011 (IMO, 2019; IMO resolution MEPC.207(62)). Moreover, in its Strategic Plan for Biodiversity 2011–2020 and the Aichi Biodiversity Targets,[7] the Conference of the Parties to the Convention on Biological Diversity called for invasive alien species and pathways to be identified and prioritized, for priority species to be controlled or eradicated and for measures to be taken to manage pathways by 2020 – a target that will be missed. The goal of the European Union Marine Strategy Framework Directive to ensure, inter alia, that, by 2020, NIS are at levels that do not adversely alter the ecosystems is also likely to be unattainable. Regulation (EU) No.1143/2014 of the European Parliament and of the Council on the prevention and management of the introduction and spread of invasive alien species, which focused only on widely spread species and those of "Union concern", is also unlikely to succeed in marine ecosystems, given that only one marine species has been listed so far. Notwithstanding the existence of some national-level regulations, including in Australia, Canada, New Zealand and the United States, there are still no legally binding and strictly monitored frameworks and tools for addressing major global and regional introduction vectors, such as biofouling, the cultivation of and trade in live organisms, and maritime canals.

5 Ibid., vol. 1833, No. 31363.
6 IMO, document BWM/CONF/36, annex.
7 United Nations Environment Programme, document UNEP/CBD/COP/10/27, annex, decision X/2, annex.

6. Other

Although NIS have long been recognized as a major threat to native biodiversity (Bax and others, 2003), they have been largely overlooked in conservation and protected area planning, regulations and management (Giakoumi and others, 2016; Mačić and others, 2018). In view of global commitments to establishing and extending conservation areas (i.e., Aichi Biodiversity Target 11, article 8 of the Convention on Biological Diversity and Sustainable Development Goal 14), that omission may undermine conservation efforts, including the effectiveness of marine protected areas, in regions overrun by NIS (Galil, 2017; Iacarella and others, 2019). In the Caribbean and the Gulf of Mexico, large populations of Indo-Pacific lionfishes (*Pterois volitans* and *P. miles*) have been documented in marine protected areas, where they have impaired native biodiversity (Ruttenberg and others, 2012; Aguilar-Perera and others, 2017). Similarly, in the Mediterranean, many Erythraean species have become the most conspicuous denizens of marine protected areas, having displaced and replaced native species, thereby reversing marine conservation efforts and hampering stock recovery of economically and ecologically important species (Jimenez and others, 2016; Galil, 2018; Stern and Rothman, 2019).

Thus far, few NIS have been reported in areas beyond national jurisdiction. It is possible that that is because survey efforts to detect NIS in those ecosystems have been limited, but it is also likely that most NIS reported globally are primarily found in coastal waters (those of all continents). In addition, as oceanic abyssal communities have been poorly described, it is possible that, even if potential NIS were to be detected, they would not be recognized as such and might be classified, at least initially, as native species. That is what occurred in South America in the case of the smooth cordgrass (*Spartina alterniflora*), where "ecological mirages" masked the true situation (Bortolus and others, 2015).

Globally, marine NIS pose significant biosecurity and biodiversity hazards, but the identification and mitigation of those hazards lag behind comparable efforts in terrestrial systems, where there has been a longer history of dealing with agricultural and forest pests. Greater efforts must be made to document NIS, their vectors and pathways, and their impacts at larger spatial scales, given that existing marine NIS data are often sparse and incomplete, possibly because of logistical and capacity constraints. Policies aimed at preventing introductions, and the development of early detection and rapid response plans, can reduce the potential impacts of NIS. Earmarked funding, political will and capacity-building related to invasion science are required to effectively understand and ultimately manage marine NIS and their vectors globally. Only then can the sustainability of marine ecosystems be ensured.

References

Aguilar-Perera, A., and others (2017). Lionfish invaded the mesophotic coral ecosystem of the Parque Nacional Arrecife Alacranes, Southern Gulf of Mexico. *Marine Biodiversity*, vol. 47, No. 1, pp. 15–16.

Anil, Arga, and others (2003). *Ballast Water Risk Assessment: Ports of Mumbai and Jawaharlal Nehru India, October 2003, Final Report*. https://doi.org/10.13140/2.1.3554.9768.

AquaNIS. Editorial Board (2019). Information system on Aquatic Non-Indigenous and Cryptogenic Species. 23 October, 2019. www.corpi.ku.lt/databases/index.php/aquanis.

Baker, Patrick, and others (2007). Range and dispersal of a tropical marine invader, the Asian green mussel, *Perna viridis*, in subtropical waters of the southeastern United States. *Journal of Shellfish Research*, vol. 26, No.2, pp. 345–356.

Barnes, David K.A., and others (2006). Incursion and excursion of Antarctic biota: past, present and future. *Global Ecology and Biogeography*, vol. 15, No.2, pp. 121–142.

Barros Pestana, Lueji, and others (2017). A century of introductions by coastal sessile marine invertebrates in Angola, South East Atlantic Ocean. *Marine Pollution Bulletin*, vol. 125, No. 1, pp. 426–32. https://doi.org/10.1016/j.marpolbul.2017.09.041

Bax, Nicholas, and others (2003). Marine invasive alien species: a threat to global biodiversity. *Marine Policy*, vol. 27, No.4, pp. 313–323.

Bergman, Kajsa C., and others (2001). Influence of algal farming on fish assemblages. *Marine Pollution Bulletin*, vol. 42, No. 12, pp. 1379–1389.

Bortolus, Alejandro, and others (2015). Reimagining South American coasts: unveiling the hidden invasion history of an iconic ecological engineer. *Diversity and Distributions*, vol. 21, No. 11, pp. 1267–1283.

Brandt, Angelika, and others (2007). First insights into the biodiversity and biogeography of the Southern Ocean deep sea. *Nature*, vol. 447, No. 7142, p. 307.

Byrne, Maria, and others (2016). From pole to pole: the potential for the Arctic seastar *Asterias amurensis* to invade a warming Southern Ocean. *Global Change Biology*, vol. 22, No. 12, pp. 3874–3887.

Campbell, Marnie L., and others (2016). Marine pests in paradise: capacity building, awareness raising and preliminary introduced species port survey results in the Republic of Palau. *Management of Biological Invasions*, vol. 7, No. 4, pp. 351–363.

Cárdenas-Calle, M., and others (2019). First report of marine alien species in mainland Ecuador: threats of invasion in rocky shores. In *Island invasives: scaling up to meet the challenge*, C.R. Veitch, and others, eds. Occasional Paper of the IUCN Species Survival Commission No. 62, Gland, Switzerland, pp. 452–457.

Carlton, James T. (1996). Biological invasion and cryptogenic species. *Ecology*, vol. 77, No.6, pp. 1653.

________ (1999). The scale and ecological consequences of biological invasions in the world's oceans. In *Invasive Species and Biodiversity Management*, Sandlund Odd Terje and others, eds., p. 431. Dordrecht: Kluwer Academic Publishers.

Carlton, James T., and others (2019). Assessing marine bioinvasions in the Galápagos Islands: implications for conservation biology and marine protected areas. *Aquatic Invasions*, vol. 14, No.1, pp. 1–20.

Carlton, James T., and Gregory M. Ruiz (2005). Vector science and integrated vector management in bioinvasion ecology: conceptual frameworks. *Invasive Alien Species: A New Synthesis*, vol. 63, pp. 36.

Carlton, James T., and others (2017). Tsunami-driven rafting: transoceanic species dispersal and implications for marine biogeography. *Science,* vol. 357, No. 3658, pp. 1402–1406.

Castilla, Juan C., and Paula E. Neill (2009). Marine bioinvasions in the southeastern Pacific: status, ecology, economic impacts, conservation and management. In *Biological Invasions in Marine Ecosystems*, pp. 439–457. Springer.

Chan, Farrah T., and others (2013). Relative risk assessment for ballast-mediated invasions at Canadian Arctic ports. *Biological Invasions*, vol. 15, No. 2, pp. 295–308.

Chandrasekaran, Sivagnanam, and others (2008). Bioinvasion of *Kappaphycus alvarezii* on corals in the Gulf of Mannar, India. *Current Science (00113891)*, vol. 94, No. 9.

Cohen, Andrew N., and James T. Carlton (1998). Accelerating invasion rate in a highly invaded estuary. *Science*, vol. 279, No. 5350, pp. 555–558.

Coles, S.L., and others (2003). *Introduced Marine Species in Pago Pago Harbor, Fagatele Bay and the National Park Coast, American Samoa*. Bishop Museum Technical Report 26.

Costello, Mark John, and others (2010). A census of marine biodiversity knowledge, resources, and future challenges. *PloS One*, vol. 5, No.8, p. e12110.

Cranfield, H.J., and others (1998). *Adventive Marine Species*. National Institute of Water and Atmospheric Research, Wellington, New Zealand.

Creed, Joel, and others (2017). The invasion of the azooxanthellate coral *Tubastraea* (Scleractinia: Dendrophylliidae) throughout the world: history, pathways and vectors. *Biological Invasions*, vol. 19, No. 1, pp. 283–305.

Crooks, Jeffrey A., and others (2011). Aquatic pollution increases the relative success of invasive species. *Biological Invasions*, vol. 13, No. 1, pp. 165–176.

Darling, John A., and others (2017). Recommendations for developing and applying genetic tools to assess and manage biological invasions in marine ecosystems. *Marine Policy*, vol. 85, pp. 54–64.

Dauvin, Jean-Claude, and others (2019). The English Channel: Becoming like the Seas around Japan. In *Oceanography Challenges to Future Earth*, pp. 105–120. Springer.

De Ramon N'Yeurt, Antoine, and Viliamu Iese (2013). Overabundant Invasive Sargassum in Funafuti, Tuvalu – Report.

Eldredge, Lucius G. (1994). Perspectives in aquatic exotic species management in the Pacific Islands. *Introductions of Commercially Significant Aquatic Organisms to the Pacific Islands*, vol. 17, p. 1.

Falk-Petersen, Jannike, and others (2011). Establishment and ecosystem effects of the alien invasive red king crab (*Paralithodes camtschaticus*) in the Barents Sea–a review. *ICES Journal of Marine Science*, vol. 68, No. 3, pp. 479–488.

Food and Agriculture Organization of the United Nations (FAO) (2019). FAO Fisheries & Aquaculture – Cultured Aquatic Species Information Programme – *Crassostrea gigas* (Thunberg, 1793). 2019. www.fao.org/fishery/culturedspecies/Crassostrea_gigas/en.

Forrest, Barrie M., and Grant A. Hopkins (2013). Population control to mitigate the spread of marine pests: insights from management of the Asian kelp *Undaria pinnatifida* and colonial ascidian *Didemnum vexillum. Management of Biological Invasions*, vol. 4, No. 4, pp. 317–326.

Galil, Bella S. (2012). Truth and consequences: the bioinvasion of the Mediterranean Sea. *Integrative Zoology*, vol. 7, No. 3, pp. 299–311.

________ (2017). Eyes wide shut: managing bio-invasions in Mediterranean marine protected areas. *Management of Marine Protected Areas: A Network Perspective*. Chichester, United Kingdom: John Wiley & Sons Ltd, 187–206.

________ (2018). Poisonous and venomous: marine alien species in the Mediterranean Sea and human health. *Invasive Species and Human Health*, vol. 10, p. 1.

Galil, Bella S., and others (2017). The enlargement of the Suez Canal—Erythraean introductions and management challenges. *Management of Biological Invasions*, vol. 8, No. 2, pp. 141–152.

Galil, Bella S., and others (2018). East is east and West is west? Management of marine bioinvasions in the Mediterranean Sea. *Estuarine, Coastal and Shelf Science*, vol. 201, pp. 7–16.

Giakoumi, Sylvaine, and others (2016). Space invaders; biological invasions in marine conservation planning. *Diversity and Distributions*, vol. 22, No. 12, pp. 1220–1231.

Goldsmit, Jesica, and others (2018). Projecting present and future habitat suitability of ship-mediated aquatic invasive species in the Canadian Arctic. *Biological Invasions*, vol. 20, No. 2, pp. 501–517.

Gómez, Fernando (2019). Comments on the non-indigenous microalgae in the European seas. *Marine Pollution Bulletin*, vol. 148, pp. 1–2.

Goulletquer, Philippe, and others (2002). Open Atlantic coast of Europe—a century of introduced species into French waters. In *Invasive Aquatic Species of Europe. Distribution, Impacts and Management*, pp. 276–290. Springer.

Gracia, Adriana, and others (2011). Guía de las especies introducidas marinas y costeras de Colombia.

Griffiths, C.L., and others (1992). Marine invasive aliens on South African shores: implications for community structure and trophic functioning. *South African Journal of Marine Science*, vol. 12, No. 1, pp. 713–722.

Halling, Christina, and others (2013). Introduction of Asian strains and low genetic variation in farmed seaweeds: indications for new management practices. *Journal of Applied Phycology*, vol. 25, No. 1, pp. 89–95.

Herbert, Roger J.H., and others (2016). Ecological impacts of non-native Pacific oysters (*Crassostrea gigas*) and management measures for protected areas in Europe. *Biodiversity Conservation*, vol. 25, pp. 2835–2865.

Hewitt, Chad L. (2002). Distribution and biodiversity of Australian tropical marine bioinvasions. *Pacific Science*, vol. 56, No. 2, pp. 213–222.

Hewitt, Chad L., and M.L. Campbell (2010). The relative contribution of vectors to the introduction and translocation of invasive marine species. Commissioned by The Department of Agriculture, Fisheries and Forestry (DAFF), Canberra. www.marinepests.gov.au/what-we-do/research/vectors-introduction-translocation.

Hughes, Kevin A., and Gail V. Ashton (2017). Breaking the ice: the introduction of biofouling organisms to Antarctica on vessel hulls. *Aquatic Conservation: Marine and Freshwater Ecosystems*, vol. 27, No. 1, pp. 158–164.

Iacarella, Josephine C., and others (2019). Non-native species are a global issue for marine protected areas. *Frontiers in Ecology and the Environment*, vol. 17, No. 9, pp. 495–501.

International Council for the Exploration of the Sea (ICES) (2005). ICES Code of Practice on the Introductions and Transfers of Marine Organisms 2005.

________ (2018). ICES Ecosystem overviews: Baltic Sea.

________ (2019). ICES VIEWPOINT: Biofouling on Vessels – What is the Risk, and What Might be Done About It? In Report of the ICES Advisory Committee, 2019, vp.2019.01.

International Council for Science (ICSU) and others (2017). *SDG14 Conserve and Sustainably Use the Oceans, Seas and Marine Resources for Sustainable Development*. International Council for Science.

Indian Ocean Commission (2016). Marine invasive species: An emerging threat in the Western Indian Ocean. Ebene: Indian Ocean Commission. https://studyres.com/doc/1377322/marine-invasive-species---commission-de-l-oc%C3%A9an-indien.

International Maritime Organization (IMO) (2019). Status of IMO Treaties. Comprehensive information on the status of multilateral Conventions and instruments in respect of which the International Maritime Organization or its Secretary-General performs depositary or other functions.

James, Kate, and others (2015). Using satellite-derived sea surface temperature to predict the potential global range and phenology of the invasive kelp *Undaria pinnatifida*. *Biological Invasions*, vol. 17, No. 12, pp. 3393–3408.

Jimenez, Carlos, and others (2016). Veni, vidi, vici: The successful establishment of the lionfish *Pterois miles* in Cyprus (Levantine Sea). *Rapport Commission International Mer Méditerranée*, vol. 41, p. 417.

Kestrup, Åsa M., and others (2015). Report of Working Group 21 on Non-indigenous Aquatic Species. *PICES Scientific Report*, No. 48, p. I.

Knapp, I.S., and others (2011). Records of non-indigenous marine species at Palmyra Atoll in the US Line Islands. *Marine Biodiversity Records*, vol. 4.

Kospartov, M., and others (2010). Non-indigenous and cryptogenic marine species in New Zealand–Current state of knowledge: Interim report. *Report Prepared for MAFBNZ Project BNZ10740. National Institute of Water and Atmospheric Research, Wellington.*

Lee, Henry, II, and Deborah Reusser (2012). *Atlas of Nonindigenous Marine and Estuarine Species in the North Pacific. Office of Research and Development, National Health and Environmental Effects Research Laboratory.* EPA/600/R/12/631.

Lee, J.E., and S.L. Chown (2007). *Mytilus* on the move: transport of an invasive bivalve to the Antarctic. *Marine Ecology Progress Series*, vol. 339, pp. 307–310.

Lewis, Patrick N., and others (2003). Marine introductions in the Southern Ocean: an unrecognised hazard to biodiversity. *Marine Pollution Bulletin*, vol. 46, No. 2, pp. 213–223.

Liu, Ruiyu (2008). *Checklist of Marine Biota of China Seas.* Science Press.

Lopes, Rubens M., and others (2009). *Informe Sobre as Espécies Exóticas Invasoras Marinhas No Brasil.* 574.5 INF.

Mačić, Vesna, and others (2018). Biological invasions in conservation planning: a global systematic review. *Frontiers in Marine Science*, vol. 5, p. 178.

Mead, Angela, and others (2011). Revealing the scale of marine bioinvasions in developing regions: a South African re-assessment. *Biological Invasions*, vol. 13, No. 9, pp. 1991–2008.

Meinesz, Alexandre, and B. Hesse (1991). Introduction et invasion de l'algue tropicale *Caulerpa taxifolia* en Méditerranée nord-occidentale. *Oceanologica Acta*, vol. 14, No. 4, pp. 415–426.

Millennium Ecosystem Assessment (MEA) (2005). *Ecosystems and Human Well-Being: Wetlands and Water.* World Resources Institute.

Miller, A. Whitman, and Gregory M. Ruiz (2014). Arctic shipping and marine invaders. *Nature Climate Change*, vol. 4, No. 6, p. 413.

Minchin, Dan, and others (2013). Alien species in British brackish and marine waters. *Aquatic Invasions*, vol. 8, No. 1, pp. 3–19.

Molnar, Jennifer L., and others (2008). Assessing the global threat of invasive species to marine biodiversity. *Frontiers in Ecology and the Environment*, vol. 6, No. 9, pp. 485–492.

Narayanaswamy, Bhavani E., and others (2013). Synthesis of knowledge on marine biodiversity in European Seas: from census to sustainable management. *PLoS One*, vol. 8, No.3, pp. e58909.

Ojaveer, Henn, and Jonne Kotta (2015). Ecosystem impacts of the widespread non-indigenous species in the Baltic Sea: literature survey evidences major limitations in knowledge. *Hydrobiologia*, vol. 750, No. 1, pp. 171–185.

Ojaveer, Henn, and others (2015). Classification of non-indigenous species based on their impacts: considerations for application in marine management. *PLoS Biology*, vol. 13, No. 4, p. e1002130.

Ojaveer, Henn, and others (2017). Dynamics of biological invasions and pathways over time: a case study of a temperate coastal sea. *Biological Invasions*, vol. 19, No. 3, p. 799–813.

Ojaveer, Henn, and others (2018). Historical baselines in marine bioinvasions: Implications for policy and management. *PloS One*, vol. 13, No. 8, p. e0202383.

Orensanz, Jose Maria Lobo, and others (2002). No longer the pristine confines of the world ocean: a survey of exotic marine species in the southwestern Atlantic. *Biological Invasions*, vol. 4, Nos. 1–2, pp. 115–143.

Paerl, Hans W., and Jef Huisman (2009). Climate change: a catalyst for global expansion of harmful cyanobacterial blooms. *Environmental Microbiology Reports*, vol. 1, No. 1, pp. 27–37.

Paulay, Gustav, and others (2002). Anthropogenic biotic interchange in a coral reef ecosystem: a case study from Guam. *Pacific Science*, vol. 56, No. 4, pp. 403–422.

Pérez, Julio, and others (2007). Especies marinas exóticas y criptogénicas en las costas de Venezuela. *Boletín Del Instituto Oceanográfico de Venezuela*, vol. 46, No. 1.

Qiao, Bing (2019). Technical methods for determining the baseline, causal relationship and degree of marine ecological environment damage + 2019 annual scientific and technological progress report. In *Science and technology report of the people's Republic of China No.400001918-2016YFC0503602/0 (to be disclosed)*, pp. 18–22.

Reise, Karsten, and others (2002). Introduced Marine Species of the North Sea Coasts. In *Invasive Aquatic Species of Europe. Distribution, Impacts and Management*, Erkki Leppäkoski, Stephan Gollasch, and Sergej Olenin, eds., pp. 260–266. Dordrecht, Netherlands: Kluwer Academic Publishers.

Richardson, David M., and others (2011). A Compendium of Essential Concepts and Terminology in Invasion Ecology. In *Fifty Years of Invasion Ecology: The Legacy of Charles Elton*, pp. 409–420. John Wiley & Sons.

Ruiz, Gregory M., and others (1997). Global invasions of marine and estuarine habitats by non-indigenous species: mechanisms, extent, and consequences. *American Zoologist*, vol. 37, No. 6, pp. 621–632.

Ruiz, Gregory M., and others (2000). Global spread of microorganisms by ships. *Nature*, vol. 408, No. 6808, p. 49.

Ruiz, Gregory M., and others (2015). Invasion history and vector dynamics in coastal marine ecosystems: A North American perspective, *Aquatic Ecosystem Health & Management*, vol. 18, No. 3, pp. 299–311.

Ruttenberg, Benjamin I., and others (2012). Rapid invasion of Indo-Pacific lionfishes (*Pterois volitans* and *Pterois miles*) in the Florida Keys, United States: evidence from multiple pre-and post-invasion data sets. *Bulletin of Marine Science*, vol. 88, No. 4, pp. 1051–1059.

Sancholle, M. (1988). Présence de *Fucus spiralis* (Phaeophyceae) en Méditerranée occidentale. *Cryptogamie Algologie*, vol. 9, No. 2, pp. 157–161.

Schlaepfer, Martin A., and others (2011). The Potential Conservation Value of Non-Native Species. *Conservation Biology*, vol. 25, No. 3, pp. 428–437.

Schröder, V., and Carlos Garcia de Leaniz (2011). Discrimination between farmed and free-living invasive salmonids in Chilean Patagonia using stable isotope analysis. *Biological Invasions*, vol. 13, No. 1, pp. 203–213.

Schwindt, Evangelina, and Alejandro Bortolus (2017). Aquatic invasion biology research in South America: Geographic patterns, advances and perspectives. *Aquatic Ecosystem Health & Management*, vol. 20, No. 4, pp. 322–333.

Schwindt, Evangelina and others (2020). Past and Future of the Marine Bioinvasions along the Southwestern Atlantic. *Aquatic Invasions*, vol. 15, No. 1, pp. 11–29.

Seaward, Kimberley, and Graeme Inglis (2018). *Long-Term Indicators for Non-Indigenous Species (NIS) in Marine Systems*. NIWA Client Report CHC2016-024.

Shatkin, Greg (1997). Considerations regarding the possible introduction of the Pacific oyster (*Crassostrea gigas*) to the Gulf of Maine: a review of global experience. *Journal of Shellfish Research*, vol. 16, pp. 463–478.

Sliwa, Cathryn, and others (2009). Marine bioinvasions in Australia. In *Biological Invasions in Marine Ecosystems*, pp. 425–437. Springer.

Spalding, Mark D., and others (2007). Marine ecoregions of the world: a bioregionalization of coastal and shelf areas. *BioScience*, vol. 57, No. 7, pp. 573–583.

Stern, Nir, and Shevy B.S. Rothman (2019). Divide and conserve the simultaneously protected and invasive species. *Aquatic Conservation: Marine and Freshwater Ecosystems*, vol. 29, No. 1, pp. 161–162.

Stiger, Valérie, and Claude Payri (1999). Spatial and seasonal variations in the biological characteristics of two invasive brown algae, *Turbinaria ornata* (Turner) J. Agardh and *Sargassum mangarevense* (Grunow) Setchell (Sargassaceae, Fucales) spreading on the reefs of Tahiti (French Polynesia). *Botanica Marina*, vol. 42, No. 3, pp. 295–306.

Tavares, Marcos, and Gustavo A.S. de Melo (2004). Discovery of the first known benthic invasive species in the Southern Ocean: the North Atlantic spider crab *Hyas araneus* found in the Antarctic Peninsula. *Antarctic Science*, vol. 16, No. 2, pp. 129–131.

Teixeira, Larissa M.P., and Joel C. Creed (2020). A decade on: an updated assessment of the status of marine non-indigenous species in Brazil. *Aquatic Invasions* vol. 15, No. 1, pp. 30–43.

Therriault, Thomas W., and others (2018). The invasion risk of species associated with Japanese tsunami marine debris in Pacific North America and Hawaii. *Marine Pollution Bulletin*, vol. 132, pp. 82–89.

Tsiamis, Konstantinos, and others (2019). Non-indigenous species refined national baseline inventories: A synthesis in the context of the European Union's Marine Strategy Framework Directive. *Marine Pollution Bulletin*, vol. 145, pp. 429–435.

Turon, Xavier, and others (2016). Too cold for invasions? Contrasting patterns of native and introduced ascidians in subantarctic and temperate Chile. *Management of Biological Invasions*, vol. 7, No. 1, pp. 77–86.

United Nations (2017). *The First Global Integrated Marine Assessment: World Ocean Assessment I*. Cambridge: Cambridge University Press.

United Nations Environment Programme (UNEP) (2002). *COP 6 Decision VI/23. Alien Species That Threaten Ecosystems, Habitats or Species. The Hague, 7–19 April 2002.*

Wan, Shuwen, and others (2009). The positive and negative effects of exotic *Spartina alterniflora* in China. *Ecological Engineering*, vol. 35, No. 4, pp. 444–452.

Wang, Xiulin, and Keqiang Li (2006). Appendix I List of Phytoplankton and Red Tide Reasons in Bohai Sea. *Marine Environmental Capacity of Major Chemical Pollutants in Bohai Sea*. Science Press. pp. 311–316.

Ware, Chris, and others (2016). Biological introduction risks from shipping in a warming Arctic. *Journal of Applied Ecology*, vol. 53, No. 2, pp. 340–349.

Williamson, Mark Herbert (1996). *Biological Invasions*. London, New York: Chapman & Hall.

Chapter 23
Developments in the exploration for and use of marine genetic resources

Contributors: Robert Blasiak and Ellen Kenchington (joint conveners), Jesús M. Arrieta, Jorge Rafael Bermúdez-Monsalve, Hilconida Calumpong (co-lead member), Shao Changwei, Sanae Chiba (lead member), Hebe Dionisi, Carlos Garcia-Soto (co-lead member), Helena Vieira and Boris Wawrik.

Keynote points

- Marine genetic resources continue to be the focus of an expanding range of commercial and non-commercial applications.
- Rapidly decreasing sequencing and gene synthesis costs and swift advances in the metabolic engineering and synthetic biology fields within the biotechnology sector have rendered scientists less reliant on physical samples and increasingly dependent on the exponentially expanding public databases of genetic sequence data.
- Sponges and algae continue to attract substantial interest for the bioactive properties of their natural compounds.
- Within the context of the Sustainable Development Goals,[1] capacity-building issues persist, with entities in a handful of countries conducting the majority of research and development associated with marine genetic resources.
- International processes and agreements with relevance to marine genetic resources include the Nagoya Protocol[2] on Access to Genetic Resources and the Fair and Equitable Sharing of Benefits Arising from Their Utilization to the Convention on Biological Diversity, and the intergovernmental conference on an international legally binding instrument under the United Nations Convention on the Law of the Sea[3] on the conservation and sustainable use of marine biological diversity of areas beyond national jurisdiction.[4]

1. Introduction

The ocean is home to a vast diversity of life forms constituting a rich source of marine genetic resources, that is, genetic material of marine origin containing functional units of heredity of actual or potential value, characterized by high biological and chemical diversity (Appeltans and others, 2012; United Nations, 2017). Over 34,000 marine natural products have been described, with recent discovery rates reaching more than 1,000 compounds each year (Lindequist, 2016; Carroll and others, 2019). A total of 188 new marine natural products from deep-sea organisms (Bryozoa, Chordata, Cnidaria, Echinodermata, Mollusca, Porifera and microbes) have been described since 2008 (Skropeta and Wei, 2014). Approximately 75 per cent of those novel products have remarkable bioactivity, with 50 per cent exhibiting moderate to high cytotoxicity towards a range of human cancer cell lines. Although the bioactivity of many marine natural products suggests high potential for drug discovery, only 13 marine-derived drugs have gained market approval to date (Liang and others, 2019; Mayer and others, 2010).[5] However, at the time of writing, 28 candidates were in clinical trials (Alves and others, 2018). Marine antifoulant research is currently focused on identifying viable non-toxic substances, and a recent review has estimated that more than 198 antifouling compounds have been obtained from marine invertebrates, specifically sponges, gorgonians and soft corals (Qi and Ma, 2017), in addition to the products derived from macroalgae and microalgae highlighted in the first *World Ocean Assessment* (United

1 See General Assembly resolution 70/1.
2 United Nations Environment Programme, document UNEP/CBD/COP/10/27, annex, decision X/1.
3 United Nations, *Treaty Series*, vol. 1833, No. 31363.
4 See General Assembly resolution 72/249.
5 See Midwestern University, "Clinical Pipeline, Marine Pharmacology".

Nations, 2017). Innovative research has also identified ingredients from discarded fish that are suitable for use in high-end cosmetics and a number of other products (Young, 2014). As of 2018, a total of 76 publicly available cosmeceutical ingredients from marine natural products had been marketed, reflecting a new growth sector (Calado and others, 2018).

At the same time, consumer demand for nutraceuticals has increased rapidly, as foreseen in the first Assessment. The global nutraceutical market is expected to reach $580 billion by 2025, more than triple the $180 billion projected for 2017 in the first Assessment, and market growth has been linked to increased innovation and consumer awareness (Grand View Research, 2017). Marine nutraceutical products such as fish oil and collagen represent a large portion of the global market, and demand for those products is expected to grow in the Asia-Pacific region, in particular in China and India (Suleria and others, 2015).

While marine genetic resources are of growing importance to the global blue economy, most commercial activity is concentrated in a comparatively small number of countries, suggesting that there is potential for technology transfer and capacity-building (Thompson and others, 2017; Blasiak and others, 2018). Several international processes addressing genetic resources, including marine genetic resources, are currently under way.

2. Trends between 2010 and 2020

Technological innovations have been key to the recent advances in the exploration for and exploitation of marine genetic resources. The discovery of new marine molecules, and their sources, has been increasing rapidly, especially since the 1970s (figure I). By November 2019, a total of 34,197 marine natural products had been documented (Carroll and others, 2019). Such growth is most likely to have been driven by modern sampling and analytical techniques that have allowed the collection of novel marine genetic resources from deeper environments and covering a wider range of chemical diversity. Approximately 11 per cent of marine genetic resources associated with patent applications are found in deep-sea and hydrothermal vent communities, reflecting increased research in remote and extreme ocean environments (Blasiak and others, 2018). However, the number of marine genetic resources collected at depths of more than 50 m remains insignificant when compared with the whole library of marine natural products (Skropeta and Wei, 2014). The discovery of enzymes from marine organisms is also accelerating thanks to the development of innovative screening methodologies (Ferrer and others, 2019). Enzymes from microorganisms adapted to extreme conditions are of particular interest for their application in industrial processes, as they are often active under challenging operational conditions (Birolli and others, 2019).

2.1. Commercial application highlights

2.1.1. Pharmaceutical applications

Thirteen drugs of marine origin have received market approval from the United States Food and Drug Administration or the European Medicines Agency, six of them since 2010. The majority of drugs of marine origin have been developed for anticancer chemotherapy (Calado and others, 2018; Liang and others, 2019; Mayer and others, 2010).[6] Since the approval of cytarabine as an anticancer agent in 1969,

[6] See Midwestern University, "Clinical Pipeline, Marine Pharmacology".

sponges have been regarded as one of the most promising sources of anticancer drugs (Hu and others, 2015; see sect. 2.3 below). Other marine invertebrates, such as tunicate and cone snail species, are also very important sources of marine natural products, as are fishes. Trabectedin (ET-743) gained United States Food and Drug Administration approval in 2015 for the treatment of soft tissue sarcoma and ovarian cancer, while Plitidepsin was approved by the Australian Therapeutic Goods Administration in 2018 for the treatment of multiple myeloma, leukemia and lymphoma (see Mayer and others, 2010).[7] Most recently, in 2020, Lurbinectedin was approved for the treatment of metastatic small cell lung cancer (see Mayer and others, 2010).[8] In all three cases, the relevant compounds were derived from tunicates. Macroalgae are also a source of pharmaceutical products. For example, OligoG, an oligoalginate with a defined structure produced from brown algae, is currently in a phase II clinical trial for the treatment of cystic fibrosis (Rye and others, 2018), and the red algae biopolymer Carragelose, which has broad anti-viral properties, is used for treating respiratory diseases (Hackl, 2017).

2.1.2. Cosmeceutical applications

Cosmeceuticals (cosmetics with pharmaceutical properties) are one of the fastest growing markets for the commercialization of marine natural products. They have a shorter development cycle than pharmaceutical and nutraceutical products, resulting in more rapid growth (Rampelotto and Trincone, 2018). Those emerging novel products with biologically active ingredients constitute an entirely new type of beauty care that will be a hallmark of coming decades. The majority of cosmeceuticals are derived from macroalgae and microalgae, but an increasing number are being generated through marine biotechnology processes based on microorganisms such as bacteria and fungi (Calado and others, 2018). There are, however, environmental concerns associated with certain cosmetic ingredients (Juliano and Magrini, 2017).

2.1.3. Food and feed applications

The consumption of omega-3 long-chain polyunsaturated fatty acids is linked to multiple positive health outcomes (Ruxton and others, 2007). However, the production of aquaculture species rich in those fatty acids remains reliant on fish-based feeds. The development of algal oils and alternative transgenic crops of omega-3 long-chain polyunsaturated fatty acids has consequently attracted substantial interest. Initial efforts have focused on oilseed crops, with reliance on enzymes from marine species (i.e., marine algae) (Ruiz-Lopez and others, 2014; Zhao and Qiu, 2018). Agro-industrial corporations have filed for patents associated with those innovations and large-scale production is envisaged by 2020 (Sprague and others, 2017). Furthermore, in addition to the direct use of macroalgae as human food, their application as feed additives is showing potential for biological methane mitigation in the cattle industry (Roque and others, 2019; Costello and others, 2019). Microalgae are also emerging as important biofuels (Fedder, 2013).

7 See Midwestern University, "Clinical Pipeline, Marine Pharmacology".
8 Ibid.

Figure I
Recent trends related to marine genetic resources

Figure I.A
Number of new marine natural products (cumulative)

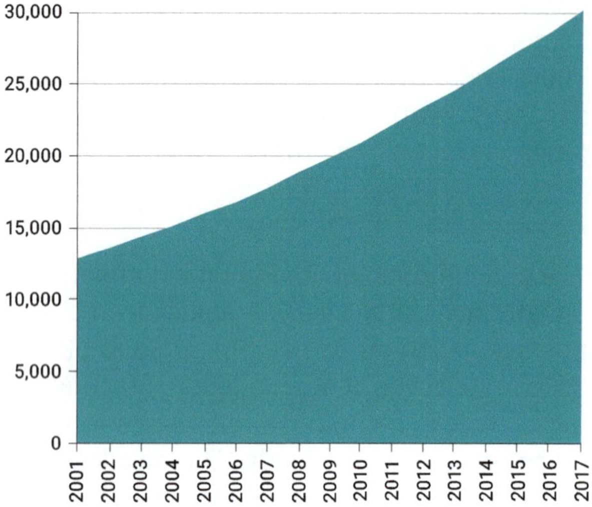

Source: Carroll and others, 2019.

Figure I.B
Number of sequences deposited in GenBank (cumulative number of base pairs)

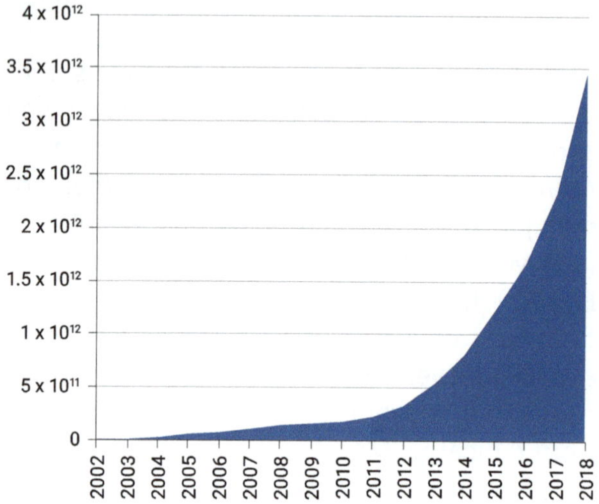

Source: United States National Institutes of Health (Wetterstrand, 2018; National Center for Biotechnology Information (NCBI), 2018).

Figure I.C
Cost of sequencing (dollars per base pair)

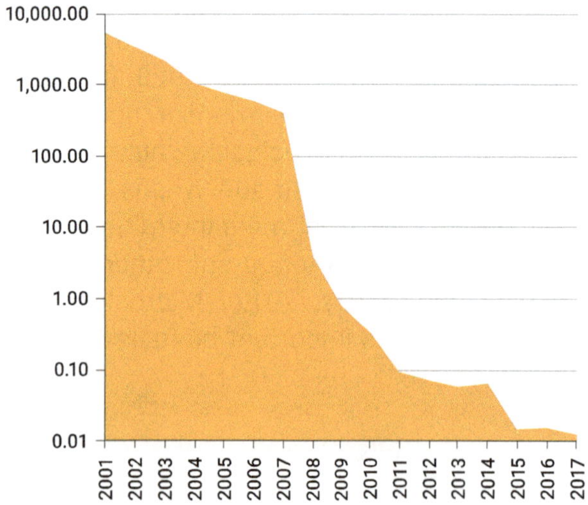

Source: National Human Genome Research Institute.

Figure I.D
Number of marine sequences associated with patent filings (cumulative)

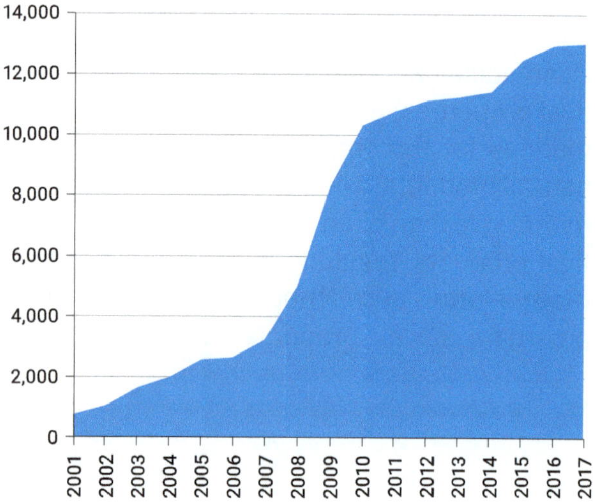

Source: Blasiak and others, 2018.

2.2. Growth in public databases of genetic sequence data

Public data archives are integral to modern biological research (Ellenberg and others, 2018; Rigden and Fernandez, 2019), in large part because rapid technological developments over the past two decades have substantially democratized the availability of nucleic acid sequencing technology. The cost per base of sequencing has fallen by more than four orders of magnitude over the past decade alone (Wetterstrand, 2018), in parallel with exponential growth in the size of publicly available repositories (see figure I). Overall, the number of bases in GenBank has doubled approximately every 18 months since 1982 (NCBI, 2018).

Although the size of public databases has grown substantially, there is good reason to believe that there are still significant gaps in the current state of knowledge regarding the extant genetic diversity in the ocean. Omics-based studies provide the best evidence for that interpretation. The most recent and comprehensive survey of marine eukaryotic genetic diversity identified approximately 53 million genes (Carradec and others, 2018), about half of which showed no similarity to existing proteins (de Vargas and others, 2015). Furthermore, estimates for oceanic plankton suggest the presence of about 150,000 eukaryotic species, which is far greater than the approximately 11,200 species that have been formally described (de Vargas and others, 2015). Large-scale initiatives such as Tara Oceans (Sunagawa and others, 2015) and Ocean Sampling Day (Kopf and others, 2015) are generating vast quantities of information that are increasing understanding of the microbial diversity that exists in the ocean at the global scale (Coutinho and others, 2018). The resulting public data sets available represent an important source of information for sequence-based research efforts (Kamble and others, 2019) and enable new directions in research, such as the use of environmental DNA in molecular ecology and in diversity assessments (Seymour, 2019).

2.3. Highlighted research

Two comprehensive volumes focused on marine biotechnology were published in 2018. The first systematically describes recent developments in the marine biotechnology sector and seeks to define its current and future economic potential (Rampelotto and Trincone, 2018), while the second goes beyond research and development aspects to delve into intellectual property law and the protections offered through patent claims (Guilloux, 2018). Previous studies on patents associated with marine genetic resources (Arrieta and others, 2010; Arnaud-Haond and others, 2011) were updated with an analysis identifying the patent filings associated with 12,998 genetic sequences from 862 marine species (Blasiak and others, 2018). Actors located or headquartered in 10 countries were responsible for patent filings covering 98 per cent of those sequences, while 165 countries were unrepresented (Blasiak and others, 2018).

SponGES,[9] a four-year project funded since 2016 through the European Union research and innovation programme Horizon 2020, is aimed at coupling exploration with bioprospecting for industrial applications, namely, drug discovery and tissue engineering. Sponges and their associated microorganisms are the richest and most prolific source of new marine natural products, accounting for nearly 30 per cent (almost 5,000) of the compounds described to date (Mehbub and others, 2014). From 2001 to 2010, more than 2,400 natural products were discovered from 671 species of sponges (Mehbub and others, 2014). SponGES research has already identified unexpected microbial diversity and resulting biotechnological potential, including unconventional C30 sterols and new barrettides with potential for antifouling activity (Lauritano and Ianora, 2018).

[9] See www.deepseasponges.org.

3. Economic and social consequences and changes

Interest in the exploration for and use of marine genetic resources is increasing, at the same time as rapid advances are being made in the global biotechnology industry and initiatives to explore the potential of the blue economy (Wynberg and Laird, 2018). Divergent views exist regarding the economic potential of marine genetic resources, in particular those from areas beyond national jurisdiction (Leary, 2018; Blasiak and others, 2020). However, the robust pipeline of marine-derived drugs in clinical trials suggests substantial interest, given that the process of bringing a new drug to market can cost as much as $2.8 billion (Wouters and others, 2020) and take 10 to 15 years (Blasiak and others, 2019).

The regulatory framework governing access to marine genetic resources and their subsequent utilization varies according to whether those resources are from areas within national jurisdiction or beyond. The former fall within the scope of the Convention on Biological Diversity[10] and the Nagoya Protocol thereto Marine genetic resources of areas beyond national jurisdiction are one of the topics in a package of issues currently under negotiation following the adoption in December 2017 of General Assembly resolution 72/249, in which the Assembly decided to convene an intergovernmental conference to elaborate the text of an international legally binding instrument under the Convention on the Law of the Sea on the conservation and sustainable use of marine biological diversity of areas beyond national jurisdiction. Three meetings of the conference were held in 2018 and 2019, and a fourth is due to take place in 2020. The conference is mandated to address the topic of marine genetic resources, including questions on the sharing of benefits, among other issues.

Discussions on whether to address and regulate the use of digital sequence data and information are taking place in the context of both the intergovernmental conference and the Convention on Biological Diversity and the Nagoya Protocol thereto. Different views have been expressed on that issue and the related terminology. In 2019, the Executive Secretary of the Convention on Biological Diversity commissioned studies covering the concept and scope of digital sequence information (Secretariat of the Convention on Biological Diversity, 2020), traceability and databases, and domestic measures. Those studies have now been published following an open review period.

Lastly, in 2017, the General Assembly of the World Intellectual Property Organization (WIPO) extended the mandate of the WIPO Intergovernmental Committee on Intellectual Property and Genetic Resources, Traditional Knowledge and Folklore and agreed that it should, inter alia, continue to expedite its work on an agreement relating to intellectual property, which would ensure the balanced and effective protection of genetic resources.[11]

All those regulatory frameworks are applicable only to the signatory countries, and they therefore govern only marine genetic resources collected in, or by, States that are parties to the relevant instruments.

[10] United Nations, *Treaty Series*, vol. 1760, No. 30619.

[11] See World Intellectual Property Organization, document WO/GA/49/21.

4. Key region-specific developments in knowledge and their consequences

In the first Assessment, the focus was on providing a general review of marine genetic resources rather than regional assessments or overviews. That was in part because regional summaries with information on trends are difficult to obtain. A brief summary of regional issues concerning the Pacific Ocean, the Southern Ocean and the Arctic Ocean, highlighting trends over the past decade, is provided below. The development of marine natural products from the Mediterranean and the Atlantic Ocean has been relatively more limited (Skropeta and Wei, 2014), but the Mediterranean, with its high biodiversity, is a potential source of new pharmaceuticals and nutraceuticals (Briand, 2010).

Skropeta and Wei (2014) conducted an update of their 2008 regional analyses of marine natural products reporting, and found that, while the proportion of marine natural products from Australia was still high (24 per cent), there had been a marked increase in reports of metabolites from deep-sea sediment sampling in the South China Sea (to 18 per cent) and in the Pacific Ocean, including maritime zones off the coast of Guam (United States of America) and Palau (to 17 per cent). That increase was attributed to the increased accessibility of remote deep-sea environments (Skropeta and Wei, 2014), with the regional pattern of discoveries of marine natural products being linked to the level of access to manned submersibles and trawling operations rather than to regional biodiversity. The increased accessibility of deep-sea environments was also reflected in the depth distribution of the discoveries since, in 2008, only 8 per cent of marine natural products were from organisms found at depths of over 1,000 m whereas, by 2013, such organisms accounted for 37 per cent of discoveries (Skropeta and Wei, 2014).

Activities in the Antarctic region are subject to the Antarctic Treaty[12] and the related agreements collectively known as the Antarctic Treaty System (Oldham and Kindness, 2020). Bioprospecting has been discussed under the Antarctic Treaty System, but the matter is very complicated owing to governance issues related to research activity, ethics and benefit-sharing. With increased scientific research taking place in Antarctica in general, biodiversity research has also increased, and a growing number of patents derived from Antarctic organisms are being filed in the United States and in Europe (Oldham and others, 2014; Oldham and Kindness, 2020).

A collaborative international research model has been established for the Arctic (Leary, 2008), although research on the biotechnology potential of Arctic genetic resources is largely being undertaken within the exclusive economic zones of the Arctic States.

5. Capacity-building gaps

Many States face challenges that hinder them from engaging directly in research on marine genetic resources. Such challenges include limited knowledge of biodiversity, limited capacity, in terms of both facilities and technological expertise, limited financial resources for research and development, a lack of experience with access and benefit-sharing mechanisms, and the need for increased collaboration across the academic, government and private sectors

12 United Nations, *Treaty Series*, vol. 402, No. 5778.

(Thompson and others, 2017). Capacity-building initiatives, such as the National Marine Biotechnology Research Network established in Brazil (Thompson and others, 2018), are key to addressing some of those limitations.

Wynberg (2016) has highlighted the rapid expansion of research activity in the Western Indian Ocean, in particular along the Eastern and Southern African coastlines, the latter of which is associated with higher biodiversity and endemism. Such research is largely being undertaken by developed countries from other regions, with few countries of the Western Indian Ocean – with the exception of South Africa and Kenya – engaged as collaborators. Comparatively few countries operate their own research vessels, and only a handful have the capacity to undertake collections from areas beyond national jurisdiction or deep-sea environments (figure II). Although public databases of genetic sequence data are available globally, many countries lack the cyberinfrastructure to gain access to such data sets or to establish and manage comparable national databases (Thompson and others, 2017).

Figure II.A
Number and distribution of research vessels by flag state as of June 2019

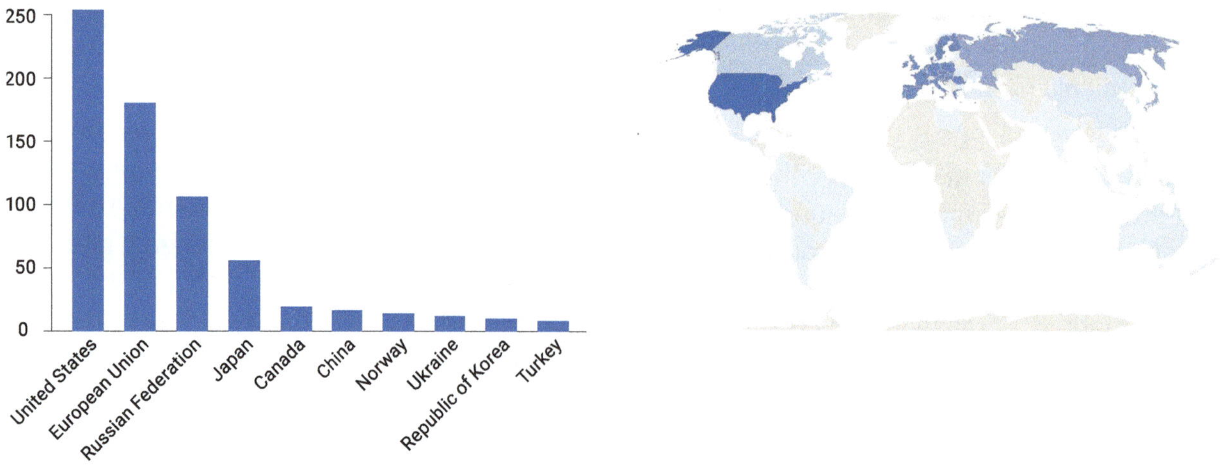

Figure II.B
Number and distribution of research vessels with offshore capacity (60 m or greater in length) as of June 2019

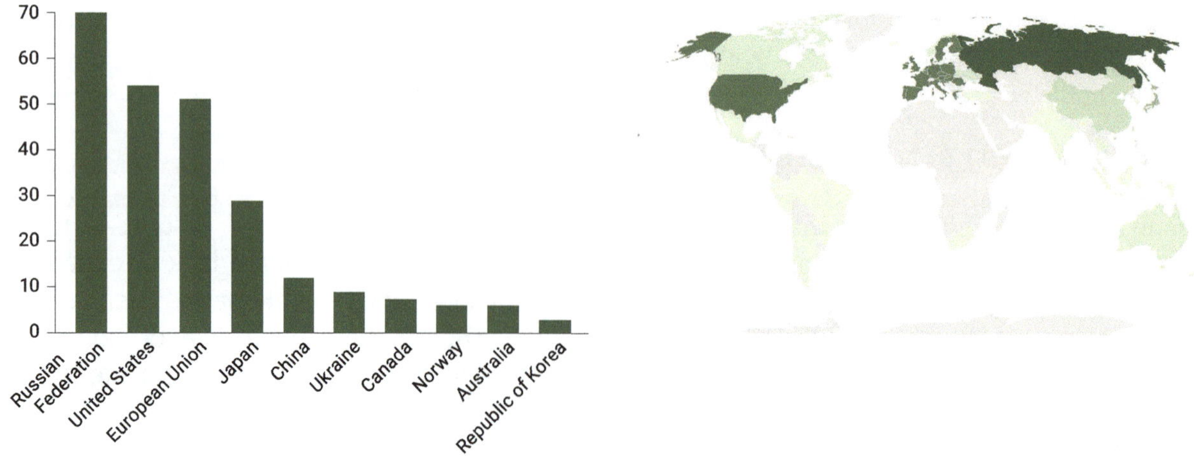

Source: International Research Vessel Database.

6. Methodological challenges and future trends

6.1. New developments in omics approaches

Over recent decades, innovations in technologies for the analysis of biomolecules have facilitated more comprehensive studies of marine organisms and their communities (Coutinho and others, 2018). Sequencing technologies with ultra-high throughput allow high coverage in the analysis of microbial communities, single-molecule sequencing technologies produce long sequence lengths from DNA and RNA, and portable real-time sequencing instruments can be used in the field (Ip and others, 2015). The current focus is on developing sequencing platforms for specific applications, and improving sequence length and output while decreasing sequencing error rates (Wuyts and Segata, 2019). Improvements in sequence length and accuracy are key to the generation of less fragmented data sets. Assembling deduced amino acid sequences, instead of DNA data, can also generate large catalogues of complete protein sequences from complex metagenomic datasets (Steinegger and others, 2019). In contrast with ecological studies, complete proteins and gene clusters are needed for biotechnological applications.

While high-throughput sequencing platforms have made the acquisition of sequence data much easier, the assignment of functions to predicted genes, proteins and pathways remains problematic (Woyke and others, 2019). Often a putative function cannot be assigned or only general functional predictions can be made, in particular, for enzymes. The experimental characterization of selected sequences with biotechnological potential is time-consuming and expensive. A combination of gene synthesis, cell-free protein expression systems and sensitive high-throughput screening methods are being developed for the discovery of novel biocatalysts and enzyme variants with improved characteristics (Rolf and others, 2019). Advances in the detection systems used in functional metagenomics – a different bioprospecting approach – are also having a positive impact on biodiscovery (van der Helm and others, 2018).

In spite of recent advances in sequencing technologies, it remains difficult to obtain high-quality near-complete genomes from uncultured microorganisms. The sequencing of the genomes of single microbial cells and the reconstruction of genomes from complex metagenome data sets have generated genomic information from thousands of uncultured marine microorganisms (Parks and others, 2017; Coutinho and others, 2018; Tully and others, 2018), creating a public resource available for bioprospecting efforts. Technological advances are needed, however, to improve the completeness of such genomes and to reduce the level of contamination that exists, prior to amplification, in the DNA cocktail generated using those culture-independent approaches (Woyke and others, 2019). The analysis of genomes from uncultured microorganisms is also being facilitated by metagenomic chromosome conformation capture (meta3C), a technique that can reveal the physical contacts in different regions of the DNA present within a cell. When applied to microbial communities, it both facilitates the assembly of genomes and allows an analysis of their tridimensional organization (Marbouty and others, 2014). Improvements in cultivation techniques for marine microorganisms are also needed, in particular in the context of the utilization of microbial marine genetic resources for industrial purposes.

The exponential growth of data generated by the different omics approaches represents a challenge, and new bioinformatic tools and platforms continue to be developed for the analysis and integration of such data to gain a better understanding of biological systems (Dihazi and others, 2018; Rohart and others, 2017). One example is the United States Department of Energy Systems Biology Knowledgebase (KBase),[13] an open-source software and data platform that enables collaborative analyses of multi-omics information, including genome or metagenome assembly, annotation, transcriptomics and metabolic modelling (Arkin and others, 2018). Through the integration of metabolomics data analysis, that is, the analysis of small biomolecules from organisms or microbial communities, it is possible to validate identified pathways, as well as to link microbial community structure, dynamics, interactions and function (Baidoo and Benites, 2019). Another example of a multi-omics integration tool, focused on data exploration and data mining, is mixOmics (Rohart and others, 2017).[14]

6.2. Marine genetic resources and synthetic biology

Given the exceptional biodiversity of marine organisms, marine genetic resources are a promising source of genes and gene clusters for the artificial redesign of organisms for industrial applications (Bloch and Tardieu-Guigues, 2014; Reen and others, 2015). Synthetic biology, in combination with enzyme and metabolic engineering, can greatly facilitate the development of high-performance strains for the production of chemicals, biomaterials and services. For instance, a synthetic biology approach can be used as an alternative to chemical synthesis for the production of marine natural products, when the extraction from the original source is not sustainable (Kiran and others, 2018). Public health and ethical considerations are important issues in synthetic biology, and the public perception of the safety of genetically modified organisms will also influence the adoption of that technology in the industrial sector (Kiran and others, 2018).

7. Marine genetic resources and the Sustainable Development Goals

Regardless of the scale of the economic benefits associated with the commercialization of marine genetic resources, capacity-building gaps remain (sect. 5), which has major implications for the achievement of the Sustainable Development Goals. The table below summarizes the relevance of marine genetic resources to the Sustainable Development Goal targets that are most applicable.

13 See http://kbase.us.
14 See http://mixomics.org.

Marine genetic resources and the Sustainable Development Goals

Relevant Sustainable Development Goal targets	Relevance of marine genetic resources
14.2 By 2020, sustainably manage and protect marine and coastal ecosystems to avoid significant adverse impacts, including by strengthening their resilience, and take action for their restoration in order to achieve healthy and productive oceans **14.5** By 2020, conserve at least 10 per cent of coastal and marine areas, consistent with national and international law and based on the best available scientific information	Ensure that the genetic diversity of populations in protected areas is taken into account, among other things to promote resilience Use marine genetic resources as tools for understanding biotic and abiotic interactions to help to manage ecosystem services Promote and focus exploitation on sustainably harvested or developed marine natural products
14.a Increase scientific knowledge, develop research capacity and transfer marine technology, taking into account the Intergovernmental Oceanographic Commission Criteria and Guidelines on the Transfer of Marine Technology, in order to improve ocean health and to enhance the contribution of marine biodiversity to the development of developing countries, in particular small island developing States and least developed countries **9.5** Enhance scientific research, upgrade the technological capabilities of industrial sectors in all countries, in particular developing countries, including, by 2030, encouraging innovation and substantially increasing the number of research and development workers per 1 million people and public and private research and development spending **9.b** Support domestic technology development, research and innovation in developing countries, including by ensuring a conducive policy environment for, inter alia, industrial diversification and value addition to commodities **17.6:** Enhance North-South, South-South and triangular regional and international cooperation on and access to science, technology and innovation and enhance knowledge-sharing on mutually agreed terms, including through improved coordination among existing mechanisms, in particular at the United Nations level, and through a global technology facilitation mechanism	Promote inclusive innovation and other mechanisms to ensure broader capacity for States to engage in the exploration for and use of marine genetic resources
3.b Support the research and development of vaccines and medicines for the communicable and non-communicable diseases that primarily affect developing countries, provide access to affordable essential medicines and vaccines, in accordance with the Doha Declaration on the TRIPS Agreement and Public Health, which affirms the right of developing countries to use to the full the provisions in the Agreement on Trade-Related Aspects of Intellectual Property Rights regarding flexibilities to protect public health, and, in particular, provide access to medicines for all	Robust pipeline of marine-derived medicines in clinical trials, and potential of marine organisms as source of new antibiotics

References

Alves, Celso, and others (2018). From marine origin to therapeutics: The antitumor potential of marine algae-derived compounds. *Frontiers in Pharmacology*, vol. 9.

Appeltans, Ward, and others (2012). The magnitude of global marine species diversity. *Current Biology*, vol. 22, No. 23, pp. 2189–2202.

Arkin, Adam P., and others (2018). KBase: the United States department of energy systems biology knowledgebase. *Nature Biotechnology*, vol. 36, No. 7.

Arnaud-Haond, Sophie, and others (2011). Marine biodiversity and gene patents. *Science*, vol. 331, No. 6024, pp. 1521–1522.

Arrieta, Jesús M., and others (2010). What lies underneath: conserving the oceans' genetic resources. *Proceedings of the National Academy of Sciences*, vol. 107, No. 43, pp. 18318–18324.

Baidoo, Edward E.K., and Veronica Teixeira Benites (2019). Mass Spectrometry-Based Microbial Metabolomics: Techniques, Analysis, and Applications. In *Microbial Metabolomics*, pp. 11–69. Springer.

Birolli, Willian G., and others (2019). Applications of marine-derived microorganisms and their enzymes in biocatalysis and biotransformation, the underexplored potentials. *Frontiers in Microbiology*, vol. 10.

Blasiak, Robert, and others (2018). Corporate control and global governance of marine genetic resources. *Science Advances*, vol. 4, No. 6, p. eaar5237.

Blasiak, Robert, and others (2019). Scientists Should Disclose Origin in Marine Gene Patents. *Trends in Ecology & Evolution*, vol. 34, No. 5, pp. 392–395.

Blasiak, Robert, and others (2020). The ocean genome and future prospects for conservation and equity. *Nature Sustainability*, pp.1–9.

Bloch, Jean-François, and Elisabeth Tardieu-Guigues (2014). Marine biotechnologies and synthetic biology, new issues for a fair and equitable profit-sharing commercial use. *Marine Genomics*, vol. 17, pp. 79–83.

Briand, Frédéric (2010). New Partnerships for Blue Biotechnology Development - Innovative solutions from the sea. Report on CIESM International Workshop. The Mediterranean Science Commission.

Calado, Ricardo, and others (2018). How to Succeed in Marketing Marine Natural Products for Nutraceutical, Pharmaceutical and Cosmeceutical Markets. In *Grand Challenges in Marine Biotechnology*, pp. 317–403. Springer.

Carradec, Quentin, and others (2018). A global ocean atlas of eukaryotic genes. *Nature Communications*, vol. 9, No. 1, p. 373.

Carroll, Anthony R., and others (2019). Marine natural products. Natural Product Reports.

Costello, Christopher, and others (2019). The Future of Food from the Sea. Washington, D.C.: World Resources Institute. www.oceanpanel.org/future-food-sea

Coutinho, Felipe Hernandes, and others (2018). Metagenomics sheds light on the ecology of marine microbes and their viruses. *Trends in Microbiology*, vol. 26, No. 11, pp. 955–965.

De Vargas, Colomban, and others (2015). Eukaryotic plankton diversity in the sunlit ocean. *Science*, vol. 348, No. 6237, pp. 1261605.

Dihazi, Hassan, and others (2018). Integrative omics-from data to biology. *Expert Review of Proteomics*, vol. 15, No. 6, pp. 463–466.

Ellenberg, Jan, and others (2018). A call for public archives for biological image data. *Nature Methods*, vol. 15, No. 11, p. 849.

Fedder, Bevis (2013). *Marine Genetic Resources, Access and Benefit Sharing: Legal and Biological Perspectives*. Routledge.

Ferrer, Manuel, and others (2019). Decoding the ocean's microbiological secrets for marine enzyme biodiscovery. *FEMS Microbiology Letters*, vol. 366, No. 1, p. fny285.

Grand View Research (2017). Nutraceuticals Market Analysis By Product (Dietary Supplements, Functional Food, Functional Beverage), By Region (North America, Asia Pacific, Europe, CSA, MEA), And Segment Forecasts, 2018–2025. Grand View Research. www.grandviewresearch.com/industry-analysis/nutraceuticals-market.

Guilloux, Bleuenn (2018). Marine Genetic Resources, R&D and the Law 1: Complex Objects of Use. Wiley Online Library.

Hackl, Christian (2017). Using red algae to fight the flu. *Les Nouvelles-Journal of the Licensing Executives Society*, vol. 52, No. 4.

Hu, Yiwen, and others (2015). Statistical research on the bioactivity of new marine natural products discovered during the 28 Years from 1985 to 2012. Marine Drugs, vol. 13, pp. 202–221.

Ip, Camilla L.C., and others (2015). MinION Analysis and Reference Consortium: Phase 1 data release and analysis. F1000Research, vol. 4.

Juliano, Claudia, and Giovanni Antonio Magrini (2017). Cosmetic ingredients as emerging pollutants of environmental and health concern. A mini-review. *Cosmetics*, vol. 4, No. 11, pp. 1–18. http://doi.org/10.3390/cosmetics4020011.

Kamble, Asmita, and others (2019). In-silico bioprospecting: finding better enzymes. *Molecular Biotechnology*, vol. 61, No. 1, pp. 53–59.

Kiran, Seghal, and others (2018). Synthetic biology approaches: Towards sustainable exploitation of marine bioactive molecules. *International Journal of Biological Macromolecules*, vol. 112, pp. 1278–1288.

Kopf, Anna, and others (2015). The ocean sampling day consortium. *Gigascience*, vol. 4, No. 1, pp. s13742–015.

Lauritano, Chiara, and Adrianna Ianora (2018). Grand Challenges in Marine Biotechnology: Overview of Recent EU-Funded Projects. In *Grand Challenges in Marine Biotechnology*, pp. 425–449. Springer.

Leary, David (2008). Bi-polar disorder? Is bioprospecting an emerging issue for the Arctic as well as for Antarctica? *Review of European Community & International Environmental Law*, vol. 17, No. 1, pp. 41–55.

________ (2018). Marine Genetic Resources in Areas beyond National Jurisdiction: Do We Need to Regulate Them in a New Agreement? *Maritime Safety and Security Law Journal*, vol. 19, pp. 22–47.

Liang, Xiao, and others (2019). Advances in exploring the therapeutic potential of marine natural products. *Pharmacological Research*, vol. 147, pp. 104373–104390.

Lindequist, Ulrike (2016). Marine-derived pharmaceuticals–challenges and opportunities. *Biomolecules & Therapeutics*, vol. 24, No. 6, p. 561.

Marbouty, Martial, and others (2014). Metagenomic chromosome conformation capture (meta3C) unveils the diversity of chromosome organization in microorganisms. *Elife*, vol. 3, p. e03318.

Mayer, A.M.S., and others (2010). The Odyssey of Marine Pharmaceuticals: A Current Pipeline Perspective. *Trends in Pharmacological Sciences*, vol. 31, pp. 255–265. www.midwestern.edu/departments/marinepharmacology/clinical-pipeline.xml, https://doi.org/10.1016/j.tips.2010.02.005

Mehbub, Mohammad Ferdous, and others (2014). Marine sponge derived natural products between 2001 and 2010: trends and opportunities for discovery of bioactives. *Marine Drugs*, vol. 12, No. 8, pp. 4539–4577.

National Center for Biotechnology Information (NCBI) (2018). GenBank and WGS Statistics. Available at www.ncbi.nlm.nih.gov/genbank/statistics.

National Human Genome Research Institute. DNA Sequencing Costs: Data. Available at www.genome.gov/about-genomics/fact-sheets/DNA-Sequencing-Costs-Data87.

Oldham, Paul, and Jasmine Kindness (2020). Biodiversity research and innovation in Antarctica and the Southern Ocean. Preprint bioRxiv 2020.05.03.074849; https://doi.org/10.1101/2020.05.03.074849.

Oldham, Paul, and others (2014). Valuing the deep: Marine genetic resources in areas beyond national jurisdiction. Defra Contract. MB, vol. 128.

Parks, Donovan H., and others (2017). Recovery of nearly 8,000 metagenome-assembled genomes substantially expands the tree of life. *Nature Microbiology*, vol. 2, No. 11, p. 1533.

Qi, Shu-Hua, and Xuan Ma (2017). Antifouling compounds from marine invertebrates. *Marine Drugs*, vol. 15, No. 9, p. 263.

Rampelotto, Pabulo H., and Trincone, Antonio (2018). *Grand Challenges in Marine Biotechnology*. Springer.

Reen, F. Jerry, and others (2015). Emerging concepts promising new horizons for marine biodiscovery and synthetic biology. *Marine Drugs*, vol. 13, No. 5, pp. 2924–2954.

Rigden, Daniel J., and Xose M. Fernandez (2019). The 27th annual Nucleic Acid Research database issue and molecular biology database collection. *Nucleic Acid Research*, vol. 48, pp. D1–D8.

Rohart, Florian, and others (2017). mixOmics: An R package for 'omics feature selection and multiple data integration. *PLOS Computational Biology*, vol. 13, No. 11, p. e1005752.

Rolf, Jascha, and others (2019). Application of cell-free protein synthesis for faster biocatalyst development. *Catalysts*, vol. 9, No. 2, p. 190.

Roque, Breanna Michell, and others (2019). Effect of the macroalgae Asparagopsis taxiformis on methane production and rumen microbiome assemblage. *Animal Microbiome*, vol. 1, No. 1, p. 3.

Ruiz-Lopez, Noemi, and others (2014). Successful high-level accumulation of fish oil omega-3 long-chain polyunsaturated fatty acids in a transgenic oilseed crop. *Plant Journal*, vol. 77, No. 2, pp. 198–208.

Ruxton, C., and others (2007). Commentary on Ruxton, C.H.S., and others (2004) The health benefits of omega-3 polyunsaturated fatty acids: a review of the evidence. Journal of Human Nutrition and Dietetics; 17, 449-459. *Journal of Human Nutrition and Dietetics: Official Journal of the British Dietetic Association*, vol. 20, No. 3, p. 286.

Rye, P.D., and others (2018). Alginate Oligomers and Their Use as Active Pharmaceutical Drugs. In *Alginates and Their Biomedical Applications*, pp. 237–256. Springer.

Secretariat of the Convention on Biological Diversity (2020). Digital Sequence Information on Genetic Resources: Concept, Scope and Current Use. Convention on Biological Diversity CBD/DSI/AHTEG/2020/1/3. www.cbd.int/doc/c/fef9/2f90/70f037ccc5da885dfb293e88/dsi-ahteg-2020-01-03-en.pdf

Seymour, Mathew (2019). Rapid progression and future of environmental DNA research. *Communications Biology*, vol. 2, No. 80, pp. 1–3.

Skropeta, Danielle, and Liangqian Wei (2014). Recent advances in deep-sea natural products. *Natural Product Reports*, vol. 31, No. 8, pp. 999–1025.

Sprague, Matthew, and others (2017). Microbial and genetically engineered oils as replacements for fish oil in aquaculture feeds. *Biotechnology Letters*, vol. 39, No. 11, pp. 1599–1609.

Steinegger, Martin, and others (2019). Protein-level assembly increases protein sequence recovery from metagenomic samples manyfold. *Nature Methods*, vol. 16, pp. 603–606.

Suleria, Hafiz Ansar Rasul, and others (2015). Marine-based nutraceuticals: An innovative trend in the food and supplement industries. *Marine Drugs*, vol. 13, No. 10, pp. 6336–6351.

Sunagawa, Shinichi, and others (2015). Structure and function of the global ocean microbiome. *Science*, vol. 348, No. 6237, p. 1261359.

Thompson, Cristiane C., and others (2017). Unlocking marine biotechnology in the developing world. *Trends in Biotechnology*, vol. 35, No. 12, pp. 1119–1121.

Thompson, Fabiano, and others (2018). Marine biotechnology in Brazil: recent developments and its potential for innovation. *Frontiers in Marine Science*, vol. 5, p. 236.

Tully, Benjamin J., and others (2018). The reconstruction of 2,631 draft metagenome-assembled genomes from the global oceans. *Scientific Data*, vol. 5, p. 170203.

United Nations (2017). *The First Global Integrated Marine Assessment: World Ocean Assessment I*. Cambridge: Cambridge University Press.

Wetterstrand, K.A. (2018). DNA Sequencing Costs: Data from the NHGRI Genome Sequencing Program (GSP). www.genome.gov/sequencingcostsdata.

Wouters, Olivier J., and others (2020). Estimated research and development investment needed to bring a new medicine to market, 2009–2018. *JAMA*, vol. 323, pp. 844–853.

Woyke, Tanja, and others (2019). Genomes From Uncultivated Microorganisms. *Encyclopedia of Microbiology*, vol. 4e, pp. 437–442.

Wuyts, Sander, and Nicola Segata (2019). At the Forefront of the Sequencing Revolution—Notes from the RNGS19 Conference. *Genome Biology*, vol. 20, No. 93, pp. 1–3.

Wynberg, Rachel (2015). Marine Genetic Resources and Bioprospecting in the Western Indian Ocean. *Western Indian Ocean*, p. 407.

Wynberg, Rachel, and Sarah A Laird (2018). Fast Science and Sluggish Policy: The Herculean Task of Regulating Biodiscovery. *Trends in Biotechnology*, vol. 36, No. 1, pp. 1–3.

Young, Lucy (2014). Marine-Derived Nutraceuticals and Cosmetics. Strategic Business Insights. www.strategicbusinessinsights.com/about/featured/2014/2014-02-marine-nutraceuticals.shtml.

Zhao, Xianming, and Xiao Qiu (2018). Analysis of the biosynthetic process of fatty acids in thraustochytrium. *Biochimie*, vol. 144, pp. 108–114.

Chapter 24
Marine hydrates – a potentially emerging issue

Contributors: Alan Simcock (convener and lead member), Aninda Mazumdar, Aaron Micallef, Katherine E.A. Segarra and Leonid Yurganov.

Keynote points

- Marine hydrates (mainly methane hydrates) exist primarily on continental slopes where there are large quantities of methane gas in the ocean, the pressure is high enough and the temperature is low enough.
- Concern has been expressed about the climatic risks resulting from the sudden release of large amounts of methane from marine hydrates. However, that hypothesis is not widely supported at present and is not mentioned in the recent special report of the Intergovernmental Panel on Climate Change on the ocean and cryosphere in a changing climate.
- Areas of gas seepage in the deep sea associated with gas hydrates host a very rich level of biodiversity supported by chemosynthetic bacteria.
- Initial successes have recently been noted by China and Japan in producing methane from marine methane hydrates.

1. Introduction

The first *World Ocean Assessment* (United Nations, 2017c) did not contain detailed material on marine hydrates. In the overall summary, it was noted that they were among the deep-water deposits that had generated continuing interest, but were not mined at that time.

It was reported in chapter 21 of the first Assessment that marine hydrates were a potential area for future offshore energy development, and an estimate was provided of the amount of marine hydrates and their carbon equivalent worldwide. While hydrates potentially represent an immense store of hydrocarbons, it was noted in the chapter that methane production from hydrates had not been documented beyond small-scale field experiments and that its relevance to the global gas supply was likely to be overshadowed by the increased development of onshore natural gas.

In chapter 35 of the first Assessment, it was noted that, because of the close relationship of gas seeps on continental margins to areas of interest for resource exploration (oil, gas and methane hydrates), an assessment of the nature of the associated rich biodiversity and its role in ecosystem functioning would be important before potential alterations and extractions could be carried out. Such biodiversity is discussed in chapter 7P of the present Assessment on hydrothermal vents and cold seeps.

The present chapter is aimed at providing a fuller assessment of the origin and estimated abundance of marine hydrates, their potential as a source of energy and the associated risks to the Earth's climate, slope stability and human society.

2. What are marine hydrates?

A marine hydrate is a crystalline solid composed of natural gas molecules retained within an ice-like cage of water molecules. The most common form of marine hydrate is methane hydrate, which has the chemical formula $(CH_4)_4(H_2O)_{23}$, or 1 mole of methane for every 5.75 moles of water, corresponding to 13.4 per cent methane by mass (Maslin and others, 2010; Chou and others, 2000). Marine hydrates are often referred to as marine or methane clathrates (from the Latin "clathri", meaning "lattice"), since the water molecules form a lattice within which the gas molecules are held. A schematic drawing of a gas hydrate is shown in figure I.

Figure I
Typical structure of a gas hydrate, with water molecules linked together to form a cage that traps gas molecules, such as methane, within

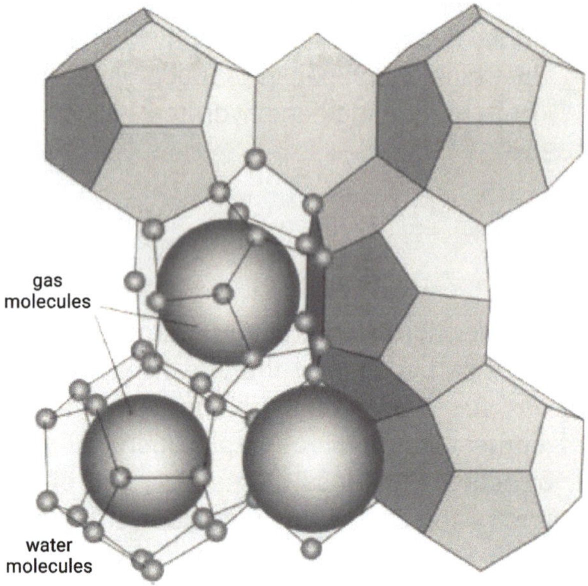

Source: Maslin and others, 2010.

Methane hydrates were first recognized in the late nineteenth century (Wróblewski, 1882; Villard, 1894). In the 1930s, they were identified in nature when their formation clogged natural gas pipelines in cold weather. In the 1950s, theoretical models for gas hydrates were developed and, in the 1960s, Russian scientists, including Vasiliev, argued that substantial marine deposits existed around the world (Vasiliev and others, 1970). That conclusion was confirmed in the early 1970s by the retrieval of samples of methane hydrates from the seabed in the Black Sea (Yefremova and Zhizhchenko, 1974). Since then, there have been similar retrievals around the world (see figure II), and countries such as Canada, China, Germany, India, Japan and the United States of America have established major hydrate research programmes (Sloan and Koh, 2007; Maslin and others, 2010; Song and others, 2014).

2.1. Location and scale of marine hydrates

Gas hydrates occur in areas of significant gas generation in which the temperature is sufficiently low and the pressure is sufficiently high to form and maintain them. The vast majority of gas hydrates occur as marine hydrates, while just over 1 per cent are located in permafrost soils (Ruppel, 2015). Most marine hydrates are formed by the accumulation of methane produced from the degradation of organic matter in buried sediments. Gas hydrate deposits (often several hundred metres thick) are embedded in sediments (Milkov and Sassen, 2002; Ruppel and Kessler, 2017). Marine hydrates are primarily driven by gas flowing through faults and channels in the sedimentary column and can be found exposed at the sea floor.

Marine hydrate distribution is driven by the combination of a gas source, water depth (usually more than 500 m, but dependent on gas composition) and temperature (geothermal gradient) to stabilize hydrates and the permeability of sediments. The most common way of inferring the presence of gas hydrates is by seismic investigation: the boundary between the gas hydrates and the underlying sediments that contain free gas reflects forms with a negative impedance contrast, which mimics the sea floor (bottom-simulating reflector) and can be interpreted to show the base of the gas hydrate stability zone. Sea floor samples can also be taken directly with cores or other sampling devices, but special steps need to be taken to maintain their stability when they are brought to the surface (Maslin and others, 2010). Seismic data indicate that methane hydrates are found in the sediments of the continental slope, while those in the Arctic Ocean are at lesser depths because of the lower temperature of the water column (Dillon and Max, 2012). In the middle of ocean basins, where the biogenic generation of gas is low owing to a lack of organic material, and in marginal seas where sea floor pressure is lower, hydrates do not form. Hydrates also form in and below the terrestrial permafrost soils of Alaska and Siberia (Maslin and others, 2010). Figure II shows a recent map of known and inferred locations of methane hydrates.

The presence of marine hydrates is constrained by the conditions in which they can persist. First, there needs to be a source of gas, generally methane of biogenic origin, derived from the decaying of organic material trapped in seabed sediments, that leads to the presence of methane in quantities greater than that which is soluble in the surrounding waters. Second, there has to be an appropriate combination of high pressure and low temperature at the sea floor. In Arctic waters, where the temperature is very low, the necessary pressure can, depending on the gas composition, be found at depths as shallow as 400 m. In warmer waters, depth as great as 1,000 m can be required. Third, there is a lower limit to the occurrence of marine hydrates: even at high pressures, the increase in temperature with regard to depth below the sea floor (geothermal gradient) will set a limit on the stability of marine hydrates at approximately 1,600 m (Kvenvolden and Lorenson, 2001; Maslin and others, 2010). The presence of methane hydrates can also act as a seal for free gas, thus retaining substantial amounts of methane in the sediments below them (Hornbach and others, 2004).

Figure II

Map showing locations where gas hydrate has been recovered, where gas hydrate is inferred to be present on the basis of seismic data and where gas hydrate drilling expeditions have been completed in permafrost or deep marine environments, also leading to the recovery of gas hydrate

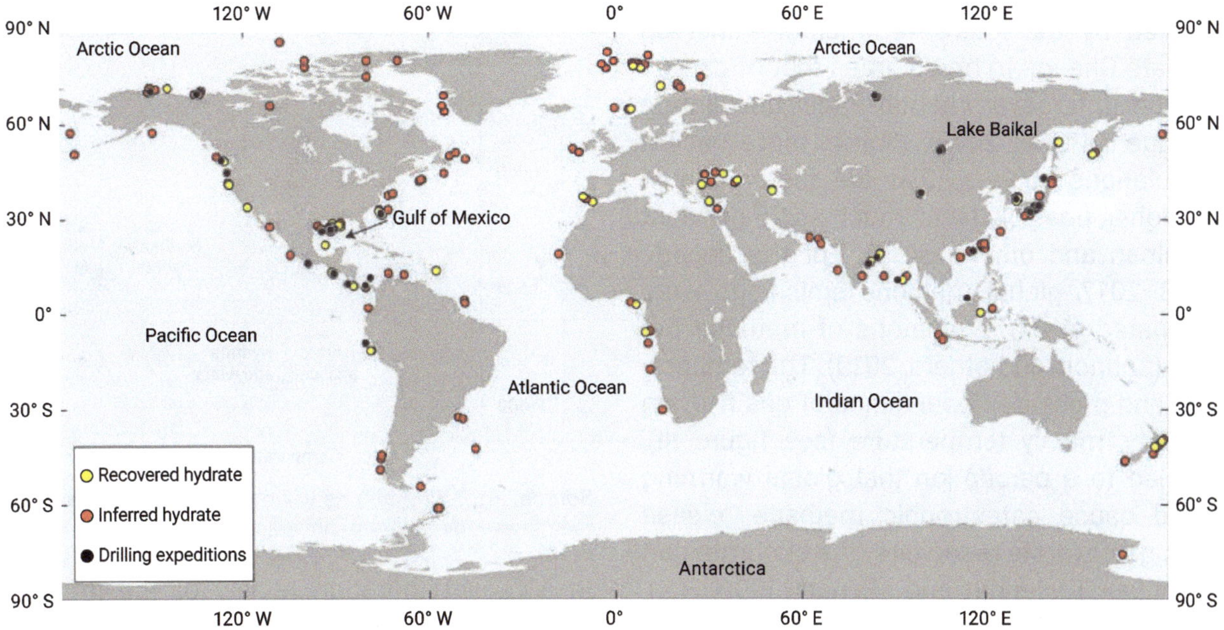

Source: Ruppel, 2018, amended to reflect Ryu and others, 2013; Minshull and others, 2020.[1]
Note: Any boundaries or names shown and designations used on the map do not imply official endorsement or acceptance by the United Nations.

1 The writing team is grateful to Chibuzo Ahaneku Valeria for his assistance in updating the map.

In 1988 and 1990, two independent estimates indicated a total global hydrate quantity of 21 × 10^{15} m^3 (MacDonald, 1990; Kvenvolden, 1999), which became a consensus view. However, in 2011, on the basis of an exhaustive review of other assessments and considering the lessons of many drilling programmes, it was estimated that there was 3 × 10^{15} m^3 of methane gas in place, calculated at the standard temperature and pressure (Boswell and Collett, 2011). That is similar to the lower end of the range (between 1–5 and 15–20 × 10^{15} m^3) calculated by Milkov (2004) and more than 30 times smaller than the 1 × 10^{17} m^3 estimated by Klauda and Sandler (2005). Some experts still support a larger estimate (Kvenvolden, 2012). The Milkov range equates to between 500–1,000 and 7,500–10,000 gigatons of carbon (Maslin and others, 2010). By way of comparison, the United States Geological Survey estimated in 2000 that the total reserves of all other fossil fuels contained 5,000 gigatons of carbon (United States Geological Survey World Energy Assessment Team, 2000). Subsequent work has supported the call for further research into the global total of marine hydrates on the basis of a wide-ranging discussion held at The Royal Society in London in 2010 (Day and Maslin, 2010).

3. Potential risks from marine methane hydrates

3.1. Risks in relation to the atmosphere

Methane is a powerful greenhouse gas with a heat-trapping potential over 100 years, estimated by the Intergovernmental Panel on Climate Change to be 25 times that of carbon dioxide (Intergovernmental Panel on Climate Change (IPCC), 2013). Some more recent calculations suggest that the factor should be higher, possibly by as much as 25 per cent (Etminan and others, 2016). For the decade 2008–2017, global methane emissions were estimated at 0.572 gigatons of methane per year (Saunois and others, 2019). The temperature and pressure dependence of gas hydrate stability, mainly temperature (see figure III), has led to a perception that global warming could cause catastrophic methane release from gas hydrate reservoirs (the clathrate gun hypothesis) (Henriet and Mienert, 1998; Haq, 1999). A similar mechanism has also been proposed as a way of explaining periods of rapid warming during the Quaternary Period (Kennett and others, 2000; Maslin and others, 2004). However, that hypothesis is not widely supported, and empirical evidence is inconclusive (Sowers, 2006; O'Hara, 2008).

Figure III
Methane hydrate stability

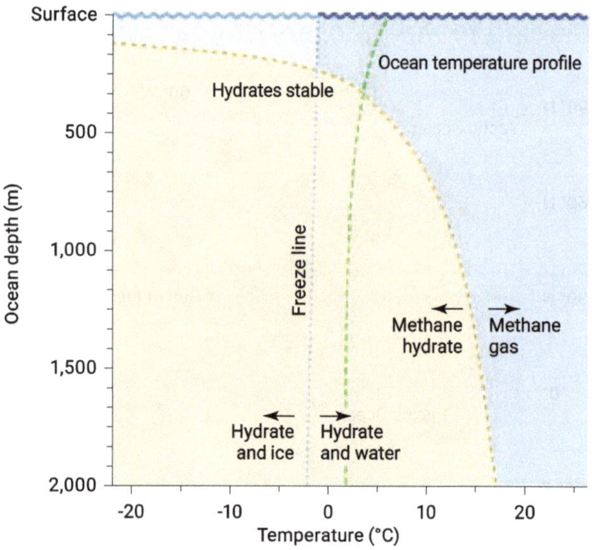

Source: https://commons.wikimedia.org/wiki/File:Undersea_methane_hydrate_phase_diagram.svg.

In a recent thorough review of the interaction between climate change and methane hydrates, it was concluded that there was no current evidence from observations that hydrate-derived methane was reaching the atmosphere or that the amounts that could potentially reach the atmosphere were significant enough to affect the overall methane budget. It was further noted that, in considering potential effects on methane flux to the atmosphere from the dissociation of marine methane hydrates, it was essential to consider the processes (sinks) that would intercept the methane before it reaches the atmosphere: in passing through sediment, methane may be broken down though anaerobic oxidation by microbes. In general, the conclusion is that methane from dissociated hydrates would not reach the atmosphere; it might dissolve in water in the sediment or in the water column, and it might be broken down further by microbial oxidation in the water column. However, more observational data and improved numerical models are needed to better characterize the climate-hydrate synergy in the future (Ruppel and Kessler, 2017).

Thus, the role of methane hydrates in contemporary and future climate change is unclear. Rather than with catastrophic, abrupt impacts, the release of methane from marine hydrates in response to rising ocean temperatures may have occurred gradually in the past and may occur over time scales of millenniums or longer (Archer, 2007; Archer and others, 2009).

However, the Arctic Ocean is warming at a faster rate than the rest of the globe (Larsen and others, 2014), and there is evidence of significant methane release into it, which may come from near-shore, submarine permafrost on the East Siberian Arctic Shelf (Shakhova and others, 2014). However, seasonal changes in the mixing of the water column appear to prevent methane from reaching the atmosphere during the summer (Yurganov and others, 2019).

In its recent special report on the ocean and cryosphere in a changing climate (IPCC, 2019), the Intergovernmental Panel on Climate Change did not mention marine hydrates, except to note (in chap. 5 of the report), with low confidence, that rising bottom temperatures or the shifting of warm currents on continental margins could increase the dissociation of buried gas hydrates on margins, potentially intensifying anaerobic methane oxidation (which produces hydrogen sulphide) and expanding the cover of methane-seep communities.

3.2. Risks in relation to seabed stability

When enclosed in sediments, and when the saturation is high enough, gas hydrates can act like cement, compacting and stabilizing the sea floor. However, if formed in deposits that are still unconsolidated, gas hydrates prevent the normal increase in compaction as the weight of the sediment increases. If destabilized by lower pressure or, in particular, by increased sea floor temperature, the gas hydrate can then dissociate. If that occurs, submarine slope failures may occur (Maslin and others, 2010). One particularly notable case in which gas hydrates are thought to have been implicated is the Storegga slide off the middle of the west coast of Norway, dated to about 8,200 years ago. According to calculations, it had a volume of 3,000 km^3 and produced a tsunami that affected Norway, the Faroe Islands (Denmark), Scotland and northern England (United Kingdom of Great Britain and Northern Ireland), with a run-up of up to 20 m. While an earthquake was probably the proximate cause, the dissociation of marine hydrates appears to have contributed significantly to the occurrence (Bondevik and others, 2005; Bryn and others, 2005; Micallef and others, 2009). In general, the consensus at present seems to be that, while the dissociation of marine hydrates can contribute to the scale, and thus the impact, of major slope failures, there is usually a separate trigger in the form of an earthquake or extreme weather event (Tappin, 2010).

4. Marine hydrates as a source of energy

Methane, as natural gas, is a well-known source of energy. Several countries have undertaken large research programmes to investigate the possibilities of using marine hydrates as a source of natural gas. Because of their lack of terrestrial natural gas resources, China and Japan are among the States that have put most effort into such exploration.

Japan established the Research Consortium for Methane Hydrate Resources in Japan (MH21) in 2002 to explore and develop energy from marine hydrates in its seas. The consortium brought together the Japan Oil, Gas and Metals National Corporation, the National Institute of Advanced Industrial Science and Technology and the Engineering Advancement Association of Japan. The work was planned in three phases. The first phase ran from 2002 to 2008 and involved cooperation with a number of other States, including Canada, Germany, India and the United States. The main outcomes were improved knowledge of the marine hydrate resources of Japan and two successful onshore methane hydrate production tests, yielding about 13,000 m^3 of methane. In the second phase, from 2008 to 2015, a successful offshore production test was run, an environmental impact assessment was developed and economic valuation and field verification was completed. The third phase, of which the major focus is to establish a technical platform for commercialization, is still in progress. The significance of the programme has increased since the great east Japan earthquake in 2011, which led to a policy of reducing planned dependence on nuclear energy (Oyama and Masutani, 2017). Through a collaborative effort between the National Energy Technology Laboratory of the Department of Energy, the Japan Oil, Gas and Metals National Corporation, the United States Geological Survey and Petrotechnical Resources of Alaska, in cooperation with Prudhoe Bay unitholders, a natural gas hydrate test well was drilled, which showed two gas hydrate reservoirs suitable for future testing. The Prudhoe well struck reservoirs at about 700 m and 844 m below the surface. According to the United States Geological Survey, gas hydrate was found to be filling 65 to >80 per cent of the spaces, or porosity, between the grains of sand and silt in the upper reservoir that comprise the rock formation. Japan is also collaborating with the United States to carry out production testing within the Prudhoe Bay unit in the fiscal year 2021/22. The experience gained through that collaboration will help Japan in its endeavour to conduct pilot testing in the fiscal year 2027/28.

Energy-related exploration for methane hydrates has been extensive in the Gulf of Mexico. The first leg of the Gulf of Mexico Gas Hydrates Joint Industry Project was undertaken in 2005 to develop technology and collect data to assist in the characterization of naturally occurring gas hydrates in the deep water of the Gulf of Mexico. The primary goal of the programme was to understand the impact of hydrate exploitation on sea floor stability and climate change, along with an assessment of the potential of methane hydrate as a future energy resource. Chevron, ConocoPhillips, Halliburton, Japan Oil, Gas and Metals National Corporation, Reliance Industries, Schlumberger, Total and the United States Mineral Management Service participated in collaboration with the Georgia Institute of Technology, Rice University and the United States Geological Survey. The investigation (Ruppel, 2018) revealed that drilling for gas hydrate in fine-grained sediments can be done safely without the expected disruption of the sea floor owing to hydrate dissociation. The results brought to light the importance of focused gas flow through localized permeability zones such as the sand body or fractures in forming hydrate deposits with a very limited lateral extent. The results also emphasized the relatively lower importance of features of the sea floor, such as mounds, and hydrates in decisions regarding coring sites for larger reserves at greater depths. Coring, drilling and wireline operations were carried out at water depths of more than 500 m, to depths of between 200 m and 459 m below the sea floor. As part of the second leg of the Gulf of Mexico Gas Hydrates Joint Industry Project, in 2009, the primary goal was to collect logging while drilling data through expected gas hydrate-bearing sand reservoirs in seven wells at three locations in the Gulf of Mexico. The findings of the second leg suggest that high-saturation gas hydrate sands free of trapped free gases are safe for exploitation since they do not present drilling hazards. The discovery of thick hydrate-bearing sands at Walker Ridge and Green Canyon validates the integrated geological and geophysical approach used in the pre-drill site selection process, and provides increased confidence in the assessment of gas hydrate volumes in the Gulf of Mexico and other marine sedimentary basins.

The National Gas Hydrate Program of India undertook the second expedition on board the drilling vessel *Chikyu* between March and July 2015 in the deep waters of the Krishna Godavari Basin, in collaboration with the Japan Agency for Marine-Earth Science and Technology and the United States Geological Survey. The objective of the expedition was to confirm the presence of sand-bearing hydrate reservoirs identified from seismic data and to calculate the reserves from the hydrate saturation percentage and sand body dimensions. Pressure coring, logging while drilling, wireline logging and formation testing operations were carried out as part of the programme. The expedition (Collett and others, 2019) confirmed the predicted slope-basin interconnected depositional model with sand-rich channel-levee facies saturated with methane hydrate in the Krishna Godavari Basin. The exceptionally detailed petrophysical information acquired through closely spaced logging while drilling and bore holes in area B of the L1 block gas hydrate accumulation has provided one of the most complete three-dimensional petrophysical-based views of any known gas hydrate reservoir system in the world.

Methane hydrate has been identified as a potential new gas source for China, and the South China Sea is believed to contain some of the world's most promising deposits. In China, a substantial number of institutions have undertaken investigations into the possibility of using marine hydrates as a source of energy, in particular the technology that would be needed to recover them. The methods considered include depressurization and thermal stimulation. Research has also been focused on the security of methane hydrate-bearing sediments during gas production and the related environmental impact (Song and others, 2014). The China Geological Survey conducted an initial production test and recovered 309,000 m^3 of methane from marine hydrates in the Shenhu area of the South China Sea between 10 May and 9 July 2017 (Li and others, 2018). China extracted 861,400 m^3 of natural gas from methane hydrate, known as "flammable ice", during a one-month trial production operation in the South China Sea, following its first experimental gas extraction from methane hydrate in 2017, during which a total of 309,000 m^3 of natural gas was produced in a 60-day period.

5. Key knowledge and capacity-building gaps

There are obvious gaps in knowledge of the global distribution and size of deposits of methane hydrates. The map in figure II shows that, for much of the world, assessments of the presence of gas hydrates are largely based on extrapolation rather than direct observation. Likewise, estimates of the global amounts of hydrate present are largely based on estimates of the volume of the methane hydrate stability zone, regardless of evidence of the presence or absence of gas to form them. Furthermore, abiogenic methane generation by ocean crust serpentinization, a major process in the ocean, has been largely ignored. A review of gas hydrates in Europe has recently been published (Minshull and others, 2020), but there is still no updated review at the global level.

There are also major gaps in understanding how methane hydrates will behave in changing circumstances, especially changes in ocean temperature, the way in which methane hydrates may dissociate and the way in which any methane released will then behave, and its impacts on climate and slope stability. Futhermore, it remains to be determined whether the oxidation of methane venting from the sea floor, presumably some sourced from dissociating hydrates, contributes significantly to ocean acidification. Such knowledge gaps have the possibility of being very significant in relation to the release of oceanic methane into the atmosphere and its consequent function as a greenhouse gas, even though the predominant opinion is that such a possibility is limited (see sect. 4 above).

Capacities are clearly being developed in China, Japan and elsewhere to enable access to methane stored as marine hydrates. At present, these are at the experimental or testing stage but could become important for States with limited access to natural gas.

6. Outlook

The outlook therefore depends very much on demand for natural gas in the context of reducing the consumption of coal and other fossil fuels, on the success of experiments in providing access to methane hydrates and on the further identification of locations of significant methane hydrate deposits that may justify their exploitation.

References

Archer, D. (2007). Methane hydrate stability and anthropogenic climate change. *Biogeosciences*, vol. 4, No. 4, pp. 521–544. https://doi.org/10.5194/bg-4-521-2007.

Archer, D., and others (2009). Ocean methane hydrates as a slow tipping point in the global carbon cycle. *Proceedings of the National Academy of Sciences of the United States of America*, vol. 106, No. 49, pp. 20596–20601.

Bondevik, Stein, and others (2005). The Storegga slide tsunami: comparing field observations with numerical simulations. *Ormen Lange: an Integrated Study for the Safe Development of a Deep-Water Gas Field within the Storegga Slide Complex, NE Atlantic Continental Margin*, vol. 22, No. 1, pp. 195–208. https://doi.org/10.1016/j.marpetgeo.2004.10.003.

Boswell, Ray, and Timothy S. Collett (2011). Current perspectives on gas hydrate resources. *Energy and Environmental Science*, vol. 4, No. 4, pp. 1206–1215.

Bryn, Petter, and others (2005). Explaining the Storegga slide. *Marine and Petroleum Geology*, vol. 22, Nos. 1–2, pp. 11–19.

Chou, I-Ming, and others (2000). Transformations in methane hydrates. *Proceedings of the National Academy of Sciences of the United States of America*, vol. 97, No. 25, pp. 13484–13487.

Collett, T.S., and others (2019). India National Gas Hydrate Program Expedition 02: Summary of Scientific Results: gas hydrate systems along the eastern continental margin of India, *Marine and Petroleum Geology*, vol. 108, pp. 39–142.

Day, S.J., and M. Maslin (2010). Gas hydrates: a hazard for the twenty-first century? *Philosophical Transactions of the Royal Society A: Mathematical, Physical and Engineering Sciences*, vol. 368, No. 1919, pp. 2579–2583.

Dillon, William, and Michael Max, (2012). Oceanic gas hydrate. In *Natural Gas Hydrate: in oceanic and permafrost environments*, M. Max, ed. Berlin: Springer Science and Business Media.

Etminan, M., and others (2016). Radiative forcing of carbon dioxide, methane, and nitrous oxide: a significant revision of the methane radiative forcing. *Geophysical Research Letters*, vol. 43, No. 24, pp. 12614–12623. https://doi.org/10.1002/2016GL071930.

Haq, Bilal U. (1999). Methane in the deep blue sea. *Science*, vol. 285, No. 5427, pp. 543–544.

Henriet, J.-P., and J. Mienert (1998). Gas hydrates: the Gent debates. outlook on research horizons and strategies. *Geological Society, London, Special Publications*, vol. 137, No. 1, pp. 1–8. https://doi.org/10.1144/GSL.SP.1998.137.01.01.

Hornbach, Matthew J., and others (2004). Critically pressured free-gas reservoirs below gas-hydrate provinces. *Nature*, vol. 427, No. 6970, pp. 142–144.

Intergovernmental Panel on Climate Change (IPCC) (2013). *Climate Change 2013: The Physical Science Basis. Contribution of Working Group I to the Fifth Assessment Report of the Intergovernmental Panel on Climate Change*. Thomas F. Stocker and others, eds. Cambridge: Cambridge University Press.

________ (2019). *Special Report on the Ocean and Cryosphere in a Changing Climate*. Geneva.

Kennett, James P., and others (2000). Carbon isotopic evidence for methane hydrate instability during Quaternary interstadials. *Science*, vol. 288, No. 5463, pp. 128–133.

Klauda, Jeffery B., and Stanley I. Sandler (2005). Global distribution of methane hydrate in ocean sediment. *Energy and Fuels*, vol. 19, No. 2, pp. 459–70. https://doi.org/10.1021/ef049798o.

Kvenvolden, Keith A. (1999). Potential effects of gas hydrate on human welfare. *Proceedings of the National Academy of Sciences of the United States of America*, vol. 96, No .7, pp. 3420–3426.

________ (2012). Natural gas hydrate: background and history of discovery. In *Natural Gas Hydrate: in oceanic and pPermafrost environments*, Michael D. Max, ed. Berlin: Springer Science and Business Media. https://doi.org/10.1007/978-94-011-4387-5_2.

Kvenvolden, Keith A., and Thomas D. Lorenson (2001). The global occurrence of natural gas hydrate. In *Natural Gas Hydrates: occurrence, distribution, and detection*, C. Paull and W. Dillon, eds., pp. 3–18. Washington, D.C.: American Geophysical Society. https://doi.org/10.1029/GM124p0003.

Larsen, J.N., and others (2014). Polar regions. In *Climate Change 2014: Impacts, Adaptation, and Vulnerability. Part B: Regional Aspects. Contribution of Working Group II to the Fifth Assessment Report of the Intergovernmental Panel on Climate Change*, V.R. Barros and others, eds., pp. 1567–1612. Cambridge: Cambridge University Press.

Li, Jin-fa, and others (2018). The first offshore natural gas hydrate production test in South China Sea. *China Geology*, vol. 1, No. 1, pp. 5–16.

MacDonald, Gordon J. (1990). Role of methane clathrates in past and future climates. *Climatic Change*, vol. 16, No. 3, pp. 247–281.

Maslin, Mark, and others (2004). Linking continental-slope failures and climate change: testing the clathrate gun hypothesis. *Geology*, vol. 32, No. 1, pp. 53–56.

Maslin, Mark, and others (2010). Gas hydrates: past and future geohazard? *Philosophical Transactions of the Royal Society A: Mathematical, Physical and Engineering Sciences*, vol. 368, No. 1919, pp. 2369–2393.

Micallef, Aaron, and others (2009). Development and mass movement processes of the north-eastern Storegga slide. *Quaternary Science Reviews*, vol. 28, Nos. 5–6, pp. 433–448.

Milkov, Alexei V. (2004). Global estimates of hydrate-bound gas in marine sediments: how much is really out there? *Earth-Science Reviews*, vol. 66, Nos. 3–4, pp. 183–197.

Milkov, Alexei V., and Roger Sassen (2002). Economic geology of offshore gas hydrate accumulations and provinces. *Marine and Petroleum Geology*, vol. 19, No. 1, pp. 1–11.

Minshull, Timothy, and others (2020). Hydrate occurrence in Europe: a review of available evidence, *Marine and Petroleum Geology*, vol. 111, pp. 1–11.

O'Hara, Kieran D. (2008). A model for late Quaternary methane ice core signals: wetlands versus a shallow marine source. *Geophysical Research Letters*, vol. 35, No. 2.

Oyama, Ai, and Stephen Masutani (2017). Review of the Methane Hydrate Program in Japan, *Energies*, vol. 10, pp. 1447–1460.

Ruppel, Carolyn (2015). Permafrost-associated gas hydrate: is it really approximately 1% of the global system? *Journal of Chemical and Engineering Data*, vol. 60, No. 2, pp. 429–436.

________(2018). The U.S. Geological Survey's Gas Hydrates Project. Report 2017–3079. Fact Sheet. Reston, Virginia. USGS Publications Warehouse. https://doi.org/10.3133/fs20173079.

Ruppel, Carolyn, and John D. Kessler (2017). The interaction of climate change and methane hydrates. *Reviews of Geophysics*, vol. 55, No. 1, pp. 126–168.

Ryu, Byong-Jae, and others (2013). Scientific results of the Second Gas Hydrate Drilling Expedition in the Ulleung Basin (UBGH2), *Marine and Petroleum Geology*, vol. 47, pp. 1–20.

Saunois, Marielle, and others (2019). The global methane budget 2000–2017. *Earth System Science Data*.

Shakhova, Natalia, and others (2014). Ebullition and storm-induced methane release from the East Siberian Arctic Shelf. *Nature Geoscience*, vol. 7, No. 1, pp. 64–70.

Sloan, E. Dendy, Jr., and Carolyn Koh (2007). *Clathrate Hydrates of Natural Gases* (3rd edition), CRC Press, Boca Raton, Florida, United States of America.

Song, Yongchen, and others (2014). The status of natural gas hydrate research in China: a review. *Renewable and Sustainable Energy Reviews*, vol. 31, pp. 778–791.

Sowers, Todd (2006). Late quaternary atmospheric CH_4 isotope record suggests marine clathrates are stable. *Science*, vol. 311, No. 5762, pp. 838–840.

Tappin, D.R. (2010). Submarine mass failures as tsunami sources: their climate control. *Philosophical Transactions of the Royal Society A: Mathematical, Physical and Engineering Sciences*, vol. 368, No. 1919, pp. 2417–2434.

United Nations (2017a). Chapter 21: Offshore hydrocarbon industries. In *The First Global Integrated Marine Assessment: World Ocean Assessment I*. Cambridge: Cambridge University Press.

________(2017b). Chapter 35: Extent of assessment of marine biological diversity. In *The First Global Integrated Marine Assessment: World Ocean Assessment I*. Cambridge: Cambridge University Press.

________(2017c). *The First Global Integrated Marine Assessment: World Ocean Assessment I*. Cambridge: Cambridge University Press.

United States Geological Survey (USGS) (2019). *Map of Gas Hydrates*. www.usgs.gov/media/images/map-gas-hydrates.

United States Geological Survey World Energy Assessment Team (2000). *US Geological Survey World Petroleum Assessment 2000: Description and Results*. USGS Digital Data Series DDS-60. US Geological Survey.

Vasiliev, V.G., and others (1970). The property of natural gases to occur in the earth crust in a solid state and to form gas hydrate deposits. *Otkrytiya v SSSR*, vol. 1969, pp. 15–17.

Villard, M.P. (1894). Sur l'hydrate carbonique et la composition des hydrates de gaz. *Comptes Rendus de l'Académie Des Sciences*, vol. 119, pp. 368–371.

Wróblewski, Zygmunt Florenty (1882). Sur la combinaison de l'acide carbonique et de l'eau. *Comptes Rendus de l'Académie Des Sciences*, vol. 94, pp. 212–213.

Yefremova, A.G., and B.P. Zhizhchenko (1974). Occurrence of crystal hydrates of gases in the sediments of modern marine basins. *Doklady Akademii Nauk SSSR*, vol. 214, No. 5, pp. 1179–1181.

Yurganov, Leonid, and others (2019). Methane increase over the Barents and Kara seas after the autumn pycnocline breakdown: satellite observations. *Advances in Polar Science*, vol. 30, pp. 82–390.

Chapter 25
Cumulative effects

Contributors: Karen Evans (convener and lead member), Roland Cormier, Piers Dunstan, Elizabeth Fulton, Jörn Schmidt (co-lead member), Vanessa Stelzenmüller, Ca Thanh Vu (co-lead member) and Skipton Woolley.

Keynote points

- Increasing pressures on marine environments from multiple sources are resulting in biodiversity loss, habitat damage and fragmentation and disease.
- Effective implementation of ecosystem-based management requires an appreciation of how, and to what extent, human activities and natural events interact and affect different ecosystem components and their functioning. It also requires the identification of solutions to prevent and mitigate the pressures being caused by those interactions.
- Over the past two decades, many frameworks for assessing the interactions, known as cumulative effects, have been developed. They have used differing approaches and terminologies and have been applied at differing levels.
- Although approaches vary, cumulative effects assessments (CEAs) conducted to date have mostly involved three main steps: (a) collation of information on the intensity and footprint of activities that may be affecting marine ecosystems; (b) identification of the responses of ecosystem components; and (c) identification of management measures that could be applied in response.
- Despite their increase in use, assessments focused on particular regions, areas or values that follow the same general steps outlined above are largely lacking from areas outside Europe and North America.
- The geographical bias in the implementation of CEAs highlights clear knowledge and capacity gaps and the need for the development of approaches that: (a) can be implemented in regions where data are sparse; (b) are easily implementable; and (c) produce outputs that can be readily understood and are translatable to decision-making processes, in particular in developing countries.

1. Introduction

The marine environment is currently subject to a number of pressures, many of which are derived from human activities. They include climate change, the extraction of resources, pollution (from land and marine sources) and invasive species, resulting in biodiversity loss, habitat damage and fragmentation and disease (e.g., Evans and others, 2017). The aim of ecosystem-based management is to balance human activities with environmental stewardship in order to maintain ecosystem properties, functions and services.[1] That requires an appreciation of how and to what extent human activities and natural events interact and affect ecosystem components and their functioning. It also requires the identification of solutions to prevent and mitigate the pressures being caused by such interactions (Halpern and others, 2008; Levin and others, 2009; Ban and others, 2010; Curtin and Prellezo, 2010). Those interactions are known as cumulative impacts or cumulative effects.

The terms "cumulative impacts" and "cumulative effects" are often used interchangeably to describe how pressures affect ecosystems. The use of standardized language is key to the transfer of knowledge, assessment approaches and expertise across management boundaries and among stakeholders and organizations. A preference for the use of the

1 See also chapter 26 for an overview of assessments associated with marine spatial planning, and chapter 27 for an overview of ecosystem-based management approaches.

term "cumulative effects" has been identified, noting that impacts are hypothesized and have been either not directly observed or attributed (Murray and others, 2015). For consistency, the term "cumulative effects" is used in the present chapter. There is as yet no universally accepted definition of cumulative effects and impacts, with definitions varying in the literature, depending on what is being assessed, and the context within which the assessment is being undertaken (e.g., Anthony, 2016; Spaling and Smit, 1993; Hegmann and others, 1999; Halpern and others, 2008; Johnson, 2016; Uthicke and others, 2016). The present chapter follows the premise that effects can be defined as a change to the environment, including its human components, while impacts represent the consequences of such change (Johnson, 2016).

There are four general types of cumulative effects: additive, synergistic, antagonistic (compensatory) and masking (Sonntag and others, 1987; Hegmann and others, 1999; Crain and others, 2008; Halpern and others, 2008). Additive effects are incremental additions to the pressures caused by an activity, with each increment adding to previous increments over time. Synergistic effects, also referred to as amplifying or exponential effects, magnify the consequences of individual pressures to produce a joint consequence that is greater than the additive effect. Antagonistic or compensatory effects produce a joint consequence that is less than additive. Masking effects produce essentially the same consequence for the ecosystem or social component as would occur with exposure to one of the pressures alone. Impacts that can be considered cumulative may result from a single activity that repeatedly produces a single pressure, a single activity that produces multiple pressures, multiple activities that produce a single pressure or multiple activities that produce multiple pressures over time (Foley and others, 2017).

The topic of assessments of cumulative effects was not included in the first *World Ocean Assessment* (United Nations, 2017a), although the range of factors affecting ecosystem services was considered in each of its regional chapters and a summary of pressures affecting the marine environment was provided in chapter 54 (United Nations, 2017b). However, in chapter 54, there was no attempt to undertake an assessment of the cumulative effects of those pressures, nor to identify the frameworks within which such assessments could be conducted. The present chapter, therefore, provides an overview of the key elements of assessments of cumulative effects, as well as an overview of approaches and their outcomes, including several regional examples of approaches in detail. The aim of the overview is to provide a baseline of the diversity of approaches and their use that can be used for establishing changes in approaches and applications in future global assessments.

2. Cumulative effects assessments

Over the past two decades, many frameworks for assessing cumulative effects on the environment have been developed using different approaches and terminologies (Stelzenmüller and others, 2018). Similarly, their focus has varied, with some taking a whole-of-system approach, in which all existing stressors and their effects on broad components of the marine environment are included in the assessment. Others have been focused on single stressors and single species or habitats (Korpinen and others, 2012; Marcotte and others, 2015; Coll and others, 2016). Of the 154 studies reviewed by Stelzenmüller and others (2018), several key conclusions regarding the various approaches used were identified, including that: (a) expert knowledge and qualitative data are sporadically or moderately used across assessments;

(b) the use of geographic information systems is almost a prerequisite for assessment; (c) large gaps exist in addressing uncertainty throughout components of each assessment; and (d) novel integrative methods, such as a combination of qualitative data and qualitative modelling to assess the ecosystem state and pressures, are increasingly being developed for use in assessments.

Although approaches may vary, several common elements have been identified that should be incorporated into cumulative effects assessments (CEAs) aimed at advising management and planning (Halpern and others, 2008; Kappel and others, 2012; International Council for the Exploration of the Sea (ICES), 2019). Those elements can be broadly categorized in terms of information on the activities that cause pressures that may be affecting marine ecosystems, information on measures that may be implemented to manage those activities and therefore pressures, and the responses of ecosystem components, which in turn depend on their resilience to and recovery potential from the pressures being exerted.

The process used to date for undertaking a CEA is fundamentally based on a mapping process that involves considering the spatial and temporal footprint of one or more pressures (including the frequency of that activity and associated pressures as a measure of intensity) in relation to components of the marine ecosystem that are being or may be affected (Elliott and others, 2020). It further considers the vulnerability of or risk to those ecosystem components (including their sensitivity), while taking into account any management measures that may be in place. That allows for the identification of residual pressures remaining after accounting for management, and the calculation of a measure of the anticipated cumulative effect (Halpern and others, 2008; Kappel and others, 2012; ICES, 2019). The various elements of information are identified in figure I. The connectivity and heterogeneity of ecosystem components, functions and processes, and the uncertainty in biophysical processes, together with varying levels of intensity of activities affecting the environment, determine the complexity of CEAs.

The key functional steps of a CEA are:

1. Definition of the values of the marine system being assessed

The first step of an assessment is to identify the values of importance in the location of the assessment and their spatio-temporal distribution within the assessment area. Values can be ecological, social, economic or cultural in nature.

2. Definition of the activities that place pressures on the marine system (stressors)

Identifying a tangible expression of the potential cumulative impacts involves confirmation that the system value and pressure do indeed interact. It is necessary to identify disturbances and activities that potentially place pressures on the marine system in the area of the assessment and to map and quantify the nature of the pressures (direct, indirect, continuous or pulse) and their spatio-temporal distribution. That is a key factor in CEAs. Many activities or disturbances concentrated in a small area over a short time can result in pressures or stressors that accumulate because of a crowding effect. An area may be resilient against some level of disturbance, but if that level is exceeded faster than the natural recovery rate, the disturbance could exceed an ecological or societal threshold for a valued component (Johnson, 2016). Furthermore, the effects of pressures can disperse from the activity area, resulting in a lagged effect on areas outside the immediate footprint of the activity. As a result, the extent, dispersal, frequency and persistence of pressures associated with an activity need to be accounted for when assessing exposure to risk (Borgwardt and others, 2019). In addition, all potential stressors within and adjacent to the area of the assessment should be considered in order to identify potential emerging risks.

Figure I
Elements of a cumulative effects assessment with a focus on quantifying effects associated with human activities on ecosystems

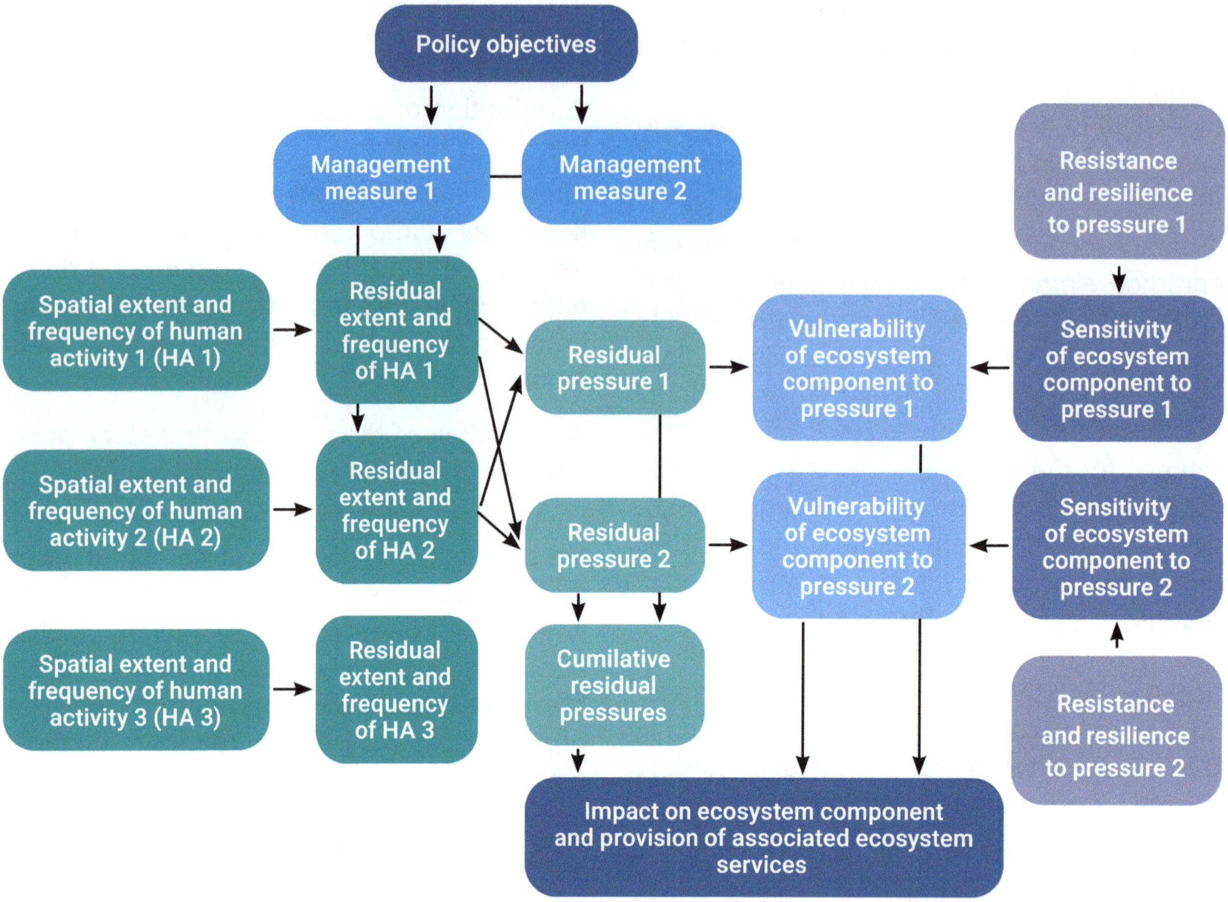

Source: Adapted from ICES (2019).

There are many approaches that can be used to map and quantify the spatial and temporal extent of both values and stressors, such as geographic information systems and spatial interpolation and dynamic models (e.g., Andersen and others, 2013; Robinson and others, 2013; Borgwardt and others, 2019; Dunstan and others, 2019). Because of the varying nature of values and stressors and their measurement, and the data or information that may be available to contribute to the mapping process, it is unlikely that a single approach will be appropriate in all circumstances. Rather, the approach undertaken should be appropriate for the data available (including their complexity), capture the spatio-temporal components of the data appropriately and address any uncertainties, biases or assumptions associated with the data.

3. Conceptual linking of pressures and values

Conceptual approaches (such as qualitative or quantitative models that identify impact pathways) can be used to link the values identified and the various potential activities and stressors in the assessment area (e.g., Dambacher and others, 2009; Anthony and others, 2013), identifying how components and processes in the marine environment are related, how natural and anthropogenic pressures can affect the system, and knowledge gaps and key uncertainties in the system. Ideally, consideration of the nature of potential interactions between pressures caused by multiple stressors is included, recognizing that interactions may be non-linear and that they may be synergistic, antagonistic or masking in nature (see sect. 1 above). Understanding of how values and pressures interact may initially be acquired by using qualitative models that allow for identification of the direction, nature and extent of interactions. Predictions of change can then be estimated through probabilistic modelling (e.g., Anthony and others, 2013). Undertaking such an approach allows for the degree of effect (i.e., severity) being placed on the values to be determined, and therefore those interactions that are most important, in order to focus efforts on better understanding, mapping and quantifying effects.

4. Assessment of risk and uncertainty

Once the pathways for the effects of pressures on values are understood, the scale of the effect on the value can be quantified, so that the level of exposure resulting from different stressors is integrated across their individual spatial extents – their "zones of influence" (Anthony and others, 2013; see also figure II). The risk to the value associated with impacts caused by the pressure and associated uncertainty can then be estimated, while noting that, often, the limited understanding of both the value and the pressures is, in itself, also a source of uncertainty. For example, often the spatial and temporal patterns of pressures are not fully known, nor are the responses of particular values to pressures that may vary over space and time (Stock and Micheli, 2016). The identification of sources of uncertainty and their influence on assessment results can be challenging in itself, so appropriate sensitivity analyses that explore the influence of all stressors and their interactions should be carried out (Stock and Micheli, 2016). Risk estimation needs to be capable of capturing the complexity of the system components and of interactions with activities and associated uncertainties, and incorporate the relevant spatial and temporal distributions of any consequences, both positive and negative in nature (e.g., Gregory and others, 2012; Stock and Micheli, 2016).

Figure II

A conceptual model illustrating zones of influence for two examples of point sources: (A) river run-off from catchments; and (B) urban or port development

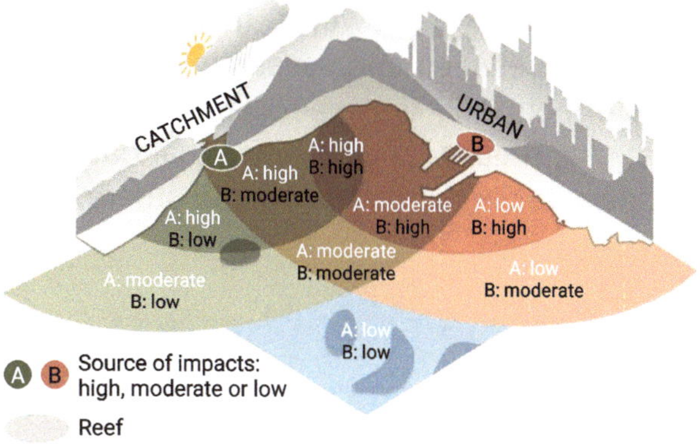

Note: The probabilities of change for each ecosystem value and the amount of ecosystem value potentially affected (with those probabilities accounting for uncertainties) are calculated within each zone of influence
Source: Anthony and others (2013).

5. Validation

Finally, where possible, the networks of interactions, maps of risk and cumulative effects should be verified observationally (though, in practice, that has occurred relatively rarely; see Halpern and Fujita, 2013). In order to facilitate such validation, risk assessments need to be reported in such a way that they can be observable; that is, measured and mapped in the field.

One framework that is useful for developing quantitative models for estimating effects and for the purpose of communication with policymakers and other decision makers is the drivers-pressures-state-impact-response framework (Smeets and Weterings, 1999; Elliott and others, 2017). It is based on the concept that drivers (underlying natural and human-caused forces) exert pressures (immediate factors) on the environment that lead to changes in the state of the environment. To be operational, a CEA should also explicitly include an evaluation of the effectiveness of management measures (Cormier and others,

2018; Stelzenmüller and others, 2018), in particular by, first, quantifying the effects of any management measures on pressures and their resulting impacts and, second, identifying how management measures may be modified to further reduce those pressures and resulting impacts. As yet, however, most CEAs lack linkages between an assessment of cumulative effects and the management measures that may regulate the activities that cause the respective pressures (Hayes and others, 2015; Cormier and others, 2017). As a result, many CEAs provide limited linkages between planning processes and regulatory frameworks that could identify where a precautionary approach may need to be implemented or where improvements to management processes are needed (ICES, 2019). Furthermore, most widely accepted CEA methods consider that the provision of ecosystem services and estimates of sociocultural effects are outside the remit of the assessment (ICES, 2019).

3. Regional applications of cumulative effects assessments on the marine environment: distribution and approaches

The implementation of CEAs within marine systems has increased rapidly over the past two decades, with applications in regional marine assessments, planning and regulatory processes (Halpern and others, 2015; ICES, 2019). However, despite their increased use, assessments following the same general steps outlined in section 2 are largely lacking from areas outside Europe and North America (Korpinen and Andersen, 2016; see also figure III and the table).

Figure III
Global distribution of cumulative effects assessments implemented in marine ecosystems (2016–2019)

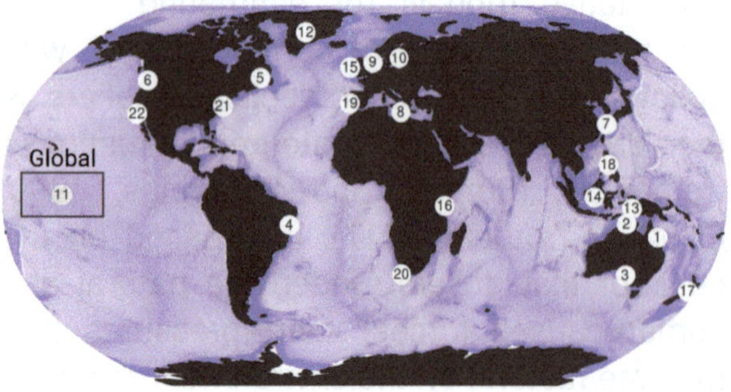

Note: Update to CEAs detailed in Korpinen and Andersen (2016). Numbers correspond to summaries provided in the table.

Korpinen and Andersen (2016) undertook a review of CEAs in an effort to provide an overview of the methods and practices associated with them in the marine environment. In particular, the review was aimed at determining if different estimate variables used in CEAs were comparable and whether the validation of the CEAs was reliable. Similar methodological approaches were identified in half of the studies reviewed, which were based on the method of Halpern and others (2008), although relatively few addressed major uncertainties in the use of such approaches as outlined in Halpern and Fujita (2013). In the review, several key areas were identified in which CEAs needed to be advanced, including the validation or benchmarking of pressures, the inclusion of accurate measures of temporal components of human activities (many assumed that activities were long-lasting and overlapped in time) and accounting for historical impacts that had already modified the marine environment.

Building upon that study, a review of the peer-reviewed literature published since 2016 was undertaken for the present Assessment by searching Scopus for the keywords "cumulative effect" and "cumulative impact", in order to provide an updated summary of assessments.

An overview of each approach and the outputs from each CEA is provided in the table.

The following key insights were gained from the updated review:

- Assessment approaches need to be contextually based. In that regard, while approaches should include the functional steps outlined in the present chapter, the development of approaches to CEAs needs to consider the scale (and resolution) at which the assessment is being conducted, the values being assessed, the data available for undertaking such an assessment, the uncertainties associated with the data, the specific management objectives in undertaking the assessment and the format of the outputs produced, in particular their suitability for informing planning and management.
- Assessments need to incorporate the extent and spatial and temporal variability in data and their associated uncertainty (including that associated with data quality), not only to ensure robust outputs from CEAs, but also to focus where knowledge gaps may lie and where future efforts should be made to improve assessments and reduce uncertainties.
- Most CEAs lack experiment-based or observation-based measures of the sensitivity of ecosystem components to the effects of stressors, and therefore the outputs of assessments should be verified observationally.
- Many assessments continue to provide single-point evaluations or temporal averages of cumulative effects, rather than repeated temporal studies that could provide information on changes in cumulative effects over time (i.e., trends).
- In order to ensure uptake in decision-making processes, CEAs should incorporate linkages between an assessment of cumulative effects and the management measures that may regulate the activities that cause the various pressures. Many assessments continue to lack such linkages. Similarly, few assessments consider an evaluation of the implementation of management measures for activities that cause the various pressures and cumulative effects on the environment.
- Assessments linked to management and regulatory processes, in most cases, consider only the effects of activities within the area under regulation and do not account for the dispersal of effects beyond the area being assessed (i.e., regional or ecosystem-scale effects). It is, therefore, important that spatial separation of the activity location and pressure effect is considered (e.g., Stephenson and others, 2019).
- Clear communication of the risk and uncertainties associated with estimates of risk is needed for the maximum uptake of CEAs in decision-making processes. Assessments need explicitly to describe the causes and consequences of deleterious effects to help managers, stakeholders, scientists and engineers to understand the causal pathways of risk (e.g., Nicol and others, 2019).
- The implementation of CEAs is geographically biased to Europe and North America, although it is encouraging that there are publications of assessments in areas in which none had been identified in Korpinen and Andersen (2016). The jurisdictions of many developing economies are yet to undergo any level of formal assessment beyond the general coverage of global analyses such as those by Halpern and others (2015). That highlights clear capacity gaps and the need for the development of approaches that: (a) can be implemented in regions where data are lacking; (b) can incorporate non-traditional data sources, such as community observations (e.g., citizen science) and traditional knowledge; (c) are easily implementable (in terms of both skills and time); (d) can be readily

updated as new information or pressures emerge; and (e) produce outputs that can be readily understood and are translatable to decision-making processes.

Detailing the many approaches to CEAs implemented in the marine environment globally is beyond the scope of the present chapter. Further details are provided below of examples of various frameworks implemented to assess cumulative effects in the southern and northern hemispheres, as well as further insights into developments in areas where CEAs have been implemented to a lesser degree.

3.1. Great Barrier Reef, Australia

The Great Barrier Reef has been identified as being under a range of pressures from the local to the global level, including those associated with climate change, (cyclonic) storms and flooding, nutrient and sediment run-off from land use, pollutants (including pesticides, marine debris, plastics, nanoparticles, noise and light), human uses of the marine environment and disease (Uthicke and others, 2016). Overall reef health has been identified as being in decline for some time (De'ath and others, 2012), and mass bleaching events that occurred in 2016, 2017 and 2020 have reduced its health even further (Smith and Spillman, 2020). According to some sources, the northern portion of the reef has been permanently altered as a result of those pressures (Hughes and others, 2017).

The development of CEAs for the reef has been an iterative process. The first formal CEA, conducted in 2012, used a combined cumulative impact and structured decision-making framework that involved both qualitative and probabilistic models to consider the influence of a subset of cumulative stressors (nutrients, turbidity and sedimentation, habitat erosion and climate change) on coral reefs and seagrass ecosystems (Anthony and others, 2013). The framework incorporated a decision-making process that allowed for the exploration of hypothetical management interventions, consequences and trade-offs.

The cumulative impact and structured decision-making approach has been further developed to incorporate statistical, ecotoxicological, conceptual, semi-quantitative and quantitative mechanistic models and structured decision analyses to assess cumulative effects on coral reef environments (Uthicke and others, 2016). Outputs from the framework include risk and exposure maps and the assessment of pressure and value thresholds for specific locations and ecological communities of interest. The application of the framework resulted in the recognition that: (a) linear changes are rare in ecosystems (i.e., change is not necessarily additive); (b) ecological thresholds and responses to multiple pressures are likely to change over ecologically relevant time frames, through acclimation (which can ameliorate effects) or dynamically compounding effects (that amplify responses); and (c) predictions without the experimental or field confirmation of responses may lead to false conclusions and suboptimal investment in management processes.

The approach of Uthicke and others (2016) goes beyond simply spatially layering the distribution of stressors and assuming that the cumulative effects are linearly additive in nature, given that it allows for mechanistic understanding of non-linear interactions (through the development of full response curves that take into account antagonistic and synergistic interactions) with management implications. That CEA framework, known as the Reef 2050 cumulative impact management policy,[2] with a proposed set of guidelines for its implementation (Dunstan and others, 2019), has only recently been applied.[3] The involvement of

2 See www.gbrmpa.gov.au/our-work/reef-strategies/Reef-2050-policies#.
3 See http://hdl.handle.net/11017/3389.

the Great Barrier Reef Marine Park Authority (the management agency responsible for the reef) in the development of the framework has seen it underwritten into all future planning and approval processes at the regional level, as well as at the level of specific development applications. That first (illustrative) application of the guidelines linked data collected on shallow coral reef systems with spatial data on the distribution of pressures through structural equation models, indicating a strong context dependence for the cumulative effects (Dunstan and others, 2019). It highlighted the role of long-term monitoring in informing assessments when evaluating cumulative effects.

3.2. North Sea

The North Sea is one of the most affected marine ecosystems in the world ocean (Halpern and others, 2008), with impacts from multiple anthropogenic stressors associated with global and regional developments, including coastal development and habitat loss, eutrophication, pollution and fishing (Emeis and others, 2015). In addition, the North Sea is a climate change hotspot (Burrows and others, 2011; Holt and others, 2012), with dramatic changes in food web structure and functioning reported in association with trends in sea level, ocean temperature and acidification (Reid and others, 2001; Beaugrand, 2003; Weijerman and others, 2005; McQuatters-Gollop and others, 2007; Kenny and others, 2009; Lynam and others, 2017). In particular, the fish community of the North Sea has been strongly affected by fishing and climate change, with rapid and substantial changes reported since 2000 (Engelhard and others, 2014; Fock and others, 2014; Sguotti and others, 2016; Frelat and others, 2017).

Assessments to investigate the effects of human activities on ecosystem components of the North Sea have largely comprised modelling studies focused on the effect of different demersal fishing practices, and derived aggregated measures of benthic disturbance by such practices (Stelzenmüller and others, 2015; Rijnsdorp and others, 2016; Hiddink and others, 2019). Only within the past decade has there been an increasing focus on assessing the combined effect of human activities other than fishing on the marine environment. (Stelzenmüller and others, 2010; Fock, 2011; Foden and others, 2011). That is not only due to limitations associated with data that may be available for CEAs, but also the complex socioecological interlinkages in the North Sea region, in no small part owing to multinational jurisdictions.

More recently, a greater emphasis has been placed on the development of approaches that not only allow for an assessment of the cumulative effects of human activities, but also assess those effects on considerably larger spatial scales than previously considered (Knights and others, 2015; Piet and others, 2019) in order to produce more targeted advice for management (Piet and others, 2017; Cormier and others, 2018). Approaches have included exposure-effect risk assessments based on sector-pressure-ecological component linkage matrices (Knights and others, 2015; Piet and others, 2019) and the spatial mapping of activities or stressors and ecosystem components, combined with linkage pathways determined by expert elicitation (Andersen and others, 2013), similar to the approach described by Halpern and others (2008). Results from such an assessment have identified key areas in which cumulative effects are greatest, as well as associated stressors. As yet, however, a North Sea-wide assessment has not been undertaken.

An emerging approach to assessing cumulative effects within the region, in particular in the context of management or regulation, comprises a framework that combines the conceptual structuring of cause-effect pathways with a quantitative assessment of effects (Cormier and others, 2018). That approach highlights the need to assess the effectiveness of management measures in reducing human pressures in order to understand the prevailing cumulative pressure load on distinct ecosystem components.

3.3. Other regions

As identified in the review described in the present chapter, few CEAs have been conducted outside North America and Europe (see table). Examples of CEAs conducted elsewhere include those in the Asian region, where an expert-based stepwise decision logic process was used to score the intensity of 10 pressures (including urbanization; coastal, anchorage and port infrastructure; sewage discharge; aquaculture; a gas platform; a saltern; and tourism) on Jiaozhou Bay in China (Wu and others, 2016). The weighted outputs were then combined with distance measures calculated using geographic information systems software to produce maps that represented the sum of the cumulative effects. In Hong Kong, China, a similar approach was taken to look at the potential implications for the survivorship of the local population of the Indo-Pacific humpback dolphin (*Sousa chinensis*) (Marcotte and others, 2015). In that case, however, the weighting was done in terms of the severity of each effect on dolphin survival.

Beyond the specific examples of CEAs outlined above, other related or precursor assessments exist for locations in Asia and Latin America, highlighting the fact that a larger number of CEAs could be undertaken in those regions. For example, the integrated fisheries risk analysis method for ecosystems approach, developed by Zhang and others (2011), used to look at the performance of fisheries management strategies in terms of the goals of an ecosystem approach to management, could easily be extended beyond fisheries to other human activities. That approach explicitly considers aspects of the local fish stocks, habitats, biodiversity measures and economic indicators for fisheries. Importantly, it considers the kinds of pressures that fisheries place on ecosystems. It is at that point that the approach could be extended to other activities as a means of creating a CEA. Dynamic process-based models that explicitly bring together multiple human activities, including fisheries, aquaculture, urban development, marine transport, mining, forestry, agriculture and tourism, to explore the implications for the future management, development or expansion of sustainable aquaculture in the Patagonian region of Chile could also be used as a basis for a CEA (Steven and others, 2019).

4. Outlook

Most CEAs undertaken to date have been focused on assessing activities and effects that have already occurred in the marine environment. There is a growing enthusiasm to move to assessments that allow foresighting or forecasting and prediction, in order to inform future planning of activities or adaptive and anticipatory management approaches (e.g., Lukic and others, 2018; see also chap. 26). The global economic contribution by marine industries is projected to double by 2030 (Organization for Economic Cooperation and Development, 2016), reaching as much as $3 trillion, with exponential (or similar) increases in their footprint and interactions (McCauley and others, 2015; Plagányi and Fulton, 2017). Avoiding undesirable outcomes and a degradation in the values of marine systems will require informative CEAs that feed into adaptive management and evidence-based decision-making. Dynamic research languages, methods and models spanning disciplines will be required in order to achieve it. Its development is not straightforward and will require substantial effort, in particular in forecasting each stressor into the future in a spatially and temporally explicit way, and accounting for the changing nature of interactions among the stressors. A unified and broadly applicable forward-looking CEA methodology is, however, probably unfeasible, at least in the near future, owing to the inherent difficulties in addressing key uncertainties in future projections. The improvement of guidelines and best practices to facilitate such CEA approaches will provide a useful step forward in that regard.

Whether forward-looking or rear-looking, there is growing agreement that the methods associated with CEAs need to be expanded from consideration of the multiple effects of single development activities or the accumulation of effects of multiple similar activities within a single industrial sector to the combined effects of all pressures on marine ecosystems. Modelling frameworks such as those detailed above have identified that the responses of marine systems are often non-linear and synergistic, and antagonistic effects play important roles in shaping the environment (Crain and others, 2008; Hunsicker and others, 2016; Uthicke and others, 2016). There is a need to improve capacity in the use of conceptual and statistical modelling approaches that allow for mechanistic understanding of non-linear interactions between stressors, non-additive effects on the marine environment and responses of the marine environment as a result. As stated above, it is recognized that the development of such approaches is not straightforward and will require substantial effort. The improvement of guidelines and best practices to facilitate such CEA approaches and commitment to building capacity in their application and use will provide a useful step forward in that regard.

Meta-analyses (e.g., Crain and others, 2008) are helping researchers to understand the prevalence of additive, synergistic and antagonistic interactions, while statistical approaches are helping to identify the presence and nature of non-additive interactions (e.g., Teichert and others, 2016). In addition, there have been significant advances in the handling of uncertainty in assessments (e.g., Rochet and others, 2010; Foster and others, 2014; Gissi and others, 2017) and some progress in defining thresholds and reference points to use in assessments, though they can be somewhat subjective as they are defined by societal objectives (e.g., Samhouri and Levin, 2012; Large and others, 2015; Samhouri and others, 2012; 2017). The incorporation of uncertainty not only allows for more robust interpretation of the outputs of assessments, but also facilitates an adaptive management process and identifies research priorities to fill knowledge gaps for the continuing improvement of management.

Ultimately, to increase the geographical spread of CEAs, future efforts have to address the development of approaches that can be applied, in particular in data-poor situations, and that produce outputs that can be readily understood and translated into decision-making processes (Stelzenmüller and others, 2020). That would better equip decision makers to deal with the dynamic nature of rapidly changing marine ecosystems, in which the mixes and relative dominance of the different pressures will change through time and space.

Summary of cumulative effects assessments (2016–2019), by country and region

Number on map[a]	Geographical region	Ocean region	Assessment approaches	Assessment objectives	Assessment results	References
1	Australia	South Pacific Ocean	Qualitative conceptual models Bayesian belief networks Statistical models Mechanistic models Index calculations Literature reviews	Map scientific understanding of coral habitats and identify gaps Identify limitations of extant assessment methods Assess impact of prawn trawling Identify impacts that affect reef habitats and communities Assess response of coral to ocean warming and sedimentation Identify cumulative effects of multiple range-shifting species under climate change scenarios and evaluate management responses	Continuing overall decline in state of Great Barrier Reef Considerations for conducting cumulative effects assessments (CEAs) (including uncertainties and biases) and recommendations for advancing them, including developing assessment frameworks for application across a range of activities and areas Identification of knowledge gaps Possibility of trophic cascades and negative impact on ecosystem dynamics and productivity caused by redistribution of multiple species	Grech and others (2011); Marzloff and others (2016); Uthicke and others (2016); Bessell-Browne and others (2017); Richards and Day (2018); Dunstan and others (2019)
2	Australia	South Pacific Ocean and Indian Ocean	Spatial mapping	Evaluate cumulative patterns in sea turtle by-catch	Identification of a by-catch hotspot in the Gulf of Carpentaria, where multiple species were affected by commercial fisheries	Riskas and others (2016)
3	Australia	Indian Ocean	Spatial mapping	Assess cumulative effects on marine environment while capturing uncertainty in expert elicitation	Increased transparency and robustness for management implementation through assessment of experts' uncertainty	Jones and others (2018)
4	Brazil	South Atlantic Ocean	Spatial mapping Index calculations	Assess exposure of coral reefs to cumulative effects of human activities	Spatial variation and variation, in terms of types of stressors to which coral reefs were exposed, caused by exposure. Areas of highest exposure were closest to population centres	Magris and others (2018)
5	Canada	North Atlantic Ocean	Species distribution models	Assess impact of ocean warming and decreases in oxygen on three marine species	Substantial change in species distributions projected over 20–30 years in varying ways	Stortini and others (2017)

6	Canada and United States	North Pacific Ocean	Spatial mapping Statistical models	Assess impacts of dissolved oxygen concentrations and bottom trawling along a depth gradient Assess impacts of shoreline armoring Assess impacts of marine noise	Influence on deep-water benthos by bottom trawling even where communities are shaped by strong environmental gradients Possible contribution to cumulative effects by shoreline armouring Predicted avoidance of or injury by loud marine noise for marine animals	De Leo and others (2017); Dethier and others (2016); Ellison and others (2016)
7	China	North Pacific Ocean	Literature reviews Statistical models Numerical models	Carry out qualitative review of potential stressors that contribute to fishery declines Assess cumulative effects of metals and polycyclic aromatic hydrocarbons on bacterioplankton communities Assess cumulative effects of restoration projects on water quality	Need for ecosystem-based management for sustainable development of fisheries Individual and cumulative effects of cadmium and phenanthrene on bacterial assemblages that are temporally variable and antagonistic in early stages of exposure Improvement in water quality due to restoration projects, but implementation was often within single objective frameworks and did not account for other activities that reduce water quality	Qian and others (2017); Zhao and others (2016); Ma and others (2017)
8	Europe and Africa	Mediterranean Sea and Black Sea	Meta-analyses Expert elicitations Estimations of uncertainty Regression models Index calculations Spatial mapping Mechanistic models Statistical models	Map and calculate cumulative impacts associated with a range of human activities Map invasive species and effects on biodiversity values	Inadequacy of current conservation initiatives to deal with cumulative threats in exclusive economic zone of Tunisia High variability in estimates of uncertainty of impacts, with impacts on only a few areas of Adriatic Sea and Ionian Sea robustly identified Overlooking of cumulative effects of extraction and dumping of marine sand No agreement in modelled importance of drivers of observed degradation on coralline outcrops with expert elicitation outputs	Coll and others (2016); Katsanevakis and others (2016); Ben Rais Lasram and others (2016); Corrales and others (2017); Depellegrin and others (2017); Gerakaris and others (2017); Gissi and others (2017); Trop (2017); Bevilacqua and others (2018); Brodersen and others (2018); Corrales and others (2018)

Number on map[a]	Geographical region	Ocean region	Assessment approaches	Assessment objectives	Assessment results	References
9	Europe	North Atlantic Ocean	Biological traits analyses Spatial mapping Expert elicitations Meta-analyses Spatial analyses Index calculations	Assess cumulative impact of five marine sectors on benthic communities Assess influence of climate change on area-based management tools in high-seas areas Assess cumulative effects of noise on two species	Variability in sensitivity of habitats to activities, with placement of hard structures on benthic habitats causing significant changes in biological and functional traits Projected reduction in usefulness of area-based management tools in high-seas areas owing to climate change Identification of high exposure risk areas for the two species	Merchant and others (2017); Johnson and others (2018); Kenny and others (2018)
10	Europe	Baltic Sea	Geographic information systems-based viewshed models	Conduct visual impact assessment of cumulative pressures caused by existing and planned anthropogenic activities	Highest potential visual impacts for sheltered coastal areas of complex geomorphological features	Depellegrin (2016)
11	Global	Global	Literature reviews Meta-analyses Spatial analyses Statistical models	Review CEAs across a range of anthropogenic activities, including social and management objectives Assess capacity of large marine protected areas to protect ecosystems from cumulative impacts Assess vulnerability of deep-sea ecosystem services to deep sea mining Assess cumulative effects on marine environment generated through oil sands production and transport	Considerations for conducting CEAs (including uncertainties and biases) and recommendations for advancing them, including developing assessment frameworks for application across a range of activities and areas Identification of knowledge gaps	Borja and others (2016); Briscoe and others (2016); Hazeem and others (2016); Lucke and others (2016); Lundquist and others (2016); Davies and others (2017); Foley and others (2017); Green and others (2017); Le and others (2017); Willsteed and others (2017); Faulkner and others (2018); Stelzenmüller and others (2018)
12	Greenland	Arctic Ocean	Spatial mapping	Assess cumulative effects of multiple stressors on biodiversity values	High level of overlap between stressors and key species along west coast of Greenland, highlighting that the area is in need of future management and protection	Andersen and others (2017)

13	Indonesia	South Pacific Ocean	Expert-elicitations Bayesian belief networks	Assess interactions between social, economic and environmental factors that influence fishing activities and the effectiveness of customary marine tenure	Social, economic and environmental outcomes of customary marine tenure influenced by complex interrelationships between community perceptions of fishing and tourism and associated conflicts	Hoshino and others (2016)
14	Indonesia	Indian Ocean	Semi-quantitative risk scoring	Assess cumulative risk to marine ecosystem of a range of human activities	Greatest risk to marine ecosystems caused by fishing, climate change and coastal development	Battista and others (2017)
15	Ireland	North Atlantic Ocean	Statistical models	Assess impact of vessel and construction-related activity on marine mammals	Reduction in occurrence of three species in association with vessel and construction-related activity	Culloch and others (2016)
16	Kenya	Indian Ocean	Statistical models	Assess cumulative effects of presence of tourist boats on a population of the Indo-Pacific dolphin	Behavioural budgets of the dolphin affected by presence of tourist boats, though cumulative effects are not significant at current levels	Pérez-Jorge and others (2017)
17	New Zealand	South Pacific Ocean	Literature reviews Meta-analyses Expert elicitations	Assess interdependencies between science, governance and society to identify risks in marine ecosystems Assess importance and magnitude of impacts of various activities and stressors on ecosystem services	Considerations for identification of risks and recommendations for risk assessments Severe total cumulative impacts for all ecosystem services considered, with climate change, commercial fishing, sedimentation and pollution contributing the most	Thrush and others (2016); Singh and others (2017)
18	Philippines	North Pacific Ocean	Semi-quantitative risk scoring	Assess cumulative risk to marine ecosystem of a range of human activities	Greatest risk to marine ecosytem posed by fishing and climate change	Battista and others (2017)
19	Portugal	North Atlantic Ocean	Spatial mapping	Assess interactions between a range of human activities and the marine environment	High level of cumulative impacts caused by human activities in Portuguese maritime space, in particular near coast	Fernandes and others (2017)

Number on map[a]	Geographical region	Ocean region	Assessment approaches	Assessment objectives	Assessment results	References
20	South Africa	Atlantic Ocean and Indian Ocean	Statistical models	Describe pelagic bioregions for defining regions of marine spatial planning	Identification through bioregional analysis of three key bioregions and a number of subregions as a framework for ecosystem reporting and systematic conservation planning	Roberson and others (2017)
21	United States	North Atlantic Ocean	Mechanistic models	Simulate effects of multiple stressors on marine living resources	Highest impacts on system productivity due to temperature increases	Ihde and Townsend (2017)
22	United States	North Pacific Ocean	Spatial mapping Statistical models	Map potential impacts of single and multiple stressors across marine protected area network Assess appropriateness of scientific activities on habitats and communities inmarine protected areas Assess cumulative effects of storm events and trampling on intertidal ecosystems	Most marine protected areas affected by intense land-based and ocean-based impacts, with climate stressors having the greatest impacts Recommendations for a decision-making framework to assess scientific activities Impacts associated with storms and trampling across similar species, thereby identifying vulnerable species, with disturbances having additive effects	Micheli and others (2016); Mach and others (2017); Saarman and others (2018)

[a] For numbers, see map in figure III.

Acknowledgements: The writing team would like to thank Nicole Stollberg of the Thünen Institute of Sea Fisheries, Bremerhaven, Germany, for developing figure I, as well as the three peer reviewers and Member States for providing comments that helped to improve the chapter.

References

Andersen, Jesper H., and others (2013). Human uses, pressures and impacts in the eastern North Sea. Technical report, Danish Centre for Environment and Energy, No. 18.

Andersen, Jesper H., and others (2017). Potential for cumulative effects of human stressors on fish, sea birds and marine mammals in Arctic waters. *Estuarine, Coastal and Shelf Science*, vol. 184, pp. 202–206.

Anthony, Kenneth R.N. (2016). Coral reefs under climate change and ocean acidification: challenges and opportunities for management and policy. *Annual Review of Environment and Resources*, vol. 41, pp. 59–81.

Anthony, Kenneth R.N., and others (2013). *A Framework for Understanding Cumulative Impacts, Supporting Environmental Decisions and Informing Resilience Based Management of the Great Barrier Reef World Heritage Area*. Final Report to the Great Barrier Reef Marine Park Authority and Department of the Environment.

Ban, Natalie C., and others (2010). Cumulative impact mapping: advances, relevance and limitations to marine management and conservation, using Canada's Pacific waters as a case study. *Marine Policy*, vol. 34, No. 5, pp. 876–886.

Battista, Willow, and others (2017). Comprehensive Assessment of Risk to Ecosystems (CARE): a cumulative ecosystem risk assessment tool. *Fisheries Research*, vol. 185, pp. 115–129.

Beaugrand, Gregory (2003). Long-term changes in copepod abundance and diversity in the north-east Atlantic in relation to fluctuations in the hydroclimatic environment. *Fisheries Oceanography*, vol. 12, Nos. 4–5, pp. 270–283.

Ben Rais Lasram, F., and others (2016). Cumulative human threats on fish biodiversity components in Tunisian waters. *Mediterranean Marine Science*, vol. 17, No. 1, pp. 190–201.

Bessell-Browne, Pia, and others (2017). Cumulative impacts: thermally bleached corals have reduced capacity to clear deposited sediment. *Scientific Reports*, vol. 7, No. 1, pp. 2716.

Bevilacqua, S., and others (2018). A regional assessment of cumulative impact mapping on Mediterranean coralligenous outcrops. *Scientific Reports*, vol. 8, No. 1, pp. 1–11.

Borgwardt, Florian, and others (2019). Exploring variability in environmental impact risk from human activities across aquatic ecosystems. *Science of the Total Environment*, vol. 652, pp. 1396–1408.

Borja, Angel, and others (2016). Bridging the gap between policy and science in assessing the health status of marine ecosystems. *Frontiers in Marine Science*, vol. 3, art. 175.

Briscoe, Dana K., and others (2016). Are we missing important areas in pelagic marine conservation? Redefining conservation hotspots in the ocean. *Endangered Species Research*, vol. 29, No. 3, pp. 229–237.

Brodersen, Maren Myrto, and others (2018). Cumulative impacts from multiple human activities on seagrass meadows in eastern Mediterranean waters: the case of Saronikos Gulf (Aegean Sea, Greece). *Environmental Science and Pollution Research*, vol. 25, No. 27, pp. 26809–26822.

Burrows, Michael T., and others (2011). The pace of shifting climate in marine and terrestrial ecosystems. *Science*, vol. 334, No. 6056, pp. 652–655.

Coll, Marta, and others (2016). Modelling the cumulative spatial–temporal effects of environmental drivers and fishing in a NW Mediterranean marine ecosystem. *Ecological Modelling*, vol. 331, pp. 100–114.

Cormier, Roland, and others (2017). Moving from ecosystem-based policy objectives to operational implementation of ecosystem-based management measures. *ICES Journal of Marine Science*, vol. 74, No. 1, pp. 406–413.

Cormier, Roland, and others (2018). The science-policy interface of risk-based freshwater and marine management systems: from concepts to practical tools. *Journal of Environmental Management*, vol. 226, pp. 340–346.

Corrales, X., and others (2017). Hindcasting the dynamics of an Eastern Mediterranean marine ecosystem under the impacts of multiple stressors. *Marine Ecology Progress Series*, vol. 580, pp. 17–36.

Corrales, X., and others (2018). Future scenarios of marine resources and ecosystem conditions in the Eastern Mediterranean under the impacts of fishing, alien species and sea warming. *Scientific Reports*, vol. 8, No. 1, pp. 14284.

Crain, Caitlin Mullan, and others (2008). Interactive and cumulative effects of multiple human stressors in marine systems. *Ecology Letters*, vol. 11, No. 12, pp. 1304–1315.

Culloch, Ross M., and others (2016). Effect of construction-related activities and vessel traffic on marine mammals. *Marine Ecology Progress Series*, vol. 549, pp. 231–242.

Curtin, Richard, and Raúl Prellezo (2010). Understanding marine ecosystem-based management: a literature review. *Marine Policy*, vol. 34, No. 5, pp. 821–830.

Dambacher, Jeffrey M., and others (2009). Qualitative modelling and indicators of exploited ecosystems. *Fish and Fisheries*, vol. 10, pp. 305–322.

Davies, T.E., and others (2017). Large marine protected areas represent biodiversity now and under climate change. *Scientific Reports*, vol. 7, No. 1, p. 9569.

De'ath, Glenn, and others (2012). The 27-year decline of coral cover on the Great Barrier Reef and its causes. *Proceedings of the National Academy of Sciences of the United States of America*, vol. 109, No. 44, pp. 17995–17999.

De Leo, Fabio C., and others (2017). Bottom trawling and oxygen minimum zone influences on continental slope benthic community structure off Vancouver Island (NE Pacific). *Deep Sea Research Part II: Topical Studies in Oceanography*, vol. 137, pp. 404–419.

Depellegrin, Daniel (2016). Assessing cumulative visual impacts in coastal areas of the Baltic Sea. *Ocean and Coastal Management*, vol. 119, pp. 184–198.

Depellegrin, Daniel, and others (2017). Multi-objective spatial tools to inform maritime spatial planning in the Adriatic Sea. *Science of the Total Environment*, vol. 609, pp. 1627–1639.

Dethier, Megan N., and others (2016). Multiscale impacts of armoring on Salish Sea shorelines: evidence for cumulative and threshold effects. *Estuarine, Coastal and Shelf Science*, vol. 175, pp. 106–117.

Dunstan, P.K., and others (2019). Draft guidelines for analysis of cumulative impacts and risks to the Great Barrier Reef. Report to the National Environmental Science Programme. Marine Biodiversity Hub. CSIRO.

Elliott, M., and others (2017). "And DPSIR begat DAPSI (W) R (M)!": a unifying framework for marine environmental management. *Marine Pollution Bulletin*, vol. 118, Nos. 1–2, pp. 27–40.

Elliott, Michael, and others (2020). Activity-footprints, pressures-footprints and effects-footprints: walking the pathway to determining and managing human impacts in the sea. *Marine Pollution Bulletin*, vol. 155, p. 111201.

Ellison, William T., and others (2016). Modeling the aggregated exposure and responses of bowhead whales Balaena mysticetus to multiple sources of anthropogenic underwater sound. *Endangered Species Research*, vol. 30, pp. 95–108.

Emeis, Kay-Christian, and others (2015). The North Sea: a shelf sea in the Anthropocene. *Journal of Marine Systems*, vol. 141, pp. 18–33.

Engelhard, Georg H., and others (2014). Climate change and fishing: a century of shifting distribution in North Sea cod. *Global Change Biology*, vol. 20, No. 8, pp. 2473–2483.

Evans, Karen, and others (2017). Australia state of the environment 2016: marine environment, independent report to the Australian Government Minister for the Environment and Energy. *Australian Government Department of the Environment and Energy, Canberra*.

Faulkner, Rebecca C., and others (2018). Guiding principles for assessing the impact of underwater noise. *Journal of Applied Ecology*.

Fernandes, Maria da Luz, and others (2017). How does the cumulative impacts approach support Maritime Spatial Planning? *Ecological Indicators*, vol. 73, pp. 189–202. https://doi.org/10.1016/j.ecolind.2016.09.014.

Fock, Heino (2011). Integrating multiple pressures at different spatial and temporal scales: a concept for relative ecological risk assessment in the European marine environment. *Human and Ecological Risk Assessment*, vol. 17, No. 1, pp. 187–211.

Fock, Heino, and others (2014). An early footprint of fisheries: changes for a demersal fish assemblage in the German Bight from 1902–1932 to 1991–2009. *Journal of Sea Research*, vol. 85, pp. 325–335.

Foden, Jo, and others (2011). Human pressures on UK seabed habitats: a cumulative impact assessment. *Marine Ecology Progress Series*, vol. 428, pp. 33–47.

Foley, Melissa M., and others (2017). The challenges and opportunities in cumulative effects assessment. *Environmental Impact Assessment Review*, vol. 62, pp. 122–134.

Foster, Scott D., and others (2014). The cumulative effect of trawl fishing on a multispecies fish assemblage in south-eastern Australia. *Journal of Applied Ecology*, vol. 52, No. 1, pp. 129–139.

Frelat, Romain, and others (2017). Community ecology in 3D: tensor decomposition reveals spatio-temporal dynamics of large ecological communities. *PLOS One*, vol. 12, No. 11. p. e0188205.

Gerakaris, V., and others (2017). Effectiveness of Posidonia oceanica biotic indices for assessing the ecological status of coastal waters in Saronikos Gulf (Aegean Sea, Eastern Mediterranean). *Mediterranean Marine Science*, vol. 18, No. 1, pp. 161–178.

Gissi, Elena, and others (2017). Addressing uncertainty in modelling cumulative impacts within maritime spatial planning in the Adriatic and Ionian region. *PLOS One*, vol. 12, No. 7, p. e0180501.

Grech, A., and others (2011). A broad-scale assessment of the risk to coastal seagrasses from cumulative threats. *Marine Policy*, vol. 35, No. 5, pp. 560–567.

Green, Stephanie J., and others (2017). Oil sands and the marine environment: current knowledge and future challenges. *Frontiers in Ecology and the Environment*, vol. 15, No. 2, pp. 74–83.

Gregory, Robin, and others (2012). *Structured Decision Making: A Practical Guide to Environmental Management Choices*. John Wiley and Sons.

Halpern, Benjamin S., and others (2008). A global map of human impact on marine ecosystems. *Science*, vol. 319, No. 5865, pp. 948–952.

Halpern, Benjamin S., and others (2015). Spatial and temporal changes in cumulative human impacts on the world's ocean. *Nature Communications*, vol. 6, p. 7615.

Halpern, Benjamin S., and Rod Fujita (2013). Assumptions, challenges, and future directions in cumulative impact analysis. *Ecosphere*, vol. 4, No. 10, pp. 1–11.

Hayes, K.R., and others (2015). Identifying indicators and essential variables for marine ecosystems. *Ecological Indicators*, vol. 57, pp. 409–419.

Hazeem, Layla J., and others (2016). Cumulative effect of zinc oxide and titanium oxide nanoparticles on growth and chlorophyll a content of Picochlorum sp. *Environmental Science and Pollution Research*, vol. 23, No. 3, pp. 2821–2830.

Hegmann, George, and others (1999). *Cumulative Effects Assessment Practitioners Guide*. Citeseer.

Hiddink, Jan Geert, and others (2019). Assessing bottom trawling impacts based on the longevity of benthic invertebrates. *Journal of Applied Ecology*, vol. 56, No. 5, pp. 1075–1084.

Holt, J.T., and others (2012). Oceanic controls on the primary production of the northwest European continental shelf: model experiments under recent past conditions and a potential future scenario. *Biogeosciences*, vol. 9, pp. 97–117.

Hoshino, Eriko, and others (2016). A Bayesian belief network model for community-based coastal resource management in the Kei Islands, Indonesia. *Ecology and Society*, vol. 21, No. 2.

Hughes, Terry P., and others (2017). Global warming and recurrent mass bleaching of corals. *Nature*, vol. 543, No. 7645, p. 373.

Hunsicker, Mary E., and others (2016). Characterizing driver–response relationships in marine pelagic ecosystems for improved ocean management. *Ecological Applications*, vol. 26, No. 3, pp. 651–663.

International Council for the Exploration of the Sea (ICES) (2019). *Workshop on Cumulative Effects Assessment Approaches in Management (WKCEAM)*, vol. 1, No. 17. ICES Scientific Reports. https://doi.org/10.17895/ices.pub.5226.

Ihde, Thomas F., and Howard M. Townsend (2017). Accounting for multiple stressors influencing living marine resources in a complex estuarine ecosystem using an Atlantis model. *Ecological Modelling*, vol. 365, pp. 1–9.

Johnson, Chris J. (2016). Defining and Identifying Cumulative Environmental, Health, and Community Impacts. In *The Integration Imperative*, pp. 21–45. Springer.

Johnson, David, and others (2018). Climate change is likely to severely limit the effectiveness of deep-sea ABMTs in the North Atlantic. *Marine Policy*, vol. 87, pp. 111–122.

Jones, Alice R., and others (2018). Capturing expert uncertainty in spatial cumulative impact assessments. *Scientific Reports*, vol. 8, No. 1, p. 1469.

Kappel, Carrie V., and others (2012). Mapping cumulative impacts of human activities on marine ecosystems. Boston, Massachusetts: SeaPlan.

Katsanevakis, Stelios, and others (2016). Mapping the impact of alien species on marine ecosystems: the Mediterranean Sea case study. *Diversity and Distributions*, vol. 22, No. 6, pp. 694–707. https://doi.org/10.1111/ddi.12429.

Kenny, Andrew J., and others (2009). An integrated approach for assessing the relative significance of human pressures and environmental forcing on the status of large marine ecosystems. *Progress in Oceanography*, vol. 81, Nos. 1–4, pp. 132–148.

Kenny, Andrew J., and others (2018). Assessing cumulative human activities, pressures, and impacts on North Sea benthic habitats using a biological traits approach. *ICES Journal of Marine Science*, vol. 75, No. 3, pp. 1080–1092.

Knights, Antony M., and others (2015). An exposure-effect approach for evaluating ecosystem-wide risks from human activities. *ICES Journal of Marine Science*, vol. 72, No. 3, pp. 1105–1115.

Korpinen, Samuli, and others (2012). Human pressures and their potential impact on the Baltic Sea ecosystem. *Ecological Indicators*, vol. 15, No. 1, pp. 105–114.

Korpinen, Samuli, and Jesper H. Andersen (2016). A global review of cumulative pressure and impact assessments in marine environments. *Frontiers in Marine Science*, vol. 3, art. 153.

Large, Scott I., and others (2015). Quantifying patterns of change in marine ecosystem response to multiple pressures. *PLOS One*, vol. 10, No. 3, p. e0119922.

Le, Jennifer T., and others (2017). Incorporating ecosystem services into environmental management of deep-seabed mining. *Deep Sea Research Part II: Topical Studies in Oceanography*, vol. 137, pp. 486–503.

Levin, Phillip S., and others (2009). Integrated ecosystem assessments: developing the scientific basis for ecosystem-based management of the ocean. *PLOS Biology*, vol. 7, No. 1, p. e1000014.

Lucke, Klaus, and others (2016). Auditory sensitivity in aquatic animals. *The Journal of the Acoustical Society of America*, vol. 139, No. 6, pp. 3097–3101.

Lukic, I., and others (2018). *Handbook for Developing Visions in MSP. Technical Study under the Assistance Mechanism for the Implementation of Maritime Spatial Planning.* www.msp-platform.eu/sites/default/files/vision_handbook.pdf.

Lundquist, Carolyn J., and others (2016). Science and societal partnerships to address cumulative impacts. *Frontiers in Marine Science*, vol. 3, art. 2.

Lynam, Christopher Philip, and others (2017). Interaction between top-down and bottom-up control in marine food webs. *Proceedings of the National Academy of Sciences of the United States of America*, vol. 114, No. 8, pp. 1952–1957.

Ma, Deqiang, and others (2017). The cumulative effects assessment of a coastal ecological restoration project in China: an integrated perspective. *Marine Pollution Bulletin*, vol. 118, Nos. 1–2, pp. 254–260.

Mach, Megan E., and others (2017). Assessment and management of cumulative impacts in California's network of marine protected areas. *Ocean and Coastal Management*, vol. 137, pp. 1–11.

Magris, Rafael, and others (2018). Cumulative Human Impacts on Coral Reefs: Assessing Risk and Management Implications for Brazilian Coral Reefs. *Diversity*, vol. 10, No. 2, pp. 26.

Marcotte, Danielle, and others (2015). Mapping cumulative impacts on Hong Kong's pink dolphin population. *Ocean and Coastal Management*, vol. 109, pp. 51–63.

Marzloff, Martin Pierre, and others (2016). Modelling marine community responses to climate-driven species redistribution to guide monitoring and adaptive ecosystem-based management. *Global Change Biology*, vol. 22, No. 7, pp. 2462–2474.

McCauley, Douglas J., and others (2015). Marine defaunation: animal loss in the global ocean. *Science*, vol. 347, No. 6219, p. 1255641.

McQuatters-Gollop, Abigail, and others (2007). A long-term chlorophyll dataset reveals regime shift in North Sea phytoplankton biomass unconnected to nutrient levels. *Limnology and Oceanography*, vol. 52, No. 2, pp. 635–648.

Merchant, Nathan D., and others (2017). Marine noise budgets in practice. *Conservation Letters*, vol. 11, No. 3, p. e12420.

Micheli, Fiorenza, and others (2016). Combined impacts of natural and human disturbances on rocky shore communities. *Ocean and Coastal Management*, vol. 126, pp. 42–50.

Murray, Cathryn Clarke, and others (2015). Advancing marine cumulative effects mapping: an update in Canada's Pacific waters. *Marine Policy*, vol. 58, pp. 71–77.

Nicol, Sam, and others (2019). Quantifying the impact of uncertainty on threat management for biodiversity. *Nature Communications*, vol. 10, No. 1, pp. 1–14.

Organization for Economic Cooperation and Development (2016). *The Ocean Economy in 2030*. https://doi.org/10.1787/9789264251724-en.

Pérez-Jorge, Sergi, and others (2017). Estimating the cumulative effects of the nature-based tourism in a coastal dolphin population from southern Kenya. *Deep Sea Research Part II: Topical Studies in Oceanography*, vol. 140, pp. 278–289.

Piet, Gerjan, and others (2017). Ecological risk assessments to guide decision-making: methodology matters. *Environmental Science and Policy*, vol. 68, pp. 1–9.

Piet, Gerjan, and others (2019). An integrated risk-based assessment of the North Sea to guide ecosystem-based management. *Science of the Total Environment*, vol. 654, pp. 694–704.

Plagányi, Éva E., and Elizabeth A. Fulton (2017). The Future of Modeling to Support Conservation Decisions in the Anthropocene Ocean. In *Conservation for the Anthropocene Ocean*, pp. 423–445. Elsevier.

Qian, Jie, and others (2017). Alteration in successional trajectories of bacterioplankton communities in response to co-exposure of cadmium and phenanthrene in coastal water microcosms. *Environmental Pollution*, vol. 221, pp. 480–490.

Reid, Philip C., and others (2001). Pulses in the eastern margin current and warmer water off the north west European shelf linked to North Sea ecosystem changes. *Marine Ecology Progress Series*, vol. 215, pp. 283–287.

Richards, Zoe T., and Jon C. Day (2018). Biodiversity of the Great Barrier Reef: how adequately is it protected? *PeerJ*, vol. 6, p. e4747.

Rijnsdorp, A.D., and others (2016). Towards a framework for the quantitative assessment of trawling impact on the seabed and benthic ecosystem. *ICES Journal of Marine Science*, vol. 73, supplement No. 1, pp. i127–i138.

Riskas, Kimberly A., and others (2016). Justifying the need for collaborative management of fisheries bycatch: a lesson from marine turtles in Australia. *Biological Conservation*, vol. 196, pp. 40–47.

Roberson, Leslie A., and others (2017). Pelagic bioregionalisation using open-access data for better planning of marine protected area networks. *Ocean and Coastal Management*, vol. 148, pp. 214–230.

Robinson, Leonie A., and others (2013). *ODEMM Pressure Assessment Userguide V.2. ODEMM Guidance Document Series No. 4.* Liverpool: University of Liverpool.

Rochet, Marie-Joëlle, and others (2010). Do changes in environmental and fishing pressures impact marine communities? An empirical assessment. *Journal of Applied Ecology*, vol. 47, No. 4, pp. 741–750.

Saarman, Emily T., and others (2018). An ecological framework for informing permitting decisions on scientific activities in protected areas. *PLOS One*, vol. 13, No. 6, p. e0199126.

Samhouri, Jameal F., and Phillip S. Levin (2012). Linking land- and sea-based activities to risk in coastal ecosystems. *Biological Conservation*, vol. 145, No. 1, pp. 118–129.

Samhouri, Jameal F., and others (2012). Sea sick? Setting targets to assess ocean health and ecosystem services. *Ecosphere*, vol. 3, No. 5, pp. 1–18.

Samhouri, Jameal F., and others (2017). Defining ecosystem thresholds for human activities and environmental pressures in the California Current. *Ecosphere*, vol. 8, No. 6, p. e01860.

Sguotti, Camilla, and others (2016). Distribution of skates and sharks in the North Sea: 112 years of change. *Global Change Biology*, vol. 22, No. 8, pp. 2729–2743.

Singh, Gerald G., and others (2017). Mechanisms and risk of cumulative impacts to coastal ecosystem services: an expert elicitation approach. *Journal of Environmental Management*, vol. 199, pp. 229–241.

Smeets, Edith, and Rob Weterings (1999). *Environmental Indicators: Typology and Overview.* Technical report No. 25. European Environment Agency.

Smith G., and C. Spillman (2020). Ocean Temperature Outlooks: Coral Bleaching Risk – Great Barrier Reef and Australian waters. Bureau Research Report No. 43, Bureau of Meteorology.

Sonntag, Nicholas C., and others (1987). *Cumulative Effects Assessment: A Context for Further Research and Development.* (No. 333.70971 C971). Canadian Environmental Assessment Research Council.

Spaling, Harry, and Barry Smit (1993). Cumulative environmental change: conceptual frameworks, evaluation approaches, and institutional perspectives. *Environmental Management*, vol. 17, No. 5, pp. 587–600.

Stelzenmüller, Vanessa, and others (2010). Quantifying cumulative impacts of human pressures on the marine environment: a geospatial modelling framework. *Marine Ecology Progress Series*, vol. 398, pp. 19–32.

Stelzenmüller, Vanessa, and others (2015). Quantitative environmental risk assessments in the context of marine spatial management: current approaches and some perspectives. *ICES Journal of Marine Science*, vol. 72, No. 3, pp. 1022–1042.

Stelzenmüller, Vanessa, and others (2018). A risk-based approach to cumulative effect assessments for marine management. *Science of the Total Environment*, vol. 612, pp. 1132–1140.

Stelzenmüller, Vanessa, and others (2020). Operationalizing risk-based cumulative effect assessments in the marine environment. *Science of the Total Environment*, vol. 724, p. 138118.

Stephenson, Robert L., and others (2019). A practical framework for implementing and evaluating integrated management of marine activities. *Ocean and Coastal Management*, vol. 177, pp. 127–138.

Steven, Andrew D.L., and others (2019). SIMA Austral: an operational information system for managing the Chilean aquaculture industry with international application. *Journal of Operational Oceanography*, vol. 12, supplement No. 2, pp. S29–S46.

Stock, Andy, and Fiorenza Micheli (2016). Effects of model assumptions and data quality on spatial cumulative human impact assessments. *Global Ecology and Biogeography*, vol. 25, No. 11, pp. 1321–1332.

Stortini, Christine H., and others (2017). Marine species in ambient low-oxygen regions subject to double jeopardy impacts of climate change. *Global Change Biology*, vol. 23, No. 6, pp. 2284–2296.

Teichert, Nils, and others (2016). Restoring fish ecological quality in estuaries: implication of interactive and cumulative effects among anthropogenic stressors. *Science of the Total Environment*, vol. 542, pp. 383–393.

Thrush, Simon, and others (2016). Addressing surprise and uncertain futures in marine science, marine governance, and society. *Ecology and Society*, vol. 21, p. 22.

Trop, Tamar (2017). An overview of the management policy for marine sand mining in Israeli Mediterranean shallow waters. *Ocean and Coastal Management*, vol. 146, pp. 77–88.

United Nations (2017a). *The First Global Integrated Marine Assessment: World Ocean Assessment I*. Cambridge: Cambridge University Press.

________ (2017b).Chapter 54: Overall assessment of human impact on the oceans. In *The First Global Integrated Marine Assessment: World Ocean Assessment I*. Cambridge: Cambridge University Press.

Uthicke, Sven, and others (2016). Multiple and cumulative impacts on the GBR: assessment of current status and development of improved approaches for management. *Final Report Project*, vol. 1.

Weijerman, Mariska, and others (2005). Regime shifts in marine ecosystems of the North Sea and Wadden Sea. *Marine Ecology Progress Series*, vol. 298, pp. 21–39.

Willsteed, Edward, and others (2017). Assessing the cumulative environmental effects of marine renewable energy developments: establishing common ground. *Science of the Total Environment*, vol. 577, pp. 19–32.

Wu, Zaixing, and others (2016). A methodology for assessing and mapping pressure of human activities on coastal region based on stepwise logic decision process and GIS technology. *Ocean and Coastal Management*, vol. 120, pp. 80–87.

Zhang, Chang Ik, and others (2011). An IFRAME approach for assessing impacts of climate change on fisheries. *ICES Journal of Marine Science*, vol. 68, No. 6, pp. 1318–1328.

Zhao, ShuJiang, and others (2016). A preliminary analysis of fishery resource exhaustion in the context of biodiversity decline. *Science China Earth Sciences*, vol. 59, No. 2, pp. 223–235.

Part six

Trends in management approaches to the marine environment

Chapter 26
Developments in marine spatial planning

Contributors: Alan Simcock (convener and lead member), Jarbas Bonetti, Louis Celliers, Karen Evans (co-lead member), Leandra Gonçalves, Marcus Polette, Julian Reyna and Ca Thanh Vu (co-lead member).

Keynote points

- The growing scale of human activities and the associated impacts on the marine environment mean that conflicts are increasingly occurring between different uses of the ocean. Marine spatial planning (MSP) is an effective way of resolving such conflicts.
- Over the past two decades, MSP has been instituted to a growing extent in many jurisdictions, in a variety of forms: some are simply zoning plans; others include more complex management systems.
- The legal status of MSP varies between jurisdictions: in some, it is guidance to be taken into account; in others, it has legal force constraining specific management decisions.
- In general, MSP has been most effective where it has been developed with the involvement of all relevant authorities and stakeholders.

1. Introduction

As noted in the summary of the first *World Ocean Assessment* (United Nations, 2017), "human activities now have so many and such great impacts on the ocean that the limits of its carrying capacity are being (or, in some cases, have been) reached". The causes of those impacts include both the intensification and extension into new areas of traditional uses of the sea, as well as the development of new uses. Increasingly, the use of ocean space cannot be taken for granted, and uses will tend to be in conflict with each other, especially in coastal zones. In the present chapter, the role of marine spatial planning (MSP) as an approach aimed at planning and managing such potential conflicts will be discussed.

The demands for goods and services from areas of the sea within national jurisdiction often exceed the capacity of those areas to meet all the demands. In the absence of special regulatory regimes, marine resources can be subject to excessive exploitation, and other uses of the sea (such as inputs of waste) can degrade the marine environment. The externalities of such exploitation and uses are often not considered within relevant market systems, and it can be necessary to identify efficient trade-offs in the allocation of sea uses (Tuda and others, 2014). A public process may, therefore, be desirable to reconcile all such factors.

At the same time, there has been increasing realization of the importance of the ocean for achieving sustainable development. Many countries have developed programmes to ensure the sustainable expansion of use of their marine resources (the blue economy) in order to achieve economic development in the context of the Sustainable Development Goals[1] (International Organization of Supreme Audit Institutions, 2019).

1.1. Marine spatial planning in the first *World Ocean Assessment*

MSP was not treated as a stand-alone topic in the first Assessment, though its relevance was noted in the chapters on ecosystem services, land-sea physical interactions, marine renewable energy and offshore hydrocarbon development, and fisheries (United Nations, 2017). It was defined as "the public process of analysing and allocating the spatial and

1 See General Assembly resolution 70/1.

temporal distribution of human activities in marine areas to achieve ecological, economic, and social objectives that are usually specified through a political process" (United Nations, 2017, chap. 15). It was noted that MSP was linked to a number of other tools and approaches that had the potential to assist in the management of conflicts among diverse stakeholders through participation, such as ecosystem-based management, marine protected areas and the ecosystem approach to fisheries (United Nations, 2017).

2. Types of marine spatial planning

There is not as yet wide agreement about the nature of MSP or how it should be evaluated (Plasman, 2008). However, the relationship between MSP and terms such as "ecosystem-based management", "sea use management" and "ocean zoning" has been clarified (Ehler and Douvere, 2009).

The concept of MSP covers a spectrum of processes. At its most basic, it may involve simply the production of a plan to allot zones for different activities. At the other end of the spectrum, it may provide a complex system for planning activities in the ocean, including elements of planning, management, licensing and enforcement (see reviews by Collie and others, 2013; Jones and others, 2016). Decisions on which type of MSP is appropriate in which areas take into account the range and intensity of pressures on the ocean, the national and local administrative frameworks and the level of economic development (Douvere and Ehler, 2009).

Many countries have already implemented some forms of development control over land, which constrain the abilities of landowners to develop and change the use of their properties. The precise extent of such controls varies (Organization for Economic Cooperation and Development (OECD), 2017). Most countries also have systems set up to regulate coastal and maritime activities. The Commission for the Protection of the Marine Environment of the North-East Atlantic, in a review, identified the following aspects as those that could be covered by MSP: coastal defences and land reclamation; dumping; fisheries; harbour works and navigational dredging; marine aquaculture; seabed minerals other than oil and gas; nature protection; navigation; offshore oil and gas; pipelines and cables; recreation (including bathing and pleasure boats); underwater cultural heritage; wind and wave energy; and wrecks and other historic features (Commission for the Protection of the Marine Environment of the North-East Atlantic, 2009).

In addition, many countries have instituted systems to promote the use of the sea and marine resources for economic development. That is especially the case for exploration for, and exploitation of, offshore hydrocarbons (see chap. 19) and for marine renewable energy installations (see chap. 21). However, the socioeconomic aspects of MSP can extend well beyond simple planning of the location of offshore installations and can also cover consideration of ways in which to enhance the maritime sectors of the coastal economy and gross domestic household incomes in the coastal communities (Jay, 2017).

It is clear from the wide range of regulatory and economic-development systems that a need arises to integrate such controls, so that they are not in conflict and allow a coherent approach. That is where MSP is useful (Ehler and Douvere, 2009).

Given the wide range of potential elements to be covered in MSP and the spectrum of types of MSP, the resulting systems vary widely, but attempts have been made to synthesize good practices (e.g., Foley and others, 2010). One such attempt is the development by the United Nations Educational, Scientific and Cultural Organization (UNESCO) of the guide *Marine Spatial Planning: a Step-by-Step Approach toward Ecosystem-based Management* (Ehler and Douvere, 2009).

3. Marine spatial planning: a step-by-step approach toward ecosystem-based management

Although initial thinking around the planning of multiple uses in coastal and ocean zones occurred during the 1980s (see sect. 5.4 on China), interest in MSP started to develop rapidly in the early 2000s. UNESCO realized that MSP could make a useful contribution to both the Programme on Man and the Biosphere and the work of the Intergovernmental Oceanographic Commission (Ehler and Douvere, 2007). A workshop was organized in 2006, which led to the development of a guide to MSP (Ehler and Douvere, 2009).

In the guide, the 10 steps described below were recommended for the MSP process. The steps are not a linear process; feedback loops, and opportunities for review and revision as the process is implemented, should be built in from the beginning (Ehler and Douvere, 2009).

Step 1: identifying need and establishing authority. This includes formulating clearly why MSP is needed and establishing the appropriate authority to plan for and implement it. In most MSP initiatives around the world, a new authority is often established for MSP, while implementation is carried out through existing authorities and institutions.

Step 2: obtaining financial support. This includes preparing a financial plan to estimate the costs involved in developing and implementing MSP and identifying sources to meet those costs. The identification of alternative sources tends to be necessary because agencies are often given responsibilities to undertake MSP activities without receiving additional funds. In many cases, some form of fee or charge on the activities authorized under the plan will be appropriate.

Step 3: organizing the process through pre-planning. MSP requires substantial preparation, including to assemble a multidisciplinary team to develop a workplan, define the boundaries, time frame, principles, goals and objectives, identify risks and develop contingency plans.

Step 4: organizing stakeholder participation. Involving key stakeholders in the development of MSP is essential, in particular because MSP is aimed at achieving a number of social, economic and ecological objectives and should therefore reflect as many expectations, opportunities or conflicts in the area under consideration. The step includes defining who should be involved in the MSP process, and when and how they should be involved.

Step 5: defining and analysing existing conditions. It is essential to know a sea area to create a useful marine spatial plan for it. The preparation of inventories of relevant information is therefore important for the creation of a plan. The inventories should include information about ecological, environmental and oceanographic conditions and human activities in the area, which would be mapped onto the area being planned. Conflicts and compatibilities among existing human uses and between those uses and the protection and preservation of the marine environment then need to be identified.

Step 6: defining and analysing future conditions. This includes evaluating the likely future development of the sea area if no changes are made ("business as usual"), estimating the effect of new demands for ocean space, and identifying alternative scenarios for the future of the area. The outcome of the step is the selection of a preferred scenario towards which MSP will work.

Step 7: preparing and approving the spatial management plan. Within this step, a marine spatial management plan should be developed to identify specific management measures that can deliver the preferred scenario, specifying criteria for selecting measures and developing a zoning plan, then evaluating and approving the spatial management plan through a formal process.

Step 8: implementing and enforcing the spatial management plan. At this stage, the planning phase ends and the implementation phase begins. Relevant institutions carry out actions towards implementation and ensure compliance with the marine spatial management plan, including through enforcement actions. The activities require new information on an ongoing basis on what is actually happening in the sea area being planned and action by a wide range of institutions to gather, evaluate and respond to that information.

Step 9: monitoring and evaluating performance. As with all policy activities, there is a need to revisit the conclusions that have been adopted and see what progress is being achieved. For MSP, an assessment of the state of the environmental system is relevant, in addition to measuring the performance of management measures.

Step 10: adapting the marine spatial management process. The results from the monitoring and evaluation would be used to adapt MSP and management so that the actions dictated by the plan have their intended effects.

MSP may need to include, or be accompanied by, an investment and development plan to provide the infrastructure, equipment and, above all, the skilled people needed to ensure the desired development of the blue economy (Schultz-Zehden and others, 2019). A review of the relevant science and technology can be helpful (Pinarbaşi and others, 2017). The involvement of stakeholders is also important. Studies are emerging on the practicalities of engaging stakeholders (e.g., Twomey and O'Mahony, 2019).

4. Tools for marine spatial planning

MSP ranges from a process to produce a plan for a given marine area to a suite of systems for managing human impacts on the ocean through the planning, management, licensing, regulation, surveillance and enforcement of human activities. Management approaches are considered in chapter 27.

As noted above, information about the ecological, environmental and oceanographic conditions of the sea area for which MSP is being developed is an essential basis for such work. Habitat mapping is therefore a necessary tool: if the current state of the natural marine environment is not known in some detail, the possible effects of both policies and individual projects can be little more than guesswork.

For the benthic layer, improvements in echo-sounding techniques – in particular in allowing whole swathes of the seabed to be explored in a single sweep – have enabled much better resolution of seabed exploration since the early 2000s. Geophysical techniques (multibeam, side-scan or seismic) may make it possible to ascertain thoroughly the nature of the seabed (mud, sand, gravel or rock), the nature of the rock and the thickness of the sedimentation. The collection of information about the plants and biota that it supports is a second level, which together with the information on the seabed will give an overall picture of the area in question. Those techniques are providing a mass of new information to support MSP and other marine policymaking (Colenutt and others, 2013). Online geospatial mapping tools are facilitating access to open source information relevant to MSP approaches (e.g., Menegon and others, 2018).

Habitat mapping does not give a comprehensive view of the ecosystem components that comprise the various habitats, including the functioning and connectivity of ecosystem components. In more developed MSP systems, therefore, an ecosystem overview is usually one of the bases of the planning system. An example is the Ecosystem Overview of the Pacific North Coast Integrated Management Area (Lucas and others, 2007), which covered geology, meteorology and climate, physical and chemical oceanography, plankton, marine plants, invertebrates, fishes, marine mammals, marine turtles and seabirds.

Similarly, where fisheries are included as part of the MSP process, it may be desirable to incorporate temporal and spatial knowledge of fish stocks and their exploitation. In France, a method for incorporating fishers' knowledge has been developed to ensure that such aspects can be brought into the MSP process (Trouillet and others, 2019).

Strategic environmental assessments are aimed at ensuring that relevant aspects are considered effectively in the development of policies, plans and programmes, because it is often at that more general level that decisions are taken that constrain specific projects. Originally focused on environmental aspects, they have broadened to cover social and sustainability issues (Fundingsland Tetlow and Hanusch, 2012).

In China, the technique grew out of the long-established administrative process of environmental impact assessments for specific projects, embodied in the revision in 2002 of the Environmental Impact Assessment Law, which provided for the assessment of integrated plans for land use and regional development and the development of drainage areas and marine areas (Zhu and others, 2005).

In Europe, the technique grew out of the Convention on Environmental Impact Assessment in a Transboundary Context (Espoo Convention)[2] and is outlined in the Protocol on Strategic Environmental Assessment of 2003.[3] The Protocol provides for six stages: screening to determine whether a strategic environmental assessment is needed to implement a plan or programme; scoping to determine what information is relevant to the environmental report; preparing an environmental report to identify, describe and evaluate the likely effects of a planned activity; informing and consulting the public, relevant authorities and any States likely to be affected; feeding the strategic environmental assessment into the decision-making process; and monitoring the effects of plans and programmes after their implementation. Stategic environmental assessments are recognized by the World Bank as a key means of integrating environmental and social considerations into policies, plans and programmes (World Bank, 2013) and are incorporated into the management of development support by a number of States in line with the Paris Declaration on Aid Effectiveness of 2005 (OECD, 2006).

At the level of an individual project, environmental impact assessments are aimed at ensuring that environmental consequences are taken into account before a decision is made to start physical changes in the environment (e.g., Morgan, 2012). A detailed description of the form adopted for States across Europe can also be found in the Espoo Convention.

If the socioeconomic aspects are to be included, surveys of the maritime industrial sectors that are local to the plan will be needed. There may be problems, however, in correlating the

2 United Nations, *Treaty Series*, vol. 1989, No. 34028.
3 Ibid., vol. 2685, No. 34028.

relevant sectors with the area for which the plan is being prepared, since fishing vessels may be based at distant ports, and other out-of-area industries may have an impact on the plan zone. To incorporate wider socioeconomic aspects, it may be appropriate to include a social survey of communities involved in the sea area to be covered by the MSP process. In addition to employment, such a survey may also (depending on the area) need to include cultural aspects, indigenous rights and traditions and other traditional involvement of the communities with the sea (Sullivan and others, 2015).

5. Progress in implementing marine spatial planning

5.1. Overview

Throughout the world, Governments have developed – or, more commonly, are developing – marine spatial plans. A joint road map to accelerate maritime and marine spatial planning processes worldwide was adopted during the second International Conference on Maritime Spatial Planning, organized by the Intergovernmental Oceanographic Commission and the European Commission in Paris in March 2017. It foresees the creation of an international forum for discussion and exchanges on cross-border MSP at the international level. Four workshops of the international MSP forum have already been conducted: in Brussels, in May 2018; in Réunion, France, in March 2019; in Vigo, Spain, in May 2019; and in Riga, in November 2019. The meetings build upon a wide exchange of good practices and interactive discussions in order to work towards the creation of international guidelines on transboundary MSP (International Oceanographic Commission (UNESCO-IOC), 2019).

A summary of the worldwide inventory of MSP provided by the International Oceanographic Commission (UNESCO-IOC, 2020) is shown in the table.

In the Baltic Sea, efforts are being made to develop transboundary MSP. The Regional Baltic MSP Road Map 2013–2020 outlines the steps planned for the development and implementation of maritime spatial plans throughout the region by 2020. In order to facilitate a coherent MSP process, the Helsinki Commission has developed guidelines for the implementation of an ecosystem-based approach in MSP in the Baltic Sea area with regard to transboundary consultations, public participation and cooperation, and the transboundary MSP output data structure (Helsinki Commission, 2016).

Further development of MSP is also being undertaken in the Republic of Korea[4] and in Peru and Ecuador.[5]

Given the varying approaches to MSP that have been implemented in different regions, further details of some are provided below, in a series of case studies selected to give an overview of different continents and different issues.

4 The Republic of Korea introduced the Marine Spatial Planning and Management Act and an associated National Marine Spatial Framework Plan in 2019.

5 See www.fao.org/in-action/coastal-fisheries-initiative/activities/latin-america/en and www.pe.undp.org/content/peru/es/home/projects/iniciativa-de-pesquerias-costeras---america-latina.html.

Countries with full or partial marine spatial planning (approved, planned, started or in progress), by region

Region	Countries with full or partial (for some aspects or some areas) marine spatial planning approved	Countries with marine spatial planning planned, started or in progress
Africa		Angola, Ghana, Kenya, Madagascar, Mauritania, Mauritius, Morocco, Namibia, Seychelles, South Africa
Asia	China, Philippines, Viet Nam	Indonesia, Myanmar, Thailand
Australia/ Oceania	Australia, Kiribati, New Zealand, Palau	Fiji, Solomon Islands, Tonga, Vanuatu
Europe	Belgium,* Germany,* Latvia,* Netherlands,* Norway, United Kingdom of Great Britain and Northern Ireland	Bulgaria,* Croatia,* Cyprus,* Denmark,* Estonia,* Finland,* France,* Greece,* Iceland, Ireland,* Italy,* Lithuania,* Malta,* Poland,* Portugal,* Romania,* Russian Federation, Slovenia,* Spain,* Sweden*
Middle East		Israel, United Arab Emirates
The Americas	Antigua and Barbuda, Belize, Canada, Mexico, United States of America	Colombia, Dominica, Grenada, Jamaica, Saint Kitts and Nevis, Saint Lucia, Saint Vincent and the Grenadines, Trinidad and Tobago

Source: UNESCO-IOC (2019).
Note: The 22 coastal States of the European Union (marked with *) are committed to the full coverage of MSP in their waters by 2021.

5.2. Case study: Australia

Australia made an impressive start on MSP with the creation of the Great Barrier Reef Marine Park in 1975. Legislation was passed to define the Great Barrier Reef region and establish the Great Barrier Reef Marine Park Authority, which manages and protects the park. The park has governance arrangements under which the Authority liaises and coordinates policies with other departments of the Government of Australia and the government of Queensland. The park is managed on the basis of ecologically sustainable principles and a zoning plan that includes multiple-use areas and provides protection for biodiversity values through a network of no-take zones for 33 per cent of its area and for at least 20 per cent of every bioregion (Vince, 2014). The Great Barrier Reef zoning plan provides the cornerstone for management within the park (Kenchington and Day, 2011; Great Barrier Reef Marine Park Authority, 2019), but many other integrated spatial and temporal management tools and strategies are also in place (Day and others, 2019; see also chap. 25). Key challenges for the management of the park are associated with global pressures, such as ocean warming as a result of climate change and resulting impacts on reef ecosystems (see chap. 7D, and the case study in sect. 3.1 of chap. 25).

Elsewhere in Australia, progress has been less straightforward. Efforts began in 1998 to develop an integrated oceans strategy, later renamed the national oceans policy. Initially, the aim was full integration between the various levels of government (in particular at the State and national levels) and across the relevant sectors. However, that would have required changes in legislative arrangements that had been settled in 1979 (Office of the Attorney-General of Australia, 1980), and therefore the model was not pursued. The national oceans policy provided a comprehensive review of each marine sector and the state of the waters. In 2004, a south-east regional marine plan was released, covering waters from southern New South Wales to eastern South Australia, including Victoria and Tasmania. It foresaw collaborative action over the following decade, leading to a review in 2014 (National Oceans Office, 2004). However, little of the specific action foreseen in the plan was developed, and the review was not conducted. In 2005, a new start was made at the national level, focusing on marine bioregional plans for national waters. The plans were based on the following conservation values: key ecological features, protected species (and habitats for such species) and protected places. They described the marine environment and conservation values of each marine region, set out broad biodiversity objectives, identified regional priorities and outlined strategies and actions to address them by bringing together scientific knowledge and information, and are intended to offer guidance for the relevant sectoral decisions (Vince and others, 2015). In taking the commitments forward, the main effort has been focused on the creation of a national representative system of marine protected areas. A review of the management plans for most of the designated marine protected areas (which cover 3.2 million km^2, about 36 per cent of the waters within the marine jurisdiction of the national Government) was completed in 2015 (Beeton and others, 2015). However, the outcome has attracted criticism from academic sources (Ocean Science Council of Australia, 2017).

5.3. Case study: Canada (Pacific coast)

Canada first developed a comprehensive approach to ocean management in the Oceans Act (*Statutes of Canada*, 1996, chap. 31). The country's oceans strategy of 2002 provided policy direction for implementing the Oceans Act on the basis of the principles of sustainable development, integrated management and a precautionary approach. Five priority areas for marine planning were identified in an Oceans Action Plan in 2005, including an area later known as the Pacific North Coast Integrated Management Area. In 2005, some of the First Nations on the Pacific Coast began to consider MSP as one of a number of issues of common interest. That eventually led to the creation of the Marine Plan Partnership for the Pacific Coast, which brought together the provincial government and (eventually) 16 First Nations. The partnership plans are not considered to have a legal function, but to set guidelines in partnership between 16 First Nations and the Province of British Columbia. The plans have a zoning regime that identifies areas of importance for biodiversity, general use and the marine industry. Four subregional plans have been synthesized into a regional action framework for the whole planning area (Rodriguez, 2017). The Department of Fisheries and Oceans of Canada organized a thorough ecological overview of the area, providing much of the basic supporting material to develop planning of the Pacific North Coast Integrated Management Area (Lucas and others, 2007). By 2010, a non-binding trilateral agreement was established between the Government of Canada, First Nations and the Province of British Columbia. The Area plan was endorsed by the Government of Canada, First Nations and the Province of British Columbia early in 2017. It provides a framework for ecosystem-based and adaptive collaborative management of marine activities and resources. A key priority of the plan currently under development is the design of a marine protected area network, which will guide the establishment of such areas and other area-based conservation measures in the future.

5.4. Case study: China

In China, marine functional zoning (MFZ) is considered to be a form of MSP and was introduced by the Government of China in 1988 (Feng and others, 2016; Kang and others, 2017). Its development can be regarded as moving through three phases and has been institutionalized through the Law on the Administration of the Use of Sea Areas, which was enacted in 2001. The law established principles for sea use authorization, user fees and MFZ systems. According to the law, MFZ is based on dividing sea areas (including islands) into different spatial areas for human activities, in the light of their geographical and ecological features, natural resources, current usage and socioeconomic development needs (Fang and others, 2018).

The first phase in the development of MFZ lasted from 1989 to 1993 and involved a pilot MFZ project implemented in the Bohai Sea in 1990. Coastal provinces then developed and implemented provincial MFZ from 1991 to 1997. The State Oceanic Administration developed the first national MFZ maps in nearshore areas of the territorial sea in 1993.

The second phase of MFZ lasted from 1997 to 2002 and began with the release of a technical directive to guide it. An initial MFZ plan was adopted in 1997 by the local government of the city of Xiamen. Based on experience in the first phase, the State Oceanic Administration organized the second phase of MFZ in 1998, which lasted until 2010. During that phase, in 1998, the Administration instructed all 11 coastal provinces of China to formulate a provincial MFZ plan. In 2001, the plans were completed and, in 2002, those of seven coastal provinces were approved. The MFZ plans of all 11 coastal provinces of China were approved by 2008 (Fang and others, 2018).

The third phase of MFZ began in 2011 and will last until 2020. In that phase, MFZ is divided into three levels (regional, provincial and local) (Huang and others, 2019; UNESCO-IOC, 2020).

MFZ has helped China to better plan the development of its seas and coasts (Fang and others, 2018; Huang and others, 2019). However, there have been a number of challenges in its implementation. Better coordination between maritime and land planning, improved resolution of conflicts between stakeholders, enhanced monitoring and evaluation and more effective participation of stakeholders have all been identified (Feng and others, 2016; Liu and Xing, 2019). In practice, MFZ is a zoning tool for multiple marine spatial users (Feng and others, 2016; Kang and others, 2017). In assessing MFZ, Huang and others (2019) found that the MFZ formulation and implementation process was essentially top-down management, which led to two issues: low applicability owing to deficiencies in marine spatial zone classification; and a lack of consistency due to the need to work, at lower (municipal) levels, within different sea use areas specified on smaller-scale maps set by the provincial authorities. Currently, MFZ lacks implementation plans and does not ensure the management of the cumulative impacts of different sectors. Its implementation does not seem to have stopped the degradation of coastal and marine natural resources and ecological systems, thus leaving the environment still polluted (Kang and others, 2017).

5.5. Case study: European Union

Following the Marine Strategy Framework Directive of 2008, the European Union decided in 2014 to adopt a directive requiring its coastal member States to develop and implement maritime spatial plans for their waters (European Union, 2014). The national legislation for the directive was to be adopted by 2016, and maritime spatial plans for all the waters covered by it are to be in place by 2021. The plans are not to include coastal waters covered by town and country planning systems and are not to deal with interactions between the land and the sea, though the results of national decisions on them are to be reflected in the plans. The planning is to take into account all the relevant human activities and uses, including aquaculture areas; fishing areas; installations and infrastructures for the exploration, exploitation and extraction of oil, gas and other energy resources, minerals and aggregates and for the production of energy from renewable sources; maritime transport routes and traffic flows; military training areas; nature and species conservation sites and protected areas; raw material extraction areas; scientific research areas; submarine cable and pipeline routes; and tourism and underwater cultural heritage areas. Member States are required to arrange for public participation in the planning process, share information and generally cooperate with each other and with relevant third countries – especially through existing regional seas organizations (European Union, 2014).

The areas to be covered by individual plans are left to the judgment of the member States. For example, in France, a high-level national strategy for the sea and coast was approved by prime ministerial decree in February 2017. That policy is to be implemented at the level of sea basins, with the development of a strategy document for the Eastern Channel – North Sea; the North Atlantic; the South Atlantic; and the Mediterranean. Each of the strategy documents is to have four parts: a situation review, challenges and a vision for the sea basin in 2030; strategic objectives defined from an economic, social and environmental perspective, together with related performance indicators; an evaluation procedure for assessing implementation of the strategy document; and an action plan. The first two parts have now been produced for each basin, and the remaining parts are due over the next few years. Taken together, the documents set the framework for all relevant decisions by national, regional and local authorities (France, Ministry for the Ecological Transition, 2017).

5.6. Case study: South Africa

The framework for MSP in South Africa was developed through an initiative of the Government of South Africa – Operation Phakisa ("phakisa" means "hurry up" in Sesotho) – aimed at unlocking the country's ocean economy as a mechanism to fulfil the National Development Plan for 2030. Within Operation Phakisa, MSP was identified as a focus area, which, in turn, fast-tracked the development of the Marine Spatial Planning Act of 2019 (South Africa, 2019). The Act provides for the development of marine spatial plans and the establishment of institutional arrangements for their implementation, and governance of the use of the ocean by multiple sectors. The accelerated pace of the development and enactment of MSP legislation in South Africa (less than three years from the first draft to promulgation) stemmed from a desire to rapidly achieve larger-scale MSP, at the level of the exclusive economic zone.

During the detailed planning and roll-out of Operation Phakisa, and while the Marine Spatial Planning Act was being drafted, the national Government also published the National Framework for Marine Spatial Planning (South Africa, 2017). The policy provided high-level direction for undertaking MSP in the context of the country's legal framework – including existing planning regimes – in order to ensure consistency in ocean space planning. The framework also highlighted the need for coordination with terrestrial and coastal planning. To simplify spatial planning, the exclusive economic zone of South Africa was divided into western, eastern and southern marine areas and the Prince Edward Islands, for which statutory marine spatial plans are to be developed. The Government aims to publish the first marine area plans by 2021. It has recognized the importance of data and information for spatial planning and has initiated projects simultaneously to fill data gaps and provide spatial data infrastructure to support MSP and ocean economy planning (South Africa, 2017).

The establishment of MSP in South Africa is built upon a legacy of environmental policies that were inherently supportive of area-based management, specifically spatial planning of environmental resources. South Africa chose a consociational democracy as the basis for its political system after apartheid (i.e., after 1994) (Karume, 2003). As a result, most post-1994 environmental legislation embraces cooperative, participatory approaches, including the need for negotiated spatial planning or zonation. That is evident in legislation for terrestrial protected areas and spatial planning for terrestrial areas (South Africa, 2004; 2013b). In 2008, the National Environmental Management: Integrated Coastal Management Act (South Africa, 2009) established cross-sectoral mechanisms to govern coastal space, thus introducing administrative (and explicitly spatial) boundaries such as coastal public property and development setback lines. In that way, spatial planning (or zoning) became a key component within the national framework for integrated coastal management, extending to the outer limit of the exclusive economic zone (e.g., South Africa, 2014, 2013a). Although considered progressive and bold (Taljaard and others, 2019; Colenbrander and Sowman, 2015), many barriers for implementation remain, including a lack of political support, a lack of resources and clarity regarding jurisdiction over private and communal land and limited civil society involvement in decision-making (Sowman and Malan, 2018).

MSP legislation in South Africa is new, and its implementation is untested and currently unchallenged in case law. Its testing in courts of law is a certainty given the intention to allocate the use of space, since it relates to highly valued marine resources that are often contested by multiple users for multiple, often conflicting uses. Operation Phakisa, with its emphasis on MSP, is also encouraging community-based, bottom-up initiatives such as the Algoa Bay project (Dorrington and others, 2018). The role of such initiatives within the national framework for MSP is not yet clear.

5.7. Case study: Viet Nam

Research on integrated coastal management and MSP began in Viet Nam in 1996. From 2010 to 2013, by implementing a regional project on coastal spatial planning, Viet Nam improved MSP capacity and undertook MSP in the coastal areas of Quang Ninh and Hai Phong Provinces. With the assistance of various donors, including Partnerships in Environmental Management for the Seas of East Asia and the National Oceanic and Atmospheric Administration, Viet Nam applied an MSP approach to the functional zoning of the Hon Mun, Bai Tu Long Bay and Cu Lao Chammarine protected areas and coastal-use zoning for integrated coastal management on the Da Nang coast (Nguyen and Hien, 2014). At the same time, through a project funded by the Global Environment Fund and executed by the Ministry of Agriculture and Rural Development of Viet Nam, seven coastal provinces (Nghe An, Thanh Hoa, Binh Dinh, Phu Yen, Khanh Hoa, Soc Trang and Ca Mau) initiated MSP between 2012 and 2018. The creation of formal institutions in Viet Nam specifically for MSP began in 2012 with the Law of the Sea. In 2015, the Law on Sea and Island Natural Resources and Environment provided for integrated planning of the sustainable use and exploitation of coastal resources. It was followed by the Planning Law, enacted in January 2017, which provides that MSP will be the basis of all relevant planning and that all other sectoral planning in the coasts and seas must follow it. The development of MSP covering all the coasts and seas of Viet Nam is under way.

References

Beeton, R.J.S., and others (2015). *Commonwealth Marine Reserves Review: Report of the Expert Scientific Panel.* Canberra: Department of the Environment.

Colenbrander, Darryl R., and Merle R. Sowman (2015). Merging socioeconomic imperatives with geospatial data: a non-negotiable for coastal risk management in South Africa. *Coastal Management*, vol. 43, No. 3, pp. 270–300.

Colenutt, Andrew, and others (2013). Nearshore substrate and marine habitat mapping to inform marine policy and coastal management. *Journal of Coastal Research*, pp. 1509–1514.

Collie, Jeremy, and others (2013). Marine spatial planning in practice. *Estuarine, Coastal and Shelf Science*, vol. 117, pp. 1–11. https://doi.org/10.1016/j.ecss.2012.11.010.

Commission for the Protection of the Marine Environment of the North-East Atlantic (2009). Overview of national spatial planning and control systems relevant to the OSPAR Maritime Area.

Day, Jon C., and others (2019). Marine zoning revisited: how decades of zoning the great barrier reef has evolved as an effective spatial planning approach for marine ecosystem-based management. *Aquatic Conservation: Marine and Freshwater Ecosystems*, vol. 29, pp. 9–32.

Dorrington, Rosemary A., and others (2018). Working together for our oceans: a marine spatial plan for Algoa Bay, South Africa. *South African Journal of Science*, vol. 114, Nos. 3–4, pp. 1–6.

Douvere, Fanny, and Charles N. Ehler (2009). New perspectives on sea use management: initial findings from European experience with marine spatial planning. *Journal of Environmental Management*, vol. 90, No. 1, pp. 77–88.

Ehler, Charles, and Fanny Douvere (2007). *Visions for a Sea Change. Report of the First International Workshop on Marine Spatial Planning.* Intergovernmental Oceanographic Commission and Man and the Biosphere Programme. IOC Manual and Guides, No. 46: ICAM Dossier, No. 3. Paris: UNESCO.

________ (2009). *Marine Spatial Planning: A Step-by-Step Approach toward Ecosystem-Based Management.* Intergovernmental Oceanographic Commission and Man and the Biosphere Programme. IOC Manual and Guides, No. 53, ICAM Dossier, No. 6. Paris: UNESCO.

European Union (2014). Directive 2014/89/EU of the European Parliament and of the Council of 23 July 2014 establishing a framework for maritime spatial planning, Official Journal L 257, 28 August 2014, pp. 135–145.

Fang, Qinhua, and others (2018). Marine functional zoning: A practical approach for integrated coastal management (ICM) in Xiamen. *Ocean and Coastal Management*, p. 104433.

Feng, Ruoyan, and others (2016). Development of China's marine functional zoning: a preliminary analysis. *Ocean and Coastal Management*, vol. 131, pp. 39–44.

Foley, Melissa, and others (2010). Guiding ecological principles for marine spatial planning. *Marine Policy*, vol. 34, pp. 955–966. https://doi.org/10.1016/j.marpol.2010.02.001.

France, Ministry for the Ecological Transition (2017). *National Strategy for the Sea and Coast.* Paris.

Fundingsland Tetlow, Monica, and Marie Hanusch (2012). Strategic environmental assessment: the state of the art. *Impact Assessment and Project Appraisal*, vol. 30, pp. 15–24. http://doi.org/10.1080/14615517.2012.666400.

Great Barrier Reef Marine Park Authority (2019). *Great Barrier Reef Outlook Report 2019.* http://elibrary.gbrmpa.gov.au/jspui/bitstream/11017/3474/10/Outlook-Report-2019-FINAL.pdf.

Helsinki Commission (2016). *MSP Guidelines.* https://helcom.fi/action-areas/maritime-spatial-planning/msp-guidelines.

Huang, Faming, and others (2019). Coordination of marine functional zoning revision at the provincial and municipal levels: a case study of Putian, China. *Journal of Marine Science and Engineering*, vol. 7, No. 12, p. 442.

Intergovernmental Oceanographic Commission (UNESCO-IOC) (2019). *MSP Around the Globe*. http://msp.ioc-unesco.org/world-applications/overview.

________ (2020). *Marine Spatial Planning Programme – China*. http://msp.ioc-unesco.org/world-applications/asia/china.

International Organization of Supreme Audit Institutions (2019). *Are Nations Prepared for Implementation of the 2030 Agenda?* www.idi.no/en/idi-library/global-public-goods/auditing-sustainable-development-goals.

Jay, S. (2017). *Marine Spatial Planning, Assessing net benefits and improving effectiveness*, Issue Paper. OECD 2017 Green Growth and Sustainable Development Forum "Greening the Ocean Economy". www.oecd.org/greengrowth/ggsd2017.

Jones, Peter, and others (2016). Marine spatial planning in reality: introduction to case studies and discussion of findings. *Marine Policy*, vol. 71, pp. 256–264. https://doi.org/10.1016/j.marpol.2016.04.026.

Kang, Min-jie, and others (2017). Discussion on marine spatial planning in China: role and prospect. *DEStech Transactions on Environment, Energy and Earth Sciences*. https://doi.org/10.12783/dteees/ese2017/14323.

Karume, Shumbana (2003). Conceptual understanding of political coalitions in South Africa: an integration of concepts and practices. Paper presented at an Electoral Institute of Southern Africa round table on Strengthening Democracy through Party Coalition Building. Cape Town. 19 June 2003.

Kenchington, R.A., and J.C. Day (2011). Zoning, a fundamental cornerstone of effective marine spatial planning: lessons learnt from the Great Barrier Reef, Australia. *Journal of Coastal Conservation*, vol. 15, No. 2, pp. 271–278.

Liu, D.H., and X. Xing (2019). Analysis of China's coastal zone management reform based on land-sea integration. *Marine Economics and Management*, vol. 2, No. 1, pp. 39–49.

Lucas, B.G., and others (2007). *Ecosystem Overview: Pacific North Coast Integrated Management Area (PNCIMA)*. Canadian Technical Report of Fisheries and Aquatic Sciences, No. 2667. Ottawa: Fisheries and Oceans Canada.

Menegon S., and others (2018). Tools4MSP: an open source software package to support Maritime Spatial Planning. PeerJ Computer Science, vol. 4, p. e165. https://doi.org/10.7717/peerj-cs.165.

Morgan, Richard (2012). Environmental impact assessment: the state of the art. *Impact Assessment and Project Appraisal*, vol. 30, pp. 5–14. http://doi.org/10.1080/14615517.2012.661557.

National Oceans Office (2004). *South-East Regional Marine Plan, Implementing Australia's Oceans Policy in the South-East Marine Region*. Hobart, Australia.

Nguyen, Chu Hoi, and Bui Thi Thu Hien (2014). Integrated Spatial Planning and Management for Marine and Coastal Sustainability in Viet Nam. Gland, Switzerland: International Union for the Conservation of Nature.

Ocean Science Council of Australia (2017). Submission to the Director of National Parks on Australian Marine Networks draft management plans. http://oceansciencecouncil.org/wp-content/uploads/2017/07/OSCA-submission-draft-management-plans-2017_09_20-1.pdf.

Office of the Attorney-General of Australia (1980). Offshore constitutional settlement: a milestone in co-operative federalism. Canberra: Australian Government Publishing Service. www.ag.gov.au/Internationalrelations/InternationalLaw/Documents/offshore-constitutional-settlement-a-milestone-in-cooperative-federalism-pages-1-10%20ocr.pdf.

Organization for Economic Cooperation and Development (OECD) (2006). *Applying Strategic Environmental Assessment: Good Practice Guidance for Development Co-Operation*. Paris.

________ (2017). *The Governance of Land Use: Policy Highlights*. Paris.

Pacific North Coast Integrated Management Area Initiative (2017). *Pacific North Coast integrated management area plan*. Canada. https://waves-vagues.dfo-mpo.gc.ca/Library/40743032.pdf.

Plasman, Cathy (2008). Implementing marine spatial planning: a policy perspective, *Marine Policy*, vol. 32, pp. 811–815. https://doi.org/10.1016/j.marpol.2008.03.016.

Pinarbaşi, Kemal, and others (2017). Decision support tools in marine spatial planning: present applications, gaps and future perspectives, *Marine Policy*, vol. 83, pp. 83–91. https://doi.org/10.1016/j.marpol.2017.05.031.

Rodriguez, Nicolas J.I. (2017). A comparative analysis of holistic marine management regimes and ecosystem approach in marine spatial planning in developed countries. *Ocean and Coastal Management*, vol. 137, pp. 185–197.

Schultz-Zehden, Angela, and others (2019). Maritime Spatial Planning and the EU's Blue Growth Policy: past, present and future perspectives. In *Maritime Spatial Planning: Past, Present, Future*, Jacek Zaucha and Kira Gee, eds., pp. 121–149. Springer International Publishing. https://doi.org/10.1007/978-3-319-98696-8_6.

South Africa (2004). *National Environmental Management: Protected Areas Act 57 of 2003*. Pretoria.

________ (2009). *National Environmental Management: Integrated Coastal Management Act 24 of 2008*. Pretoria.

________ (2013a). *National Estuarine Management Protocol (10 May 2013)*. Pretoria.

________ (2013b). *Spatial Planning and Land Use Management Act 16 of 2013*. Pretoria.

________ (2014). *The National Coastal Management Programme of South Africa*. Cape Town.

________ (2017). *Marine Spatial Planning Framework (26 May 2017)*. Pretoria.

________ (2019). *Marine Spatial Planning Act 16 of 2018*. Pretoria.

Sowman, M., and N. Malan (2018). Review of progress with integrated coastal management in South Africa since the advent of democracy. *African Journal of Marine Science*, vol. 40, No. 2, pp. 121–136.

Sullivan, Colleen, and others (2015). Combining geographic information systems and ethnography to better understand and plan ocean space use. *Applied Geography*, vol. 59, pp. 70–77. https://doi.org/10.1016/j.apgeog.2014.11.027.

Taljaard, S., and others (2019). The legal landscape governing South Africa's coastal marine environment: helping with the 'horrendogram'. *Ocean and Coastal Management*, vol. 178, p. 104801.

Trouillet, Brice, and others (2019). More than maps: providing an alternative for fisheries and fishers in marine spatial planning. *Ocean and Coastal Management*, vol. 173, pp. 90–103.

Tuda, Arthur, and others (2014). Resolving coastal conflicts using marine spatial planning, *Journal of Environmental Management*, vol. 133, pp. 59–68.

Twomey, S., and C. O'Mahony (2019). Stakeholder processes in Marine Spatial Planning: ambitions and realities from the European Atlantic experience. In *Maritime Spatial Planning*, Zaucha J. and Gee K., eds. Springer International Publishing. https://doi.org/10.1007/978-3-319-98696-8_13.

United Nations (2017). *The First Global Integrated Marine Assessment: World Ocean Assessment I*. Cambridge: Cambridge University Press.

Vince, Joanna (2014). Oceans governance and marine spatial planning in Australia. *Australian Journal of Maritime and Ocean Affairs*, vol. 6, No. 1, pp. 5–17.

Vince, Joanna, and others (2015). Australia's oceans policy: past, present and future. *Marine Policy*, vol. 57, pp. 1–8.

World Bank (2013). *Brief: Strategic Environmental Assessment.* www.worldbank.org/en/topic/environment/brief/strategic-environmental-assessment.

Zhu, Tan, and others (2005). Requirements for strategic environmental assessment in China. *Journal of Environmental Assessment Policy and Management*, vol. 7, No. 1, pp. 81–97.

Chapter 27
Developments in management approaches

Contributors: Piers Dunstan (convener), Louis Celliers, Valerie Cummins, Michael Elliott, Karen Evans (co-lead member), Antony Firth, Frédéric Guichard, Quentin Hanich, Ana Cristina de Jesus, Manuel Hildago, Hector Manuel Lozano-Montes, Chanda L. Meek, Marcus Polette, Jemma Purandare, Anita Smith, Anastasia Strati (lead member) and Ca Thanh Vu (co-lead member)

Keynote points

- The ecosystem approach is one of the most significant approaches to ocean management, consisting of the environmental, social and economic management of human interactions with oceans and coasts at multiple levels (transboundary, regional, national and local).
- While there is general agreement that the ecosystem approach provides an effective framing of ocean management, further research and capacity-building are needed to realize its full potential benefits across the oceans.
- Management has two different levels of governance, namely: decision-making processes that provide a framework for making decisions and implementing policy focused on the conservation and sustainable use of marine resources; and management tools (area-based and non-area-based) that can be used to regulate and modify human activity in a particular system.
- The implementation of the 2030 Agenda for Sustainable Development[1] requires management grounded in the ecosystem approach in order to achieve the integrated set of global priorities and objectives set out in the Sustainable Development Goals. That will allow for the integration of interactions, benefits and trade-offs between the Goals and support the achievement of each of the ocean-related targets.
- There is a growing trend towards incorporating the cultural values of the ocean into management.

1. Introduction

1.1. Need for management of the marine environment

The past decade has seen a step change in the development of management approaches for ocean resource management and sustainability. The present chapter is aimed at providing an overview of the nature of that change, as well as examples of selected good practices worldwide, including decision-making processes and tools. To understand those changes, it is important to recognize that approaches to ocean management have deep roots in local and indigenous communities, as well as in science, having evolved incrementally from initial attempts to deal with specific environmental issues, such as pollution from land-based sources in the 1960s, to more integrated approaches, such as integrated coastal zone management starting in the 1970s. Modern approaches to ocean management cover many different tools, tailored to regionally specific issues at various scales. The needs and nature of ocean management are influenced by social, cultural, economic and governance contexts, including the norms and value systems that have an impact on approaches to decision-making between government, industry and civil society at various levels. In general, ocean management is expanding from coasts and regional seas to include the regulation of increasing human activities in deeper waters of exclusive economic zones and continental shelves, such as through marine spatial planning (see chap. 26). Areas beyond national jurisdiction are currently the focus of negotiations at the United Nations in the context of the intergovernmental conference on an international legally binding instrument under the United Nations Convention on the Law of the

1 See General Assembly resolution 70/1.

Sea[2] on the conservation and sustainable use of marine biological diversity of areas beyond national jurisdiction (see chap. 28). In applying the many different forms of management, an understanding of the approaches and their success to date is, therefore, necessary.

The present chapter commences with an introduction to one of the most significant emerging paradigms for ocean management, the ecosystem approach, which is now universally accepted at the global, regional and national levels (Secretariat of the Convention on Biological Diversity, 2004) as a strategy for integrated management. The ecosystem approach embraces the need for the engagement of all relevant sectors of society and has motivated increasing levels of support for bottom-up, community-led approaches to ocean management that take into consideration traditional rights and social justice and apply participatory processes. Those trends are juxtaposed in a stocktaking of global approaches to management, organized according to area-based and non-area-based examples. The bottom-up approaches are complemented by top-down approaches, developed through international, regional and national governance initiatives. That shows a diversity of ocean management interventions designed to address a wide range of issues, from global wetlands conservation to networks of marine protected areas. Adaptive management to integrate flexible strategies that mitigate and adapt to shifts in marine ecosystems associated with climate change is also analysed in the context of region-specific issues, capacity-building, gaps and future research.

1.2. Summary of the first *World Ocean Assessment*

The first *World Ocean Assessment* (United Nations, 2017) did not explicitly include management approaches in a stand-alone chapter, but rather provided a high-level commentary on management approaches integrated into individual chapters. Recognizing the importance of providing a consolidated overview of the many approaches to marine management and their application, a chapter specifically focused on ocean management has been included in the present Assessment.

1.3. Overlaps and interactions with other chapters

Management tools broadly apply across all marine uses and users; therefore the present chapter is relevant to all other chapters in the present Assessment, in particular chapter 15 on capture fisheries, chapter 16 on aquaculture, chapter 21 on renewable energy, chapter 25 on cumulative effects and chapter 26 on marine spatial planning.

2. Management approaches

2.1. Introduction to the ecosystem approach

The ecosystem approach consists of an integrated approach with three main pillars, namely the environmental, social and economic management of human interactions with oceans and coasts at multiple levels (i.e., transboundary, regional, national and local), incorporating both top-down and bottom-up perspectives. The Conference of the Parties to the Convention on Biological Diversity (United Nations Environment Programme (UNEP), 2000), in its decision V/6, described the ecosystem approach as "a strategy for the integrated management of land, water and living resources that promotes conservation and sustainable use of biodiversity in an equitable way". As such, the approach has been widely

2 United Nations, *Treaty Series*, vol. 1833, No. 31363.

accepted and implemented as an effective management mechanism (see, for example, the European Union Marine Strategy Framework Directive[3] and the integrated ecosystem assessment implemented by the National Oceanic and Atmospheric Administration[4]).

There is a plethora of legislative instruments covering all aspects of marine use and requiring both vertical and horizontal integration (Boyes and Elliott, 2014). Top-down management approaches generally include policy and legislative instruments focused on implementing international conventions, agreements and instruments and meeting national priorities for marine spaces. Bottom-up management tools, including customary or indigenous ecosystem-based and stakeholder-based approaches to resource management (Thornton and Maciejewski Scheer, 2012; Turner and Berkes, 2006), are generally driven by a local-level need to implement effective management on a local scale. Bottom-up management tools can be motivated by social, economic or environmental aspects specific to an area, such as the need to address point source pollution impacts through targeted management.

Principles of the ecosystem approach adopted by the Conference of the Parties to the Convention on Biological Diversity (see decisions V/6 (2000) and VII/11 (2004))

Principle 1: the objectives of management of land, water and living resources are a matter of societal choice.

Principle 2: management should be decentralized to the lowest appropriate level.

Principle 3: ecosystem managers should consider the effects (actual or potential) of their activities on adjacent and other ecosystems.

Principle 4: recognizing potential gains from management, there is usually a need to understand and manage the ecosystem in an economic context. Any such ecosystem-management programme should:

(a) Reduce those market distortions that adversely affect biological diversity;

(b) Align incentives to promote biodiversity conservation and sustainable use;

(c) Internalize costs and benefits in the given ecosystem to the extent feasible.

Principle 5: conservation of ecosystem structure and functioning, in order to maintain ecosystem services, should be a priority target of the ecosystem approach.

Principle 6: ecosystems must be managed within the limits of their functioning.

Principle 7: the ecosystem approach should be undertaken at the appropriate spatial and temporal scales.

Principle 8: recognizing the varying temporal scales and lag effects that characterize ecosystem processes, objectives for ecosystem management should be set for the long term.

Principle 9: management must recognize that change is inevitable.

Principle 10: the ecosystem approach should seek the appropriate balance between, and integration of, conservation and use of biological diversity.

Principle 11: the ecosystem approach should consider all forms of relevant information, including scientific and indigenous and local knowledge, innovations and practices.

Principle 12: the ecosystem approach should involve all relevant sectors of society and scientific disciplines.

---

3 See https://ec.europa.eu/environment/marine/eu-coast-and-marine-policy/marine-strategy-framework-directive/index_en.htm.

4 See www.integratedecosystemassessment.noaa.gov.

The Conference of the Parties to the Convention on Biological Diversity acknowledges in its implementation guidelines (see box) that there are often limitations in current understanding and, in such cases, a precautionary approach should be followed.[5] The precautionary approach, as reflected in principle 15 of the Rio Declaration on Environment and Development of 1992[6] – in which it is stated that, where there are threats of serious or irreversible damage, a lack of full scientific certainty should not be used as a reason for postponing cost-effective measures to prevent environmental degradation – has been incorporated into an increasing number of international treaties and other instruments, reflecting a trend towards making the precautionary approach part of customary international law (see, for example, the advisory opinion of the Seabed Disputes Chamber of the International Tribunal of the Law of the Sea, 2011, para. 135).

2.2. Implementation of the ecosystem approach to management

The ecosystem approach can be operated and implemented in a single sector, as in the case of ecosystem-based fisheries management (Cowan and others, 2012), ecosystem approaches to fisheries and aquaculture (Brugère and others, 2019), or in multiple sectors, as with integrated coastal zone management (UNEP, 2018). Over the past decade, specific cases of implementation of the ecosystem approach have resulted in management mechanisms moving towards establishing methods for operation and implementation (Zhang and others, 2011; Link and Browman, 2017). Despite this, there are still large gaps in implementation and incomplete uptake across sectors and regions. For example, there are still significant differing opinions on the implementation of ecosystem-based fisheries management from different stakeholders, such as policymakers, managers, scientists, conservationists and ecologists (Trochta and others, 2018). It is, therefore, necessary to create frameworks and criteria for ecosystem assessment (Harvey and others, 2017; Zador and others, 2017), in particular on the basis of demonstrated best practices. Developing methods to increase stakeholder engagement is also essential to ensuring successful implementation (Oates and Dodds, 2017).

Management is generally conducted at two different levels of governance: (a) decision-making processes that provide a framework for making decisions and implementing policy focused on the conservation and sustainable use of marine resources, such as marine spatial planning, an ecosystem approach to fisheries and integrated coastal zone management; and (b) management tools (area-based and non-area-based) that can be used to manage or regulate human activity in particular systems, such as marine protected areas and zoning (Maestro and others, 2019), fisheries closures (Hall, 2002), particularly sensitive sea areas (Basiron and Kaur, 2009) and fisheries management tools (Pope, 2002) (see also sect. 3 below). Numerous approaches have been developed to facilitate the implementation of ecosystem approaches through management mechanisms. The figure below illustrates a typology of approaches to ocean management.

---

5 See Conference of the Parties to the Convention on Biological Diversity decision VII/11 (2004), annex I, implementation guideline 6.2.

6 Report of the United Nations Conference on Environment and Development, Rio de Janeiro, 3–14 June 1992, vol. I, Resolutions Adopted by the Conference (United Nations publication, Sales No. E.93.I.8 and corrigendum), resolution 1, annex I. See also www.cbd.int/doc/ref/rio-declaration.shtml.

Illustrative typology of approaches to ocean management

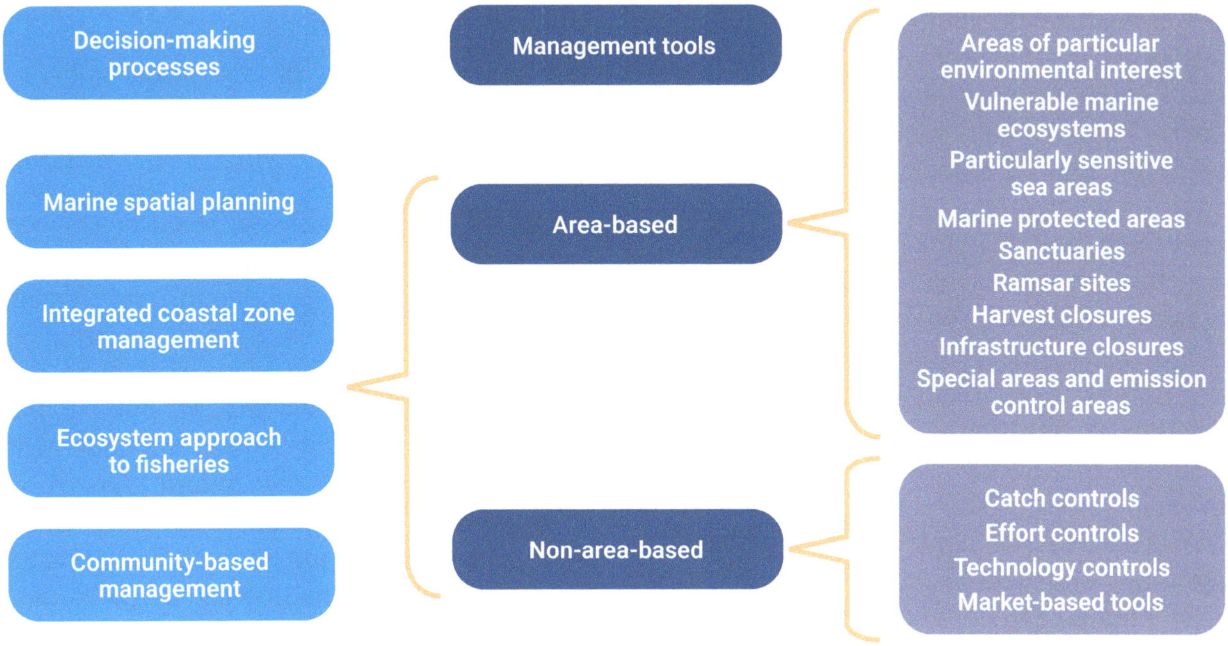

2.3. Community-based and culture-based management

One area in which ocean management based on ecosystem approaches continues to develop is the way that it supports engagement with communities and their culture. The Millennium Ecosystem Assessment identified cultural ecosystem services as the non-material benefits that people obtain from ecosystems through spiritual enrichment, cognitive development, reflection, recreation and aesthetic experiences (Milcu and others, 2013; Díaz and others, 2018). As already noted, the principles of the ecosystem approach include the decentralization of management to the lowest appropriate level and the involvement of all relevant sectors of society. Furthermore, management approaches should recognize that the cultural services provided by the marine environment also include specific values and benefits derived from sites of anthropogenic origin, including archaeological and historical sites (such as shipwrecks and prehistoric submerged sites, known as underwater cultural heritage). Such sites or locations can exhibit a variety of values, including those of historical and archaeological significance, of a sacred nature (war graves or tombs) or of cultural importance (myths and folklore). They are benefits provided by the cultural footprint within the marine ecosystem. Hence, there is growing recognition that many marine ecosystem services are hybrids of culture and nature, appreciated holistically by coastal communities. For example, management of the Papahānaumokuākea Marine National Monument in Hawaii, United States of America, is framed by native Hawaiian understanding of the ocean as a cultural seascape, where all natural resources are cultural resources, connected through ancestral stories and perpetuated through traditional practices, including wayfinding and voyaging (Kikiloi and others, 2017). Notwithstanding an anthropogenic emphasis, community-based and culture-based management approaches respect the intrinsic value of nature for its own sake.

Equally, recognition of the limitations of top-down management approaches and increased understanding of the rights, tenures and traditional and indigenous customary uses of inshore marine environments have catalysed widespread recognition of the strength and sustainability of community-based management, or bottom-up, approaches to marine conservation. Community-based management recognizes local community stewardship, knowledge and practices in monitoring, assessing and managing marine resources and through participatory, collaborative governance structures led by or involving local communities and systems of authority (Turner and Berkes, 2006). Many such schemes often develop from long-standing local institutions, such as the Alaska Eskimo Whaling Commission (Meek, 2013) and its self-organized aboriginal whaling captain associations that are now engaged in cross-scale (local to international) and community-based management. In the southern hemisphere, dugong management is shared by State and territorial agencies and communities in the Torres Strait between Australia and Papua New Guinea, through a system of indigenous rangers and Papua hunters (Miller and others, 2018). Such systems of shared management may be framed by a general understanding of the ecosystem approach but, at the local level, communities shape management approaches within their social and cultural values and the cultural benefits of their traditional practices (Delisle and others, 2018). As another example, networks of locally managed marine areas in the Pacific are building community resilience by supporting village-level management and sustainable use of marine resources (Govan, 2009; Veitayaki, 2003).

Growing recognition of the importance of marine ecosystem services to coastal communities and culture will undoubtedly intensify as those communities face pressures associated with climate change, in particular sea level rise and both temporary and permanent coastal inundation (Goodhead and Aygen, 2007; see also chap. 9). Cultural information is increasingly regarded as an integral part of ecosystem-based management, both in the context of community-based management and for safeguarding the cultural dimension of the marine environment. Such information may be very diverse and intangible, relating to, for example, traditional marine resource use, sea routes, ancient navigational skills, maritime identities, legends, rituals, beliefs and practices, aesthetic and inspirational qualities, cultural heritage and places of spiritual, sacred and religious importance.[7] That may make it challenging to incorporate such cultural values and practices into planning and management. Nonetheless, the cultural dimension of the sea can be integrated and mapped as a precursor to management. Once taken on board, culture can be powerful, not simply as a factor to be managed and monitored, but as the foundation upon which ecosystem approaches to management may be developed in the context of sustainable development.

3. Advances in ocean management approaches

The past decade has been characterized by the proliferation and expansion of new and existing approaches to the management of the oceans and seas. That has been manifested by the regulation of human activity in specific areas to achieve conservation or resource management policy objectives. Although all areas of the marine environment may be managed in some way (e.g., fisheries, tourism, oil and gas extraction), it often consists of a patchwork of policies and legislation that results in piecemeal approaches to protection

7 A number of cultural practices relating to the sea have been inscribed in the Representative List of the Intangible Cultural Heritage of Humanity of the United Nations Educational, Scientific and Cultural Organization. See https://ich.unesco.org/en/lists.

(Boyes and Elliott, 2014). While the management processes and tools described in the present section tend to have a spatial dimension, they share the following set of common characteristics:

- Scale: from global to regional and to local
- Driving factors: motivated by conservation, economic development, environmental, social/cultural concerns
- Sectoral dimensions: single sector, multi-sector or cross-sector
- Implementation measures: hard measures (legally binding), soft measures (voluntary)
- Approaches to management: top-down, bottom-up or both

The present Assessment is focused on management approaches that alter some aspect of human use. Other tools, such as the description of ecologically or biologically significant marine areas[8] under the Convention on Biological Diversity,[9] do not change use but provide information that may play a role in decision-making processes. Such areas, however, should be distinguished from decision-making processes, such as fisheries stock assessments, integrated ecosystem assessments and strategic environmental assessments, as they are a purely scientific and technical process exercise and do not include management measures, even though they have the potential to inform policy and management decisions. The same applies to other tools, such as important marine mammal areas.

3.1. Decision-making processes for management

Decision-making processes are used to identify the most appropriate policy and management objectives of competent authorities tasked with developing and implementing management approaches or strategies (see table 1). Governments, industry, communities and civil society identify the outcomes that they wish to achieve (i.e., management objectives) and use one of the potential approaches to identify how and where to achieve those outcomes. The outcomes described cover different aspects of sustainable development, including environmental, economic and social aspects. They may be global, regional, national, subnational or community-led. Common examples are marine spatial planning, integrated ecosystem assessments, strategic environmental assessments, an ecosystem approach to fisheries, ecosystem-based fisheries management, systematic conservation planning (McIntosh and others, 2017), community-based resource management (see sect. 2.3), source-to-sea approaches[10] and integrated coastal zone management.

At the regional level, examples of such approaches can be found in the context of the Convention for the Protection of the Marine Environment of the North-East Atlantic,[11] the Convention on the Protection of the Marine Environment of the Baltic Sea Area,[12] the Convention for the Protection of the Marine Environment and the Coastal Region of the Mediterranean (Barcelona Convention)[13] and the Convention on the Protection of the Black Sea against Pollution. The conventions use an area-based approach to assess the status of the environment and control activities, aimed at ensuring the good environmental status of marine assets. The organizations established under the conventions have working groups that focus on marine spatial planning, fisheries management and integrated coastal zone management.

8 See www.cbd.int/ebsa.
9 United Nations, *Treaty Series*, vol. 1760, No. 30619.
10 See www.siwi.org/publications/implementing-the-source-to-sea-approach-a-guide-for-practitioners.
11 United Nations, *Treaty Series*, vol. 2354, No. 42279.
12 Ibid., vol. 2099, No. 36495.
13 Ibid., vol. 1102, No. 16908.

The concept of adaptive management or adaptive resource management is shared across the decision-making processes listed (Dunstan and others, 2016), but the actual process used is often determined by the policy objectives (see also sect. 4). Within adaptive management frameworks, management measures or actions are implemented sequentially over time, taking into account future conditions and uncertainties associated with the responses of the resource being managed (Schultz and others, 2015). Conservation objectives are often met by using systematic conservation planning and community-based approaches at the local level in order to support local communities in the sustainable use and conservation of marine resources (Berkes and others, 2000; Nguyen and others, 2016). In contrast, the ecosystem approach to fisheries is aimed at providing a holistic approach to managing fisheries and other living marine resources by taking into account relevant human activities and their interactions with the ecosystem, with the purpose of maintaining health, productivity and resilience in order to ensure the continued delivery of ecosystem services and societal goods and benefits (Cowan and others, 2012). However, even with the more holistic processes, issues regarding the integration of multiple sectors remain (Jones and others, 2016).

Table 1
Decision-making processes and their associated attributes, including primary drivers, sectors, implementation measures, direction and scale

Decision-making processes		Management approach														
		Primary driver			Sector			Measures		Direction			Spatial scale			
Example in practice	Relevant authority	Economic	Environmental	Social well-being/cultural	Single-sector	Multi-sector	Cross-sector	Legally binding	Voluntary	Top-down	Bottom-up	Both	Global	Regional	National	Subnational
Marine spatial planning (zoning, consenting, licensing, policy-led mechanisms)	Competent national or local authorities	x	x	x		x	x	x				x		x	x	x
Integrated coastal zone management		x	x	x		x	x	x	x			x		x	x	x
Systematic conservation planning			x		x	x		x		x					x	
Integrated ecosystem assessment		x	x		x	x		x		x					x	
Ecosystem approach to fisheries		x	x		x			x		x				x	x	x
Community-based management plans		x	x	x	x	x		x	x		x					x
Strategic environmental assessment		x	x	x	x	x		x		x				x	x	x

3.2. Area-based management tools

Area-based management tools provide a spatial context to management approaches, whereby, usually, the area is defined as having distinctive characteristics that warrant measures that are different from the management of surrounding sea areas. Examples of area-based management tools that change or regulate aspects of human use of the marine environment include marine protected areas, particularly sensitive sea areas, areas of particular environmental interest, world heritage sites, fisheries closures, infrastructure closures and designations under the Convention on Wetlands of International Importance especially as Waterfowl Habitat (Ramsar Convention).[14] The application of the tools worldwide and the use of terminology is highly variable, owing in part to local hazards, risk and vulnerability and the need for resilience-building (Fanini and others, 2020). Notwithstanding such variability, there is general consistency in overall goals to improve pathways towards sustainability, and some of the tools could be used as other effective area-based conservation measures.[15] Examples (by no means exhaustive) of area-based management tools currently in use are highlighted below.

Marine protected areas provide specific protection mechanisms for specific areas of the ocean. They have been identified as one of the tools that should be implemented to achieve Aichi Biodiversity Target 11[16] and target 5 of Sustainable Development Goal 14.[17] The indicators and global targets for marine protected areas as identified under the Convention on Biological Diversity are currently undergoing revision through the process of negotiation of the post-2020 global biodiversity framework of the Convention. The areas can take many forms, covering varying spatial scales and providing varying levels of marine environmental protection. Examples of such areas include the 94,000 km^2 South Orkney Islands Southern Shelf area (established in 2009) and the 1.5 million km^2 Ross Sea area (established in 2017) designated by the Commission for the Conservation of the Antarctic Marine Living Resources;[18] the network of areas under the Convention for the Protection of the Marine Environment of the North-East Atlantic, with a total surface area of 864,337 km^2;[19] the specially protected areas of Mediterranean importance under the Protocol concerning Specially Protected Areas and Biological Diversity in the Mediterranean to the Barcelona Convention, including the 87,500 km^2 Pelagos Sanctuary for the Conservation of Marine Mammals established by a tripartite agreement between France, Italy and Monaco (established in 2001),[20] and the Natura 2000 network of the European Union, the largest coordinated

14 United Nations, *Treaty Series*, vol. 996, No. 14583. See also www.ramsar.org.

15 A definition and voluntary guidance for other effective area-based conservation measures was adopted by the Conference of the Parties to the Convention on Biological Diversity, at its fourteenth session. See www.cbd.int/doc/decisions/cop-14/cop-14-dec-08-en.pdf.

16 See United Nations Environment Programme, document UNEP/CBD/COP/10/27, annex, decision X/2, target 11: "By 2020, at least 17 per cent of terrestrial and inland water areas and 10 per cent of coastal and marine areas, especially areas of particular importance for biodiversity and ecosystem services, are conserved through effectively and equitably managed, ecologically representative and well-connected systems of protected areas and other effective area-based conservation measures, and integrated into the wider landscapes and seascapes".

17 See General Assembly resolution 70/1, Sustainable Development Goal 14, target 5: "By 2020, conserve at least 10 per cent of coastal and marine areas, consistent with national and international law and based on the best available scientific information".

18 See www.ccamlr.org/en/science/marine-protected-areas-mpas.

19 As at 1 October 2018, the network of marine protected areas under the Convention for the Protection of the Marine Environment of the North-East Atlantic comprised 496 such areas, including 7 collectively designated in areas beyond national jurisdiction. See 2018 Status Report on the OSPAR Network of MPAs, Commission for the Protection of the Marine Environment of the North-East Atlantic, 2019. See also https://ospar.org.

20 See www.rac-spa.org/spami.

network of protected areas in the world, spanning the marine territory of 23 European Union countries and, as at the end of 2018, covering more than 551,000 km^2.[21] Marine protected areas have increased rapidly in both number and size in recent years, largely in response to internationally agreed targets under the Convention on Biological Diversity and the 2030 Agenda, and are an important tool for marine conservation (Humphreys and Clark, 2020). Currently, global coverage in areas within national jurisdiction has reached 18 per cent, which amounts to 8 per cent coverage of the entire ocean. In contrast, only 1 per cent of areas beyond national jurisdiction have been established as protected areas (World Conservation Monitoring Centre and International Union for Conservation of Nature, 2019).

With regard to incorporating community and indigenous values into area-based management, examples can be found in the marine protected areas of Canada in the Arctic (including Anguniaqvia niqiqyuam in the Amundsen Gulf, Tarium Niryutait in the Beaufort Sea and Tuvaijuittuq off the north-west coast of Ellesmere Island in Nunavut). Anguniaqvia niqiqyuam was the first marine protected area in Canada with conservation objectives based on traditional and indigenous knowledge. The sites were identified as ecologically important areas that provide habitat for species of cultural importance and contribution to social and cultural values.[22]

Other examples of area-based management tools are provided for under conventions that seek to protect specific areas of diversity, habitat or heritage. In areas designated under the Ramsar Convention, for example, the broad aim is to halt the worldwide loss of wetlands and to conserve those that remain through wise use and management. As at February 2019, 2,341 sites had been designated under the Convention, comprising 252.48 million ha of internationally significant wetlands. A recently designated site is the Qurm Nature Reserve in Oman, which has successfully protected 106.83 ha of coastal wetland ecosystems through specific planning and management, as a result of its designation as a site listed under the Convention. Programmes include encouraging the development of nature-based tourism and community engagement in active management of the wetlands, which has resulted in an increased economic value of the reserve to the community.[23]

Other mechanisms that use area-based management include the implementation of offshore exclusion zones or closures to facilitate infrastructure installation and operation, such as pipelines, offshore wind farms and telecommunications cables. Those areas are restricted primarily for public health and safety although, indirectly, they have resulted in the protection of marine habitats and biodiversity.

The area-based management tools of particular sectors, such as shipping, encompass the 17 areas designated by the International Maritime Organization as particularly sensitive sea areas,[24] including the Great Barrier Reef, the Torres Strait, the Florida Keys, the Papahānaumokuākea Marine National Monument, the Galapagos Islands, the Wadden Sea and Western European waters. The protection afforded in those areas includes routing measures and anchoring bans, mandatory reporting requirements and the strict application of discharge and equipment requirements for ships, such as oil tankers, as set out under the International Convention for the Prevention of Pollution from Ships of 1973, as modified by the Protocol of 1978 and the Protocol of

21 See www.eea.europa.eu/data-and-maps/dashboards/natura-2000-barometer.
22 See https://cases.open.ubc.ca/the-cultural-and-conservation-significance-of-anguniaqvia-niqiqyuam-marine-protected-area-mpa-north-west-territories-canada.
23 See https://rsis.ramsar.org/ris/2144.
24 See www.imo.org/en/OurWork/Environment/Pages/PSSAs.aspx.

1997.[25] Four of the areas (the Great Barrier Reef, the Papahānaumokuākea Marine National Monument, the Galapagos Islands and the Wadden Sea) are also protected as marine world heritage sites (see below).

The regional environmental management plan adopted by the International Seabed Authority for the Clarion-Clipperton Fracture Zone in the eastern central Pacific included the establishment of an initial set of nine areas of particular environmental interest as "no-mining areas", on the basis of expert recommendations. Those areas were intended to protect the biodiversity and ecosystem structure and functioning of the Zone from the potential impacts of seabed mining (Jones and others, 2019; see also chap. 18).

Marine protected areas may also be used in combination with fisheries management tools and sanctuaries (no-take zones, which may be within such areas). Sanctuary areas, seasonal and year-round fisheries closures[26] and exclusion zones provide area-based management mechanisms that seek to improve species population and biodiversity recovery. For example, the International Whaling Commission has established two sanctuaries, both of which prohibit commercial whaling: the Indian Ocean Whale Sanctuary, which was established in 1979 and covers the whole of the Indian Ocean as far south as 55° S; and the Southern Ocean Whale Sanctuary, which was established in 1994 and covers the waters around Antarctica.[27]

Seasonal and year-round fisheries closures support the maintenance or recovery of overexploited species, preserve the livelihoods of local communities, protect habitats and key ecological processes, such as spawning, and prevent the exploitation of living resources in areas beyond national jurisdiction prior to specific rule setting as a precautionary measure. Examples include the identification of vulnerable marine ecosystems and spatial closures by regional fisheries management organizations or associations, no-trawl zones in the United Kingdom of Great Britain and Northern Ireland to protect fish stocks and habitats, dynamic spatio-temporal closures in Australia to manage catches associated with migratory species and the closure of Arctic waters to commercial fishing under the Agreement to Prevent Unregulated High Seas Fisheries in the Central Arctic Ocean, pending a scientific assessment of the sustainability of such fisheries.

Area-based management is also used to safeguard marine sites of significance owing to their cultural value or the way in which the marine seascape combines cultural and natural attributes. World heritage sites under the United Nations Educational, Scientific and Cultural Organization (UNESCO) Convention for the Protection of the World Cultural and Natural Heritage of 1972 (UNESCO, 1972) provide an international example. Since the inscription of the first marine site on the UNESCO World Heritage List in 1981, 50 marine sites in 37 countries have been recognized for their unique marine biodiversity, ecosystems, geological processes or incomparable beauty.[28] The largest is the French Austral Lands and Seas, designated in 2019, covering 67,296,900 ha, followed by the Phoenix Islands Protected Area in Kiribati, at

25 See www.imo.org/en/About/Conventions/Pages/International-Convention-for-the-Prevention-of-Pollution-from-Ships-(MARPOL).aspx.

26 See, for example, European Union Regulation No. 2019/1022 establishing a multiannual plan for the fisheries exploiting demersal stocks in the western Mediterranean, which provides, inter alia, for the establishment of three-month closures of areas for the protection of juveniles, to be determined spatially and temporally by each member State. See www.consilium.europa.eu/en/press/press-releases/2019/06/06/first-ever-multi-annual-management-plan-for-fisheries-in-the-western-mediterranean-becomes-reality.

27 See www.iwc.int/sanctuaries.

28 See whc.unesco.org/en/marine-programme.

408,250 km², inscribed in 2010.[29] Four of the sites (the Papahānaumokuākea Marine National Monument in the United States; Saint Kilda in the United Kingdom; Ibiza in Spain; and Rock Islands Southern Lagoon in Palau) are internationally recognized for their mixed cultural and natural outstanding universal value. In a national context, all of the national marine sanctuaries in the United States include protections for historical, archaeological and cultural resources throughout the sanctuary system, and there are several sanctuaries designated specifically for their collections of historic shipwrecks (e.g., Thunder Bay sanctuary, Monitor sanctuary and Mallows Bay sanctuary).[30] In Scotland, the marine protected area concept has been developed to introduce areas around significant historic wreck sites (Historic Environment Scotland, 2019). Similarly, many national heritage laws provide for the designation of protection zones around underwater archaeological and historical sites, including measures such as the prohibition of fishing, anchoring and scuba diving without special authorization (e.g., Law No. 3028/2002 of Greece on the protection of antiquities and cultural heritage in general). Finally, special reference should be made to the recognition of the wreck site of RMS *Titanic* as an international maritime memorial under United States law and the international agreement between the United Kingdom and the United States that entered into force in 2019.[31]

29 In addition, the number of marine world heritage sites declared as being "in danger" has been reduced from three to two sites. The Belize-Barrier Reef Reserve System was removed from the List of World Heritage in Danger in 2018 owing to the effective implementation of a national policy specifically relating to the adoption of forests (protection of mangroves) regulations, a moratorium on oil exploration and other petroleum operations within the entire maritime zones of Belize, and further revision and amendment of the environmental impact assessment checklist and the corresponding ongoing revision of the assessment regulations.

30 See https://sanctuaries.noaa.gov.

31 See www.gc.noaa.gov/gcil_titanic.html. See also the International Maritime Organization circular (MEPC.1/Circ.779, dated 31 January 2012) on pollution prevention measures in the area surrounding the wreckage of RMS *Titanic*. Since 2012, the wreck site has fallen within the scope of protection of the Convention on the Protection of the Underwater Cultural Heritage of 2001 (United Nations, *Treaty Series*, vol. 2562, No. 45694), which applies to all traces of human existence having a cultural, historical or archaeological character that have been under water for at least 100 years. See www.unesco.org/new/en/culture/themes/underwater-cultural-heritage/the-heritage/did-you-know/titanic.

Table 2

Area-based management tools and their associated attributes, including primary drivers, sectors, implementation measures, direction and scale

Area-based tools		Management approach														
		Primary driver			Sector			Measures		Direction			Spatial scale			
Example in practice	Relevant authority	Economic	Environmental	Social well-being/cultural	Single-sector	Multi-sector	Cross-sector	Legally binding	Voluntary	Top-down	Bottom-up	Both	Global	Regional	National	Subnational
Areas of particular environmental interest	International Seabed Authority	x	x		x			x		x			x			
Vulnerable marine ecosystems	Regional fisheries management organizations or associations, competent national authorities		x		x			x		x				x		
Particularly sensitive sea areas and areas to be avoided	International Maritime Organization	x	x		x			x		x			x			
Fisheries closures and fisheries restricted areas	Food and Agriculture Organization of the United Nations, regional fisheries management organizations or associations, European Union, competent national authorities		x		x			x		x				x	x	x
Whale sanctuaries	International Whaling Commission		x		x			x		x			x			
Infrastructure closures: pipeline (e.g., oil, gas, waste, freshwater) and cable closures (e.g., telecommunications, grid)	International Maritime Organization, competent national authorities	x			x			x		x					x	

National marine conservation zones and priority areas for conservation	Competent national authorities		X		X			X		X			X	X
Aquaculture closures	Competent national authorities	X	X		X			X		X			X	X
World heritage sites, including those recognized for their mixed cultural and natural outstanding universal value	United Nations Educational, Scientific and Cultural Organization		X	X		X		X		X			X	
Marine protected areas	Aichi Biodiversity Targets, regional seas conventions, competent national authorities		X		X			X		X		X	X	
Protection zones around archaeological and historical sites	Competent national authorities			X				X		X			X	
Sites listed under the Convention on Wetlands of International Importance especially as Waterfowl Habitat (Ramsar Convention)	Ramsar Convention		X		X			X		X		X	X	
Species-specific sanctuaries (e.g., shark or dugong)	Competent national authorities		X		X			X		X		X	X	
Co-location (e.g., ocean energy or aquaculture)	Competent national authorities	X	X			X			X	X			X	X
Special areas and emission control areas	International Maritime Organization		X		X			X		X		X	X	X
Community-based spatial closures	Local government or communities		X	X	X		X		X		X			X
Traditional management approaches, including indigenous ranger programmes	Community leadership or authority, competent national or local authorities	X	X	X	X	X		X	X		X			X

3.3. Non-area-based management tools

The management of the ocean is not limited to area-based approaches, although, paradoxically, all management measures are applied across a spatial area even if they are required or sanctioned at larger scales. Many activities are dealt with through a range of other measures, such as the regulation of chemicals and pollution events, the management of transboundary migratory species and the application of technical measures in fisheries management (see chap. 15).

Non-area-based tools are primarily sectoral in nature and regulate specific sectoral activities to achieve a specific outcome. For example, global emissions controls are applied to international shipping vessels (global sulphur cap),[32] while catches within fisheries can be restricted through catch limits and limits on effort (such as through quota-based systems, hook limits and capacity limits). Technology-based measures can also be applied to fisheries to restrict catches of non-target species (e.g., turtle exclusion devices), and market-based approaches (e.g., accreditation schemes, seafood sustainability or eco-labelling) can be applied across an entire fishery, at the global, regional, national or subnational level.

Non-area-based tools are also widely used in national law for managing cultural heritage at sea, such as the requirement to report discoveries and obtain a licence before carrying out any activities directed towards the excavation, removal or disturbance of underwater cultural heritage.

At the international level, the United Nations Convention on the Law of the Sea sets out the jurisdictional framework for the duty to protect objects of an archaeological and historical nature at sea (see art. 303 of the Convention; Strati, 1995). The Convention on the Protection of the Underwater Cultural Heritage of 2001 elaborates on that duty in specific rights and obligations within the various maritime zones as defined by the United Nations Convention on the Law of the Sea, by providing for, inter alia, a system of reporting or notification and consultation for the protection of underwater cultural heritage found in the exclusive economic zone and on the continental shelf, as well as in the Area. In addition, the rules annexed thereto concerning activities directed at underwater cultural heritage contain general principles of protection along with technical rules, such as standards for conservation and management.

---

32 See www.imo.org/en/MediaCentre/HotTopics/Pages/Sulphur-2020.aspx.

Table 3
Non-area-based management tools and their associated attributes, including primary drivers, sectors, implementation measures, direction and scale

Non-area-based tools		Management approach														
		Primary driver			Sector			Measures		Direction			Spatial scale			
Example in practice	Relevant authority	Economic	Environmental	Social well-being/cultural	Single-sector	Multi-sector	Cross-sector	Legally binding	Voluntary	Top-down	Bottom-up	Both	Global	Regional	National	Subnational
Catch and effort controls	Regional authorities; Competent national authorities	X	X		X			X		X				X	X	X
Technology controls		X	X		X			X		X				X	X	X
Market-based tools		X	X		X			X	X		X			X	X	X
Underwater cultural heritage protection mechanisms			X	X	X			X		X					X	X

4. Management tools to support mitigation of and adaptation to climate change, including building resilience

In undertaking the ecosystem approach, decision-making processes are also required to consider knowledge of climate impacts and mitigation and adaptive responses. In that regard, identifying the adaptation pathways that may be undertaken to advance climate resilience is important in determining which management processes and tools can incorporate the uncertainty and unpredictability of environmental impacts and responses across spatio-temporal scales (Holsman and others, 2019; Wise and others, 2014). The choice of different adaptive measures that may be implemented to achieve greater resilience can vary greatly and is contingent upon the decision-making processes that frame them. As an example, ecosystem-based disaster risk reduction contributes to the adaptability of integrated coastal zone management and protected area management, in particular in the case of vulnerable communities and countries (Ferrario and others, 2014; Satta and others, 2017). Alternative strategies may apply mitigation and compensation measures,[33]

[33] These follow a hierarchy of management measures: preventative measures (e.g., stopping pollutants from entering the sea); mitigation measures (e.g., reducing direct impacts); and compensation (e.g., the user (such as fishers for loss of catch), the resource (by restocking fishes or replanting mangroves) or the habitat (creating new habitats to replace those lost to new infrastructure)) (Elliott and others, 2016).

such as the blue carbon initiative. Effective mitigation approaches should also enhance linkages with adaptation finance, technology transfer and capacity-building, while adaptive responses should consider environmental, social and economic aspects in order to identify effective mechanisms that balance the needs and maximize benefits to all.

The global application of marine protected area networks helps to promote mitigation and adaptation to climate change (Dudley and others, 2010; Roberts and others, 2017) by supporting ecosystem resilience. By building resilience, ecosystems have a greater ability to cope with perturbations and recover from adverse circumstances, thereby maintaining ecosystem functions and the provision of services necessary for human well-being (Chong, 2014).

Resilience-based management (alongside area-based management tools) uses knowledge of current and future drivers that influence ecosystem function (e.g., coral disease outbreaks and changes in land use, trade or fishing practices) to prioritize, implement and adapt management actions that sustain ecosystems and human well-being (Mcleod and others, 2019). To support the maintenance of ecosystem resilience, managers must reduce local stressors (e.g., pollution and destructive fishing pressures), while fostering key resilience processes (e.g., recovery, reproduction, recruitment and connectivity) (Anthony and others, 2015; Graham and others, 2013). That requires managing the causes and consequences of the endogenic (local) pressures and responding to the consequences of the exogenic (global) pressures, given that responding to the causes of the latter requires global action (Elliott, 2011). For example, marine protected area networks can be designed for climate resilience by maintaining a diversity and redundancy of species, habitats and functional groups and pathways of connectivity and reducing stressors, and by including adaptive processes to accommodate uncertainty and change (Mcleod and others, 2019). Coral resilience and the associated ability to recover from bleaching events across the marine protected area network of Hawaii is supported through active management of herbivore fish aimed at maintaining and increasing herbivore biomass, abundance and functional diversity (Chung and others, 2019).

Along with marine protected area networks, there is a diversity of adaptation measures that can be carried out at the community and institutional levels. They include tools such as cross-sectoral coordination, flexible fishing licences, seasonal rights, transboundary management and enhanced institutional cooperation that can be applied in conjunction with market and livelihood diversification and resilience-building tools such as emergency preparedness, early-warning systems, remittances and post-disaster recovery plans (Poulain and others, 2018). In applying specific management tools, trade-offs should also be considered, as the tools can trigger contrasting effects on different sectors or countries. In the Arctic, for example, transboundary cooperation engages new actors and sectors, such as polar tourism, but also brings new risks, such as shipping and mineral exploration and exploitation. In the Mediterranean, transcontinental cooperation (between Africa and Europe) is needed to embrace regional adaptation measures to deal with the contrasting local needs and adaptive capacities of African and European countries (Karmaoui, 2018; Hidalgo and others, 2018).

5. Key region-specific issues

The implementation of the ecosystem approach through decision-making processes and management tools in the marine environment has progressed at different paces in different regions. Regions with higher skill levels, financial capacities and resources have experienced considerable progress in the implementation of the ecosystem approach. For example, rapid environmental change in the Arctic Ocean, driven by large-scale warming, has necessitated a shift by the Arctic Council from a focus on soft-policy scientific assessments to legally binding agreements negotiated by member countries. Those agreements have also become necessary as a result of the increasing opportunities for industrial uses of the Arctic Ocean and their attendant risks, including shipping activities, Arctic tourism, the transfer of alien species and mineral exploitation on the continental shelf of Arctic coastal States. Such rapid changes have prompted countries to adjust their policies to better respond to fast-emerging social, economic and environmental challenges resulting from climate change. Canada, for example, amended its Oceans Act in 2019 to be able to apply precautionary principles and allow interim protection for an area for up to five years, through the use of a ministerial order provision that freezes the footprint of human activities, meaning that no new or additional human activities will be allowed in the area for the duration of the order. In 2019, the Tuvaijuittuq marine protected area, the first created through the use the ministerial order provision, was established to protect the oldest and thickest sea ice in the Arctic Ocean as an important summer habitat for species as ice cover continues to decline in the Arctic.

In regions with more limited capacity, it is more difficult to implement the ecosystem approach. Many marine and coastal areas in such regions are confronting decades, if not centuries, of degradation as a result of a lack of management practices or controls and owing to the fact that restoration approaches are being implemented reactively. In South America (Gianelli and others, 2018; Reis and D'Incao, 2000), the implementation of ecosystem approaches to fisheries has been challenging, with restrictions on both institutional and scientific capacity, which has limited success to areas with favourable enabling conditions. Similar capacity challenges can be seen in the management of marine protected areas (Gerhardinger and others, 2011), although engagement with local knowledge holders has led to improvements in outcomes (Gerhardinger and others, 2009).

Much of the recent growth in the surface areas of marine protected areas can be accounted for by a small number of countries that have established large marine protected areas. Although the data reflect progress towards the conservation of biodiversity and marine resources, protection is still focused on waters under national jurisdiction and countries with the capability and capacity to identify and implement marine protected area networks. However, the designation of a marine protected area is not necessarily reflective of active management and protection, since many of them lack adequate management plans and associated enforcement measures (World Conservation Monitoring Centre and International Union for Conservation of Nature, 2019; Maestro and others, 2019). Similarly, the uneven geographical distribution of the areas limits their effectiveness, connectivity, coherence and representativeness.

Finally, climate change is becoming a key driver in prioritizing restoration approaches in many parts of the world, including the restoration of mangrove forests in Indonesia and in a number of small island developing States in the Pacific, which are aimed at protecting local communities from coastal inundation (Food and Agriculture Organization of the United Nations, 2016) and increasing resilience to future changes, as well as restoring parts of the Great Barrier Reef in Australia following multiple bleaching events (Reef Restoration and Adaptation Program Consortium, 2018). The restoration of coral reefs in the Caribbean and oyster reefs worldwide employed small-scale techniques, such as micro-fragmentation, to address local-scale damage (Gilby and others, 2018). However, such approaches are often still limited in their scale. Further examples of climate adaptation and disaster risk reduction include the measures taken by Colombia, Ecuador and Grenada for mangrove restoration and coastal protection; the United Kingdom for coastal realignment; Mexico for sustainable fishing and mangrove rehabilitation; and Vanuatu for coral reef restoration (Secretariat of the Convention on Biological Diversity, 2019). The forthcoming United Nations Decade on Ecosystem Restoration (2021–2030)[34] is aimed at accelerating that trend (Waltham and others, 2020).

6. Capacity-building

Most management approaches require information that cuts across the natural and social sciences. In many regions, especially in developing countries, scientists and practitioners are simply not sufficiently trained to implement existing or new approaches to management, in particular those involving the ecosystem approach. Increased capacity, not only in understanding management approaches, but also in having the tools to implement them, will support Governments and other stakeholders in understanding the suite of options available for marine management and governance in their jurisdictions. Hence, there are several key capacity-building and technology-transfer requirements in that field. First, there is a need for training and expertise in marine management and governance linked to the relevant science, including training in policy drivers, as well as in policy-relevant science and policy repercussions of science – that is, how relevant science can be used in developing policy and what adaptations or revisions need to be made to policy as new scientific information becomes available. Second, there is a large scope for learning within and between nations and regions (i.e., knowledge and technology transfer), especially since some approaches have worked well in certain conditions, such as marine spatial planning programmes under the Convention for the Protection, Management and Development of the Marine and Coastal Environment of the Western Indian Ocean. In that regard, increased capacity in transboundary cooperation is needed, with science-based management as a core element. Third, there is also a large scope for learning across the breadth of different policies, including how policy was derived, especially for new practitioners but also as continuing professional development for more experienced professionals.

Knowledge of the key stages in implementing the planning and policy process for marine management, as well as the metrics for measuring and monitoring the effectiveness of management measures, are key requirements for countries that are starting to implement management approaches. An understanding by scientists and other stakeholders (including the public) of policymaking and the management of public behaviour, including related economic aspects, is also important in that regard. To achieve those goals, both formal and non-formal approaches to education are required. In addition, the transfer of knowledge regarding decision-making processes and tools across sectors should be promoted in order to ensure that the ecosystem approach can be applied holistically across marine sectors.

[34] See General Assembly resolution 73/284.

Gill and others (2017) indicated that staff and budget capacity were the strongest predictors of conservation impacts. Their study indicated that marine protected areas with adequate staff capacity had ecological benefits 2.9 times greater than those with inadequate capacity. Creating such areas without adequate investment will, therefore, result in suboptimal conservation outcomes. Limited resources in some cases may increase the need for citizen science programmes that can complement or support monitoring limitations (e.g., in the United Kingdom for shore biota and beach litter monitoring and clean-up programmes, as well as the Reef Check Foundation, MangroveWatch and the Manta Trust) in their global programmes (see also sect. 7.1 below). The techniques can be deployed worldwide as best practices for greater benefit.

7. Gaps and future perspectives

7.1. Data and information for management needs

Marine management approaches, processes and tools are often hampered by a lack of data of appropriate quality and quantity (Borja and others, 2017). Recent developments in the use of big-data methods, the innovative use of data and information in policy approaches and the linking of databases help to provide information in such situations. However, understanding of ecological causes and effects related to socioeconomic priorities, as reflected in modelling expertise and scientific support systems for decision-making (recognizing the complexity of coastal and marine systems) is still limited across many regions. The sharing of knowledge (e.g., the Ocean Biodiversity Information System) and open access to information and data streams, in particular across sectors, should be encouraged in order to ensure that the data collected are made available to all (e.g., "collect once, use many times"). The enhanced collaboration and connectivity of monitoring programmes will assist, not only in the sharing of capacity across sectors and institutions, but also in providing for more efficient approaches to the monitoring and provision of data and information. Data from citizen science are increasingly becoming an important source of monitoring information, where they are validated and accepted by the academic community (Bennett, 2019) to provide key information on environmental states and trends (e.g., Edgar and Stuart-Smith, 2014).

Challenges that still need to be addressed include the gathering of data for marine management in a cost-effective manner. The role of technology in marine conservation and management will become increasingly important, especially the collection and use of data from remote sensing and satellites. In sectoral and spatial management, for example, automatic identification system and vessel monitoring system data are used to manage shipping and fishing activities, in particular for mapping. Novel analytical approaches, such as machine learning, are increasingly being applied to the identification of illegal activities in those sectors (Longépé and others, 2018) and to monitor fishery catches (Lee and others, 2008).

7.2. Management requirements

Marine management requires the best available science for maintaining and protecting the natural system, while also providing benefits for the private sector and for society. More research is needed on ecological adaptation and resilience, inter alia, and the prediction of ecosystem response trajectories. Those variables should be built into management approaches that cover the scale of both the impact and response of marine ecosystems, which implies the need for a greater recognition of human intervention in the marine environment as measured against baselines and for using thresholds and targets for unacceptable change. However, it is a major challenge and there is often no baseline or, because of climate change, baselines are moving. Better interconnected monitoring programmes across institutions also need to be established. Areas beyond national jurisdiction present a major challenge in that regard, in particular in deep sea ecosystems that are poorly surveyed.

Management approaches are underpinned by detailed governance mechanisms, such as policies, polities, administration and legislation. Improving the science-policy interface by enhancing capacity is necessary and particularly important where the knowledge base of informing decision-making is quickly expanding and emerging. Greater coordination is needed in that regard between social and natural sciences, between scientists and policymakers and between science and civil society, including industry, as is the inclusion of traditional knowledge, culture and social history in management. Such cross-sectoral understanding is important for management that is truly holistic.

7.3. Incorporating multiple values into management

The present chapter has shown an evident trend in management approaches from being focused on predominantly ecological aspects to the inclusion of diverse links between ecological aspects and societal, economic and cultural aspects of the marine environment. Management would be better equipped to achieve the fundamental goal of protecting and maintaining natural systems if it also recognized the wide range of ecosystem services and benefits derived from the oceans. Protecting and preserving the marine environment depends on engaging those who live or work with the sea and who gain benefits from it, in order to address deleterious behavious, restore inadvertently damaged systems and mitigate the impacts of a changing climate.

However, the values that people place on the marine environment and its services vary not only in quantity but also in character. Challenging to most management systems is the need to accommodate the multiplicity of values, for which real or perceived benefits cannot be equated with each other or reconciled. The best opportunities to understand and address multiple values are those that engage affected communities in the management approach, resulting in the need to combine ecosystem-based management with community-based management that is sensitive to the cultural dimensions of the sea. Such hybrid systems are more capable of balancing all three pillars of sustainable development (environmental, economic and social) and, as such, are likely to be more successful.

8. Outlook

While the present chapter has identified a plethora of approaches to the management of the marine environment, there is still much that can be done to improve and enhance progress, including with regard to the successful integration of the Sustainable Development Goals, especially Goal 14, into management objectives and programmes. There is also a need for the increased integration of measures to manage anthropogenic pressures that are not currently the focus of management measures, such as anthropogenic noise.

The implementation of the 2030 Agenda requires management grounded in the ecosystem approach in order to achieve the integrated set of global priorities and objectives set out in the Goals. That will allow for the integration of interactions, benefits and trade-offs between the Goals and support the achievement of each of the ocean-related targets. Overall, the progress made to date, notwithstanding existing actions for the implementation of Goal 14, is insufficient. Accelerated action, in particular with regard to the targets of Goal 14 that mature in 2020, is necessary as a matter of urgency, including for targets 14.2, 14.4, 14.5 and 14.6. Although Goal 14 does not explicitly include any reference to marine cultural aspects, the outcome of the United Nations Conference to Support the Implementation of Sustainable Development Goal 14, entitled "Our ocean, our future: call for action", includes the need to develop comprehensive strategies to raise awareness of the natural and cultural significance of the ocean.[35] Similarly, the SIDS Accelerated Modalities of Action (SAMOA) Pathway recognizes the cultural connection of the communities of small island developing States to the ocean and the importance of traditional knowledge in the sustainable development of ocean-based economies.[36]

The outputs of the United Nations Decade of Ocean Science for Sustainable Development (2021–2030)[37] and the concurrent United Nations Decade on Ecosystem Restoration will support the implementation of Goal 14 and provide many of the necessary data sources to apply management processes and tools, and will also increase ocean literacy.[38] The initiatives have the potential to advance the tools needed for current and future decision-making, improve overall understanding of issues and solutions for ocean management and increase societal engagement in decision-making and solution applications. Integrating the protection of the underwater cultural heritage into the United Nations Decade of Ocean Science for Sustainable Development[39] is also pertinent to support the tangible and intangible resources and cultural benefits provided by the oceans (UNESCO, 2019; Trakadas and others, 2019).

While it is implicit in the context of marine management, the present chapter has not covered the detailed nature of marine governance, nor the challenges associated with the sectoral and often fragmented nature of administrative bodies (e.g., Boyes and Elliott, 2014; 2015). In order to be effective across wider scales and for species that span large scales, both area-based and non-area-based management approaches will need to overcome the often fragmented and complex governance regimes worldwide.

The effective management of marine resources will also need to extend to areas beyond national jurisdiction, where challenges are greater owing to the complexities of the legal regime. That gives added significance to the current negotiations at the United Nations on an international legally binding instrument under the United Nations Convention on the Law of the Sea for the conservation and sustainable use of marine biological diversity of areas beyond national jurisdiction (see chap. 28). Similar discussions have been initiated at UNESCO on expanding the scope of application of the Convention for the Protection of the World Cultural and Natural Heritage to provide for the protection and management of marine sites of outstanding universal value in high seas areas (UNESCO, 2016; 2019).

35 See General Assembly resolution 71/312, annex.
36 See General Assembly resolution 69/15, annex. See also https://sidsnetwork.org/samoa-pathway.
37 See General Assembly resolution 72/73.
38 See https://oceanconference.un.org/commitments/?id=15187 and http://ioc-unesco.org/index.php?option=com_oe&task=viewEventAgenda&eventID=2200.
39 See General Assembly resolution 72/73, para. 292. See also www.oceandecadeheritage.org.

References

Anthony, Kenneth R.N., and others (2015). Operationalizing resilience for adaptive coral reef management under global environmental change. *Global Change Biology*, vol. 21, No. 1, pp. 48–61. https://doi.org/10.1111/gcb.12700.

Basiron, Mohd, and Cheryl Kaur (2009). *Designating a Particularly Sensitive Sea Area: Specifics, Processes and Issues*. Proceedings of the 6th MIMA International Conference on the Straits of Malacca. Kuala Lumpur.

Bennett, Nathan J. (2019). Marine social science for the peopled seas. *Coastal Management*, vol. 47, No. 2, pp. 244–252. https://doi.org/10.1080/08920753.2019.1564958.

Berkes, Fikret, and others (2000). Rediscovery of traditional ecological knowledge as adaptive management. *Ecological Applications*, vol. 10, No. 5, pp. 1251–1262.

Borja, Angel, and others (2017). Bridging the gap between policy and science in assessing the health status of marine ecosystems. Editorial. Lausanne: *Frontiers in Marine Science*. https://doi.org/10.3389/978-2-88945-126-5.

Boyes, Suzanne J., and Michael Elliott (2014). Marine legislation: the ultimate 'horrendogram' – international law, European directives and national implementation. *Marine Pollution Bulletin*, vol. 86, Nos. 1–2, pp. 39–47.

________(2015). The excessive complexity of national marine governance systems: has this decreased in England since the introduction of the Marine and Coastal Access Act 2009? *Marine Policy*, vol. 51, pp. 57–65.

Brugère, Cecile, and others (2019). The ecosystem approach to aquaculture 10 years on: a critical review and consideration of its future role in blue growth. *Reviews in Aquaculture*, vol. 11, No. 3, pp. 493–514. https://doi.org/10.1111/raq.12242.

Chong, J. (2014). Ecosystem-based approaches to climate change adaptation: progress and challenges. *International Environmental Agreements: Politics, Laws and Economics*, vol. 14, pp. 391–405. https://doi.org/10.1007/s10784-014-9242-9.

Chung, Anne E., and others (2019). Building coral reef resilience through spatial herbivore management. *Frontiers in Marine Science*, vol. 6, art. 98. https://doi.org/10.3389/fmars.2019.00098.

Cowan, James H., Jr., and others (2012). Challenges for implementing an ecosystem approach to fisheries management. *Marine and Coastal Fisheries*, vol. 4, No. 1, pp. 496–510.

Delisle, Aurélie, and others (2018). The socio-cultural benefits and costs of the traditional hunting of dugongs (*Dugong dugon*) and green turtles (*Chelonia mydas*) in Torres Strait, Australia. *Oryx*, vol. 52, No. 2, pp. 250–261. https://doi.org/10.1017/S0030605317001466.

Díaz, Sandra, and others (2018). Assessing nature's contributions to people. *Science*, vol. 359, No. 6373, p. 270. https://doi.org/10.1126/science.aap8826.

Dudley, Nigel, and others (2010). The revised IUCN protected area management categories: the debate and ways forward. *Oryx*, vol. 44, No. 4, pp. 485–490.

Dunstan, Piers K., and others (2016). Using ecologically or biologically significant marine areas (EBSAs) to implement marine spatial planning. *Ocean and Coastal Management*, vol. 121, pp. 116–127.

Edgar, Graham J., and Rick D. Stuart-Smith (2014). Systematic global assessment of reef fish communities by the reef life survey program. *Scientific Data*, vol. 1, pp. 140007–140007. https://doi.org/10.1038/sdata.2014.7.

________(2016). Ecoengineering with ecohydrology: successes and failures in estuarine restoration. *Estuarine, Coastal and Shelf Science*, vol. 176, pp. 12–35. https://doi.org/10.1016/j.ecss.2016.04.003.

Elliott, M. (2011). Marine science and management means tackling exogenic unmanaged pressures and endogenic managed pressures: a numbered guide. *Marine Pollution Bulletin*, vol. 62, pp. 651–655.

Elliott, M., and others (2016). Ecoengineering with ecohydrology: successes and failures in estuarine restoration. *Estuarine, Coastal and Shelf Science*, vol. 176, pp. 12–35. http://doi.org/10.1016/j.ecss.2016.04.003.

Fanini, Lucia, and others (2020). Advances in sandy beach research: local and global perspectives. *Estuarine, Coastal and Shelf Science*, vol. 234, p. 106646.

Ferrario, Filippo, and others (2014). The effectiveness of coral reefs for coastal hazard risk reduction and adaptation. *Nature Communications*, vol. 5, p. 3794.

Food and Agriculture Organization of the United Nations (FAO) (2016). *The State of World Fisheries and Aquaculture*. Rome.

________ (2018). *The State of World Fisheries and Aquaculture 2018: Meeting the Sustainable Development Goals*. Rome.

Gerhardinger, Leopoldo C., and others (2009). Local ecological knowledge and the management of marine protected areas in Brazil. *Ocean and Coastal Management*, vol. 52, Nos. 3–4, pp. 154–165.

Gerhardinger, Leopoldo C., and others (2011). Marine protected dramas: the flaws of the Brazilian national system of marine protected areas. *Environmental Management*, vol. 47, No. 4, pp. 630–643.

Gianelli, I., and others (2018). Operationalizing an ecosystem approach to small-scale fisheries in developing countries: the case of Uruguay. *Marine Policy*, vol. 95, pp. 180–188.

Gilby, Ben L., and others (2018). Maximizing the benefits of oyster reef restoration for finfish and their fisheries. *Fish and Fisheries*, vol. 19, No. 5, pp. 931–947. https://doi.org/10.1111/faf.12301.

Gill, David A., and others (2017). Capacity shortfalls hinder the performance of marine protected areas globally. *Nature*, vol. 543, p. 665.

Goodhead, Tim, and Zeynep Aygen (2007). Heritage management plans and integrated coastal management. *Marine Policy*, vol. 31, No. 5, pp. 607–610. https://doi.org/10.1016/j.marpol.2007.03.005.

Govan, Hugh (2009). Achieving the potential of locally managed marine areas in the South Pacific. *SPC Traditional Marine Resource Management and Knowledge Information Bulletin*, vol. 25.

Graham, Nicholas A.J., and others (2013). Managing resilience to reverse phase shifts in coral reefs. *Frontiers in Ecology and the Environment*, vol. 11, art. 10, pp. 541–548. https://doi.org/10.1890/120305.

Hall, Stephen (2002). Chapter 3: the use of technical measures in responsible fisheries: area and time restrictions. In *A Fishery Manager's Guidebook: Management Measures and Their Application*, Kevern L. Cochrane, ed. Fisheries Technical Paper, No. 424. Rome: FAO. www.fao.org/3/y3427e/y3427e00.htm.

Harvey, C.J., and others (2017). Implementing "the iea": using integrated ecosystem assessment frameworks, programs, and applications in support of operationalizing ecosystem-based management. *ICES Journal of Marine Science*, vol. 74, No. 1, pp. 398–405. https://doi.org/10.1093/icesjms/fsw201.

Hidalgo, Manuel, and others (2018). Climate change impacts, vulnerabilities and adaptations: Mediterranean Sea and the Black Sea marine fisheries. In *Impacts of Climate Change on Fisheries and Aquaculture*, pp. 139–158. FAO.

Historic Environment Scotland (2019). *Scotland's Historic Marine Protected Areas*. Historic Environment Scotland. www.historicenvironment.scot/archives-and-research/publications/publication/?publicationId=fe248e27-0c19-4e4e-8d65-a62d00a2ce6a.

Holsman, Kirstin K., and others (2019). Towards climate resiliency in fisheries management. *ICES Journal of Marine Science*, vol. 76, No. 5, pp. 1368–1378. https://doi.org/10.1093/icesjms/fsz031.

Humphreys, John, and Robert W.E. Clark (2020). *Marine Protected Areas: Science, Policy and Management*. Elsevier.

International Tribunal of the Law of the Sea (Seabed Disputes Chamber) (2011). *Case No. 17, Advisory Opinion on Responsibilities and Obligations of States with Respect to Activities in the Area, 1 February 2011*.

Jones, Daniel O.B., and others (2019). Existing environmental management approaches relevant to deep-sea mining. *Marine Policy*, vol. 103, pp. 172–181.

Jones, Peter J.S., and others (2016). Marine spatial planning in reality: introduction to case studies and discussion of findings. *Marine Policy*, vol. 71, pp. 256–264. https://doi.org/10.1016/j.marpol.2016.04.026.

Karmaoui, Ahmed (2018). Environmental vulnerability to climate change in Mediterranean Basin: socio-ecological interactions between North and South. In *Hydrology and Water Resource Management: Breakthroughs in Research and Practice*, pp. 61–96. Hershey, Pennsylvania, United States of America: IGI Global. https://doi.org/10.4018/978-1-5225-3427-3.ch003.

Kikiloi, Kekuewa, and others (2017). Papahānaumokuākea: integrating culture in the design and management of one of the world's largest marine protected areas. *Coastal Management*, vol. 45, No. 6, pp. 436–451.

Lee, Dah-Jye, and others (2008). Contour matching for fish species recognition and migration monitoring. In *Applications of Computational Intelligence in Biology*, Tomasz G. Smolinski, and others, eds., vol. 122, pp. 183–207. Studies in Computational Intelligence. Springer.

Link, Jason S., and Howard I. Browman (2017). Operationalizing and implementing ecosystem-based management. *ICES Journal of Marine Science*, vol. 74, No. 1, pp. 379–381.

Lo, Veronica (2018). Voluntary Guidelines for the Design and Effective Implementation of Ecosystem-Based Approaches to Climate Change Adaptation and Disaster Risk Reduction. CBD/SBSTTA/22/INF/1.

Longépé, Nicolas, and others (2018). Completing fishing monitoring with spaceborne Vessel Detection System (VDS) and Automatic Identification System (AIS) to assess illegal fishing in Indonesia. *Marine Pollution Bulletin*, vol. 131, pp. 33–39. https://doi.org/10.1016/j.marpolbul.2017.10.016.

Maestro, María, and others (2019). Marine protected areas in the 21st century: current situation and trends. *Ocean and Coastal Management*, vol. 171, pp. 28–36.

McIntosh, Emma J., and others (2017). The impact of systematic conservation planning. *Annual Review of Environment and Resources*, vol. 42, pp. 677–697.

Mcleod, Elizabeth, and others (2019). The future of resilience-based management in coral reef ecosystems. *Journal of Environmental Management*, vol. 233, pp. 291–301.

Meek, Chanda L. (2013). Forms of collaboration and social fit in wildlife management: a comparison of policy networks in Alaska. *Global Environmental Change*, vol. 23, No. 1, pp. 217–228.

Milcu, Andra Ioana, and others (2013). Cultural ecosystem services: a literature review and prospects for future research. *Ecology and Society*, vol. 18, No. 3. https://doi.org/10.5751/ES-05790-180344.

Miller, Rachel L., and others (2018). Protecting migratory species in the Australian marine environment: a cross-jurisdictional analysis of policy and management plans. *Frontiers in Marine Science*, vol. 5, art. 229.

Nguyen, K.D., and others (2016). The Vietnamese state and administrative co-management of nature reserves. *Sustainability*, vol. 8, No. 3, p. 292.

Oates, Jennifer, and Lyndsey A. Dodds (2017). An approach for effective stakeholder engagement as an essential component of the ecosystem approach. *ICES Journal of Marine Science*, vol. 74, No. 1, pp. 391–397. https://doi.org/10.1093/icesjms/fsw229.

Pope, John G. (2002). Chapter 4: input and output controls: the practice of fishing effort and catch management in responsible fisheries. In *A Fishery Manager's Guidebook: Management Measures and Their Application*, Kevern L. Cochrane, ed. Fisheries Technical Paper, No. 424. Rome: FAO. https://doi.org/10.1002/9781444316315.ch9.

Poulain, Florence, and others (2018). Methods and tools for climate change adaptation in fisheries and aquaculture. In *Impacts of Climate Change on Fisheries and Aquaculture. Synthesis of Current Knowledge, Adaptation and Mitigation Options*, Manuel Barange and others, eds., pp.535–566. FAO Fisheries and Aquaculture Technical Paper, No. 627. Rome: FAO.

Reef Restoration and Adaptation Program Consortium (2018). *Reef Restoration and Adaptation Program*. Australian Marine Science Association. www.aims.gov.au/documents/30301/0/RRAP+Brochure/909e6dea-c7e9-4125-bece-0f10b639da5b.

Reis, Enir G., and Fernando D'Incao (2000). The present status of artisanal fisheries of extreme Southern Brazil: an effort towards community-based management. *Ocean and Coastal Management*, vol. 43, No. 7, pp. 585–595.

Roberts, Callum M., and others (2017). Marine reserves can mitigate and promote adaptation to climate change. *Proceedings of the National Academy of Sciences of the United States of America*, vol. 114, No. 24, pp. 6167–6175. https://doi.org/10.1073/pnas.1701262114.

Satta, Alessio, and others (2017). Assessment of coastal risks to climate change related impacts at the regional scale: the case of the Mediterranean region. *International Journal of Disaster Risk Reduction*, vol. 24, pp. 284–296.

Schultz, Lisen, and others (2015). Adaptive governance, ecosystem management, and natural capital. *Proceedings of the National Academy of Sciences of the United States of America*, vol. 112, No. 24, pp. 7369–7374. https://doi.org/10.1073/pnas.1406493112.

Secretariat of the Convention on Biological Diversity (2004). *The Ecosystem Approach*. Montreal. www.cbd.int/ecosystem.

________ (2019). *Voluntary Guidelines for the Design and Effective Implementation of Ecosystem-Based Approaches to Climate Change Adaptation and Disaster Risk Reduction and Supplementary Information*. Technical Series, No. 93. Montreal.

Strati, Anastasia (1995). *The Protection of the Underwater Cultural Heritage: An Emerging Objective of the Contemporary Law of the Sea*, vol. 23. Martinus Nijhoff Publishers.

Thornton, Thomas F., and Adela Maciejewski Scheer (2012). Collaborative engagement of local and traditional knowledge and science in marine environments: a review. *Ecology and Society*, vol. 17, No. 3. https://doi.org/10.5751/ES-04714-170308.

Trakadas, Athena, and others (2019). The Ocean Decade Heritage Network: Integrating Cultural Heritage Within the United Nations Decade of Ocean Science 2021–2030. *Journal of Maritime Archaeology*, vol. 14, No. 2, pp. 153–165.

Trochta, John T., and others (2018). Ecosystem-based fisheries management: perception on definitions, implementations, and aspirations. *PLOS One*, vol. 13, No. 1, p. e0190467.

Turner, Nancy J., and Fikret Berkes (2006). Coming to understanding: developing conservation through incremental learning in the Pacific Northwest. *Human Ecology*, vol. 34, No. 4, pp. 495–513.

United Nations (2017). *The First Global Integrated Marine Assessment: World Ocean Assessment I*. Cambridge: Cambridge University Press. https://doi.org/10.1017/9781108186148.

United Nations Conference on Environment and Development (1992). *Annex I, Rio Declaration on Environment and Development, Principle 15: Precautionary Approach, Rio de Janeiro, 3–14 June 1992*. www.unesco.org/education/pdf/RIO_E.PDF.

United Nations Educational, Scientific and Cultural Organization (UNESCO) (1972). Convention Concerning the Protection of the World Cultural and Natural Heritage, Adopted by the General Conference at its seventeenth session, Paris, 16 November 1972. https://whc.unesco.org/archive/convention-en.pdf.

________ (2016). *World Heritage in the High Seas: An Idea Whose Time Has Come*. World Heritage Report, No. 44. UNESCO.

________ (2019). *Report on the Evaluation of 2001 Convention on the Protection of Underwater Cultural Heritage*. UNESCO.

United Nations Environment Programme (UNEP) (2018). Conceptual guidelines for the application of marine spatial planning and integrated coastal zone management approaches to support the achievement of sustainable development goal targets 14.1 and 14.2. United Nations Regional Seas Reports and Studies, No. 207. www.unep-wcmc.org/system/dataset_file_fields/files/000/000/548/original/Final_ConceptualGuidelines_240918.pdf?1538124788.

United Nations Environment Programme (UNEP) and Secretariat of the Convention on Biological Diversity (2000). Annex III, Decisions Adopted by the Conference of the Parties to the Convention on Biological Diversity at Its Fifth Meeting. www.cbd.int/doc/decisions/COP-05-dec-en.pdf.

Veitayaki, Joeli, and others (2003). Empowering local communities: case study of Votua, Ba, Fiji. *Ocean Yearbook Online*, vol. 17, No. 1, pp. 449–463.

Waltham, Nathan J., and others (2020). United Nations decade on ecosystem restoration 2021–2030: what chance for success in restoring coastal ecosystems? *Frontiers in Marine Science*, vol. 7, art. 71. https://doi.org/10.3389/fmars.2020.00071.

Wise, Russell M., and others (2014). Reconceptualising adaptation to climate change as part of pathways of change and response. *Global Environmental Change*, vol. 28, pp. 325–336.

World Conservation Monitoring Centre and International Union for Conservation of Nature (2019). Protected Planet: World Database on Protected Areas. Cambridge, United Kingdom of Great Britain and Northern Ireland. www.protectedplanet.net/marine.

Zador, S.G., and others (2017). Ecosystem considerations in Alaska: the value of qualitative assessments. *ICES Journal of Marine Science*, vol. 74, No. 1. https://doi.org/10.1093/icesjms/fsw144.

Zhang, C. I., and others (2011). An IFRAME approach for assessing impacts of climate change on fisheries. *ICES Journal of Marine Science*, vol. 68, No. 6, pp. 1318–1328.

Chapter 28
Developments in the understanding of overall benefits from the ocean to humans

Contributors: Luciano Hermanns (convener), Denis Worlnanyo Aheto, Adem Bilgin, Robert Blasiak, Cecile Brugere, Karen Evans, Antony Firth, Deborah Greaves, Osman Keh Kamara (co-lead member), Wenhai Lu, Essam Yassin Mohamed (lead member), Iryna Makarenko, Stale Navrud, Marinez Eymael Garcia Scherer, Jörn Schmidt (co-lead member), Anita Smith, Anastasia Strati (co-lead member), Rashid Sumaila, Kateryna Utkina, Hans Van Tilburg and Wojciech Wawrzynski

Keynote points

- Ocean resources provide the main sources of livelihoods to millions of people across the globe, as well as a wide range of ecosystem services and benefits, including oxygen production, food provision, carbon storage, minerals, genetic resources and cultural and general life support services. However, the ecosystem services from marine and coastal ecosystems are deteriorating at an alarming rate, owing to several human pressures, including climate change.
- Human activities are directly or indirectly affecting ecosystem services and can thus reduce or erase benefits that would otherwise be provided. As human activities in the marine environment are expected to increase in the future, in particular in areas beyond national jurisdiction, not only will they exert growing pressure on natural resources, but they may also threaten marine biodiversity and therefore the benefits that people obtain from ecosystem services.
- International law as reflected in the United Nations Convention on the Law of the Sea[1] plays a crucial role in the conservation and sustainable use of the ocean and its resources and in safeguarding the many ecosystem services that the ocean provides for both current and future generations. Actions and efforts should be primarily focused on implementation and regulatory gaps, especially in areas beyond national jurisdiction.
- That gives added significance to the current negotiations at the United Nations on the elaboration of an international legally binding instrument under the United Nations Convention on the Law of the Sea on the conservation and sustainable use of marine biological biodiversity of areas beyond national jurisdiction.
- The distribution around the world of the benefits drawn from the ocean is still very uneven. Efforts by less developed countries to take advantage of what the ocean can offer them are hampered by gaps in capacity-building and resource and financial constraints.
- Capacity-building, shared scientific knowledge and collaboration to develop and transfer innovative marine technology will empower States to fully participate in and benefit from the conservation and sustainable use of the ocean and its resources and assist them in meeting their obligations.

1. Introduction

Ocean resources provide the basis for the livelihoods of millions of people across the globe, as well as a range of critical ecosystem services, including oxygen production and carbon storage, several biodiversity-related services, such as the harvesting of living resources, coastal protection and genetic resources (Mohammed, 2012) and cultural and amenity services (Whitmarsh, 2011). The most commonly valued services are tourism and recreation, as well as storm protection (Mehvar and others, 2018). Fisheries alone provide multiple benefits to millions of people, including those living in poverty in the coastal communities of low-income countries. Fishes and other seafood are a major source of food, protein and micronutrients for many vulnerable communities. It is estimated that, in 2016, 59.6 million

1 United Nations, *Treaty Series*, vol. 1833, No. 31363.

people were employed in the primary sector of capture fisheries and aquaculture, with a great majority in low-income countries (although that figure includes some inland activities). With the addition of those who work in associated processing, marketing, distribution and supply industries, it is estimated that fisheries and aquaculture support nearly 250 million livelihoods (Food and Agriculture Organization of the United Nations (FAO), 2018).

Benefits from marine and coastal ecosystems can be categorized in several ways. Traditionally, they have been understood in terms of goods (i.e., products, resources and harvests from nature with a market value), services (i.e., processes that sustain all forms of life but do not have a market value) and cultural benefits (i.e., spiritual and religious heritage, with no explicit market value). While goods have a direct use (consumptive) value, determined through market prices, services and cultural benefits have an indirect use (non-consumptive) value that can be determined through the application of a variety of valuation techniques (see figure).

How marine ecosystems generate economic benefits

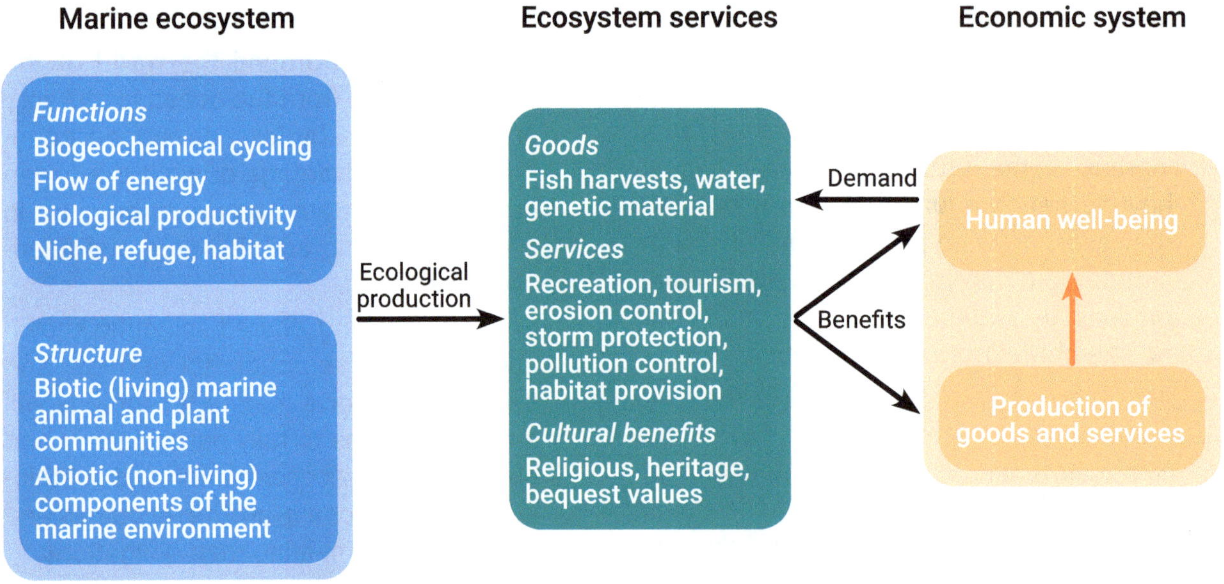

Note: The structure and functioning of marine ecosystems lead to the ecological production of ecosystem services. Some of the goods, services and cultural benefits have a direct impact on human well-being, whereas others indirectly affect the welfare of humans by supporting or protecting valuable economic assets and production activities.
Source: adapted from Barbier (2017).

In the seminal Millennium Ecosystem Assessment of 2005, a different classification was proposed for the benefits obtained by humans from ecosystems, otherwise known as ecosystem services. They are classified as provisioning, regulating, cultural and supporting services (the latter category needed for the existence of the others). Provisioning services, such as the food, fuel and fibre extracted from ecosystems, are similar to consumptive benefits and have a use value, while other services, such as regulating the climate, absorbing carbon dioxide, maintaining life cycles and landscapes and creating income and employment opportunities and cultural identity, are mostly immaterial (i.e., non-consumptive in nature, with a non-use value).

1.1. Provisioning services of marine and coastal ecosystems

The ocean provides a multitude of direct and indirect benefits of value to humans. The most direct benefit that marine and coastal ecosystems provide is through their primary productivity and the resulting products, such as fishes, plants, animals, fuel, timber (e.g., mangroves), biochemicals, natural medicines, pharmaceuticals, raw materials (sand and corals) and, to a lesser extent, fresh water and fibre. In 2016, 79.3 million tons of marine fishes[2] were caught, and 28.7 million tons of marine aquaculture species were farmed, supplying together an average of 14.6 kg of seafood per person on earth (FAO, 2018). Seafood is essential for food security: it provides more than 20 per cent of the average per capita animal protein intake for 3 billion people, and more than 50 per cent in some developing countries (FAO, 2018).

1.2. Regulating services of marine and coastal ecosystems

Oceans perform fundamental regulating services. They influence biologically mediated processes, such as carbon fixation and oxygen release, enabling climate mitigation and regulation. Similarly, coastal fringes perform a key role in sequestering carbon. Those services have an indirect use value for humans as they enable the maintenance of favourable and stable climate conditions (e.g., temperatures and precipitation) to which livelihood activities have adapted (e.g., crop cultivations), the preservation of human health, and infrastructure and other assets on which livelihoods depend. The role of coastal ecosystems in controlling pests and animal populations through trophic-dynamic relations and supporting pollination helps to keep at bay pests and diseases that can have an impact on cultivations, aquaculture activities and, potentially, human health.

Coastal ecosystems play an important role in the prevention of coastal erosion and can act as both shoreline stabilization and protection against storms, attenuating the strength of the waves and reducing the vulnerability of coastal settlements to sea surges and flooding events. For example, it was estimated that the Indian Ocean tsunami of 2004 caused greater damage to areas that had been converted to shrimp ponds and other uses than those where the mangrove had remained intact (FEnvironmental Justice Foundation, 2006) and that, overall, the thicker the mangrove fringes were, the greater the protection to economic activity that they offered (Hochard and others, 2019). Albeit to a lesser extent, coral reefs, seagrass beds and other vegetated coastal ecosystems can also have a significant impact in dissipating wave action and offering shoreline protection (Spalding and others, 2014) provided that they are in a healthy state themselves.

1.3. Supporting services of marine and coastal ecosystems

Photosynthesis occurring in marine and coastal ecosystems enables the conversion of solar energy into plants and animals and the maintenance of the net primary productivity of the ecosystems. Coastal ecosystems perform a key role in maintaining biodiversity and suitable reproductive habitats and nursery grounds for aquatic species. The ecological niches and refuge for wild animals and plants that they provide directly support the provisioning services of marine and coastal ecosystems. For example, seagrass beds in the Mediterranean are estimated to contribute 30 to 40 per cent of the value of commercial fisheries landings and approximately 29 per cent of recreational fisheries expenditure (Jackson and others,

---

2 This excludes aquatic mammals, crocodiles and related species, seaweed and other aquatic plants.

2015). Coastal ecosystems also act as pollution sinks, enable the storage and recycling of nutrients and support water cycling.

1.4. Cultural services and other social benefits of marine and coastal ecosystems

The aesthetic, cultural, religious and spiritual services from the ocean (cultural services) cover a wide range of practices. The services are essential to the maintenance and creation of social capital, education, cultural identity and traditions (human and social capital). Around the world, many beliefs and rituals are rich in references to the sea. Research on marine and coastal cultural ecosystem services is, however, still limited (Garcia Rodrigues and others, 2017; Blythe and others, 2020; Diaz and others, 2018).

Some cultural practices form integral parts of the traditional use of the ocean (such as ways of building boats or harvesting shellfish, and stone fish traps found across the coast of South-East Asia, Australia and the Pacific Islands). The diversity and technological sophistication of such structures attest to indigenous traditional knowledge of the ocean and its resources (Jeffery, 2013; Rowland and Ulm, 2011) Traditional watercraft such as the Hawaiian voyaging canoe *Hōkūle`a* provide an active platform for the restoration and maintenance of Pacific non-instrument navigation and cultural identity. Numerous other voyaging canoes have been constructed in the Pacific and, in many places, knowledge of traditional wayfinding has been preserved. Fautasi races in Samoa and dragon boat races in China merge history and cultural traditions with health, fitness and competition. People have long incorporated water-related activities as habitual or significant parts of their lives. Other non-consumptive ocean activities are swimming, diving, kayaking, surfing, sailing and wildlife viewing.

Finally, for many indigenous communities, fishing and the sharing of fishes form essential parts of traditional foodways, which support sociocultural cohesion and identity as well as linked ceremonial and cultural practices (Loring and others, 2019; Leong and others, 2020).

Other cultural activities represent ways of reacting to the ocean (such as dances to celebrate the ocean or religious practices to safeguard against danger on the ocean). Such practices can constitute an important part of the cultural heritage of a people. One example is the role of whale hunting for the indigenous peoples of the western seaboard of Canada and the United States of America, as discussed in the first *World Ocean Assessment*. In Washington State, United States, one tribe, the Makah, has been pursuing special authorization to resume some whale hunting since 2005. In November 2019, a hearing was held for the tribe's request and, in February 2020, a revised environmental impact assessment was published. The Makah fear that, without the special authorization, that particular element of their culture would remain connected to the past without any present reinforcement (A National Oceanic and Atmospheric Administration, 2015; 2020).

Heritage is also part of the cultural services provided by the ocean, with significant, though often unquantified, social and economic benefits (Firth, 2015). The iconic nature of underwater cultural heritage, such as historic shipwrecks, captures archaeological and historical information, revealing unique aspects of past human seafaring and behaviour, to be shared through museums, documentaries and public research. Shipwrecks can also yield valuable information about the sociocultural, historical, economic and political contexts on various scales of reference (local, regional or global) between the date of the vessel's construction (e.g., hull design, rig, materials used or purpose) and the reason for its eventual demise in the sea (e.g., warfare, piracy, privateering, intentional abandonment or natural weather events) (Gould, 1983). The remains of prehistoric and historic landscapes submerged by changing sea levels and the continuing destruction of important coastal sites by exposure and erosion are important reminders of climate change in the human past and of the impact of the climate crisis today (Harkin and others, 2020).

Wreck site tourism plays a role in the recreational diving industry. Services to memorialize vessel losses, such as wreath-laying ceremonies at submerged warship gravesites, are an expression of the deep connection to sacrifice at sea. The diversity of cultural services arising from shipwrecks and other historic structures in the sea is complemented by the role that underwater cultural heritage can play as artificial reef, providing habitats that are important for nature conservation, sea angling and commercial fishing, for example (Firth, 2018).

Finally, there is a sense of place engendered in onlookers by the ocean. The sense of openness and exposure to the elements can be very important to those who live by the sea or visit it as tourists. As discussed in chapter 8B on human health and the ocean, there is growing evidence that the sense of openness engendered by the ocean can improve human health. The ocean has also been an important source of inspiration to artists, composers and writers, often reflecting economically important aspects of society. Some studies reveal the deep emotional attachment of people to the marine environment (e.g., the Black Sea in Fletcher and others (2014) and the North Sea in Gee and Burkhard (2010)), as well as the importance of maintaining that relationship to preserve both nature and culture (Fletcher and others, 2014). However, despite progress to date, marine research and management have until recently largely neglected the critically important role of the sense of place, including how it influences the success and efficacy of management interventions (Van Putten and others, 2018; Hernandez and others, 2007).

Opportunities for income generation and employment opportunities, for education and recreation and for scientific and artistic information and inspiration are also part of the wider range of social benefits that marine and coastal ecosystems provide and upon which the well-being of populations, regardless of their distance from the shore, hinges directly and indirectly.

2. Benefits and their distribution

While some benefits from the ocean are very central and ensure the existence of life on earth, including the production of oxygen and the uptake of carbon dioxide and heat, most services are related to specific ecosystems or elements therein and are thus not evenly distributed. Moreover, not all States have the capacity to participate fully in and benefit from the ocean and its resources. That may be because they either do not have access to the ocean, such as landlocked States, or do not have the financial means to develop maritime industries, which is the case for many developing countries. Some States do not have the capacity for access to areas beyond national jurisdiction or even parts of their own exclusive economic zone. For example, in areas beyond national jurisdiction, the collection of marine genetic resources, their sequencing and potential commercialization are currently concentrated in a small number of countries (Blasiak and others, 2018; 2019; Harden-Davies, 2019; Levin and Baker, 2019).

One of the main provisioning services, living resources, is not only unevenly distributed, with productivity hotspots concentrated in the upwelling areas of the world (Kämpf and Chapman, 2016), but a very substantial proportion of capture fisheries is carried out by relatively few fishing vessels from few States. Vessels from 25 States took 42 per cent of the global catch in 2016 (FAO, 2018). Thus, profits are not necessarily going to the countries with the exclusive economic zone in which the fishes are produced. McCauley and others (2018) found that vessels flagged to higher-income nations, for example, are responsible for 97 per cent of trackable industrial fishing on the high seas and 78 per cent within the national waters of lower-income countries.

Economic assessments of the cultural benefits of ecosystem services are increasingly undertaken by applying environmental valuation methods to recreational uses such as tourism, marine recreational fishing, whale watching, and enjoying the seascape (Hanley and others, 2015; Aanesen and others, 2015; Spalding and others, 2017), as well as non-use values (i.e., existence and bequest values) of coral reefs and other marine biodiversity (Aanesen and others, 2015; Navrud and others, 2017). Tourism relies particularly on specific characteristics such as coral reefs (Brander and others, 2007) and specific activities such as cruise tourism, and are concentrated in certain areas such as the Caribbean and the Mediterranean but increasingly in polar areas too (see chap. 8A).

The International Seabed Authority has been established as the organization through which States parties to the United Nations Convention on the Law of the Sea organize and control activities in the Area (i.e., all exploration and exploitation activities for the mineral resources of the seabed and ocean floor and subsoil thereof beyond the limits of national jurisdiction) for the benefit of humankind as a whole and to provide for the equitable sharing of financial and other economic benefits from activities in the Area (see art. 140 of the Convention). However, in addition to economic benefits from deep seabed mining, the benefits from leaving ecosystems intact should also be considered in the context of article 140, thus integrating redistribution (international solidarity) with ecological preservation (intergenerational solidarity) (Tladi, 2015; Feichtner, 2019).

Specific revenue-sharing obligations are also included in article 82 of the Convention, which provides for a system of payments or contributions in kind by coastal States with regard to exploiting the non-living resources of the continental shelf beyond 200 nautical miles. Such payments or contributions are to be made through the International Seabed Authority for distribution to States parties to the Convention, on the basis of equitable sharing criteria (Spicer and McIsaac, 2016).

3. Disbenefits to humans

In its conceptual framework, the Intergovernmental Science-Policy Platform on Biodiversity and Ecosystem Services defines nature's contributions to people as all the positive contributions or benefits (including ecosystem services) and occasionally the negative contributions, losses or detriments that people obtain from nature (Pascual and others, 2017). The negative contributions of nature, which in some fields have become known as ecosystem disservices, are beginning to be incorporated into ecosystem service assessments (Campagne and others, 2018).

Ecosystem disservices or disbenefits are functions or properties of ecosystems that are unpleasant or cause harm, by either reducing ecosystem services or directly affecting humans (Lyytimäki, 2014; Shackleton and others, 2016). Direct effects on humans are, for example, caused by floods and storm surges, which can lead to economic losses exceeding $100 billion annually (Kousky, 2014). Direct harm to humans can also come through seafood-borne diseases mainly caused by harmful algal blooms, such as amnesic shellfish poisoning, paralytic shellfish poisoning, diarrhetic shellfish poisoning, neurotoxic shellfish poisoning and ciguatera fish poisoning. In addition to the negative impact on human well-being, the diseases also cause economic losses owing to hospitalization expenses and lost productivity (Sanseverino and others, 2016). Harmful algal blooms can also cause losses in terms of fisheries and aquaculture production. Natural sedimentation can negatively affect human activities, including shipping.

Despite recognition of the occurrence of negative contributions from nature, the increase in those occurrences and the magnification of such events are, most of the time, related to anthropogenic activities and pressures. For instance, coastal flooding normally affects human settlements misallocated in low and susceptible coastal areas. Likewise, some algal blooms are attributable to contaminants from human activities.

4. Threats to ocean ecosystem services

Human activities are directly or indirectly affecting ecosystem services and can thus reduce or erase benefits that would otherwise be provided. Those threats are the pressures that are detailed in chapters 9–25. As human activities in the marine environment are expected to increase in the future, in particular in areas beyond national jurisdiction, they will not only exert growing pressure on natural resources, but may also threaten marine biodiversity and thus the benefits that people obtain from ecosystem services (Altvater and others, 2019). Relatively little is understood about how social and ecological processes interact to determine marine ecosystem benefits (Outeiro and others, 2017). While the co-production process can sustain desirable ecosystem service flows, it can also produce trade-offs that constrain flows of ecosystem services or exacerbate the provision of disservices, with negative impacts on human well-being at various levels (Pope and others, 2016). Those impacts, which can be categorized as extractive threats (e.g., fishing, mining, offshore hydrocarbon exploration and extraction, offshore and marine renewable energy installation and mangrove exploitation) and non-extractive threats (e.g., ocean warming and acidification, eutrophication, pollution and habitat destruction and conversion), interact, often with compounded effects (McCauley and others, 2015; Sumaila and others, 2016; Simas and others, 2015; O'Hagan and others, 2015; Greaves and others, 2016).

5. Safeguarding ocean benefits through regional and international cooperation and improved implementation of international law as reflected in the United Nations Convention on the Law of the Sea

5.1. United Nations Convention on the Law of the Sea, its implementing agreements and related instruments

The United Nations Convention on the Law of the Sea, which sets out the legal framework within which all activities in the oceans and seas must be carried out, plays a crucial role in the conservation and sustainable use of the ocean and its resources and in safeguarding the many ecosystem services that the ocean provides for both current and future generations.

The integration of environmental, social and economic dimensions is at the core of the Convention, which establishes a delicate balance between the need for economic and social development through the use of the ocean and its resources and the need to conserve and manage those resources in a sustainable manner and to protect and preserve the marine environment. It also provides a framework for international cooperation in the conservation and sustainable use of the ocean and its resources, which can take place through intergovernmental institutions or bilaterally among States (United Nations, 2017b).

The Convention is a factor of stability, peace and progress and holds special importance in a difficult international context. It provides a solid basis and the legal tools for effectively dealing with issues related to sustainable development and with new challenges raised in that field. Full respect for and implementation of its provisions, in particular the duty to protect and preserve the marine environment and to cooperate therefor, is of pivotal importance in this respect.

The integrated approach to ocean management as reflected in the Convention is essential for promoting sustainable development, given that sectoral and fragmented approaches lack coherence and may lead to solutions that have a limited impact on the conservation and sustainable use of the ocean and its resources. At the international level, it is important that the integrated approach guides the regulatory work and capacity-building activities of international organizations in the framework of their competences and that such organizations effectively respond to the increasing need for coordination and cross-sectoral cooperation. At the same time, at the national level, the integrated approach requires that a comprehensive legal framework for ocean matters be put in place and that institutional mechanisms enabling inter-agency cooperation be set up and improved.

The Convention is, in many fields, supplemented by more specific, sectoral instruments. In addition to its two implementing agreements, the 1994 Agreement relating to the Implementation of Part XI of the Convention[3] and the 1995 Agreement for the Implementation of the Provisions of the Convention relating to the Conservation and Management of Straddling Fish Stocks and Highly Migratory Fish Stocks,[4] there are numerous international legal instruments at the global and regional levels that

[3] United Nations, *Treaty Series*, vol. 1836, No. 31364.
[4] Ibid., vol. 2167, No. 37924.

cover many aspects of ocean use, including the conservation and sustainable use of oceans and their resources.

These instruments include global treaties relating to sustainable fisheries management (including the FAO Agreement on Port State Measures to Prevent, Deter and Eliminate Illegal, Unreported and Unregulated Fishing of 2009), pollution from ships, maritime safety, atmospheric pollution, the release of hazardous substances into the environment, the protection of certain species or habitats, the conservation and sustainable use of biodiversity, the working conditions of seafarers, fishers and other maritime workers and the protection of underwater cultural heritage. Several regional treaties are also included in the framework, including those that establish regional fisheries management organizations and arrangements, as well as regional seas conventions and action plans. In addition, a number of soft law instruments also address relevant issues, including technical guidelines on fisheries, such as the FAO International Guidelines for the Management of Deep-Sea Fisheries in the High Seas, the FAO Code of Conduct for Responsible Fisheries and the International Oceanographic Commission guide *Marine Spatial Planning: a Step-by-Step Approach toward Ecosystem-based Management*. While the guidelines are universal, they highlight best practices and regional specificities and, therefore, support individual countries in the implementation of global ocean agendas. Components of a working system for global ocean management are also supported by soft law mechanisms that provide guidance for international action, such as the Rio Declaration on Environment and Development[5] and the 2030 Agenda for Sustainable Development[6] and its Sustainable Development Goals, in particular Goal 14 on life below water.

The effective conservation and sustainable use of the ocean and its resources will be achieved only with the full and effective implementation of that body of international law. Actions and efforts should be focused primarily on implementation gaps. All States are challenged by the implementation of such a comprehensive legal framework, in particular, developing countries. Many small island developing States and least developed countries lack the detailed knowledge and skilled workforce needed for ocean management, in particular in the light of their limited resources and capacity compared with the large ocean areas under their jurisdiction. Capacity and technologies for planning and managing land-based activities that have impacts on coastal and marine environments, as well as activities that occur in coastal and marine environments, will ensure that economic benefits can be maximized in an environmentally sustainable manner.

It was noted in the first Assessment that capacity-building, shared scientific knowledge and the transfer of marine technology, taking into account the Criteria and Guidelines on the Transfer of Marine Technology of the Intergovernmental Oceanographic Commission, would empower States to fully participate in and benefit from the conservation and sustainable use of the ocean and its resources and assist them in meeting their obligations (United Nations, 2017a).

The situation has not changed drastically since then. Human, institutional and systemic capacities, as well as financing, continue to be the primary limiting factors, in particular for developing countries. Resource capacity, including financial capacity, remains a significant constraint in relation to the protection and preservation of the marine environment

5 Report of the United Nations Conference on Environment and Development, Rio de Janeiro, 3–14 June 1992, vol. I, Resolutions Adopted by the Conference (United Nations publication, Sales No. E.93.I.8 and corrigendum), resolution 1, annex I.

6 See General Assembly resolution 70/1.

and marine scientific research, while technology constraints are often an impediment to effective implementation of a State's obligations (United Nations, 2017b; see also chap. 27).

Gaps also exist with regard to the material or geographical scope of relevant instruments. For example, while some aspects of marine debris, plastics and microplastics are covered by global, regional and national instruments, none, other than some regional action plans on marine litter and sector-specific measures such as annex V to the International Convention for the Prevention of Pollution from Ships of 1973, are specifically dedicated to those issues. At the same time, while there is widespread coverage by regional instruments relevant to the implementation of aspects of the United Nations Convention on the Law of the Sea and the Fish Stocks Agreement (United Nations, 2017b), some gaps remain.

Specific challenges are encountered in the enforcement of effective management measures in areas beyond national jurisdiction, primarily owing to a lack of cross-sectoral coordination, but also to regulatory gaps (Altvater and others, 2019; see also chap. 27). These issues are currently being discussed at the United Nations in the context of the intergovernmental negotiations on the elaboration of a legally binding instrument under the United Nations Convention on the Law of the Sea on the conservation and sustainable use of marine biodiversity in areas beyond national jurisdiction.

Sustainable Development Goal 14 can be a strong driver for strengthening ocean governance and enhancing policy coherence, while providing an impetus for collective global accountability for the oceans under the 2030 Agenda. In target 14.c, States pledged that they would "enhance the conservation and sustainable use of oceans and their resources by implementing international law" as reflected in the Convention.[7] Increasing participation in international instruments and addressing the challenges of implementation, including resource and capacity constraints, strengthening intersectoral cooperation, coordination and information-sharing at all levels and developing new instruments to address emerging challenges in a timely fashion will be key elements in accelerating the implementation of the target (United Nations, 2019).

5.2. Third implementing agreement of the United Nations Convention on the Law of the Sea on the conservation and sustainable use of marine biodiversity in areas beyond national jurisdiction, currently under consideration

Efforts to strengthen the international legal framework through the elaboration of new instruments include, in particular, the intergovernmental conference convened by the General Assembly to elaborate the text of an international legally binding instrument on the conservation and sustainable use of marine biological diversity of areas beyond national jurisdiction. More specifically, following a decade of work in the framework of a working group

7 See General Assembly resolution 71/313, annex. In the indicator to monitor progress against target 14.c, indicator 14.c.1, the Assembly called for an assessment of the number of countries making progress in ratifying, accepting and implementing through legal, policy and institutional frameworks, ocean-related instruments that implement international law, as reflected in the United Nations Convention on the Law of the Sea, for the conservation and sustainable use of the oceans and their resources. Recently, a new methodology for the measurement of such progress has been developed. Data to be collected on the basis of the approved methodology will provide, for the first time, a baseline of the current state of implementation of the Convention and its implementing agreements with regard to the conservation and sustainable use of the oceans and their resources. See also Division for Ocean Affairs and the Law of the Sea, Information note: development of a methodology for Sustainable Development Goal indicator 14.c.1, 4 October 2019.

and a Preparatory Committee, the Assembly decided in its resolution 72/249 of 24 December 2017 to convene an intergovernmental conference, under the auspices of the United Nations, to consider the recommendations of the Preparatory Committee established by its resolution 69/292 of 19 June 2015 on the elements and to elaborate the text of an international legally binding instrument under the United Nations Convention of the Law of the Sea on the conservation and sustainable use of marine biological diversity of areas beyond national jurisdiction, with a view to developing the instrument as soon as possible.

The conference convened three substantive sessions from 2018 to April 2020 to address the topic identified in the package agreed upon in 2011, namely the conservation and sustainable use of marine biological diversity of areas beyond national jurisdiction, in particular, together and as a whole, marine genetic resources, including questions of the sharing of benefits, measures such as area-based management tools, including marine protected areas, environmental impact assessments, capacity-building and the transfer of marine technology. Negotiations are at a critical juncture. Regrettably, however, in accordance with General Assembly decision 74/543 of 11 March 2020, the fourth session, initially scheduled for March and April 2020, was postponed owing to the COVID-19 pandemic.

References

Aanesen, M., and others (2015). Willingness to pay for unfamiliar public goods: preserving cold-water corals in Norway. *Ecological Economics*, vol. 112, pp. 53–67.

Altvater, Susanne, and others (2019). The need for marine spatial planning in areas beyond national jurisdiction. In *Maritime Spatial Planning: Past, Present, Future*, Jacek Zaucha and Kira Gee, eds., pp. 397–415. Cham, Switzerland: Springer International Publishing. https://doi.org/10.1007/978-3-319-98696-8_17.

Barbier, Edward B. (2017). Marine ecosystem services. *Current Biology*, vol. 27, No. 11, pp. R507–R510. https://doi.org/10.1016/j.cub.2017.03.020.

Blasiak, Robert, and others (2018). Corporate control and global governance of marine genetic resources. *Science Advances*, vol. 4, No. 6.doi: 10.1126/sciadv.aar5237.

Blasiak, Robert, and others (2019). Scientists should disclose origin in marine gene patents. *Trends in Ecology and Evolution*, vol. 34, No. 5, pp. 392–395.

Blythe, J., and others (2020). Frontiers in coastal well-being and ecosystem services research: a systematic review. *Ocean and Coastal Management*, vol. 185, p. 105028. www.sciencedirect.com/science/article/abs/pii/S0964569119304703.

Brander, L., and others (2007). The recreational value of coral reefs: a meta-analysis. *Ecological Economics*, vol. 63, No. 1, pp. 209–218.

Campagne, Carole Sylvie, and others (2018). Looking into Pandora's box: ecosystem disservices assessment and correlations with ecosystem services. *Ecosystem Services*, vol. 30, pp. 126–136.

Diaz, Sandra, and others (2018). Assessing nature's contributions to people: recognizing culture, and diverse sources of knowledge, can improve assessments. *Science*, vol. 359, No. 6373.

Environmental Justice Foundation (2006). *Mangroves: Nature's Defence against Tsunamis: A Report on the Impact of Mangrove Loss and Shrimp Farm Development on Coastal Defences*. London: Environmental Justice Foundation.

Feichtner, Isabel (2019). Sharing the riches of the sea: the redistributive and fiscal dimension of deep seabed exploitation. *European Journal of International Law*, vol. 30, No. 2, pp. 601–633.

Firth, Antony (2015). *The Social and Economic Benefits of Marine and Maritime Cultural Heritage*. London: Fjordr Ltd., for the Honor Frost Foundation. ISBN 978-0-9933832-1-2.

________ (2018). *Managing Shipwrecks*. London: Honor Frost Foundation. ISBN 978-0-9933832-3-6.

Fletcher, Ruth, and others (2014). Revealing marine cultural ecosystem services in the Black Sea. *Marine Policy*, vol. 50, pp. 151–161.

Food and Agriculture Organization of the United Nations (FAO) (2018). *The State of World Fisheries and Aquaculture 2018: Meeting the Sustainable Development Goals*. Rome.

Garcia Rodrigues, João, and others (2017). Marine and coastal cultural ecosystem services: knowledge gaps and research priorities. *One Ecosystem*, vol. 2, p. e12290. https://doi.org/10.3897/oneeco.2.e12290.

Gee, Kira, and Benjamin Burkhard (2010). Cultural ecosystem services in the context of offshore wind farming: a case study from the west coast of Schleswig-Holstein. *Ecological Complexity*, vol. 7, No. 3, pp. 349–358.

Gould, Richard A. (1983). Looking below the surface: shipwreck archaeology as anthropology. In *Shipwreck Anthropology*, pp. 3–22.

Greaves, D., and others (2016). Environmental impact assessment: gathering experience at wave energy test centres in Europe. *International Journal for Marine Energy*, http://doi.org/10.1016/j.ijome.2016.02.003.

Hanley, N., and others (2015). Economic valuation of marine and coastal ecosystems: is it currently fit for purpose? *Journal of Ocean and Coastal Economics*, vol. 2, No. 1, art. 1. https://doi.org/10.15351/2373-8456.1014.

Harden-Davies, Harriet R. (2019). Research for regions: strengthening marine technology transfer for Pacific Island countries and biodiversity beyond national jurisdiction. In *Conserving Biodiversity in Areas beyond National Jurisdiction*, David Freestone, ed., pp. 298–323. Leiden, The Netherlands: Brill-Nijhoff.

Harkin, K., and others (2020). Impacts of climate change on cultural heritage. *Marine Climate Change Impacts Partnership Science Review*, vol. 16, pp. 24–39.

Hernandez B., and others (2007). Place attachment and place identity in natives and non-natives. *Journal of Environmental Psychology*, vol. 27, pp. 310–319.

Hochard, Jacob P., and others (2019). Mangroves shelter coastal economic activity from cyclones. *Proceedings of the National Academy of Sciences of the United States of America*, vol. 116, No. 25, pp. 12232–12237.

Jackson, Emma L., and others (2015). Use of a seagrass residency index to apportion commercial fishery landing values and recreation fisheries expenditure to seagrass habitat service. *Conservation Biology*, vol. 29, No. 3, pp. 899–909. https://doi.org/10.1111/cobi.12436.

Jeffery, Bill (2013). Reviving community spirit: furthering the sustainable, historical and economic role of fish weirs and traps. *Journal of Maritime Archaeology*, vol. 8, No. 1, pp. 29–57.

________ (2019). Global observing needs in the deep ocean. *Frontiers in Marine Science*, vol. 6, art. 241.

Kämpf, J., and P. Chapman (2016). *Upwelling Systems of the World: A Scientific Journey to the Most Productive Marine Ecosystems*. Cham, Switzerland: Springer. https://doi.org/10.1007/978-3-319-42524-5.

Kousky, Carolyn (2014). Informing climate adaptation: a review of the economic costs of natural disasters. *Energy Economics*, vol. 46, No. C, pp. 576–592.

Leong, K.M., and others (2020). Beyond Recreation: When Fishing Motivations Are More than Sport or Pleasure. NOAA Administrative Report, No. H-20-05. https://doi.org/10.25923/k5hk-x319.

Levin, L.A., and Maria Baker (2019). Grand challenge from the deep: opinion. *Ecomagazine*.

Loring, P.A., and others (2019). Fish and food security in small-scale fisheries. In *Transdisciplinarity for Small-Scale Fisheries Governance*, pp. 55–73. Springer. https://link.springer.com/chapter/10.1007/978-3-319-94938-3_4.

Lyytimäki, Jari (2014). Bad nature: newspaper representations of ecosystem disservices. *Urban Forestry and Urban Greening*, vol. 13, No. 3, pp. 418–424.

McCauley, Douglas J., and others (2015). Marine defaunation: animal loss in the global ocean. *Science*, vol. 347, No. 6219, p. 1255641.

McCauley, Douglas J., and others (2018). Wealthy countries dominate industrial fishing. *Science Advances*, vol. 4, No. 8, https://doi.org/10.1126/sciadv.aau2161.

Mehvar, Seyedabdolhossein, and others (2018). Quantifying economic value of coastal ecosystem services: a review. *Journal of Marine Science and Engineering*, vol. 6, No. 1, p. 5.

Millennium Ecosystem Assessment (2005). *Ecosystems and Human Well-Being: Synthesis*. Washington, D.C.: Island Press.

Mohammed, Essam Yassin (2012). *Payments for Coastal and Marine Ecosystem Services: Prospects and Principles*. London: International Institute for Environment and Development.

National Oceanic and Atmospheric Administration (2015). *Draft Environmental Impact Statement on the Mikah Tribe Request to Hunt Grey Whales*. Silver Spring, Maryland, United States of America: NOAA.

________ (2020). *Makah Tribal Whale Hunt*. www.fisheries.noaa.gov/west-coast/makah-tribal-whale-hunt-chronology.

Navrud, S., and others (2017). Chapter 5: Valuing marine ecosystem services loss from oil spills for use in cost-benefit analysis of preventive measures, pp. 124–137. In *Handbook on the Economics and Management of Sustainable Oceans*, P. Nunes, and others, eds. Cheltenham, United Kingdom: Edward Elgar Publishing.

O'Hagan, A.M., and others (2015). Wave energy in Europe: views on experiences and progress to date, *International Journal for Marine Energy*, http://doi.org/10.1016/j.ijome.2015.09.001.

Outeiro, Luis, and others (2017). The role of non-natural capital in the co-production of marine ecosystem services. *International Journal of Biodiversity Science, Ecosystem Services and Management*, vol. 13, No. 3, pp. 35–50.

Pascual, Unai, and others (2017). Valuing nature's contributions to people: the IPBES approach. *Current Opinion in Environmental Sustainability*, vol. 26, pp. 7–16.

Pope, Kevin L., and others (2016). Fishing for ecosystem services. *Journal of Environmental Management*, vol. 183, pp. 408–417.

Rowland, Michael J., and Sean Ulm (2011). Indigenous fish traps and weirs of Queensland. *Queensland Archaeological Research*, vol. 14, pp. 1–58.

Sanseverino, Isabella, and others (2016). *Algal Bloom and Its Economic Impact*. EUR 27905 EN. http://doi.org/10.2788/660478.

Shackleton, Charlie M., and others (2016). Unpacking Pandora's box: understanding and categorising ecosystem disservices for environmental management and human wellbeing. *Ecosystems*, vol. 19, No. 4, pp. 587–600.

Simas, T., and others (2014), Review of consenting processes for ocean energy in selected European Union member States, *International Journal for Marine Energy*, vol. 9, No. 2015, pp. 41–59.

Spalding, Mark, and others (2014). The role of ecosystems in coastal protection: adapting to climate change and coastal hazards. *Ocean and Coastal Management*, vol. 90, pp. 50–57.

Spalding, Mark, and others (2017). Mapping the global value and distribution of coral reef tourism. *Marine Policy*, vol. 82, pp. 104–113.

Spicer, W., and E. McIsaac (2016). *A Study of Key Terms in Article 82 of the United Nations Convention on the Law of the Sea*, International Seabed Authority technical study, No. 5.

Sumaila, U. Rashid, and others (2016). Fishing for the future: an overview of challenges and opportunities. *Marine Policy*, vol. 69, pp. 173–180.

Tladi, D. (2015). The Common Heritage of Mankind and the Proposed Treaty on Biodiversity in Areas beyond National Jurisdiction: The Choice between Pragmatism and Sustainability, *Yearbook of International Environmental Law*, vol. 25, No. 1, p. 113.

United Nations (2017a). Chapter 23: Offshore mining industries. In *The First Global Integrated Marine Assessment: World Ocean Assessment I*. Cambridge: Cambridge University Press Cambridge.

________ (2017b). Concept Paper: Partnership Dialogue 7: Enhancing the Conservation and Sustainable Use of the Oceans and Their Resources by Implementing International Law as Reflected in the United Nations on the Law of the Sea. The Ocean Conference 2017. https://sustainabledevelopment.un.org/content/documents/14402Partnershipdialogue7.pdf.

________ (2017c). The Ocean and Sustainable Development Goals under the 2030 Agenda for Sustainable Development. A Technical Abstract of the First Global Integrated Marine Assessment. New York.

________ (2019). Preparatory process of the 2020 United Nations Conference to Support the Implementation of Sustainable Development Goal 14: Conserve and sustainably use the oceans, seas and marine resources for sustainable development. Note by the Secretary-General, A/74/630, 24 December 2019.

Van Putten, I.E., and others (2018). A framework for incorporating sense of place into the management of marine systems. *Ecology and Society*, vol. 23, No. 4, pp. 42–65. www.jstor.org/stable/26796884?seq=1#metadata_info_tab_contents.

Whitmarsh, David (2011). *Economic Management of Marine Living Resources: A Practical Introduction*. Routledge.

Annexes

Chapter 1

Maria João Bebianno, Hilconida Calumpong, Sanae Chiba, Karen Evans, Carlos Garcia-Soto, Osman Keh Kamara, Enrique Marschoff, Essam Yassin Mohammed, Henn Ojaveer, Chul Park, Ylenia Randrianarisoa, Renison Ruwa, Jörn Schmidt, Alan Simcock, Anastasia Strati, Joshua T. Tuhumwire, Ca Thanh Vu, Juying Wang and Tymon Zielinski (Group of Experts of the Regular Process for Global Reporting and Assessment of the State of the Marine Environment, including Socioeconomic Aspects).

Chapter 2

Maria João Bebianno, Hilconida Calumpong, Sanae Chiba, Karen Evans, Carlos Garcia-Soto, Osman Keh Kamara, Enrique Marschoff, Essam Yassin Mohammed, Henn Ojaveer, Chul Park, Ylenia Randrianarisoa, Renison Ruwa, Jörn Schmidt, Alan Simcock, Anastasia Strati, Joshua T. Tuhumwire, Ca Thanh Vu, Juying Wang and Tymon Zielinski (Group of Experts of the Regular Process for Global Reporting and Assessment of the State of the Marine Environment, including Socioeconomic Aspects).

Chapter 3

Bing Qiao (convener), Carlos Francisco Andrade, Paulo Antunes Horta, Nene Bi Trace Boniface, Sanae Chiba (co-lead member), Mohammad Zahedur Rahman Chowdhury, Antonio Di Natale, Karen Evans (co-lead member), Carlos Garcia-Soto (co-lead member), Enrique Marschoff (co-lead member), Colin Moffat, Jocelyne Mpemba Kazadi, Henn Ojaveer, Renison Ruwa (lead member), Jörn Schmidt (co-lead member), Hoinsoude Segniagbeto, Sekou Tidiane Bangoura, Kedong Yin, Chang-Ik Zhang and Tymon Zielinski (co-lead member).

Chapter 4

Chang-Ik Zhang (convener), Karen Evans (co-lead member), Andrew F. Johnson, Osman Keh Kamara (co-lead member), Ben S. Malayang, Renison Ruwa (lead member), Jörn Schmidt (co-lead member) and Thomas W. Therriault.

Chapter 5

Carlos Garcia-Soto (convener and lead member), Levke Caesar, Anny Cazenave, Lijing Cheng, Alicia Cheripka, Paul Durack, Karen Evans (co-lead member), David Halpern, Libby Jewett, Sung Yong Kim, Guancheng Li, Ignatius Rigor, Sunke Schmidtko, Juying Wang (co-lead member) and Tymon Zielinski (co-lead member).

Chapter 6

Chul Park (chapter lead member).

Chapter 6A

Thomas Malone (convener), Maurizio Azzaro, Russell Hopcroft, Chul Park (chapter lead member), Kazuaki Tadokoro, Michael Thorndyke and Sinjae Yoo.

Chapter 6B

Lis Lindal Jørgensen (convener), Christos Arvanitidis, Silvana N.R. Birchenough, Malcolm R. Clark, Igor Cristino Silva Cruz, Marina Cunha, Alan Deidun, Judith Gobin, Maruf Hossain, Ana C.M. de Jesus, Carmen Mifsud, Khac Bat Nguyen, Chul Park (chapter lead member), Rachel Przeslawski, Jake Rice, Lennert Schepers, Paul Snelgrove, Natalia Strelkova and Leen Vandepitte.

Chapter 6C

Thomas J. Webb (convener), Fernanda de Oliveira Lana, Maria José Juan-Jordá, Hiroyuki Motomura, Francisco Navarrete-Mier, Khac Bat Nguyen, Henn Ojaveer (subchapter lead member), Hazel Oxenford, Chul Park (chapter lead member), Clive Roberts, V.N. Sanjeevan, Mudjekeewis D. Santos, Tracey Sutton, Michael Thorndyke and Burcu Bilgin Topcu.

Chapter 6D

David Lusseau (convener), Luciano Dalla Rosa, Karen Evans (subchapter lead member), Iryna Makarenko, André Silva Barreto, Mette Skern-Mauritzen, Chul Park (chapter lead member) and Marta Soeffker.

Chapter 6E

Qamar Schuyler (convener), Karen Evans (subchapter lead member), Maximillian Hirschfeld, Carmen Mifsud, Chul Park (chapter lead member), Gabriel Hoinsoude Segniagbeto, André Silva Barreto and Vinay Udyawer.

Chapter 6F

Martin Cryer (convener), Igor Debski, Maria Dias, Karen Evans (subchapter lead member), Carolina Hazin, Chul Park (chapter lead member), Cleo Small, Graeme Taylor and Ross McLeod Wanless.

Chapter 6G

Hilconida Calumpong (joint convener and co-lead member), Hugh Kirkman (joint convener), Nair Sumie Yokoya (joint convener), Paula Bontempi, Sanae Chiba (co-lead member), Phillip Da Silva, Jason M. Hall-Spencer, Honghui Huang, Carmen Mifsud, Nahid Abdel Rahim Osman, Chul Park (chapter lead member), Franciane Pellizzari, Elizabeth Sinclair, Mario Soares, John A. West and Isabel Sousa Pinto.

Chapter 7

Hilconida Calumpong (chapter lead member).

Chapter 7A

Julia Sigwart (convener), Hilconida Calumpong (chapter lead member), Ronaldo Christofoletti, Karen Evans (subchapter lead member), Judith Gobin and Patricia Miloslavich.

Chapter 7B

Ronaldo Christofoletti (joint convener), Judith Gobin (joint convener), Mohammed Abdallah, Hilconida Calumpong (chapter lead member), Frédéric Guichard, Fahad Ibrahim, Sergiy Medinets, Anna Metaxas, José Souto Rosa Filho, Evangelina Schwindt, Julia Sigwart and Nicole Webster.

Chapter 7C

Colin D. Woodroffe (convener), Hilconida Calumpong (chapter lead member), Fernanda de Oliveira Lana, Karen Evans (subchapter lead member), David Obura and Arthur P. Webb.

Chapter 7D

Ian Butler (convener), Hilconida Calumpong (chapter lead member), Karen Evans (subchapter lead member), Hazel Oxenford, Kemraj Parsram, Alex Rogers, Raquel Silva Peixoto and Hiroya Yamano.

Chapter 7E

Erik Cordes (convener), Hilconida Calumpong (chapter lead member), Malcolm R. Clark, Karen Evans (subchapter lead member), Sebastian Hennige and Georgios Kazanidis.

Chapter 7F

Colin D. Woodroffe (convener), Hilconida Calumpong (chapter lead member), Ronaldo Christofoletti, Dana E. Hunt, José H. Muelbert, Pablo Muniz, Bing Qiao, Moriaki Yasuhara and Tymon Zielinski (subchapter lead member).

Chapter 7G

Hugh Kirkman (convener), Hilconida Calumpong (chapter lead member), Sanae Chiba (co-lead member), Kiho Kim, Paul Lavery, Nahid Abdel Rahim Osman, Elizabeth Sinclair and Konstantinos Topouzelis.

Chapter 7H

José Souto Rosa Filho (convener), Hilconida Calumpong (chapter lead member), Sanae Chiba (co-lead member), Mohammad Zahedur Rahman Chowdhury, Phillip Da Silva, Seon Hamer, Nahid Abdel Rahim Osman, Mario Soares and Colin D. Woodroffe.

Chapter 7I

Judith S. Weis (convener), Hilconida Calumpong (chapter lead member), Oscar O. Iribarne, Luis M. Pinheiro, Katherine E.A. Segarra and Alan Simcock (subchapter lead member).

Chapter 7J

Lisa A. Levin (convener), Peter Auster, Hilconida Calumpong (chapter lead member), Malcolm R. Clark, Jason M. Hall-Spencer, Russell Hopcroft, Jeroen Ingels, Anna Metaxas, J. Murray Roberts, Bhavani E. Narayanaswamy, Joshua T. Tuhumwire (subchapter lead member) and Moriaki Yasuhara.

Chapter 7K

Grant R. Bigg (convener), Maurizio Azzaro, Hilconida Calumpong (chapter lead member), Karen Evans (subchapter lead member), Huw Griffiths and Moriaki Yasuhara.

Chapter 7L

Malcolm R. Clark (convener), Angelo F. Bernardino, Hilconida Calumpong (chapter lead member), Jason M. Hall-Spencer, J. Murray Roberts, Bhavani E. Narayanaswamy, Paul Snelgrove and Joshua T. Tuhumwire (subchapter lead member).

Chapter 7M

Jeroen Ingels (convener), Diva Amon, Angelo F. Bernardino, Punyasloke Bhadury, Holly Bik, Hilconida Calumpong (chapter lead member), Malcolm R. Clark, Thomas Dahlgren, Daniel O.B. Jones, Craig McClain, Clifton Nunnally, Paul Snelgrove, Joshua T. Tuhumwire (subchapter lead member) and Moriaki Yasuhara.

Chapter 7N

Peter Croot (convener), Hilconida Calumpong (chapter lead member), Fernanda de Oliveira Lana, Osman Keh Kamara (co-lead member), Joseph Montoya, Tracey T. Sutton, Michael Vecchione and Tymon Zielinski (co-lead member).

Chapter 7O

Ana Colaço (convener), Angelika Brandt, Hilconida Calumpong (chapter lead member), Ana Hilario, Tomo Kitahashi, Nuno Lourenço, Bhavani E. Narayanaswamy, Imants George Priede, Joshua T. Tuhumwire (subchapter lead member), Michael Vecchione and Hiromi Watanabe.

Chapter 7P

Nadine Le Bris (convener), Hilconida Calumpong (chapter lead member), Sanae Chiba (subchapter lead member), Ana Colaço, Elva Escobar, Anna Metaxas, Paraskevi Nomikou, Julia Sigwart, Verena Tunnicliffe and Hiromi Watanabe.

Chapter 7Q

Howard S.J. Roe (convener), Hilconida Calumpong (chapter lead member), David Freestone, Laurence Kell, Brian E. Luckhurst, Chul Park (co-lead member) and Tammy Warren.

Chapter 8A

Alan Simcock (convener and chapter lead member), Austin Becker, Marcelo Bertellotti, Joan Bondareff, Robert Boysen, Anthony Charles, Leandra Gonçalves, Miguel Iñíguez, Osman Keh Kamara (co-lead member), Paula Keener, Jenna Lamphere, Candace May, Angeliki N. Menegaki, Ishmael Mensah, Essam Yassin Mohammed (co-lead member), Tanya O'Gara, Christina Pita, Jean Edmond Randrianantenaina, Maria Sahib, Regina Salvador, Anastasia Strati (co-lead member), Jean-Claude Tibe and Gregory Wetterau.

Chapter 8B

Michael Moore (convener), Martin Edwards, Bella S. Galil, Osman Keh Kamara (co-lead member), Essam Yassin Mohammed (co-lead member), Alan Simcock (lead member), Anastasia Strati (co-lead member) and Dick Vethaak.

Chapter 9

Carlos Garcia-Soto (convener and lead member), Denise Breitburg, Monica Campillos, Patricia Castillo-Briceno, Sanae Chiba (co-lead member), Matthew Collins, Ganix Esnaola, Karen Evans (co-lead member), Louise B. Firth, Thomas Frölicher, Jason M. Hall-Spencer, David Halpern, Karen L. Hunter, Gabriel Ibarra, Sung-Yong Kim, Roxy M. Koll, Kathleen McInnes, Jon Saenz, Ca Thanh Vu (co-lead member), Bess Ward and Tymon Zielinski (co-lead member).

Chapter 10

Thomas Malone (convener), Archis Ambulker, Maria João Bebianno (co-lead member), Paula Bontempi, Michael Krom, Harri Kuosa, Joseph Montoya, Alice Newton, Yapo Ossey, João Sarkis Yunes, Walker Smith, Lars Sonesten, Georgios Sylaios, Juying Wang (lead member) and Kedong Yin.

Chapter 11

Ralf Ebinghous (convener: pharmaceuticals and personal care products), Bjørn Einar Grøsvik (convener: hydrocarbons), Ida-Maja Hassellöv (convener: ships), Colin F. Moffat (convevner: persistent organic pollutants), Alan Simcock (convener: radioactivity; and co-lead member), Lars Sonesten (convener: atmospheric inputs), Penny Vlahos (convener: metals), Eric P. Achterberg, Babajide Alo, Robin Anderson, Carlos Francisco Andrade, Michael Angelidis, Maria João Bebianno (lead member), Arsonina Bera, Nene Bi Trace Boniface, Miguel Caetano, Isabel Natalia Garcia Arevalo, Kissao Gnandi, Julio Esteban Guerra Massón, Gi Hoon Hong, Suk Hyun Kim, Rainer Lohmann, Kida Rose Ninsemon, Jae Ryoung Oh, Bing Qiao, Monika Stankiewicz, Joshua T. Tuhumwire (co-lead member), Juying Wang (co-lead member) and Judith Weis.

Chapter 12

François Galgani (convener: marine debris), Aleke Stöfen-O'Brien (convener: dumping), Archis Ambulkar, Maurizio Azzaro, Maria João Bebianno (lead member), Arsonina Bera, Joan Bondareff, Alan Deidun, Fernanda de Oliveira Lana, Huw Griffiths, Bjørn Einar Grøsvik, Martin Hassellöv, Christos Ioakeimidis, Jenna Jambeck, Ahmed M. Kawser, Paula Keener, Iryna Makarenko, Chelsea Rochman, Qamar Schuyler, Paula Sobral, Konstantinos Topouizelis, Joshua T. Tuhumwire (co-lead member), Dick Vethaak, Ca Thanh Vu (co-lead member), Penny Vlahos, Juying Wang (co-lead member) and Judith Weis.

Chapter 13

Ca Thanh Vu (convener and lead member), Paulette Bynoe, Trang Minh Duong, Matt Eliot, Frank Hall, Sylvain Monde, Tuan Le Nguyen, Roshanka Ranasinghe, Matthieu de Schipper and Joshua T. Tuhumwire (co-lead member).

Chapter 14

Ca Thanh Vu (convener and lead member), Matchonnawe Hubert Bakai, Sam Bentley, Nene Bi Trace Boniface, Lionel Carter, Catherine Creese, Robert Dapa, Hugo Masson Fiallos, Regina Folorunsho, Gheorghe Ftadeev-Brat, Alan Simcock (co-lead member) and Alix Willemez.

Chapter 15

Porter Hoagland (convener), Megan Bailey, Lena Bergström, Alida Bundy, Fernanda de Oliveira Lana, Karen Evans (co-lead member), Manuel Hidalgo, Andrew Johnson, Melina Kourantidou, Hector Lozano-Montes, Enrique Marschoff (lead member), Essam Yassin Mohammed (co-lead member), Henn Ojaveer (co-lead member), Franklin Ormaza-Gonzalez, Imants George Priede, Ylenia Randrianisoa (co-lead member), Jörn Schmidt (co-lead member), Zacharie Sohou, Burcu Bilgin Topçu, Lynn Waterhouse and Chang-Ik Zhang.

Chapter 16

Rohana Subasinghe (convener), Pedro Barón, Malcolm Beveridge, Enrique Marschoff (lead member), Doris Oliva and Renison Ruwa (co-lead member).

Chapter 17

Hilconida Calumpong (convener and lead member), Paula Bontempi, Adam Hughes, Franciane Pellizzari, Isabel Sousa Pinto, Renison Ruwa (co-lead member), Jörn Schmidt (co-lead member) and Noemí Solar-Bacho.

Chapter 18

James R. Hein (joint convener), Pedro Madureira (joint convener), Maria João Bebianno (co-lead member), Ana Colaço, Giorgio de la Torre, Paraskevi Nomikou, Luis M. Pinheiro, Richard Roth, Pradeep Singh, Anastasia Strati (co-lead member) and Joshua T. Tuhumwire (lead member).

Chapter 19

Amardeep Dhanju (convener), Arsonina Bera, Hans-Peter Damian, Robert Dapa, Giorgio de la Torre, Kacou Yeboue Seraphim, Alan Simcock (co-lead member) and Joshua T. Tuhumwire (lead member).

Chapter 20

Ana Širović (convener), Karen Evans (lead member), Carlos Garcia-Soto (co-lead member), John A. Hildebrand, Sérgio M. Jesus and James H. Miller.

Chapter 21

Takvor Soukissian (convener), Joan Bondareff, Valerie Cummins, Amardeep Dhanju, Carlos Garcia-Soto (co-lead member), Lars Golmen, Osman Keh Kamara (co-lead member), Jimmy Murphy, Eric Mwangi Njoroge, Anastasia Strati (lead member) and Georges Vougioukalakis.

Chapter 22

Thomas W. Therriault (convener), Marnie L. Campbell, Alan Deidun, Bella S. Galil, Chad L. Hewitt, Graeme Inglis, Henn Ojaveer (lead member), Chul Park (co-lad member), Bing Qiao, Renison Ruwa (co-lead member) and Evangelina Schwindt.

Chapter 23

Robert Blasiak (joint convener), Ellen Kenchington (joint convenor), Jesús M. Arrieta, Jorge Rafael Bermúdez-Monsalve, Hilconida Calumpong (co-lead member), Shao Changwei, Sanae Chiba (lead member), Hebe Dionisi, Carlos Garcia-Soto (co-lead member), Helena Vieira and Boris Wawrik.

Chapter 24

Alan Simcock (convener and lead member), Carlos Garcia-Soto (co-lead member), Aninda Mazumdar, Aaron Micallef, Joseph Montoya, Katherine E.A. Segarra, Joshua T. Tuhumwire (co-lead member) and Leonid Yurganov.

Chapter 25

Karen Evans (convener and lead member), Roland Cormier, Piers Dunstan, Elizabeth Fulton, Essam Yassin Mohammed (co-lead member), Jörn Schmidt (co-lead member), Alan Simcock (co-lead member), Vanessa Stelzenmüller, Ca Thanh Vu (co-lead member) and Skipton Woolley.

Chapter 26

Alan Simcock (convener and lead member), Jarbas Bonetti, Louis Celliers, Karen Evans (co-lead member), Leandra Gonçalves, Ståle Navrud, Marcus Polette, Julian Renya and Ca Thanh Vu (co-lead member).

Chapter 27

Piers Dunstan (convener), Hilconida Calumpong (co-lead member), Louis Celliers, Valerie Cummins, Ana Cristina de Jesus, Michael Elliott, Karen Evans (co-lead member), Antony Firth, Frédéric Guichard, Quentin Hanich, Manuel Hildago, Hector Manuel Lozano-Montes, Chanda L. Meek, Essam Yassin Mohammed (co-lead member), Marcus Polette, Jemma Purandare, Anita Smith, Anastasia Strati (lead member) and Ca Thanh Vu (co-lead member).

Chapter 28

Luciano Hermanns (convener), Denis Worlnanyo Aheto, Adem Bilgin, Robert Blasiak, Cecile Brugere, Karen Evans, Antony Firth, Marinez Eymael Garcia Scherer, Deborah Greaves, Osman Keh Kamara (co-lead member), Wenhai Lu, Iryna Makarenko, Juan Ramon Martinez, Essam Yassin Mohammed (lead member), Ståle Navrud, Jörn Schmidt (co-lead member), Anita Smith, Anastasia Strati (co-lead member), Rashid Sumaila, Kateryna Utkina, Hans Van Tilburg, Wojciech Wawrzynski and Vladimir Žulkus.